The
Physiological Measurement
Handbook

Series in Medical Physics and Biomedical Engineering

Series Editors: John G Webster, E Russell Ritenour, Slavik Tabakov, and Kwan-Hoong Ng

Other recent books in the series:

Radiosensitizers and Radiochemotherapy in the Treatment of Cancer
Shirley Lehnert

The Physiological Measurement Handbook
John G Webster (Ed)

Diagnostic Endoscopy
Haishan Zeng (Ed)

Medical Equipment Management
Keith Willson, Keith Ison, and Slavik Tabakov

Targeted Muscle Reinnervation: A Neural Interface for Artificial Limbs
Todd A Kuiken; Aimee E Schultz Feuser; Ann K Barlow (Eds)

Quantifying Morphology and Physiology of the Human Body Using MRI
L Tugan Muftuler (Ed)

Monte Carlo Calculations in Nuclear Medicine, Second Edition: Applications in Diagnostic Imaging
Michael Ljungberg, Sven-Erik Strand, and Michael A King (Eds)

Vibrational Spectroscopy for Tissue Analysis
Ihtesham ur Rehman, Zanyar Movasaghi, and Shazza Rehman

Webb's Physics of Medical Imaging, Second Edition
M A Flower (Ed)

Correction Techniques in Emission Tomography
Mohammad Dawood, Xiaoyi Jiang, and Klaus Schäfers (Eds)

Physiology, Biophysics, and Biomedical Engineering
Andrew Wood (Ed)

Proton Therapy Physics
Harald Paganetti (Ed)

Practical Biomedical Signal Analysis Using MATLAB®
K J Blinowska and J Żygierewicz (Ed)

Physics for Diagnostic Radiology, Third Edition
P P Dendy and B Heaton (Eds)

Series in Medical Physics and Biomedical Engineering

The Physiological Measurement Handbook

Edited by
John G. Webster
University of Wisconsin
Madison, USA

CRC Press
Taylor & Francis Group
Boca Raton London New York

CRC Press is an imprint of the
Taylor & Francis Group, an **informa** business

CRC Press
Taylor & Francis Group
6000 Broken Sound Parkway NW, Suite 300
Boca Raton, FL 33487-2742

First issued in paperback 2020

© 2015 by Taylor & Francis Group, LLC
CRC Press is an imprint of Taylor & Francis Group, an Informa business

No claim to original U.S. Government works

ISBN-13: 978-1-4398-0847-4 (hbk)
ISBN-13: 978-0-367-78366-2 (pbk)

Visit the Taylor & Francis Web site at
http://www.taylorandfrancis.com

and the CRC Press Web site at
http://www.crcpress.com

Contents

Series Preface, vii

Preface, ix

Editor, xi

Contributors, xiii

CHAPTER 1 ▪ Overview of Physiological Measurements   1

P. ÅKE ÖBERG AND TATSUO TOGAWA

CHAPTER 2 ▪ Cardiology: Blood Pressure   13

ALBERTO AVOLIO, MARK BUTLIN, AND DEAN WINTER

CHAPTER 3 ▪ Cardiology: Electrocardiography   43

JASON NG AND JEFFREY J. GOLDBERGER

CHAPTER 4 ▪ Cardiology: Hemodynamics   75

SÁNDOR J. KOVÁCS

CHAPTER 5 ▪ Phonocardiography: Recording of Heart Sounds and Murmurs   101

HOMER NAZERAN

CHAPTER 6 ▪ Dermatology   123

JOHN G. WEBSTER

CHAPTER 7 ▪ Gastroenterology   131

JOHN WILLIAM ARKWRIGHT AND PHIL DINNING

CHAPTER 8 ▪ Measuring Glomerular Filtration Rate in Clinical Practice   161

A. MICHAEL PETERS

CHAPTER 9 ■ Neurology: Central Nervous System   171

RICHARD B. REILLY

CHAPTER 10 ■ Neurology: Peripheral Nerves and Muscles   213

ERNEST NLANDU KAMAVUAKO AND DARIO FARINA

CHAPTER 11 ■ Obstetrics and Gynecology   237

MICHAEL R. NEUMAN

CHAPTER 12 ■ Ophthalmology   267

AYYAKKANNU MANIVANNAN AND VIKKI MCBAIN

CHAPTER 13 ■ Orthopedics   287

MIRIAM HWANG, UBONG UDOEKWERE, AND GERALD HARRIS

CHAPTER 14 ■ Physiologic Measures in Otology, Neurotology, and Audiology   317

IL JOON MOON, E. SKOE, AND JAY T. RUBINSTEIN

CHAPTER 15 ■ Pathology: Chemical Tests   355

ARLYNE B. SIMON, BRENDAN M. LEUNG, TOMMASO BERSANO-BEGEY, LARRY J. BISCHOF, AND SHUICHI TAKAYAMA

CHAPTER 16 ■ Pediatric Physiological Measurements   393

MICHAEL R. NEUMAN

CHAPTER 17 ■ Measurements in Pulmonology   419

JASON H.T. BATES, DAVID W. KACZKA, AND G. KIM PRISK

CHAPTER 18 ■ Radiology   453

MELVIN P. SIEDBAND AND JOHN G. WEBSTER

CHAPTER 19 ■ Rehabilitation   477

JOHN G. WEBSTER

CHAPTER 20 ■ Urology   485

MICHAEL DRINNAN AND CLIVE GRIFFITHS

CHAPTER 21 ■ Data Processing, Analysis, and Statistics   537

GORDON SILVERMAN AND MALCOLM S. WOOLFSON

INDEX, 589

Series Preface

THE SERIES IN MEDICAL PHYSICS AND BIOMEDICAL ENGINEERING deals with the applications of physical sciences, engineering, and mathematics in medicine and clinical research.

The Series in Medical Physics and Biomedical Engineering is the official book series of the International Organization for Medical Physics. The series seeks (but is not restricted to) publications in the following topics:

- Artificial organs
- Assistive technology
- Bioinformatics
- Bioinstrumentation
- Biomaterials
- Biomechanics
- Biomedical engineering
- Clinical engineering
- Imaging
- Implants
- Medical computing and mathematics
- Medical/surgical devices

- Patient monitoring
- Physiological measurement
- Prosthetics
- Radiation protection, health physics, and dosimetry
- Regulatory issues
- Rehabilitation engineering
- Sports medicine
- Systems physiology
- Telemedicine
- Tissue engineering
- Treatment

This is an international series that meets the need for up-to-date texts in this rapidly developing field. Books in the series range in level from introductory graduate textbooks and practical handbooks to more advanced expositions of current research.

THE INTERNATIONAL ORGANIZATION FOR MEDICAL PHYSICS

The International Organization for Medical Physics (IOMP) represents over 18,000 medical physicists worldwide and has a membership of 80 national and 6 regional organizations, together with a number of corporate members. Individual medical physicists of all national member organizations are also automatically members of the IOMP.

The mission of IOMP is to advance medical physics practice worldwide by disseminating scientific and technical information, fostering the educational and professional development of medical physics, and promoting the highest quality medical physics services for patients.

A world congress on medical physics and biomedical engineering is held every three years in cooperation with the International Federation for Medical and Biological Engineering (IFMBE) and International Union for Physics and Engineering Sciences in Medicine (IUPESM). A regionally based international conference, the International Congress of Medical Physics (ICMP), is held between world congresses. The IOMP also sponsors international conferences, workshops, and courses.

The IOMP has several programs to assist medical physicists in developing countries. The joint IOMP Library Programme supports 75 active libraries in 43 developing countries, and the Used Equipment Programme coordinates equipment donations. The Travel Assistance Programme provides a limited number of grants to enable physicists to attend the world congresses.

The IOMP cosponsors the *Journal of Applied Clinical Medical Physics*. It publishes an electronic bulletin, *Medical Physics World*, twice a year. It also publishes *e-Zine*, an electronic newsletter, about six times a year. The IOMP has an agreement with Taylor & Francis Group for the publication of textbooks in The Series in Medical Physics and Biomedical Engineering. IOMP members receive a discount.

The IOMP collaborates with international organizations, such as the World Health Organization (WHO), the International Atomic Energy Agency (IAEA), and other international professional bodies such as the International Radiation Protection Association (IRPA) and the International Commission on Radiological Protection (ICRP), to promote the development of medical physics and the safe use of radiation and medical devices.

Guidance on education, training, and professional development of medical physicists is issued by the IOMP, which is collaborating with other professional organizations in the development of a professional certification system for medical physicists that can be implemented on a global basis.

The IOMP website (www.iomp.org) contains information on all of its activities, policy statements 1 and 2, and the *IOMP: Review and Way Forward*, which outlines its activities and plans for the future.

Preface

INTRODUCTION

The purpose of *The Physiological Measurement Handbook* is to provide a reference that is both concise and useful for biomedical engineers in universities, the medical device industry, scientists, designers, managers, research personnel, and students, as well as health care personnel such as physicians, nurses, and technicians who make physiological measurements. The handbook covers an extensive range of topics that comprise the subject of measurement in all departments of medicine.

The handbook describes the use of instruments and techniques for practical measurements required in medicine. It includes sensors, techniques, hardware, and software. It also includes information on processing systems, automatic data acquisition, reduction and analysis, and their incorporation for diagnosis.

The chapters include descriptive information for professionals, students, and workers interested in physiological measurement. Equations also assist biomedical engineers and health care personnel who seek to discover applications and solve diagnostic problems that arise in medical fields not in their specialty. The chapters also include specialized information needed by informed specialists who seek to learn advanced applications of the subject, evaluative opinions, and possible areas for future study. By including all of this descriptive information, the handbook serves the reference needs of the broadest group of users—from the advanced high school science student to health care and university professionals.

ORGANIZATION

The handbook is organized according to each *medical specialty*. Chapter 1: Overview of Physiological Measurements. Chapter 2: Cardiology: Blood Pressure. Chapter 3: Cardiology: Electrocardiography. Chapter 4: Cardiology: Hemodynamics. Chapter 5: Phonocardiography: Graphic Recording of Heart Sounds and Murmurs. Chapter 6: Dermatology. Chapter 7: Gastroenterology. Chapter 8: Measuring Glomerular Filtration Rate in Clinical Practice. Chapter 9: Neurology: Central Nervous System. Chapter 10: Neurology: Peripheral Nerves and Muscles. Chapter 11: Obstetrics and Gynecology. Chapter 12: Ophthalmology. Chapter 13: Orthopedics. Chapter 14: Physiologic Measures in Otology, Neurotology, and Audiology. Chapter 15: Pathology—Chemical Tests. Chapter 16: Pediatric Physiological

Measurements. Chapter 17: Measurements in Pulmonology. Chapter 18: Radiology. Chapter 19: Rehabilitation. Chapter 20: Urology. Chapter 21: Data Processing, Analysis, and Statistics.

LOCATING YOUR PHYSIOLOGICAL VARIABLE

Select a physiological variable for measurement, skim the contents and peruse the chapter that describes different methods of making that physiological measurement. Consider alternative methods and their advantages and disadvantages prior to selecting the most suitable one.

For more detailed information, consult the index, as certain principles of physiological measurement may appear in more than one chapter.

John G. Webster

Editor

John G. Webster earned his BEE from Cornell University, Ithaca, New York, in 1953, and MSEE and PhD from the University of Rochester, Rochester, New York, in 1965 and 1967, respectively.

He is currently professor emeritus of biomedical engineering at the University of Wisconsin-Madison. In the field of medical instrumentation, he teaches undergraduate and graduate courses and conducts research on intracranial pressure monitoring, ECG dry electrodes, tactile vibrators, a visual voiding device, and apnea.

Professor Webster is the author of *Transducers and Sensors*: an IEEE/EAB individual learning program (Piscataway, NJ: IEEE, 1989). He is coauthor, with B. Jacobson, of *Medicine and Clinical Engineering* (Englewood Cliffs, NJ: Prentice-Hall, 1977), *Analog Signal Conditioning* (New York: Wiley, 1999) and *Sensors and Signal Conditioning, Second Edition* (New York: Wiley, 2001). He is the editor of *Tactile Sensors for Robotics and Medicine* (New York: Wiley, 1988), *Electrical Impedance Tomography* (Bristol, U.K.: Adam Hilger, 1990), *Teaching Design in Electrical Engineering* (Piscataway, NJ: Educational Activities Board, IEEE, 1990), *Prevention of Pressure Sores: Engineering and Clinical Aspects* (Bristol, U.K.: Adam Hilger, 1991), *Design of Cardiac Pacemakers* (Piscataway, NJ: IEEE Press, 1995), *Design of Pulse Oximeters* (Bristol, U.K.: IOP Publishing, 1997), *Encyclopedia of Electrical and Electronics Engineering* (New York, Wiley, 1999), *Encyclopedia of Medical Devices and Instrumentation, Second Edition* (New York: Wiley, 2006), and *Medical Instrumentation: Application and Design, Fourth Edition* (New York: Wiley, 2010). He is the co-editor, with A. M. Cook, of *Clinical Engineering: Principles and Practices* (Englewood Cliffs, NJ: Prentice-Hall, 1979) and *Therapeutic Medical Devices: Application and Design* (Englewood Cliffs, NJ: Prentice-Hall, 1982), with W. J. Tompkins, of *Design of Microcomputer-Based Medical Instrumentation* (Englewood Cliffs, NJ: Prentice-Hall, 1981) and *Interfacing Sensors to the IBM PC* (Englewood Cliffs, NJ: Prentice Hall, 1988), and with A. M. Cook, W J. Tompkins, and G. C. Vanderheiden, of *Electronic Devices for Rehabilitation* (London, U.K.: Chapman & Hall, 1985).

Dr. Webster has been a member of the IEEE-EMBS Administrative Committee and the National Institutes of Health (NIH) Surgery and Bioengineering Study Section. He is a fellow of the Institute of Electrical and Electronics Engineers, the Instrument Society of America, the American Institute of Medical and Biological Engineering, the Biomedical Engineering Society, and the Institute of Physics. He is a recipient of the AAMI Foundation Laufman-Greatbatch Prize, the ASEE/Biomedical Engineering Division's Theo C. Pilkington Outstanding Educator Award, and the IEEE-EMBS Career Achievement Award.

Contributors

John William Arkwright
Materials Science and Engineering
Australian National University
Commonwealth Scientific and Industrial
    Research Organisation
Lindfield, New South Wales, Australia

Alberto Avolio
Australian School of Advanced Medicine
Macquarie University
Sydney, New South Wales, Australia

Jason H.T. Bates
Department of Medicine
University of Vermont
Burlington, Vermont

Tommaso Bersano-Begey
Department of Biomedical Engineering
University of Michigan
Ann Arbor, Michigan

Larry J. Bischof
Department of Pathology
University of Michigan Health System
Ann Arbor, Michigan

Mark Butlin
Australian School of Advanced Medicine
Macquarie University
Sydney, New South Wales, Australia

Phil Dinning
Department of Gastroenterology
and
Department of Surgery
Flinders University
Adelaide, South Australia, Australia

Michael Drinnan
Regional Medical Physics Department
Freeman Hospital
Newcastle upon Tyne, United Kingdom

Dario Farina
Department of Health Science and
    Technology
Aalborg University
Aalborg, Denmark

Jeffrey J. Goldberger
Feinberg School of Medicine
Northwestern University
Evanston, Illinois

Clive Griffiths
Department of Urology
Freeman Hospital
Newcastle upon Tyne, United Kingdom

Gerald Harris
Orthopaedic & Rehabilitation Engineering
   Center
Medical College of Wisconsin
Marquette University
Milwaukee, Wisconsin

Miriam Hwang
Orthopaedic & Rehabilitation Engineering
   Center
Medical College of Wisconsin
Marquette University
Milwaukee, Wisconsin

David W. Kaczka
Department of Anesthesia
The University of Iowa Hospital and Clinics
Iowa City, Iowa

Ernest Nlandu Kamavuako
Department of Health Science and
   Technology
Aalborg University
Aalborg, Denmark

Sándor J. Kovács
Cardiovascular Division
Cardiovascular Biophysics Laboratory
Department of Medicine
School of Medicine
Washington University
St. Louis, Missouri

Brendan M. Leung
Department of Biomedical Engineering
University of Michigan
Ann Arbor, Michigan

Ayyakkannu Manivannan
Department of Bio-Medical Physics
NHS Grampian and University of Aberdeen
Aberdeen, United Kingdom

Vikki McBain
Eye Outpatient Department
Aberdeen Royal Infirmary
NHS Grampian
Aberdeen, United Kingdom

Il Joon Moon
Department of Otorhinolaryngology—
   Head and Neck Surgery
Samsung Medical Center
School of Medicine
Sungkyunkwan University
Seoul, South Korea

Homer Nazeran
Department of Electrical and Computer
   Engineering
University of Texas at El Paso
El Paso, Texas

Michael R. Neuman
Department of Biomedical Engineering
Michigan Technological University
Houghton, Michigan

Jason Ng
Feinberg School of Medicine
Northwestern University
Chicago, Illinois

P. Åke Öberg
Department of Biomedical Engineering
Linköping University
Linköping, Sweden

A. Michael Peters
Brighton and Sussex Medical School
University of Sussex
Brighton, United Kingdom

G. Kim Prisk
Department of Medicine
and
Department of Radiology
University of California, San Diego
La Jolla, California

Richard B. Reilly
Trinity Centre for Bioengineering
Trinity College Dublin
Dublin, Republic of Ireland

Jay T. Rubinstein
Virginia Merrill Bloedel Hearing Research
  Center
University of Washington
Seattle, Washington

Melvin P. Siedband
Department of Medical Physics
University of Wisconsin–Madison
Madison, Wisconsin

Gordon Silverman
Department of Electrical & Computer
  Engineering
Manhattan College
Riverdale, New York

Arlyne B. Simon
Macromolecular Science and Engineering
  Center
University of Michigan
Ann Arbor, Michigan

E. Skoe
Virginia Merrill Bloedel Hearing Research
  Center
University of Washington
Seattle, Washington

Shuichi Takayama
Biomedical Engineering Department
Macromolecular Science and Engineering
  Center
University of Michigan
Ann Arbor, Michigan

Tatsuo Togawa
School of Human Sciences
Waseda University
Tokyo, Japan

Ubong Udoekwere
Orthopaedic & Rehabilitation Engineering
  Center
Medical College of Wisconsin
Marquette University
Milwaukee, Wisconsin

John G. Webster
Department of Biomedical Engineering
University of Wisconsin–Madison
Madison, Wisconsin

Dean Winter
AtCor Medical, Inc.
West Ryde, New South Wales, Australia

Malcolm S. Woolfson
Department of Electrical and Electronic
  Engineering
Faculty of Engineering
University of Nottingham
Nottingham, United Kingdom

CHAPTER 1

Overview of Physiological Measurements

P. Åke Öberg and Tatsuo Togawa

CONTENTS

1.1	Introduction	1
1.2	Historical Perspective	3
1.3	Measurands and Requirements for Physiological Measurements	4
1.4	Common Physiological Measurements Used for Diagnosis	4
	1.4.1 Electrophysiological Measurements	5
	1.4.2 Pressure Measurements	6
	1.4.3 Blood Flow Measurements	7
	1.4.4 Temperature Measurements	8
	1.4.5 Respiratory Measurements	8
1.5	Common Physiological Measurements Used for Research	9
	1.5.1 Intracellular Measurements	9
	1.5.2 Impedance Measurements	9
1.6	Possible Future Instruments	9
	1.6.1 Optical Biopsies	10
	1.6.2 MEMS and Lab-on-a-Chip	10
References		11

1.1 INTRODUCTION

The information required to present the diagnosis of a particular disease or illness can be extracted from a number of sources. The past history of the disease, the present symptoms and their developments, a thorough physical examination of the patient, and a number of laboratory tests are all important sources of information. Samples for laboratory tests can be gases from the lungs, blood samples, urine, feces, and tissue samples taken by biopsies. In laboratories, these samples are analyzed in a variety of ways, for instance, cell counts, cell staining for analysis of signs of cancer, or cell cultures to detect the presence of bacteria. Physiological measurements of these kinds of samples can be classified using Figure 1.1.

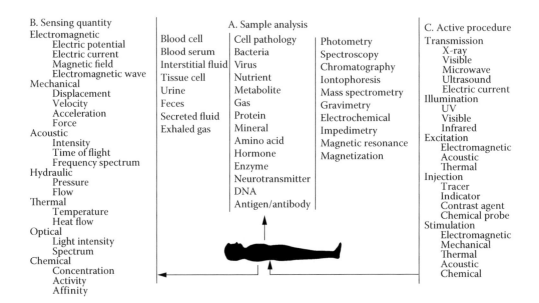

B. Sensing quantity
Electromagnetic
　　Electric potential
　　Electric current
　　Magnetic field
　　Electromagnetic wave
Mechanical
　　Displacement
　　Velocity
　　Acceleration
　　Force
Acoustic
　　Intensity
　　Time of flight
　　Frequency spectrum
Hydraulic
　　Pressure
　　Flow
Thermal
　　Temperature
　　Heat flow
Optical
　　Light intensity
　　Spectrum
Chemical
　　Concentration
　　Activity
　　Affinity

A. Sample analysis
Blood cell
Blood serum
Interstitial fluid
Tissue cell
Urine
Feces
Secreted fluid
Exhaled gas

Cell pathology
Bacteria
Virus
Nutrient
Metabolite
Gas
Protein
Mineral
Amino acid
Hormone
Enzyme
Neurotransmitter
DNA
Antigen/antibody

Photometry
Spectroscopy
Chromatography
Iontophoresis
Mass spectrometry
Gravimetry
Electrochemical
Impedimetry
Magnetic resonance
Magnetization

C. Active procedure
Transmission
　　X-ray
　　Visible
　　Microwave
　　Ultrasound
　　Electric current
Illumination
　　UV
　　Visible
　　Infrared
Excitation
　　Electromagnetic
　　Acoustic
　　Thermal
Injection
　　Tracer
　　Indicator
　　Contrast agent
　　Chemical probe
Stimulation
　　Electromagnetic
　　Mechanical
　　Thermal
　　Acoustic
　　Chemical

FIGURE 1.1　Variety of techniques utilized in physiological measurements. A column lists samples taken from the patient, B column lists observable signals generated by the body, and C column lists extrinsic energy probes. (From Rushmer, R.F., *Medical Engineering: Projections for Health Care Delivery*, Academic Press, London, U.K., 1972.)

In Figure 1.1, the A column lists samples taken from the patient and analyzed remotely using chemical, cellular, and structural methods. Today, laboratory analysis is automated to reduce analysis time, increase quality and reduce cost. In a major hospital, the number of these type of samples can be as many as 100,000 per day.

However, many of the sample collection methods involve a slight risk or inconvenience for the patient. The introduction of needles, catheters, or biopsy instruments through the skin or into the body orifices can result in infections, pain, or other hazards.

A preferred way to understand the status of a particular disease is to study the intrinsic signals, generated by the body itself, without interaction with sampling methods. Examples of such methods are listed in the B column in Figure 1.1. Most of them are well known and as old as the medical science itself. The electrophysiological methods to investigate the heart (electrocardiogram [ECG]), brain (electroencephalogram [EEG]), and muscles (electromyogram [EMG]) are all familiar. Pressure recordings from the vascular system have been used since the beginning of the eighteenth century. Sounds from the airways, lungs, fetus, and the heart have been used as diagnostic tools long before the invention of the stethoscope in the early nineteenth century. Body heat production and temperature changes can be used to understand infectious diseases or inflammatory disorders.

The beauty of intrinsic energy analysis is that these methods do not interfere with the physiological systems under study, nor do they cause the patient discomfort.

An alternative to extract diagnostic information from the human body is to utilize external energy probes. The common denominator here is that we *inject* external energies in the human body and study how these energies are modified when interacting with the organ under study. Optical beams, x-ray, ultrasound, and contrast agents are all examples of this type of diagnostic methodology. The physical phenomena utilized are absorption, scattering, Doppler shift, and reflection.

In the C column in Figure 1.1, the extrinsic energy probes are listed. These methods all utilize external probes or energies.

1.2 HISTORICAL PERSPECTIVE

Physiological measurements have been used for a very long time for the diagnosis and treatment of illnesses. In ancient days, the physicians were referred to visual observations like skin color changes like paleness, flush, rashes from infectious diseases, urine and feces abnormalities, and phlegm from the airways. During the eighteenth century, more quantitative methods came into use. van Leeuwenhoek and Hooke developed the first microscopes that opened the microbiological world to diagnostic questions. Stephen Hales made the first measurements of arterial blood pressure in the early eighteenth century. Poiseuille used a mercury manometer for the same purpose and Ludwig improved this device and was able to record the movements of the mercury column on a revolving smoked drum (the kymograph), and dynamic blood pressure measurements became possible for the first time. Physiological measurements have performed essential roles in the progress of science and technology in medicine for at least five centuries (Gedeon 2006).

At the turn of the twentieth century, Riva-Rocci and Korotkoff invented the noninvasive blood pressure cuff technique, still in use today. An inflatable pressure cuff surrounds the upper arm, and the cuff pressure is increased until the palpated pulse in the radial artery disappears, which defines the systolic pressure.

In the Korotkoff method, the blood pressure cuff is inflated to a pressure above systolic pressure and then slowly deflated while listening to the sounds from the artery using a stethoscope. Systolic and diastolic pressures can be determined from the sounds.

The inventions of Riva-Rocci and Korotkoff formed the basis of modern automatic blood pressure monitors for diagnosis and monitoring 150 years later.

Animal cells generate electric currents when stimulated. Volta and Galvani discovered the excitability of nerves and muscles around 1790. These early findings together with the development of the string galvanometer by Einthoven led to the possibility to record electrical activity from the heart (ECG), the brain (EEG), and the skeletal muscles (EMG), all of them having a substantial diagnostic impact. The recording techniques have undergone a tremendous development. Einthoven recorded the ECG from *electrodes* utilizing buckets, filled with salt solution in which the extremities were placed. Today's electrodes for all the electrophysiological signals are small metal plates, fixed at the skin surface.

Blood flow measurements are an important group of physiological measurements that has a long and interesting history. Plethysmography is one of the earliest methods, frequently used even today for blood flow measurements in the extremities. The term is derived from the Greek words *plethysmos* (volume) and *grafien* (to record). Glisson and Swammerdam were the first to use plethysmography to study muscular contraction, but the same method was later used to record blood flow. Francois-Frank published the first blood flow studies in 1876 using the venous occlusion technique. Brodie and Russell studied renal flow by enclosing the kidney in a closed chamber while occluding venous outflow. The same method can be used to study blood flow in a number of organs like the heart, liver, kidney, and the limbs.

The most common application is to diagnose obstructions in the limb vascular system. The early influx of arterial blood to the limb is recorded when the venous drainage is stopped temporarily by inflating a cuff. There are numerous ways to record the volume changes like water-filled cuffs, air-filled cuffs, strain gages, and photoelectric probes.

1.3 MEASURANDS AND REQUIREMENTS FOR PHYSIOLOGICAL MEASUREMENTS

Measurands in physiological measurements are quantities on which physiological states and functions are reflected. Sometimes, human senses can be used to detect such a quantity. For example, a fever can be detected by the touch of a hand to the forehead. The use of a thermometer provides objective and quantitative information and is thus preferable for diagnosis. Very many sensing techniques are now available so that accurate, less invasive or noninvasive, continuous measurements can be achieved if an adequate device is employed. However, a physiologically significant quantity is not simply a physical quantity of a matter. Although the temperature of a human body is not uniform, single-temperature values are always used as the index of body temperature that is usually defined by the measurement technique such as oral, rectal, or esophageal temperature. Blood pressure is also a physiological quantity defined by measurement technique. Common blood pressure measurement at the upper arm using a cuff is often regarded as the gold standard.

Performances required for a physiological measurement system should thus be determined by the quality of obtained physiological information. For commonly used physiological measurements, adequate measurement techniques and their performances are shown in recommendations provided by authorized societies. In such a recommendation, necessary performances such as accuracy and response characteristics required for clinical diagnostic applications are described.

1.4 COMMON PHYSIOLOGICAL MEASUREMENTS USED FOR DIAGNOSIS

A number of principles exist for diagnostic measurements of functional parameters from the human body. Electrophysiological measurements can record and analyze signals from the heart, the brain, skeletal muscles, and the nervous system. Blood flow in the major arteries and veins can be recorded with ultrasound and blood flow in the skin and minor vessels with laser Doppler flowmetry or heat dissipation probes. In addition, various indicators exist, the dilution of which in blood can estimate blood flow.

Blood pressure measurements can be performed with inflatable cuffs and a stethoscope or palpation. Invasive small sensors can be introduced via the vascular system into the heart and its chambers. Body temperature is an important variable for diagnostic work.

Airflow in the airways and flow distribution over the lung surfaces can detect respiratory dysfunctions. Various external sensors (Fleisch sensors, hot wire techniques,

Venturi tubes, etc.) and radioactive isotopes are used for these. But spirometers and body boxes are used for volume displacement measurements.

A short presentation of the various techniques for diagnosis of physiological conditions is presented below. For a more extensive review the reader is referred to Togawa *et al* (2011) in which various measurement techniques are systematically reviewed according to the physiological quantities to be measured.

1.4.1 Electrophysiological Measurements

The ECG is one of the most common diagnostic measurements in modern health care. Millions of ECGs are taken every day around the world. The ECG is also used for monitoring the conditions of critically ill patients in intensive care. The frequency ranges and signal magnitudes of the most common electrophysiological signals are presented in Figure 1.2.

When cardiac cells depolarize, small currents are spread in the whole body, and corresponding potentials about 1 mV in magnitude can be recorded on the skin surface. The potentials change rapidly during each heartbeat, and their shape varies with the position on the body.

In diagnostic work, electrodes are fixed on the skin surface in a standardized way.

The electrodes are attached to the extremities and to the chest. Usually, 12 different leads are used. In the bipolar extremity leads, the voltages between the arms and the left leg are recorded. From these potentials, the so-called augmented unipolar extremity leads can be calculated. In the chest leads, six electrodes are positioned on the rib cage above the heart.

The potentials from the heart are recorded in various ways. The most common way is the use of *electrocardiographs* or ECG machines that print out signals from the various leads on a strip of paper. Monitors that display the ECG on a screen are commonly used during surgery and in intensive care. Computer software is used for the analysis of the ECG, and the programs can identify deviations from the normal pattern and even suggest diagnostic statements.

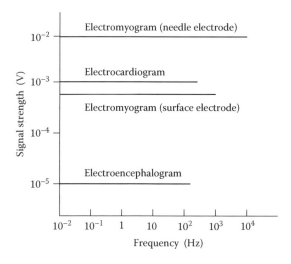

FIGURE 1.2   The frequency ranges and signal magnitudes of the most common electrophysiological signals.

Electrodes for ECG recording are seemingly simple devices but are important *translators* of ionic currents in the body to currents of electrons in the electrocardiograph.

Both the time course and the amplitudes of the ECG carry information about the physiological conditions of the heart.

The EEG records electric signals from the brain using an array of 21 surface electrodes arranged like a *hat* of elastic strips between the electrode sites. The EEG has an amplitude of only 100 μV that increases the demands on the electrodes, electrode site preparation, and recording technology.

The EEG is a sensitive mirror of alertness and gives also specific diagnostic information in case of epileptic attacks.

Today, the magnetic correspondence to the EEG, magnetoencephalography (MEG), can be recorded with ultrasensitive detectors and very advanced technology. The MEG gives a better spatial resolution of the position of an ectopic focus generating an epilepsy attack. With this information, invasive surgery or radiation can eliminate the problem.

The EMG records the electric activity from a working muscle. Often, the same type of skin electrodes are used as in ECG recordings. However, sometimes, the potentials from a single motor unit are of interest to record. Then, a thin needle electrode is introduced into the muscle that picks up the electrical activity from a much smaller volume than does skin electrodes.

A variant is the recording from individual muscle fibers, *single-fiber electromyography* (*SFEMG*). In this method, very tiny electrodes are used and the activity comes from single muscle cells.

EMG is an important method used for investigations of neuromuscular disorders and also used to differentiate between muscle problems and nerve malfunction.

1.4.2 Pressure Measurements

Pressures in the human body are kept within certain limits by regulatory systems. Very often, the proper function of an organ or the whole body requires that these limits are not exceeded.

Blood pressure can be used as an example. If the systemic pressure drops to too low levels, the brain will become seriously underperfused with blood, which leads to unconsciousness. On the other hand, if blood pressure becomes too high (extreme hypertension), this leads to injuries to the arterial system vessel walls or even stroke.

Increased intraocular pressure can result in retinal malfunction.

Thus, it is important to monitor pressure levels in the body.

The methods for pressure measurements can be *indirect* or *direct*. Indirect methods usually mean that it is not necessary to puncture an organ for the measurement. If the organ has soft walls and an external pressure can be applied, it is possible to obtain pressure readings from the interior of the organ.

Direct methods usually mean introduction of a sensor into the organ under study. Direct methods usually mean that higher levels of accuracy are met compared to the indirect method.

The *indirect* Riva-Rocci cuff method from 1896 means that a blood pressure cuff is placed around the upper arm. The cuff is inflated by a pump to a pressure level that exceeds the systolic pressure. The cuff pressure is monitored by a pressure gage. The cuff is then slowly deflated to a level at which the blood flow under the cuff is reestablished. The return of the flow can be monitored by a stethoscope placed over the elbow and by listening to the sounds produced in the partly occluded artery. This level represents the systolic blood pressure. The cuff pressure is then reduced further, and the sound from the vessel becomes longer and louder. When the flow no longer is obstructed by the cuff pressure, the sounds disappear. This pressure level equals the diastolic pressure.

The systolic pressure can also be measured by sensing the pulsation of the artery and when they reappear after the occlusion.

The intraocular pressure can also be measured with indirect methods. The classical Goldman applanometer utilizes a force measurement in which the probe is pressed against the cornea until the contact area is a circle of 3 mm, which is monitored through a prism. At this point, the force is recorded by a spring balance. A number of variants of the principle exist.

1.4.3 Blood Flow Measurements

To get a good estimate of the perfusion of an organ, blood flow measurements are important. A variety of methods exist. Some of them measure average flow and others give information about the momentary flow.

The method used in each diagnostic problem is selected from what part of the circulatory system (central or peripheral circulation) is investigated.

To diagnose the central circulation, *indicator dilution methods*, *Fick's principle*, and *ultrasound* methods are used. In peripheral circulation, *Doppler ultrasound*, *plethysmography*, *thermal probes*, and *laser Doppler flowmetry* are used.

Fick's method is used to determine the cardiac output average over several heartbeats.

Oxygen is used as an indicator, and the patient inhales this gas from a bag. The arteriovenous oxygen concentration difference is determined by sampling arterial blood and mixed venous blood that require cardiac catheterization. If the oxygen uptake is known, the cardiac output can be calculated. Fick's method gives an accurate estimation of cardiac output but requires cardiac catheterization and gas laboratory facilities.

Indicators, often a colored dye, can be injected into the blood and used for flow measurements.

The degree of dilution in blood is an estimate of the blood flow rate. The dye is often injected into the right atrium of the heart, and the dye concentration is measured in the pulmonary artery during the period after the injection.

A problem with indicators is that they remain in the vascular system and make repeated measurements impossible. However, cold saline (cooler than blood) can be used as indicator that permits frequent measurements, because the temperature difference is equalized quite rapidly through heat losses from the vascular walls.

Ultrasound can also be used for cardiac output measurements. The Doppler principle is applied, and knowledge about descending aorta dimensions and the ejection time of the heart allows the calculation of cardiac output. The ultrasound beam is directed along the descending aorta from the suprasternal notch. Esophageal probes are also used.

In the peripheral part of the vascular system, a variety of methods are used for flow estimation. The classical plethysmographic method is used for recording flow to the extremities.

An inflatable cuff is placed proximal to the tissue volume under study. The cuff is inflated to above venous pressure, preventing venous blood to leave the extremity, whereas influx of arterial blood is maintained. The increased blood volume in the extremity can be measured with sensors surrounding the actual volume. The volume increase per time unit during the first heartbeats is an estimate of the blood flow.

Continuous Doppler ultrasound is used to qualitatively monitor flow in peripheral arteries. The Doppler frequencies occur in the audible spectrum and can be easily listened to during the investigation of the patient.

To estimate flow in tissue and skin require different techniques than what is used in the central circulation. Thermal probes in which a metal ring and a central plate heated at a constant power are arranged in a concentric manner are common. The temperature difference between the ring and the plate is a measure of the blood flow underneath the probe.

Laser Doppler flowmetry measures the Doppler shift of laser light scattered in moving erythrocytes and stationary skin structures. The flow signal is proportional to the product mean velocity of cells times their number in a small volume (typically 1 mm^3).

Specially designed instruments with scanning facilities produce a color-coded image of the flow pattern in a smaller area (a few cm^2) (Shepherd and Öberg 1990, Strömberg *et al*).

1.4.4 Temperature Measurements

Body temperature is controlled by a center in the hypothalamus. Today, a number of body sites are used for temperature measurement. Rectum, the eardrum, and esophageal measurements are the most reliable because here, the temperature is stable. Less accurate are the oral cavity and the armpit. Electronic thermometers in various forms are now used for body temperature measurements. Mercury thermometers are no longer in use because of the negative environmental effects of mercury. In most of these, a thermistor (a temperature-sensitive resistor) is used as the sensing element. In eardrum thermometers, the infrared radiation from the eardrum is sensed. In this type of measurement, it is important that the instrument really *sees* the eardrum and not the wall of the external auditory meatus.

1.4.5 Respiratory Measurements

Respiratory measurements include air volume and flow rate measurements. Volume measurements utilize spirometers in which the gas from the lungs is accumulated and the volume is recorded. Body plethysmographs are closed chambers in which the patient is placed during the measurements. The patient inhales air from the outside of the chamber through a tube. The pressure in the chamber varies with the respiratory work of the patient, and lung volumes can be recorded.

For flow rate measurements, a number of instrumental principles exist. A Fleisch tube is a part of a flow channel in which a small flow resistance in the form of a mesh is introduced. By monitoring the pressure drop across the flow resistance, the volume flow can be calculated.

A Venturi tube is a constriction of a flow tube through which the flow of interest is passing and over which the pressure drop is recorded. The geometry of the constriction is such that it gives a quadratic relation between the passing flow and the pressure drop. Methods for determining forced vital capacity (FVC) and forced expiratory volume in 1 s (FEV_1) require a sensor solution with a low mechanical inertia.

1.5 COMMON PHYSIOLOGICAL MEASUREMENTS USED FOR RESEARCH

1.5.1 Intracellular Measurements

As an example of physiological measurements for research, we have chosen intracellular recordings with glass microelectrodes, a technique that has contributed enormously to the understanding of the excitable cell and its neurophysiological significance.

The glass microelectrode with a tip diameter of a few micrometers is manufactured from glass tubes that are heated and drawn to very small diameters. The sharp tip facilitates the puncture of the cell membrane with minor trauma. To make the electrode a conductor, its interior is often filled with three molar KCl. The electrodes can be introduced into the cell membrane without disturbing the membrane or cell function.

An amplifier connected to the microelectrode measures the transmembrane potential and its variations.

1.5.2 Impedance Measurements

In a number of problem areas, electrical impedance measurements have been suggested as the type of measurement to enlighten a physiological problem. For instance, cardiac output can be recorded from an impedance measurement. Electrodes in contact with the skin and surrounding the body are attached to the thorax and the neck region. Four electrodes are used. The two inner electrodes are used for monitoring the voltage value, and the outer electrodes are used for feeding a current through the thorax. The stroke volume can be calculated if the inner electrode distance and the resistivity of blood are known. Impedance has also been used as a detection method in occlusion plethysmography and hence is called impedance plethysmography.

1.6 POSSIBLE FUTURE INSTRUMENTS

The development of future sensor systems is very diversified and based on many techniques and manufacturing processes. This wide, growing field cannot be summarized in a short presentation like this. However, a few examples can be given to mirror some important steps taken toward future sensor systems.

The future of medical and biological sensors is likely to be associated with the developments in optics, *microelectromechanical systems* (MEMS) and extremely sensitive biological sensing principles.

As examples of future sensor technologies, we have chosen to describe the fields of optical biopsies and MEMS-based sensors.

1.6.1 Optical Biopsies

Today, a biopsy usually means to take a tissue sample with some surgical technique from the site of interest. The sample is then analyzed with cytological, histological, or biochemical methods in order to understand if the sample shows normal, benign, or malignant changes. The analysis time can be long before a diagnosis can be secured. Removal of tissue means discomfort or even pain for the patient, and there is a risk of infection in the area from which the sample is taken. In addition, there is an administrative risk that samples from many patients are interchanged.

The development of optical components like optical fibers, photodetectors, and light sources with small dimensions makes it possible to design instruments that can be introduced into the human body through the vascular system or the gastrointestinal channel. A specific area of interest in most parts of the body can be illuminated through the optical fibers, and diffuse reflected light can be analyzed by the photodetector. Absorption, reflection, or scattering can be used to provide optical signatures of morphological, chemical, or physiological changes that may help the physician to determine if the tissue is normal, benign, or malignant.

There are many examples of successful attempts to utilize these new techniques in diagnostic work. Optical mammography research (Alfano 1998) developed sensors for the detection of cancerous lesions of humans breasts at an early stage. The sensor utilizes noninvasive and nonionizing near-infrared light. The number of possible applications of similar setups is numerous. Esophageal cancers and dysplasia, bladder tumors, and head and neck cancers are areas in which optical biopsies have been evaluated so far.

Optical coherence tomography (OCT) is expected to become a strong tool for optical biopsy. It is an interferometric technique using low-coherent light beams. By scanning in two or three directions, 2D or 3D images of scattering objects can be obtained with submicrometer resolution. Although the observational region is limited to a few millimeters beneath the tissue or organ surface, it is already applied to obtain cross-sectional images of the retina noninvasively.

Changes in tissue vascularity and the presence of metabolic cofactors are examples of optical markers of a disease that can support the use of optical biopsy techniques.

Minimally invasive diagnosis, even at the bedside, can be possible after future development of optical biopsies techniques that today require laboratory handling with time delays as a result.

1.6.2 MEMS and Lab-on-a-Chip

MEMS can probably become a key technology for the development of new very small sensors of different kinds for medical problem areas.

Photolithography in combination with etching processes (wet and dry etching) makes it possible to manufacture advanced tools and sensors with very small dimensions.

Components can be batch fabricated, enabling miniaturization, low-cost, and integrated solutions to advanced sensor problems. Silicon is very often used as material in the MEMS processes.

Silicon probes for neural activity monitoring, microcutters for eye surgery, probes for atomic force microscopy, and electrodes for electrophysiological recording of skin potentials are all examples of medical applications of MEMS technology.

The application of this technology in the medical field is often called Bio-MEMS (Tay 2001). Bio-MEMS differs from traditional MEMS in a number of ways. Bio-MEMS is often a common term for analytical techniques used in biochemistry, biology, and medicine. The number of manufacturing materials used is much wider in Bio-MEMS than in MEMS to achieve biocompatibility in applications where long-term implantations in tissue are considered. Glass, polypropylene, and polymeric materials as well as rigid plastic are all examples of such materials.

A major bio-MEMS subgroup of techniques is called lab-on-a-chip (LOC) (Herold and Rasooly 2009). An LOC is a device that can integrate laboratory functions on a single chip. The geometrical dimensions are often on the order of a millimeter to a few square centimeters. An LOC handles very small fluid volumes down to a picoliter. The development started with applications like pressure sensors, airbag sensors, and fluid-handling devices like microvalves, mixers, micropumps, and capillary systems.

A growing interest for LOC applications is seen in various single-cell analysis, detection of bacteria and viruses based on antigen–antibody reactions, biochemical assays, and real-time polymerase chain reaction (PCR).

The LOC techniques can become an important step toward improvement of global health care. In countries with limited health-care resources, the future new chips can provide a diagnostic, low-cost tool for diseases where drugs for treatment already exist. For instance, around 40 million people in the world are infected with HIV, but only 1.3 million receive antiretroviral treatment. Around 90% of the persons having HIV have never been tested for the disease.

REFERENCES

Alfano R R (ed.) 1998 Advances in biopsy and optical mammography *Ann. N. Y. Acad. Sci.* **838**(ix–xii) 1–197

Gedeon A 2006 *Science and Technology in Medicine* (New York: Springer)

Herold K E and Rasooly A (eds.) 2009 *Lab-on-a-Chip Technology: Biomolecular Separation and Analysis* (Caister: Academic Press)

Rushmer R F 1972 *Medical Engineering: Projections for Health Care Delivery* (London: Academic Press)

Shepherd A P and Öberg P Å 1990 *Laser-Doppler Blood Flowmetry* (London: Kluwer Academic)

Strömberg T, Wårdell K, Larsson M and Salerud E G 2014 Laser Doppler perfusion monitoring and imaging in T Vo-Dinh (ed.), *Biomedical Photonics Handbook*, 2nd edition, Part 1 (Boca Raton: CRC Press)

Tay F E H 2001 *Microfluidics and Bio-MEMS Applications* (New York: Kluwer Academic)

Togawa T, Öberg P Å and Tamura T 2011 *Biomedical Sensors and Instruments* (Boca Raton: CRC Press)

CHAPTER 2

Cardiology

Blood Pressure

Alberto Avolio, Mark Butlin, and Dean Winter

CONTENTS

2.1	Introduction	14
2.2	Brief History	15
2.3	Direct Measurement of Blood Pressure	17
	2.3.1 Rational for Measurement	17
	2.3.2 Fluid-Filled Manometers	17
	2.3.2.1 Manometer Systems	17
	2.3.2.2 Frequency Response	18
	2.3.3 Catheter-Tip Sensors	20
	2.3.3.1 Piezoresistive Sensors	21
	2.3.3.2 Capacitive-Based Sensors	22
	2.3.3.3 Other Sensor Technologies	22
2.4	Indirect Measurement of Blood Pressure	23
	2.4.1 Basis for Measurement: The Occluding Cuff	23
	2.4.2 Technical Issues with Cuff-Based Measurements	24
	2.4.3 Korotkoff Sounds	25
	2.4.4 Oscillometric Technique	26
	2.4.4.1 Oscillogram	26
	2.4.4.2 Problems with Estimation of Pressure from the Oscillogram	28
	2.4.5 Ambulatory Blood Pressure Measurement	29
	2.4.6 Home Blood Pressure Monitoring Systems	29
	2.4.7 Guidelines, Standards, and Validation	30
	2.4.7.1 Systems for Validation	30
	2.4.7.2 Evaluation Standards and Protocols	30
2.5	Applications Using the Arterial Pulse Waveform	31
	2.5.1 Rationale	31
	2.5.1.1 Pulse Wave Analysis and the Importance of the Pulse Waveform	31
	2.5.1.2 Significance of Difference between Central and Peripheral Pressures	33

	2.5.2	Techniques for Detection of the Arterial Pulse	33
		2.5.2.1 Applanation Tonometry	33
		2.5.2.2 Finger Pressure Methods (Vascular Unloading)	35
		2.5.2.3 Brachial Cuff-Based Methods for Detection of the Arterial Pulse	36
	2.5.3	Noninvasive Measurement of Central Aortic Pressure	36
		2.5.3.1 Carotid Pressure Pulse	36
		2.5.3.2 Transfer Function Methods	37
		2.5.3.3 Central Systolic Pressure from Radial Systolic Inflection (P_i)	38
2.6	Future Directions		39
	2.6.1	Current Problems	40
		2.6.1.1 Improvement of Accuracy of Cuff-Based Systems	40
		2.6.1.2 Validation Protocols of Instruments	40
References			41

2.1 INTRODUCTION

The presence and vital significance of the pulse in blood vessels has been known since antiquity, but the measurement of the pressure exerted by the flowing blood in the circulation due to cardiac contraction is relatively recent. For accurate and true values of arterial blood pressure, there is no real substitute for direct measurements, with the accuracy being determined by the physical characteristics of the sensor system. However, direct measurements can be done only in limited conditions in humans and animals, and indirect measurements are necessary for the majority of clinical assessments. A major hallmark of the noninvasive sphygmomanometric measurement of arterial blood pressure is that it has undergone little change, since the cuff technique was proposed as the means of obtaining quantifiable values of arterial blood pressure and has been the main instrument in assessing cardiovascular risk related to levels of blood pressure. While the brachial cuff technique is embedded in clinical practice, there are advances being made where this is combined with information present in the arterial pulse so as to enhance the measurement of arterial blood pressure beyond the conventional systolic pressure (SP) and diastolic pressure (DP) values.

This chapter addresses the measurement of blood pressure in relation to the standard techniques, as well as new developments made in the areas of sensors and pulse wave analysis. There are many excellent reference texts on medical instrumentation that describe the fundamental principles in considerable depth. This chapter will include the basic concepts but will also canvas emerging fields and examine changes that are taking place to existing techniques.

The development of techniques for blood pressure measurement will be placed in the recent historical context, and detailed aspects of both direct and indirect measurement techniques will be described, with emphasis on the limitations due to system characteristics. These will also encompass ambulatory and home blood pressure measurement, as well as system validation and make reference to the continuously updated online resources. Methodologies incorporating pulse wave analysis will be described in relation to the use of the cuff sphygmomanometer as a calibration device to obtain noninvasive measurements of aortic pressure, as a means of enhanced assessment of cardiovascular function. Future directions will be examined in relation to the current problems that need to be addressed.

Current blood pressure monitoring in clinical practice faces a number of daunting challenges that have not yet been adequately addressed. The elimination of the mercury manometer to measure pressure has brought these challenges into sharp focus with the need to determine the best replacement pressure measurement system. Oscillometric-based systems appear to present the best opportunity for accurate measurement of SP, DP, and mean arterial pressure (MAP). However, the method presents both challenges (e.g., the discrepancies among manual sphygmomanometer measurements, oscillometry-based measurements, and invasive measures of blood pressure) and opportunities (e.g., the possibility of reducing or eliminating the confounding effects of the presence of observers during a measurement). The increased interest in home blood pressure monitoring as a more representative measurement of an individual's blood pressure and health risk is based, in part, on eliminating observer bias from the measurement. In addition, a new appreciation of the clinical importance of other features of the blood pressure waveform is expanding the measurement of blood pressure beyond the current parameters of SP, DP, and MAP.

2.2 BRIEF HISTORY

Johannes Müller, one of the most renowned physiologists of the nineteenth century, stated that "the discovery of blood pressure was more important than the discovery of blood." Indeed, both discoveries have produced highly significant and major advances in the fundamental understanding of cardiovascular physiology and in the practice of medicine. While the modern understanding of the circulation of blood is attributed to the exquisitely reasoned work of William Harvey published in 1628, that of blood pressure is attributed to the careful observations by Reverend Stephen Hales of the forces responsible for fluid movement in plants described in *Vegetable Staticks* in 1727 and extended to the forces due to the movement of blood as described in *Hemo-Staticks* in 1733. Hales not only demonstrated that a column of blood in a tube inserted in a major artery of a horse rose to "8 feet 3 inches perpendicular above the level of the left ventricle of the heart" but also quantified the presence of pulse pressure (PP) "when it was at its full height it would rise and fall at and after each pulse 2, 3, or 4 inches."

Following the work of Hales, the advances made in the eighteenth and nineteenth centuries were essentially in mechanical devices that would register directly the arterial pressure in human patients, with the first mercury manometer introduced by Poisseuille in 1827 leading to Ludwig's kymograph in 1847. While these and other devices gave measurements through arterial puncture, they had limited clinical use, and noninvasive means of measuring the force as registered by the arterial pulse were developed. The first sphygmomanometer was invented by Vierdort in 1854 using levers and weights leading to Marey's sphygmograph in 1860, a device used extensively to describe the change in the arterial pulse with various diseases. While attempts were made to quantify the forces due to arterial pulsations, it was realized that the resistance of the arterial wall and other tissue to displacement of mechanical levers and springs would produce errors. The first clinically useful instrument was developed by von Basch in 1881 where water-filled bags were used to transmit the force of the radial pulse using the tambours developed earlier by Marey. This was modified by Potain in 1889 by replacing fluid-filled chambers with air and aneroid manometers.

Although the progression of modifications since the early Ludwig instruments produced gradual improvements, it was the simplified modifications by Riva-Rocci that produced the single most useful advance in the clinical application of the sphygmomanometer. While early devices attempted to measure the force associated with the directly palpable pulse, Riva-Rocci's invention was to apply the cuff entirely around the upper arm and compress it to the point where the radial pulse becomes obliterated. The pressure in the cuff corresponding to the reappearance of the radial pulse judged by palpation by the fingers of the operator corresponded to SP. The circumferential compression of the brachial artery by the Riva-Rocci cuff method was a significant improvement to the von Basch techniques of essentially asymmetric lateral compression. However, the relatively small cuffs (5 cm) used by Riva-Rocci could not be reliably applied to a wide range of adults. Later improvements by von Recklinghausen in 1901 showed that 12 cm cuffs gave more reliable measurements.

Even though the brachial cuff produced useful measurements of SP, the palpatory method could not be used to measure DP, since the pulse was present at all levels of cuff pressure below arterial SP. The search for a method to measure DP produced the second most significant advance in sphygmomanometry, that of the auscultatory method proposed by Korotkoff in 1905. By placing a stethoscope distal to the brachial cuff at the cubital fossa over the brachial artery, appearance and disappearance of sounds could be heard. Korotkoff reasoned that the cuff pressure at the point of appearance and disappearance of the sounds corresponded to arterial SP and DP, respectively.

The introduction of the brachial cuff enabled the application of other methodologies for measurement of arterial blood pressure. The application of the Korotkoff sounds enabled measurement of PP, where sounds can be detected by acoustic microphones or ultrasound methods. The oscillogram, the recording of the pressure oscillations in the cuff, was observed in various forms as early as Marey's work. However, its application to measuring blood pressure did not come until a century later when the advent of microprocessors and small, sensitive pressure sensors led to the development of practical oscillometric blood pressure monitors. By the late 1970s, several companies had independently developed oscillometric devices, expanding the application of blood pressure measurement to the wider community. Although oscillometric devices now dominate the automated blood pressure measurement market, the specific advances made by Riva-Rocci and Korotkoff that developed from the pioneering principles of Hales, Ludwig, von Basch, and Marey have stood the test of time, and the manual detection of the Korotkoff sounds using a stethoscope and cuff/mercury sphygmomanometer remains the most accurate device for the noninvasive measurement of arterial blood pressure.

2.3 DIRECT MEASUREMENT OF BLOOD PRESSURE

2.3.1 Rational for Measurement

The direct measurement of blood pressure is necessary when absolute values of true pressure are required. This is in situations where any inaccuracies or artifact can result in serious adverse outcomes, such as in operating theaters where anesthetic agents are titrated according to blood pressure levels or in intensive care monitoring. Throughout the circulatory system, the only location where pressure can be obtained by indirect means is the large arteries; all other locations including cardiac chambers, pulmonary circulation, and veins essentially require direct measurement.

The American Heart Association (AHA) recommends direct blood pressure measurements using fluid-filled catheters or catheter-tip sensors for most experimental applications in laboratory animals. Indirect measurements are also possible in some animals (e.g., mice and rats) using a tail cuff and oscillometric pressure techniques. Several techniques have been developed to detect flow or pressure pulsations during cuff deflation, but these methods do not appear to correlate well with direct pressure measurements. Algorithms based on analysis of the oscillogram have been shown to provide satisfactory estimates of SP, DP, and MAP. Indirect methods are recommended by the AHA for animal experiments mainly in which a high throughput of measurements is required.

Accuracy of measurement, however, depends on both the sensor and sensor system and whether the sensor is mounted on a catheter or externally connected via fluid-filled tubes.

2.3.2 Fluid-Filled Manometers

There is a large body of work addressing the theory and practice of fluid-filled manometers for measurement of blood pressure, and much of it is found in the texts and references listed at the end of this chapter. The following sections will treat the quantitative aspects of the sensor and the manometer system so as to be able to determine the essential measurement characteristics related to the dynamic response of the system in terms of natural resonant frequency and damping coefficients.

2.3.2.1 Manometer Systems

The models used for fluid-filled manometer system are based on a lumped parameter representation of second-order systems. The models and their application to manometer systems are extensively described in standard engineering and medical instrumentation texts (Webster 2010). Essentially, the manometer system is represented by a lumped parameter circuit where electrical analogs correspond to the catheter fluid resistance (R_c), fluid inertance due to the liquid mass contained within the catheter, and the sensor casing (L_c) and the capacitive element (C_c) of the volume storage compartment, which consists of the elastic properties of the catheter tubing and the diaphragm of the sensor (Figure 2.1). For a manometer system with known dimensions (L, length of the catheter, m; r, radius, m), physical constants (η, fluid viscosity, Pa·s; ρ, fluid density, kg/m^3), and measured change in volume (ΔV) and pressure (ΔP), the calculations for the values for the analog parameters are

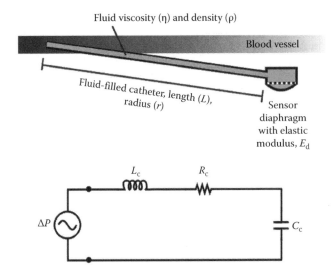

FIGURE 2.1 The physical manometer system with catheter of length L and radius r, filled with a fluid of viscosity η and density ρ and attached to a pressure sensor with a diaphragm of stiffness E_d. The electrical analog of the manometer system is schematically represented, with the catheter fluid resistance (R_c), fluid inertance (L_c), and the compliance of the sensor diaphragm (C_c) linked in series to the pressure source (ΔP). Model parameters are determined from the catheter and sensor physical properties as shown in Equations 2.1 through 2.3.

given in Equations 2.1 through 2.3. The calculation of the elastic constant of the sensor diaphragm (E_d) assumes that the majority of the volume change is taken up by the displacement of the sensor diaphragm, which is then transformed into a voltage change. The quantities of resistance, inertance, and capacitance determine the frequency response of the system.

$$R_c = \frac{8\eta L}{\pi r^4} \tag{2.1}$$

$$L_c = \frac{\rho L}{\pi r^2} \tag{2.2}$$

$$C_c = \frac{\Delta V}{\Delta P} \propto \frac{1}{E_d} \tag{2.3}$$

2.3.2.2 Frequency Response

For catheter systems, the frequency response can be obtained by introducing a step change in pressure and analyzing the transient response. The step change in pressure is obtained by the *pop test* method, where a rubber balloon is inflated to a constant pressure and then suddenly opened to atmospheric pressure by bursting with a lighted match or a pinprick (Nichols *et al* 2011, Webster 2010). Depending on the combination of values of the circuit that constitutes a model of a second-order system, the response can be underdamped, overdamped, or critically damped.

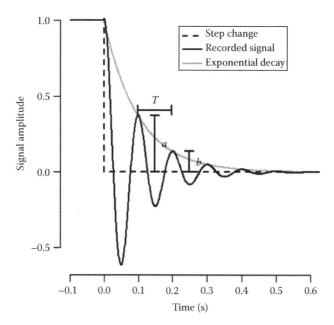

FIGURE 2.2 An example of an underdamped system response (solid line) to a step change (dashed line) in pressure. The logarithmic decrement (Λ) in pressure can be calculated by the damping ratio (ζ) and the underdamped natural frequency (ω_n), where T is the period between peaks of the pressure in the system response (Equations 2.4 through 2.6). The response is mathematically defined by Equation 2.9, where the peak amplitude (A) is dependent upon the magnitude of the step change (G), which is normalized to 1 in this example.

An underdamped system will oscillate about a step change in the signal, with the magnitude of the oscillations decreasing exponentially (Figure 2.2). The magnitude of the damping of the oscillations, the logarithmic decrement (Λ), and the frequency of the oscillations, the natural frequency (ω_n), characterize the system. The logarithmic decrement can be calculated from the amplitude of two adjacent peaks in the oscillations (Figure 2.2, Equation 2.4) and the natural frequency calculated from the damping ratio (ζ, Equation 2.5) and the period (T) of the oscillations (Equation 2.6). The response with respect to time ($P(t)$) to a step change of magnitude G is therefore described by Equation 2.9:

$$\Lambda = \ln(a) - \ln(b) \tag{2.4}$$

$$\zeta = \frac{\Lambda}{\sqrt{4\pi^2 + \Lambda^2}} \tag{2.5}$$

$$\omega_n = \frac{2\pi}{\left(T\sqrt{1-\zeta^2}\right)} \tag{2.6}$$

$$A = \frac{G}{\sqrt{1-\zeta^2}}$$

(2.7)

$$\phi = \arcsin\left(\sqrt{1-\zeta^2}\right)$$

(2.8)

$$P(t) = \frac{Ae^{-\zeta\omega t}}{\sqrt{1-\zeta^2}\sin\left(\omega t\sqrt{1-\zeta^2}+\phi\right)}$$

(2.9)

From the theory of second-order systems, the parameters of the response defined by Equations 2.4 through 2.9 are determined by the equivalent analog parameters in Figure 2.1 and defined by catheter and sensor physical characteristics defined by Equations 2.1 through 2.3 (Nichols *et al* 2011, Webster 2010).

While a qualitative evaluation of the response of the system can be gauged by visual inspection of the response in the time domain, a much more readily quantifiable evaluation can be made in the frequency domain where the extent of amplification of the signal $(X(j\omega)/Y(j\omega))$ is expressed for each frequency (ω) as a function of the underdamped natural frequency (ω_n) and the damping coefficient (ζ):

$$\frac{X(j\omega)}{Y(j\omega)} = \frac{K}{\sqrt{\left[1-\left(\omega/\omega_n\right)^2\right]^2 + 4\zeta^2\omega^2/\omega_n^2}}$$

(2.10)

This analysis has been applied to catheters attached to external sensors for the measurement of intra-arterial pressure, as routinely used in catheterization laboratories. Figure 2.3 shows the frequency response of a pigtail and JL4 catheter (6F and 5F, respectively; Cordis) when directly attached to a low-compliance pressure sensor (Braun Angiotrans® manifold and pressure sensor for angiography) and when attached to the sensor, as conventionally done, via a 90 cm fluid-filled line of 3 mm inner diameter. The effect of the fluid-filled line is to reduce the resonant peak from greater than 25 Hz to around 17 Hz. This has the effect of increasing the amount of distortion in the lower frequency band (<10 Hz), which contains the major part of the energy of the arterial pressure pulse. This would have the effect of overestimating SP compared to the true value.

2.3.3 Catheter-Tip Sensors

Placement of the sensing element on the distal tip of an invasive catheter has allowed the more precise and higher fidelity measurement of arterial pressure than can generally be achieved using a fluid-filled catheter. This is especially true in recording blood pressure in small animals where the high heart rates may produce frequency components (50–70 Hz) that exceed the frequency response of fluid-filled catheters (up to 20 Hz). Catheter-tip pressure sensors have other advantages over fluid-filled catheters. Firstly, they are easier to zero

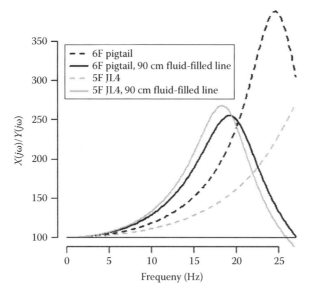

FIGURE 2.3 Actual frequency responses of catheters (dashed lines) and catheters with a 90 cm fluid-filled line of 3 mm inner diameter (solid lines) used in a clinical setting. Each catheter was characterized using a *pop test* (Figure 2.2) to ascertain the underdamped natural frequency (ω_n) and damping ratio (ς). The resultant frequency response is given by Equation 2.10. The underdamped natural frequency corresponds to the peak, and the damping ratio determines the decay from this peak. The addition of a fluid-filled line to the catheter decreases the underdamped natural frequency but increases the damping ratio.

to a voltage equivalent to atmospheric pressure and do not require periodic flushing to removed clots or air bubbles. Secondly, they are generally less susceptible to catheter whip, the catheter motion induced by unstable or accelerating fluid velocities.

Several types of sensors have been applied to invasive blood pressure measurement, including capacitance-based sensors, fiber-optic sensors, surface acoustic wave (SAW) sensors, and piezoelectric-based sensors. However, the most commonly used technology by far is the piezoresistive sensor on a silicon substrate.

2.3.3.1 Piezoresistive Sensors

In this design, two arms of a Wheatstone bridge are typically embedded in the silicon substrate at the catheter tip, while two other arms, along with a compensation resistor, are in a signal processing unit connected to the proximal end of the catheter. Thus, the electrical connections to the catheter tip consist only of the wires connected to the two bridge arms. The catheter also typically contains a vent from atmospheric pressure to the face of the silicon not subjected to blood pressure so that the measured pressure is always relative to atmospheric (gauge) pressure.

The excellent physical and thermal stability of piezoresistive sensors allows them to be zeroed and calibrated using set voltages programmed into the supporting electronics. These sensors have an excellent frequency response (>1000 Hz), well above that required for blood pressure waveforms, so that they are most frequently applied when a high-fidelity waveform is required. Piezoresistive sensor-tipped catheters as small as 1F (0.3 mm diameter) are commercially available.

Catheter whip is an issue in all pressure catheters and is caused by the flexible catheter being subjected to the rapidly varying blood velocities in locations such as the ascending aorta. However, this condition can be readily recognized by inspecting the recorded waveform.

2.3.3.2 Capacitive-Based Sensors

Capacitive sensors register pressure by measuring the capacitance between two electrode surfaces, one on a deformable diaphragm exposed to the sensed pressure and the other on a rigid substrate. As the diaphragm deforms under a pressure load, the distance between the two electrodes, and therefore the capacitance of the circuit, changes. The distance between electrodes and the area of electrodes define the initial capacitance of the capacitive pressure sensor and, together with the geometry and flexibility of the diaphragm, define the sensitivity of the sensor (single-plate designs are also available).

Capacitive-based sensors have not found a significant market in catheter-tip manometers, in large part because of their higher inherent electrical noise compared to piezoresistive sensors and the resultant reduced frequency response. They also tend to be more complex in their signal processing electronics compared to piezoresistive sensors because of the need for dynamic excitation of the capacitors. This requires an internal oscillator, along with a signal demodulator, which increases their overall cost. However, capacitive-based sensors have been incorporated into prototype implantable systems to measure blood pressure because of their low power requirements and very small size. They can also be mass-produced through silicon micromachining techniques.

2.3.3.3 Other Sensor Technologies

2.3.3.3.1 Piezoelectric Sensors: While piezoresistive sensors dominate in the application to blood pressure measurement (with capacitive sensors a distant second), other technologies have been investigated. Piezoelectric sensors have the disadvantage of being an AC-coupled sensor, and therefore, mean blood pressure cannot be measured. This characteristic makes piezoelectric sensors impractical for blood pressure measurement, except when only the time-varying portion of the signal is required.

2.3.3.3.2 Surface Acoustic Wave (SAW) Sensors: SAW sensors have also been investigated for use in measuring blood pressure. These sensors are simple, have low power requirements, and are very stable, showing little or no drift over long periods of time (months or longer) when temperature compensated. These sensors operate on the principle that the deformation of a diaphragm, usually a piezoelectric material, changes the resonance frequency of the SAW. While this type of sensor has not yet found commercial application in blood pressure measurement, their characteristics indicate that they may be suitable, as are the capacitive sensors, in implantable applications.

2.3.3.3.3 Fiber-Optic Sensors: Fiber-optic pressure sensors have a number of potentially attractive features, such as small size and immunity from electromagnetic interference.

These sensors operate on the principle that an incident light beam on a deformable diaphragm is reflected from the diaphragm to create interference patterns. Early versions of fiber-optic pressure catheters were subject to drift, which has limited their application in blood pressure measurement.

2.4 INDIRECT MEASUREMENT OF BLOOD PRESSURE

2.4.1 Basis for Measurement: The Occluding Cuff

The basis for almost all indirect blood pressure measurement systems is an inflatable cuff wrapped around the upper arm. The cuff is inflated to above the pressure required to collapse and completely occlude the brachial artery before the cuff is slowly deflated. Blood pressure measurement is based on two basic assumptions:

1. During deflation, certain landmarks (described below) identify SP and DP and, in the case of the oscillometric technique, MAP.

2. The pressure in the cuff at those landmarks corresponds to the arterial pressure associated with the landmarks.

This technique has not fundamentally changed in more than 100 years following its introduction. The original landmark used by Riva-Rocci was the reappearance of the palpated radial pulse during cuff deflation, identifying SP. Landmarks to identify DP are more subtle and have been described as a "thin, very abrupt…snapping" when the brachial artery is palpated during cuff deflation. This *snapping* disappears at the point of diastole. The technique, however, requires considerable skill and training on the operator's part, and so has not found wide application. The description of the Korotkoff sounds several years after Riva-Rocci provided a means to identify DP, along with an easier and more reliable identification of SP, and is still used today.

Riva-Rocci's original occluding cuff was a narrow (5 cm in width) rubber bladder encased in a leather sheath. The modern version of the cuff follows the same basic design as Riva-Rocci's original cuff, that of an inflatable bladder inside of an inelastic covering. However, for the cuff pressure to match the arterial pressure, the artery must be subjected to a uniform pressure along the length of the artery under the cuff. A narrow cuff results in longitudinal wall and tissue stresses contributing to the balance of forces from the cuff and blood pressure so that the cuff and blood pressures are not equivalent. Thus, the occluding cuff pressure must overcome a component of the longitudinal wall stress as well as the blood pressure, resulting in, for example, an overestimation of SP. This overestimation was quickly noted by Riva-Rocci and others and led to the use of wider cuffs for more accurate measurements. Conversely, cuffs that are too large may give incorrectly low systolic values, especially in children. Today, the upper arm circumference is used to determine the required width of the brachial cuff. Current guidelines recommend a bladder length of 80% and width of 40% of the arm circumference, but this subject continues to be an area of discussion.

Only limited studies have been reported that compare invasive brachial pressure measurements with auscultatory measurements. Though such studies are technically challenging as ideally the two measurements should be taken simultaneously at the same location (the brachial artery), several studies have been completed. These studies indicate that the auscultatory method underestimates invasive SP by approximately 10 mmHg and overestimates invasive DP by approximately the same amount. This finding appears to be consistent in hypertensive adults over a wide age range, not just in the elderly.

2.4.2 Technical Issues with Cuff-Based Measurements

Choosing an appropriately sized cuff is a major factor in obtaining accurate blood pressure measurement with the occluding cuff. However, there are a number of other technical and protocol issues that should be taken into account or adhered to. Technical issues include the following:

1. *Rate of cuff deflation.* The AHA guidelines recommend a deflation rate of 2 mmHg/s. A faster deflation rate may lead to an underestimation of SP, since the initial opening of the artery at systole can occur at any time between pulses; as the deflation rate increases, the range of pressure between the pulses increases (Figure 2.4). Protocol problems with occluding cuff systems are many and widespread. Various societies, including the AHA, publish protocols for blood pressure measurement, but adherence to these protocols does not tend to be widespread.

2. *Pressure measurement systems.* The mercury column sphygmomanometer, considered to be the most accurate pressure measurement method, is being phased out of clinical use because of safety concerns over mercury, a potent neurotoxin. Alternatives

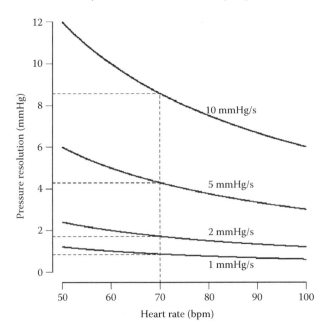

FIGURE 2.4 Dependency of resolution of pressure readings on the linear deflation rate (mmHg/s) for given values of heart rate. The dashed line shows the pressure resolution for a heart rate of 70 bpm.

to the mercury sphygmomanometer are aneroid pressure sensors, typically used in auscultatory measurements, and solid-state pressure sensors typically used in oscillometric measurements. All three measurement methods have been shown to provide reliable and comparable pressure measurements, provided that they are adequately maintained and regularly (at least annually) calibrated. However, aneroid manometers have been shown to require the most care and are prone to breakage and calibration drift. Solid-state sensors are much less susceptible to such problems. Errors with mercury manometers tend to be due to incorrect zeroing of the mercury column or blockages that may prevent the mercury column from falling freely.

3. *Body posture.* Errors in blood pressure measurement can be reduced by quietly sitting upright with one's back supported and feet flat on the floor.

4. *Arm position.* If the occlusive cuff is not at the level of the heart, a hydrostatic pressure gradient, either positive or negative, will be introduced in the measurement. In addition, if the cuff is applied over a shirt or other clothing, an artificially high blood pressure may be measured.

5. *Transiently altered physiology.* Transient changes in blood pressure may occur due to prescribed or over-the-counter medications, recent smoking, caffeine consumption, or anxiety, often due to the presence of the operator or the measurement itself.

6. *Recent physical activity.* If the subject does not sit quietly for at least 5 min, sometimes even longer, then the measured blood pressure can be at an artifactual, nonresting, elevated level.

7. *Operator error.* This problem includes wrong choice of cuff size, under- or overinflation of the cuff, digit or zero preference, or too rapid cuff deflation.

The technical problems can lead to inaccurate measurements, while the protocol violations can lead to artifactual or transient changes in blood pressure that do not reflect the patient's cardiovascular health.

Some patients become more hypertensive (white-coat) or less hypertensive (masked) when their blood pressure is measured in a clinical setting. The incidence of both conditions is the subject of much dispute, but both appear to be in the range of 10%–20% for the general population and as high as 50% in some populations, for example, those treated for hypertension. Ambulatory blood pressure monitoring, and possibly home blood pressure monitoring, has been shown to be useful in accurately diagnosing these conditions.

2.4.3 Korotkoff Sounds

It was the Korotkoff association of the sounds heard through the stethoscope placed over the brachial artery, which change with phases of the arterial pressure pulse that provides the information for estimation of DP. The sounds appear and disappear as the brachial cuff is gradually deflated from levels of suprasystolic cuff pressure. The cuff pressure at which the sounds are initially heard corresponds to arterial SP and, when they disappear, to DP. However, in some individuals, the Korotkoff sounds (or K-sounds) may continue for some time during cuff deflation below DP. Hence, a more formal description of the phases of the K-sounds is employed to standardize measurement. The spectrum of the K-sounds is divided into five different phases:

Phase I: The appearance of the sound with a *snapping* characteristic

Phase II: Continuous persisting murmurs

Phase III: Increasing of sound intensity and *sharpness* above that of Phase II

Phase IV: The muffling of sounds

Phase V: The cessation of sounds

Recommendations for determination of DP are to use Phase V; however, if sounds persist throughout cuff deflation, Phase IV should be used. The sounds can be detected with a stethoscope in manual recordings, and the operator makes the decision to associate the cuff pressure reading with the specific sounds. In automated devices, the sounds are detected by microphones and computerized algorithms are used to classify the phase and associate these with cuff pressures.

A variety of mechanisms have been speculated for the origin of the K-sounds, including flow-based mechanisms, such as the water-hammer phenomenon and wall vibrations induced by a turbulent jet, and pressure-based mechanisms, such as arterial wall instabilities or shock waves in the blood. However, none of these have survived close analysis. The sounds instead have been shown to be a complex interaction between the nonlinear pressure–flow relationships in the portion of the brachial artery under the cuff and the elasticity of that same portion of the brachial artery (Drzewiecki *et al* 1989).

The occurrence of an auscultatory gap, where the sounds in Phases I and II disappear and reappear, can be a source of error, especially for SP determination. In these cases, the original palpatory method of Riva-Rocci can be used to determine SP. The causes of the auscultatory gap have been attributed primarily to the effects of respiration. However, studies have also shown a significant association with vascular wall properties, with suggestions that the presence of the auscultatory gap, in itself, may have a prognostic significance.

2.4.4 Oscillometric Technique

2.4.4.1 Oscillogram

The oscillometric method involves the analysis of the oscillations in the cuff pressure (the oscillogram) during deflation of the cuff. Compared to the mean cuff pressure, the oscillations are relatively small, typically only a few mmHg. When the oscillogram is plotted as a function of mean cuff pressure, the oscillation amplitude increases from zero to a maximum and then decreases to a smaller value or zero.

The pressure at maximum amplitude corresponds to the vascular unloading condition of minimal (or zero) transmural arterial pressure. At this pressure, there is maximal wall expansion and the cuff pressure is equal to the MAP (Geddes *et al* 1982). To obtain SP and DP, the oscillogram is analyzed to obtain characteristic points that can be identified to correlate with these pressures. Early embodiments of the oscillometric technique assumed that SP and DP occurred at the onset and cessation, respectively, of the cuff pressure pulsations, but more sensitive pressure measurements showed that these pulsations occur well before and continue well after the occurrence of SP and DP (Figure 2.5). Currently, both SP and DP are identified as corresponding to a predetermined fraction of the maximum

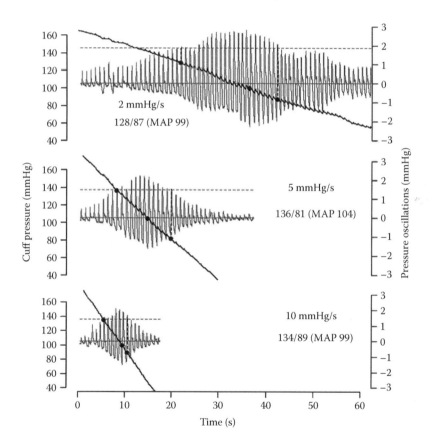

FIGURE 2.5 The oscillogram and the effect of different linear cuff deflation rates (2, 5, and 10 mmHg/s, in black, left axis) on the oscillogram (in gray, right axis) and the estimation of SP, DP, and MAP (black circles) in a 31-year-old male. A simplified algorithm finding points where the oscillogram first and last exceeded two-thirds (horizontal dashed line) of the maximum positive pulse amplitude were used to define SP and DP, respectively. MAP was defined as the point of maximum oscillations. The brachial blood pressure measured with a commercial oscillometric device (Omron M4, with a deflation rate between 2 and 5 mmHg/s) was 135/82 mmHg.

PP in the cuff. In general, SP has been found to correspond to the cuff pressure when the pressure oscillations reach 50% of their maximum value, and DPs found to correspond to the cuff pressure when the pressure oscillations decrease to 80% of their maximum value. However, various manufacturers of oscillometric-based devices have developed proprietary algorithms to determine both SP and DP.

2.4.4.2 Problems with Estimation of Pressure from the Oscillogram
Comparative studies between the oscillometric and auscultatory methods have shown that in normal subjects, oscillometric-based devices may systematically overestimate SP and underestimate DP compared to auscultatory measurements. A number of investigators have developed theoretical analyses of the oscillometric method of determining blood pressure (Baker *et al* 1997). Baker *et al* investigated sources of error in the mean pressure estimation. Their analysis showed that the differences between oscillometric and auscultatory measures are due to a number of factors, including local arterial and arm tissue mechanics, high PP, mechanical properties of the inflating cuff, the shape of the arterial pressure waveform, and heart rate. These effects generally result in an overestimation of MAP. The overestimation can be substantial, as much as 10–15 mmHg, and is most prominent at the extremes of the different factors. For example, in diabetic patients who typically have increased arterial stiffness, oscillometric methods overestimate both SP and DP compared with auscultatory measurements. This may also be a factor in elderly people with very stiff arteries that more strongly resist collapse under the pressure of the occluding cuff.

The proprietary algorithms used by each manufacturer to determine SP and DP from the oscillogram make it difficult to analyze systematic errors in measurement results. However, these algorithms are generally dependent on the MAP value, so that any error in MAP adversely affects SP and DP values. Such an error may be due to the difficulty in identifying the peak of the oscillogram when the oscillogram is only gradually increasing and decreasing in amplitude (Figure 2.5). Also, in a small number of patients, spurious oscillogram shapes can occur, leading to an inaccurate determination of the peak and, therefore, the MAP. In addition, while almost all manufacturers determine the MAP from the oscillogram, some *display* the MAP as computed from the widely reported formula (Equation 2.11) using the DP and PP. However, this approximation depends on the shape of the pressure waveform, and its use may produce significant error in the displayed MAP:

$$\mathrm{MAP} = \mathrm{DP} + \frac{1}{3} \times \mathrm{PP} \tag{2.11}$$

As with the auscultatory technique, deflation rate is an important parameter, since, in combination with heart rate, it determines the sampling rate of the cuff pressure and, therefore, the resolution of the blood pressure measurement. For example, at the AHA recommended deflation rate of 2 mmHg/s and a heart rate of 70 bpm (or a period of 60/70 = 0.86 s), the pressure resolution is 1.7 mmHg. However, if the deflation rate is increased to 10 mmHg/s, the pressure resolution is 8.6 mmHg (Figures 2.4 and 2.5). A linear deflation rate is also desirable, since a nonlinear rate would result in a varying pressure resolution.

2.4.5 Ambulatory Blood Pressure Measurement

Devices for ambulatory blood pressure measurement (ABPM) are lightweight (often less than 300 g), portable systems that can be worn on a 24 h basis and set to record blood pressures at regular intervals, typically every 15–30 min, or on demand. ABPM devices are typically used when clinic blood pressure measurements are thought to be inadequate, such as in cases of suspected white-coat or masked hypertension. However, they are increasingly used in a variety of hypertension-related conditions and are becoming a significant means of hypertension diagnosis and management. These devices are now available from a variety of medical device manufacturers.

Early ABPM systems used automated auscultatory techniques to measure blood pressure to reduce the effect of motion artifact. But with improvements in signal processing and sensor technology, almost all current ABPM systems are based on oscillometric techniques, with a number of systems offering a choice between the two. Volume oscillometry using a finger cuff has also been used as a measurement technique, but these systems have shown only variable correlation with brachial cuff-based systems and so are not widely used.

Data logging is an important capability for ABPM devices, and the availability of these data over 24 h has introduced considerable new insights into our knowledge of blood pressure. One example is a better understanding of the diurnal cycle of blood pressure and the association of abnormalities in the cycle with cardiovascular diseases. Data are recorded on a solid-state memory device and can then be downloaded to a computer to display the 24 (or 48) h record of blood pressure and ancillary information.

2.4.6 Home Blood Pressure Monitoring Systems

Home blood pressure monitors have come into increasingly widespread use, in large part because they eliminate some of the artifacts associated with office blood pressure measurements, such as the influence of the presence of a nurse of physician. Some studies indicate that home blood pressure measurement helps to identify the presence of white-coat and masked hypertension, and recent studies have shown that home blood pressure readings tend to be lower than office blood pressure readings. The units are relatively inexpensive, but systems that employ an upper arm cuff still can be reliable and accurate, provided the users are adequately trained. Two main problems with the systems, however, are adherence to protocol and accurate data recording. The AHA has published guidelines for home blood pressure monitoring, but compliance is difficult to determine. Electronic data logging that makes blood pressure readings available to a physician greatly improves the accuracy of the recorded data.

Virtually all home blood pressure monitors are based on oscillometry. Wrist cuff systems are increasing in popularity because of their ease of use, but their accuracy has been widely questioned. One problem with the wrist cuff is that there are no guidelines for cuff width. As with upper arm cuffs, the width of the cuff can significantly alter the measured blood pressure. Commercially available wrist cuff systems have cuff widths that range from 6 to 10 cm, and adult wrist circumferences range from approximately 15 to 30 cm. If the guidelines for choosing a cuff size for the upper arm are applied to the wrist, the ratio of cuff width to wrist circumference is less than the minimum 40% recommended for upper arm cuffs. This means that pressures measured with wrist cuffs will tend to be higher than the actual SP and DP. In addition, the effect of radius and ulna bones on arterial collapse under the wrist cuff is not known. If these bones tend to counter the effect of cuff inflation on arterial collapse, they may contribute to inaccurately high-pressure measurements.

2.4.7 Guidelines, Standards, and Validation

2.4.7.1 Systems for Validation

The measurement of arterial pressure is a simple and practical means to obtain one of the most basic quantitative parameters in clinical medicine. However, it is subject to a range of possible errors due to the interaction of the operator with all components of the sphygmomanometer system, such as proper application of the cuff, selection of cuff size, fiducial auscultation of sounds, cuff deflation rates, and reliable analysis of the oscillogram. In addition, systematic validation studies indicate that not all devices meet the recommended criteria for equivalence with intra-arterial measurements (dabl® Educational Trust Ltd 2011).

2.4.7.2 Evaluation Standards and Protocols

Standards for evaluation of blood pressure measurement devices are set by professional societies. At present, there are three principal societies that have issued recommendations: the Association for the Advancement of Medical Instrumentation (AAMI) (American Association for the Advancement of Medical Instrumentation 2008), the British Hypertension Society (BHS), and the European Society of Hypertension (ESH). The original standard for validation of electronic or automated sphygmomanometers as used with an occluding cuff was published by the AAMI in 1986, with subsequent revisions in 2008. The standards have now been consolidated under several IEC standards: IEC 60601-2-34 *Invasive blood pressure monitoring equipment*, IEC 80602-1 *Validation of non-automated sphygmomanometers*, IEC 80602-2-30 *Basic safety and essential performance on automated sphygmomanometers*, and IEC 81060-2 *Clinical validation of automated, non-invasive sphygmomanometers.*

Information regarding all aspects of sphygmomanometric measurement of blood pressure, ranging from specifications and testing of commercial devices to standardization of measurement techniques, along with a continuously updated list of relevant publication techniques, is contained in a comprehensive online resource maintained by the dabl Educational Trust Ltd (2014). This website was established in 2003 by Professor Eoin O'Brien and William J. Rickard to provide open access to information on a wide range of blood pressure monitors. The content is regularly updated and is rich in quantitative data. Devices are classed in specific categories:

1. *Devices for clinical use*: Manual, mercury, and aneroid sphygmomanometers and automated devices for clinical use

2. *Self-measurement devices*: Upper arm cuff devices, wrist devices, and devices for community use

3. *Ambulatory blood pressure measurement (ABPM) devices*: Devices for intermittent and continuous measurement of blood pressure

All of these classes of devices have linked tables of specific devices and manufacturers classified as *recommended*, *questionable*, and *not recommended*, based on an evaluation to a set of standardized criteria as per the International protocol (IP) approved by the ESH, the BHS, and the American Association for the Advancement of Medical Instrumentation Standards.

Recommendations have also been compiled for the use of various techniques in clinical blood pressure measurement and the exact procedures through which blood pressure should be measured (Pickering *et al* 2005). For example, the measurement of brachial blood pressure using a cuff should be taken with a pillow under the arm if in the supine position and supported if in the seated position with a deflation rate of 2–3 mmHg/s and with a minimum of two readings taken with a time between the measurements of at least 1 min. The AHA has also developed recommendations for blood pressure measurement in animals (Kurtz *et al* 2005), covering both indirect and direct methods of measuring blood pressure and the use of fluid-filled and sensor-tipped catheters.

2.5 APPLICATIONS USING THE ARTERIAL PULSE WAVEFORM

2.5.1 Rationale

There has been recent interest in combining the early methods of *sphygmography* (noninvasive detection of the arterial pressure pulse waveform) with s*phygmomanometry* (noninvasive measurement of the arterial pressure using the arterial pulse) so as to obtain information beyond the conventional SP and DP values. The combination of measurement or arterial blood pressure and pulse wave analysis is being accepted as a novel potential means of enhancing cardiovascular assessment (Avolio *et al* 2010). The principal additional features of this novel methodology are the noninvasive quantification of the central aortic pressure in relation to the conventional measurement of brachial cuff pressure and waveform features that relate to qualitative and quantitative assessment of arterial stiffness and peripheral wave reflection.

2.5.1.1 Pulse Wave Analysis and the Importance of the Pulse Waveform
The periodic cardiac ejection can be analyzed by Fourier analysis of oscillatory time-dependent signals of arterial pressure and flow and relationships expressed in the

frequency domain (Nichols *et al* 2011). Assuming linearity, a specific component of pressure is uniquely related to the same harmonic component of flow determined by the vascular impedance at that frequency. Similarly, a specific harmonic component of distal pressure is uniquely related to the same harmonic component of proximal pressure as determined by the vascular transmission values of modulus and phase at that frequency.

The time course of the pressure pulse waveform at the aortic root is determined by the pattern of ventricular ejection and the elastic and geometric properties of the arterial tree (Nichols *et al* 2011). During the cardiac cycle, the waveform exhibits prominent features that can be used as descriptors of the pressure pulse. In the ascending aorta (Figure 2.6), peak pressure (P_s) occurs after peak flow due to the capacitive effects of the ascending aortic segment. The first inflection generally coincides with time of peak flow velocity (at pressure value of P_i), at time T_i, at approximately 30% of the ejection duration. The ratio of the augmented component of pressure to the PP (Equation 2.12) is defined as the augmentation index (AIx). The AIx has been found to have a significant heritability factor and shows changes with age. It is generally negative in young individuals and increases to positive values with increasing age due to significant changes in the waveform morphology brought about by changes in structural components of the arterial system affecting pulse wave propagation. It was conceived as a parameter associated with ventricular–vascular coupling:

$$\text{AIx}\,(\%) = 100 \times \frac{(P_s - P_i)}{(P_s - P_d)} \tag{2.12}$$

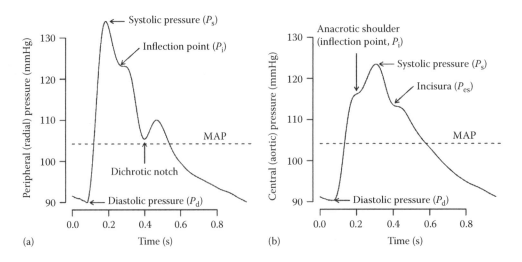

FIGURE 2.6 Features of the radial pulse wave (a) and corresponding central aortic pressure wave (b). Radial pressure wave was recorded using tonometry from a 50-year-old male, BMI of 24 kg/m², and the aortic pressure wave derived with an aortic to radial transfer function employed in the SphygmoCor device (AtCor Medical). MAP, mean arterial pressure.

The radial pulse is also characterized by a number of waveform features (Figure 2.6). The features correspond to the following descriptors:

1. A sharp ascent from the lowest point corresponding to DP (P_d). This point is often equated to that in the aorta (or in the carotid artery) and, together with mean pressure, is used as a calibration reference.

2. A steep rise of pressure corresponding to continued left ventricular ejection.

3. A first peak in the waveform corresponding to SP (P_s).

4. An inflection occurring late in systole. The pressure at this inflection point (P_i) is usually lower than P_s but may form a local maximum higher than P_s in cases of high arterial stiffness as seen with advancing age. It is also related to P_s in the aorta.

5. A local minimum near the end of systole, the dicrotic notch, often incorrectly called the incisura. Unlike the incisura, the dicrotic notch is not causally related to the closing of the aortic valve. However, there is a strong correlation between the timing of the dicrotic notch (radial) and the timing of the incisura (aortic), and thus, it can be used to obtain systolic duration.

6. A gradual decline in pressure during diastole after closure of the aortic valve.

7. The lowest point in the wave corresponding to DP, after which the cycle recommences.

2.5.1.2 Significance of Difference between Central and Peripheral Pressures
A characteristic feature of the pressure pulse is that it changes shape as it travels away from the heart. In large arteries, mean pressure is essentially constant. Hence, shape changes are such that the total waveform area over one cardiac period is constant. The resultant alteration to the pressure waveform is that the pulse height changes; that is, the pulse becomes distorted with a concomitant increase in pulse height. The pulse amplification with distance away from the heart is due to elastic and geometric nonuniformities of the arterial tree as well as effects of wave reflection at discontinuities (Nichols *et al* 2011). These changes were documented by the early studies of Kroeker and Wood in 1955 and 1956, in which simultaneous measurements of arterial pressure were obtained using intra-arterial catheters. These studies are significant as they illustrate the different time delay and the marked difference in PP at different anatomical sites. The importance of these observations is in the large differences in peak peripheral pressure as surrogate values of aortic pressure and thus the peak load on the ejecting ventricle. The difference is exaggerated in exercise, as in the 1968 studies of Rowell and Murray, where radial PP can be more than double the aortic PP.

2.5.2 Techniques for Detection of the Arterial Pulse
2.5.2.1 Applanation Tonometry
Of all the many and varied techniques employed for noninvasive detection of the arterial pressure pulse, applanation tonometry has the widest application in devices that perform pulse wave analysis. The strict definition of a tonometer is essentially any instrument that

measures pressure or tension. The specific application of *applanation* tonometry, however, is one where a curved surface is flattened, such that the normal wall stress is effectively reduced to zero and there is transmission of the internal force to the external sensor. This principle found specific application in the field of ophthalmology, where intraocular pressure (IOP) can be determined by a force sensor applanating the cornea. The governing relationships are based on the Imbert–Fick principle (Equation 2.13) that states the internal pressure (P_i) in a spherical body consisting of an infinitely thin, dry, and elastic membrane wall is the ratio of the applied force (F_a) and the area of the applanated surface (A). Although the Imbert–Fick principle requires ideal conditions such as a thin wall, with correct applanation of the corneal surface and with accurate calibration, it is possible to determine IOP and quantify changes with an applanation diameter of approximately 3 mm:

$$P_i = \frac{F_a}{A} \tag{2.13}$$

The theoretical basis for arterial applanation tonometry, which was developed from the ocular application, was also applied specifically to arteries. The models developed by Pressman and Newgard in 1963 and 1965 included properties of the sensor elements and the overlying tissues where the uniform compressible tissue surrounding the elastic artery can be represented by linear springs. Furthermore, the deflections caused by the application of the tonometer were assumed to be small, allowing nonlinearities to be disregarded. The elastic artery was also represented by a spring model, where the artery was represented by a cylindrical tube instead of a spring model. The overlying tissue and skin layer was neglected and the artery assumed to be thin walled and with isotropic and homogeneous wall properties. By assuming a uniform deformation along the length of the arterial segment, the contact stress, as a function of distance from the center of the circular artery, was calculated using curved beam mechanics. The elaborate analysis performed by Drzewieki et al. in 1983 indicates the importance of correct placement of the sensor to obtain zero contact stress so that only the internal blood pressure is detected with sufficient hold-down force, but no additional deformational stress.

The application of the applanation tonometry principle to arteries followed directly from ocular tonometry, given the propensity for a circular arterial segment to be flattened by an external force. However, this requires a rigid support for the artery. The most accessible and suitable anatomic location is the wrist, where the radial pulse can be readily palpated against the ulna bone.

The main sensor types are the single sensor at the end of a handheld pencil-type holder, where the element consists of a piezoresistive sensor with dimensions much smaller than the arterial diameter or an array of sensors strapped over the radial artery where the optimum signal is selected using computer-based algorithms.

As in the ocular application, the original aim was to obtain actual calibrated measurements of arterial pressure using the principles of applanation tonometry. In theory, this was deemed to be possible by eliminating the contact stress. However, the practical application of this has not proved entirely successful. As yet, there is no reliable and reproducible

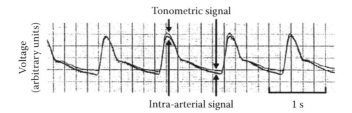

FIGURE 2.7 Similarity of the radial artery pressure pulse recorded noninvasively by tonometry and the pressure measured with an intra-arterial, 1F microtip piezoelectric sensor located directly beneath the tonometer.

tonometric technique able to quantify the intra-arterial pressure in a way to match the sphygmomanometer with respect to the ease of use. Notwithstanding the inability to obtain calibrated pressure values, the applanation principle is highly effective in recording the time-related change in intra-arterial pressure. Thus, for an uncalibrated signal, the non-invasive tonometric pulse is similar to the intra-arterial pressure pulse. Figure 2.7 shows the strong correspondence between the noninvasive tonometric radial artery pulse and the pressure measured with an intra-arterial, 1F microtip piezoresistive sensor located directly beneath the tonometer. This property of arterial applanation tonometry has enabled the accurate measurement of the peripheral pulse. This has resulted in the development of devices that combine pulse wave analysis and the cuff sphygmomanometer.

Although the tonometric sensor is a solid state device and has a high-frequency response in the kHz range, the overall response is determined by the sensor and the coupling to the artery through the surrounding tissue. For a multisensor array device, the frequency response of the sensor is given as being flat for the range 0–50 Hz for a pressure applied directly to the sensor, although a much lower-flat-frequency response exists for the complete system. However, there is good agreement with intra-arterial pressure waves, with the largest discrepancies occurring in early systole.

2.5.2.2 Finger Pressure Methods (Vascular Unloading)

The vascular unloading method introduced by the Czech physiologist, Jan Peñáz, in 1967, measures continuous peripheral arterial blood pressure, specifically in the finger. Changes in the vascular volume of the phalanges are detected by photoplethysmography. The optical signal detecting diameter changes of the finger cuff is used as a compensatory signal for changes in volume through a servonulling principle by continuous adjustment of the pressure in the cuff around the phalanges (Figure 2.8). The time-dependent finger cuff pressure required to maintain the constant blood volume is the instantaneous arterial pressure in the finger and describes a calibrated finger pressure pulse. The device can be uncomfortable when used for long periods as the finger cuff pressure is greater than venous pressure and occludes venous return. The Portapres and other similar devices (e.g., BMV, CNSystems) attempt to alleviate this problem with alternate measurement on two fingers. Recent developments in the technology have shown an improvement in the association between finger pressure and brachial cuff pressure values.

The finger cuff device gives a calibrated continuous recording of arterial pressure. It has been used extensively for applications requiring beat-to-beat information such as

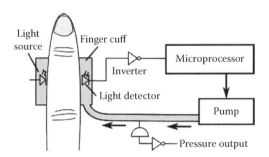

FIGURE 2.8 Servonulling vascular unloading technique for continuous measurement of the pressure pulse in the finger.

quantification of baroreceptor function and obstructive sleep apnea as well as in microgravity simulations for space missions. Modern devices now have pressure compensation for hydrostatic pressure due to changing positions of the hand in relation to the heart level.

2.5.2.3 Brachial Cuff-Based Methods for Detection of the Arterial Pulse

The arterial pulse can be detected using a volume displacement technique in a statically inflated limb pressure cuff, where small oscillations in the cuff caused by arterial pulsations within the area covered by the cuff are detected by a piezoelectric pressure sensor. The Vicorder device (Skidmore Medical, United Kingdom) inflates pressure cuffs to subdiastolic levels for the detection of the arterial pulse at the brachial, femoral, and carotid regions. The Arteriograph (TensioMed) measures the terminal arterial pulse against a brachial pressure cuff inflated to supersystolic pressure, associating features of the terminal pulse with arterial stiffness parameters such as AIx and pulse wave velocity. Similar techniques of suprasystolic cuff inflation are employed in the Pulsecor device using a wideband external pulse (WEP) measurement with a broad bandwidth piezoelectric sensor located over the brachial artery under the distal edge of a sphygmomanometer cuff. The WEP signal is similar to the first derivative of the intra-arterial pressure signal and tracks changes following the use of vasoactive agents. Recent developments in cuff based detection of the arterial pulse have been made in the Mobil-O-Graph (IEM) and SphygmoCor (AtCor Medical) devices.

2.5.3 Noninvasive Measurement of Central Aortic Pressure

2.5.3.1 Carotid Pressure Pulse

With the detection of the pulse in the carotid artery (as used in the PulsePen (Diatecne) device) using applanation tonometry and measurement of brachial pressure with a cuff sphygmomanometer, the values corresponding to the carotid SP can be obtained noninvasively. The trough of the carotid voltage signal is assigned the value of DP obtained from the brachial cuff, and the integrated mean value is assigned the estimated mean pressure from the brachial cuff systolic and diastolic values. Hence, the corresponding systolic value can then be determined for the carotid pressure pulse.

While it is possible to obtain a calibrated carotid pulse using the aforementioned technique, the pulse itself may not always be obtained as readily as the radial pulse, since applanation tonometry requires essentially a hard surface behind the artery to obtain a certain degree of applanation. Movement of the carotid artery with the pressure applied to the probe can result in artifact and inconsistent waveform.

2.5.3.2 Transfer Function Methods

Transfer functions are mathematical entities used to describe a system in terms of the relationship between input and output signals. In essence, they are a mathematical relationship between an input and an output. Various functions have been applied to estimate central or aortic pressure parameters from peripheral or brachial blood pressure values and waveforms. This method is employed in the SphygmoCor device (AtCor Medical, Sydney, New South Wales, Australia).

A transfer function can take the form of a complex quantity expressed as modulus and phase as a function of frequency, autoregressive functions, and closed-form filter functions such as neural networks and Chebyshev filters or be expressed in the time domain using autoregressive models. Applications extend to using peripheral pressure signals with and without wave velocity, multichannel adaptive algorithms, and moving average filters.

If the input is a flow signal and the output is pressure, the transfer function between flow and pressure measured at the same location and at the same time is the input impedance of the system (Z_{in}), described as the ratio of the frequency components of pressure and flow (Nichols *et al* 2011). If the input is a pressure signal (P_0) and the output is another pressure signal (P_L) measured at the same time but at a different location separated by a finite distance (L), the transfer function is the ratio of the frequency (ω) components of the output and input pressure signals. The following transfer function ($P(\omega)_L/P(\omega)_0$) is a function of the distance (L); the propagation coefficient ($\gamma(\omega)$), which depends on arterial geometric and elastic properties defining pulse wave velocity; and the reflection coefficient ($\Gamma(\omega)$):

$$\frac{P(\omega)_L}{P(\omega)_0} = \frac{\left(1+\Gamma(\omega)\right)}{\left(e^{\gamma(\omega)L}+\Gamma(\omega)e^{-\gamma(\omega)L}\right)} \tag{2.14}$$

Devices are increasingly using transfer functions to noninvasively estimate aortic pressure pulse values and waveform shape. The ratio of the aortic pressure pulse to the radial pressure pulse in the frequency domain, averaged from invasive pressure measurements in a cohort of subjects, will give a generalized transfer function that, when applied to an individual's peripheral pulse, will give, with reasonable accuracy, the aortic waveform by noninvasive means. The premise of this is displayed in Figure 2.9. The inverse transfer function between the radial artery and the ascending aorta takes the form of a low-pass filter. As such, a simple moving average filter of suitable width (e.g., 25% of the pulse period) also acts as a low-pass filter transfer function and will give a good approximation of central aortic PP values, though it loses detail in the aortic waveform shape (Figure 2.10). The Fourier transform can also be expressed in parametric terms in an autoregressive model, where the aortic waveform is defined in terms of the previous time domain data of both the peripheral ($P(t-i)$) and aortic ($A(t-i)$) pulse waveforms, as expressed in the following

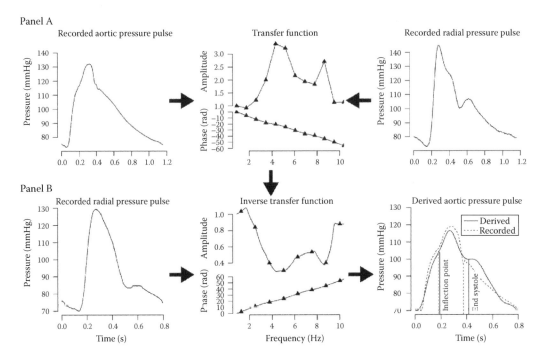

FIGURE 2.9 The generation and application of a transfer function in the frequency domain. Panel A: an invasively recorded aortic and tonometrically recorded radial waveform (48-year-old male with a heart rate of 52 bpm) was used to generate the transfer function in the frequency domain. Panel B: the inverse of this function is applied to a tonometrically recorded radial waveform (71-year-old female with a heart rate of 75 bpm) to generate the aortic waveform noninvasively. The derived aortic waveform has measurable features, such as the inflection (augmentation) point and the end of systole. The accuracy of the derived aortic pressure wave would be improved by *generalizing* the transfer function, by incorporating a greater number of subjects in the formation of the transfer function (Panel A). However, the robustness of the application of this method for estimation of PP is essentially due to the low-pass filter characteristics of the inverse transfer function (Panel B). The coefficients of the autoregressive function (Equation 2.16) representing this transfer function in the time domain are given in Table 2.1.

equations, where $a_1 \ldots a_n$ and $b_1 \ldots b_m$ are the weighting coefficients for the time sampled data points of the series $A(t)$ and $P(t)$:

$$P(t) = -a_1 P(t-1) - a_2 P(t-2) - \cdots a_n P(t-n) + b_1 A(t-1) + \cdots + b_m A(t-m) \qquad (2.15)$$

$$A(t-1) = -\left(\frac{b_2}{b_1}\right) A(t-2) - \cdots \left(\frac{b_m}{b_1}\right) A(t-m) + \left(\frac{1}{b_1}\right) P(t) + \left(\frac{a_1}{b_1}\right) P(t-1) + \cdots \left(\frac{a_n}{b_1}\right) P(t-n)$$

$$(2.16)$$

2.5.3.3 Central Systolic Pressure from Radial Systolic Inflection (P_i)

The propagation characteristics of the aortic to brachial artery system result in a high degree of correlation between the value of SP in the ascending aorta and the second shoulder of the radial pulse. From Figure 2.6, P_s in the aorta is at the same level as P_i in the radial artery.

TABLE 2.1   Coefficients of a 10-Order, $n = 2$ Delay, Autoregressive Model
for Aortic to Radial Pressure Pulse

n	a_n	b_n
1	1	0
2	−1.038	0
3	−0.122	2.069
4	−0.006	−2.421
5	0.021	−1.650
6	0.002	2.282
7	0.082	1.859
8	−0.064	−5.216
9	−0.004	2.352
10	−0.037	1.403
11	−0.081	0.282
12	—	−0.960

Note: Based on tonometric (radial) and invasive (aortic) measurements of the pulses
(Figure 2.9, Panel A, of the form defined in Equation 2.16).

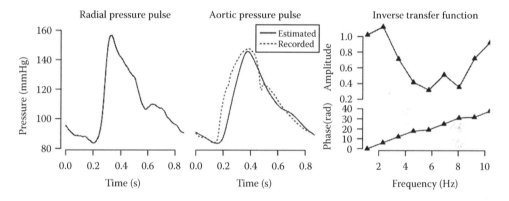

FIGURE 2.10   A moving average filter, with window width of 25% of the pulse period, generates a pressure pulse that, when calibrated to mean pressure and DP, suitably predicts the aortic systolic peak value. If the moving average filter is expressed as an inverse transfer function (radial to estimated aortic pressure pulse) in the frequency domain, it can be seen that the moving average filter acts as a low-pass filter, similar to that in Figure 2.9. However, the moving average filter removes detail from the waveform shape, and features such as the inflection point and end of systole cannot be located. Pulse waveforms recorded in a 62-year-old male.

With a peripheral radial pulse calibrated with the brachial cuff measurements, using SP and DP values, or diastolic and mean values, this method is suitable to estimate central aortic SP when the central aortic pressure waveform is not required. This technique is used in the HEM 9000AI (Omron) device.

2.6 FUTURE DIRECTIONS

Blood pressure measurement is the most frequent measurement performed by physicians. It is widely considered as a first indicator of the well-being of a patient. And yet it is often done casually, with little regard for protocols that produce an accurate measurement. This culture of casual inattention to following protocol guidelines is due to factors not necessarily related to the measurement technique: the dynamic and situational nature of blood pressure and the lack of agreement as to hypertension management guidelines. Thus, when treating individual patients, physicians may tolerate a significant amount of variability, and therefore inaccuracy, in assessing any given patient. However, especially in research to understand the dynamic aspects of blood pressure and to develop a consensus on hypertension management guidelines, noninvasive brachial blood pressure monitors with improved precision and reduced systematic errors must be developed. When it has been demonstrated that more accurate and precise cuff-based blood pressure monitors are necessary to improve patient health and reduce health-care costs, then more attention to protocol compliance and careful measurement of blood pressure will follow.

A second area for improving blood pressure measurement is in the more sophisticated assessment of blood pressure itself. Ambulatory blood pressure monitors have highlighted chronological changes in blood pressure, such as nighttime versus daytime patterns, that are clinically important. More recently, it has been possible to determine the blood pressure in the ascending aorta noninvasively using a number of different techniques. This has highlighted the importance of arterial stiffness in the assessment of cardiovascular risk and the importance of aortic blood pressure on major organs such as the heart, brain, and kidney. Improvements in the accuracy, precision, and ease of use of such systems will provide researchers with tools to better measure blood pressure and understand its impact on health and disease and clinicians with tools to manage the health of their patients.

2.6.1 Current Problems

2.6.1.1 Improvement of Accuracy of Cuff-Based Systems

The area that potentially offers the most opportunity for improvement in blood pressure measurement using an occluding cuff is in protocol compliance. Some areas of compliance require closer attention to the protocol by the operator, such as allowing the patient a 5–10 min period of relaxing prior to beginning the measurement, choosing the appropriate cuff size, and ensuring that the patient's posture is correct. More fully automated systems address the issue of the influence of the operator on the measurement and the standardization of the measurement while reducing the training necessary for an accurate measurement. Systems such as the BpTRU, which enforces the premeasurement relaxation period and also allows the operator to leave the room during the measurement, are a step in the right direction.

The algorithms used in oscillometric systems are another area of active research and an opportunity to improve measurement accuracy. However, the evaluation, research, and improvement of these algorithms continue to be hampered by the proprietary nature of commercial algorithms.

2.6.1.2 Validation Protocols of Instruments

The diagnosis and treatment of one of the most severe silent killers, hypertension, or the progression and development of high blood pressure depends on the accuracy of measurement of arterial blood pressure. In addition to the integrity of the instrumentation and its components, systematic errors can lead to severe underdetection or overdetection of hypertensive individuals. A systematic error of 5 mmHg (underreading) would not detect some 8% of hypertensives (Turner 2010). This, of course, relates not only to the validation of instruments but also to the quality control of the validation systems and protocols.

The validation of devices is treated extensively with updated information in online resources (dabl® Educational Trust Ltd 2011). However, what is not available is the same quality control and traceability as applied to other measurement systems in a range of industries. Recent recommendations involve the establishment of quality control systems where laboratories that validate blood pressure measurement devices should be a part of the general international metrology framework and to be accredited by external agencies that are members of the International Laboratory Accreditation Cooperation. Similar to Medical Testing Laboratories and Medical Imaging facilities being externally accredited to international standards ISO 15189 (*Medical laboratories—Particular requirements for quality and competence*) and ISO 17025 (*General requirements for the competence of testing and calibration laboratories*) by the same agencies that accredit industrial calibration and testing laboratories, it is recommended that laboratories that validate automatic sphygmomanometers also be accredited to ISO 17025 with demonstrable competence in performing the validation protocols (Turner 2010).

REFERENCES

Avolio A P, Butlin M and Walsh A 2010 Arterial blood pressure measurement and pulse wave analysis—Their role in enhancing cardiovascular assessment *Physiol. Meas.* **31** R1–47

Baker P D, Westenskow D R and Kück K 1997 Theoretical analysis of non-invasive oscillometric maximum amplitude algorithm for estimating mean blood pressure *Med. Biol. Eng. Comput.* **35** 271–8

dabl® Educational Trust Ltd. 2014 Blood pressure monitors—Validations, papers and reviews. Retrieved from www.dableducational.org (Accessed August 12, 2014).

Drzewiecki G M, Melbin J and Noordergraaf A 1989 The Korotkoff sound *Ann. Biomed. Eng.* **17** 325–59

Geddes L A, Voelz M, Combs C, Reiner D and Babbs C F 1982 Characterization of the oscillometric method for measuring indirect blood pressure *Ann. Biomed. Eng.* **10** 271–80

Kurtz T W, Griffin K A, Bidani A K, Davisson R L and Hall J E 2005 Recommendations for blood pressure measurement in humans and experimental animals. Part 2: Blood pressure measurement in experimental animals: A statement for professionals from the subcommittee of professional and public education of the American Heart Association Council on High Blood Pressure Research *Hypertension* **45** 299–310.

Nichols W W, O'Rourke M F and Vlachopoulos C 2011 *McDonald's Blood Flow in Arteries: Theoretical, Experimental and Clinical Principles*, 6th edition (London: Hodder Arnold)

Pickering T G, Hall J E, Appel L J, Falkner B E, Graves J, Hill M N, Jones D W *et al* 2005 Recommendations for blood pressure measurement in humans and experimental animals: Part 1: Blood pressure measurement in humans: A statement for professionals from the Subcommittee of Professional and Public Education of the American Heart Association Council on High Blood Pressure Research *Hypertension* **45** 142–61

Turner M J 2010 Can we trust automatic sphygmomanometer validations? *J. Hypertens.* **28** 2353–6

Webster J G ed 2010 *Medical Instrumentation: Application and Design*, 4th edition (Hoboken: John Wiley & Sons) p 713

CHAPTER 3

Cardiology

Electrocardiography

Jason Ng and Jeffrey J. Goldberger

CONTENTS

3.1	Origin of Cardiac Potentials	44
3.2	Electrodes	46
3.3	Lead Systems	47
	3.3.1 12-Lead ECG	47
	3.3.2 Mason–Likar Lead Configuration	49
	3.3.3 Expanded 12 Leads	49
	3.3.4 Orthogonal Lead Configuration	50
	3.3.5 Lead Configurations for Bedside and Ambulatory Monitoring	51
	3.3.6 Body Surface Mapping	51
3.4	Amplifiers	52
3.5	Displays	53
3.6	Arrhythmias	54
3.7	Automatic Diagnosis	54
	3.7.1 QRS Detection	54
	3.7.2 QRS Duration	55
	3.7.3 QT Interval	58
	3.7.4 PR Interval	59
	3.7.5 ST Height	59
	3.7.6 QRS Axis	59
3.8	Advanced ECG Signal Processing	59
	3.8.1 Median Beat Analysis	59
	3.8.2 Pacemaker Detection	60
	3.8.3 Heart Rate Variability	60
	3.8.4 Signal-Averaged ECG	62
	3.8.5 T-Wave Alternans	62

3.9   Ambulatory Monitoring   64
      3.9.1   Holter Monitoring   64
      3.9.2   Long-Term Ambulatory/Event Recorders   65
      3.9.3   Implantable Loop Recorders   65
3.10 Telemetry   65
3.11 Cardiac Stress Testing   66
3.12 Cardiac Mapping   67
      3.12.1 Electrograms   67
      3.12.2 Effective Refractory Periods   69
      3.12.3 Electroanatomic Mapping   69
      3.12.4 Noncontact Mapping   71
3.13 Radio-Frequency Lesion Measurements   71
      3.13.1 Contact Sensing and Force Measurement   71
      3.13.2 Tissue Temperature Sensing   72
3.14 Possible Future Instruments   73
Bibliography   73

3.1 ORIGIN OF CARDIAC POTENTIALS

The heart is a muscular organ responsible for circulating blood through the body's circulatory system. The pumping action of the heart is coordinated by electrical activity that propagates through the heart's four chambers. This same electrical activity makes it possible to noninvasively evaluate some aspects of heart function from recordings known as the electrocardiogram (ECG). The ECG is currently the fundamental technology used by physicians to evaluate the electrical activity of the heart and diagnose conditions such as abnormal rhythms (arrhythmias), abnormal conduction, damage due to heart attacks, and abnormal chamber sizes.

From a functional perspective, the basic operation of the heart consists of coordinated contractions of its four chambers: the right atrium, the left atrium, the right ventricle, and the left ventricle. A diagram is shown in Figure 3.1. The right atrium receives

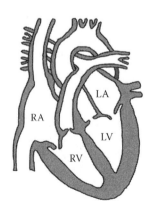

FIGURE 3.1   Diagram of the four chambers of the heart. RA, right atrium; LA, left atrium; RV, right ventricle; LV, left ventricle.

deoxygenated blood, which then fills the right ventricle. The right ventricle then pumps this blood to the lungs via the pulmonary artery. The left atrium receives oxygenated blood from the pulmonary veins, which then fills the left ventricle. The left ventricle then pumps the oxygenated blood to all parts of the body via the aorta. The pumping of the heart occurs with rhythmic contractions that occur at rates typically between 60 and 80 beats/min. Heart rates vary depending on need (e.g., higher heart rates with physical activity but lower rates at rest).

On the cellular level, the heart is made up of electrically active cells called myocytes. Each cardiac myocyte in its resting state has a potential difference from the inside to the outside of the cell due to concentration gradients of sodium, calcium, and potassium ions. Activation of a cardiac myocyte is invoked either by an activation of a neighboring cell, through an applied electrical impulse, or even by an automatic mechanism. The response of an activated cell consists of inward and outward flows of the sodium, calcium, and potassium ions, which in sum produce a rapid depolarization of the intracellular potential and the subsequent gradual repolarization. The typical waveform of the intracellular potential during this process, known as the action potential, is shown in Figure 3.2. The refractory period in which the cardiac cell is not excitable after activation is related to the duration of the action potential. The refractory period tends to decrease with an increase in heart rate.

In a normal heart rhythm, a cluster of specialized pacemaker cells in the right atrium, known as the sinus node, initiates each heart beat. Activation wavefronts then propagate into the right and left atria, resulting in contraction of the atria. The atria and the ventricles are electrically connected by a specialized pathway called the atrioventricular node. Propagation through the atrioventricular node is slow compared to conduction in the atria and ventricles, allowing the atria time to contract before contraction of the ventricles is initiated.

The summed strength of the electrical activity from the large number of cardiac cells that make up the heart and the coordination of the electrical activity allows electric potentials to be detected using electrodes placed in contact with the wall of the heart or even

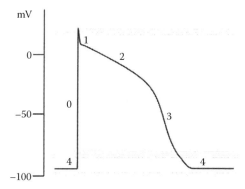

FIGURE 3.2   Representation of a ventricular myocyte action potential. Phase 4 is the resting potential. Phase 0 is the rapid depolarization phase. Phase 1 is the early repolarization phase. Phase 2 is the plateau phase of the action potential. Phase 3 is the repolarization phase.

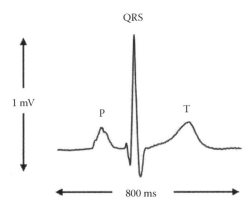

FIGURE 3.3   Example of an ECG waveform for a single beat in normal sinus rhythm.

externally from the surface of the body. Recordings made from electrodes placed directly on the heart are known as electrograms, while body surface recordings of cardiac activity are known as ECGs.

The ECG during a normal sinus rhythm has three main components, as illustrated in Figure 3.3. The first sign of activity is the P-wave. This typically low amplitude hump reflects atrial depolarization. The next waveform, the QRS complex, occurs during the ventricular depolarization process. The third waveform is the T-wave, which reflects ventricular repolarization. The waveforms can vary in size, timing, and morphology, depending on the rate, rhythm, heart condition, and electrode locations.

3.2 ELECTRODES

The electrical activity of the heart has strong enough magnitude to be detected on the surface of the body. However, if one were to attempt to measure the electrical activity of the heart by simply touching the positive and negative ends of an oscilloscope by the left and right hands, no potentials relating to cardiac activity would be observed. The reason for this is that the oscilloscope requires an electrical current (i.e., a flow of electrons). In contrast, biopotentials from cardiac activity are the result of ionic currents. Thus, an interface between the oscilloscope probes and the skin that converts ionic currents to electrical currents is needed for the oscilloscope to display the body surface cardiac potentials. The same is true when an ECG recording system is used to record surface potentials.

Electrodes are transducers that convert the ionic currents to an electron current from which the ECG potentials can be amplified, recorded, and displayed. Electrodes are attached to the skin with an electrolyte gel serving as an interface between the electrode and the skin. The electron current is created by a chemical reaction between the electrolyte and the electrode.

There are two classifications of ideal electrodes: polarizable and nonpolarizable. With a polarizable electrode, the behavior of the electrode is similar to that of a capacitor. Current flowing through the electrode will charge the electrode causing "polarization." The current responsible for charging (or discharging) the electrode is known as "capacitive current." In a polarizable electrode, there is no "faradaic" current, which is the current

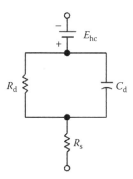

FIGURE 3.4   Circuit equivalent of a biopotential electrode and the electrolyte interface. E_{hc} is the electrode half potential. R_d and C_d are the resistance and capacitance components of the electrode, respectively. R_s is the series resistance due to the electrode–electrolyte interface.

due to chemical reaction within the electrode, and no charge actually passes through the electrode–electrolyte interface. The potential of nonpolarizable electrodes, on the other hand, will not change from its equilibrium potential even with large current through the electrode. This behavior is attributed to the extremely fast electrode reaction (has an almost infinite exchange current density). Polarizable electrodes are thus more suitable for recording, while nonpolarizable electrodes are more suited for pacing and stimulation. Figure 3.4 shows the circuit equivalent of a biopotential electrode and the electrolyte interface. With a polarizable electrode, the resistance of R_d is infinite, and capacitor C_d is responsible for the polarization. With a nonpolarizable electrode, the capacitance of C_d is negligible.

3.3 LEAD SYSTEMS

Activation and repolarization of the heart is a 4D process. Therefore, multiple leads are needed to capture directional information of activation and repolarization. In this section, the common electrode configurations used for ECG recordings will be described.

3.3.1 12-Lead ECG

The 12-lead ECG is the most widely used configuration by cardiologists for diagnostic surface ECG evaluation. The 12-lead ECG tracings in this configuration are derived from electrodes placed in 10 standardized locations. The 12 leads are also commonly classified into 3 categories: the limb leads, the augmented limb leads, and the precordial leads. The limb leads (leads I, II, and III) are recorded from electrodes placed on the arms and legs, as shown in Figure 3.5a. The bipolar recording measuring the difference across the left arm (+) and the right arm (−) is designated as lead I. The bipolar recording across the left leg (+) and the right arm (−) is designated as lead II. Lead III is recorded between the left leg (+) and the left arm (−). The right leg is used as a reference for the amplifier circuitry. Lead III can be obtained by subtracting lead I from lead II. The vector representation of the limb leads known as Einthoven's triangle is shown in Figure 3.5b.

The augmented limb leads (leads aVL, aVR, and aVF) are considered unipolar recording, because the positive electrode is recorded in reference to a combination of other electrodes. For aVL, the left arm potential is recorded in reference to the average of the right

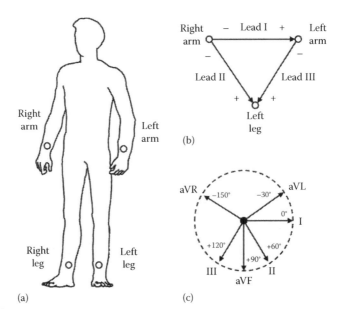

FIGURE 3.5 (a) Location of the limb lead. (b) Einthoven's triangle. (c) Vector directions of the limb leads and augmented limb leads.

arm and left leg potentials. For aVR, the right arm potential is recorded in reference to the average of the left arm and left leg potential. For aVF, the left leg potential is recorded in reference to the average of the right and left arm potentials. The augmented limb leads can also be derived from leads I and II with the following equations:

$$aVL = \frac{Lead\,I - Lead\,III}{2} = Lead\,I - \frac{Lead\,II}{2}$$

$$aVR = -\frac{Lead\,I + Lead\,II}{2}$$

$$aVF = \frac{Lead\,II + Lead\,III}{2} = Lead\,II - \frac{Lead\,I}{2}$$

The combined vector representation of the limb leads and augmented limb leads is shown in Figure 3.5c. Propagation of cardiac potentials traveling in the direction of the vector of a lead will result in a positive change in amplitude in that lead. Propagation in the opposite direction will result in a negative change in amplitude. Propagation perpendicular to the lead vector will result in zero potential.

The precordial leads of the 12-lead system consist of leads V1, V2, V3, V4, V5, and V6. The precordial leads are unipolar leads that use a common reference known as Wilson's central terminal. Wilson's central terminal is essentially the average potential of the right arm, left arm, and left leg electrodes. This average is obtained by the

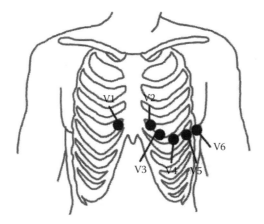

FIGURE 3.6   Placement of the precordial leads.

interconnection of the three electrodes with resistors. The positive electrodes of the precordial leads are placed in the following locations (Figure 3.6):

- V1: right fourth intercostal space

- V2: left fourth intercostal space

- V3: halfway between V2 and V4

- V4: left fifth intercostal space, midclavicular line

- V5: horizontal to V4, anterior axillary line

- V6: horizontal to V5, midaxillary line

V1, V2, and V3 are used to detect electrical propagation traveling in the anterior/posterior directions (i.e., front/back). V4, V5, and V6 are used to detect lateral propagation (i.e., left/right).

3.3.2 Mason–Likar Lead Configuration

Although the 12-lead ECG is considered the standard for diagnostic ECG, the limb lead locations on the arms and leg make the recordings susceptible to motion artifact. Not only are the limb leads and augmented limb leads affected by movement of the arm and legs, the precordial leads are also affected, because Wilson's central terminal is derived from the limb electrodes. In the Mason–Likar lead configuration, the arm electrodes are moved to the shoulders, and leg electrodes are moved to the hips to minimize movement of these electrodes. Thus, this system is often used in ambulatory (Holter) monitoring and exercise testing. Precordial leads remain in the standard positions if used with the Mason–Likar configuration.

3.3.3 Expanded 12 Leads

In the standard 12-lead system, the majority of the unipolar precordial leads are positioned in the left torso. Although not commonly used, the mirrored locations of V3, V4, V5, and

V6 on the right side of the torso are labeled lead V3R, V4R, V5R, and V6R. Additionally, V7 (the posterior axillary line), V8 (below the scapula), and V9 (paravertebral border) can be used to record from the posterior chest wall. These additional leads may be useful in the detection of myocardial ischemia and/or infarction.

3.3.4 Orthogonal Lead Configuration

Orthogonal lead systems were developed for the ability to record and display vectorcardiograms (VCGs). VCGs are created by plotting two or three orthogonal leads as a function of the other. The result is a display that shows the progression of dipole vectors as a series of loops. The magnitude of the vectors allows evaluation of the ECG waveform independent of the specific cardiac axis of the patient. Although the VCG is rarely used today in clinical practice, orthogonal ECG lead systems are still used in specialized ECG tests, such as signal-averaged ECG and microvolt T-wave alternans testing.

The Frank lead system, the most widely used orthogonal lead configuration, has the following seven electrode locations (shown in Figure 3.7a):

- A: left midaxillary line

- E: midsternum

- C: anterior left chest wall midway between A and E

- I: right midaxillary line

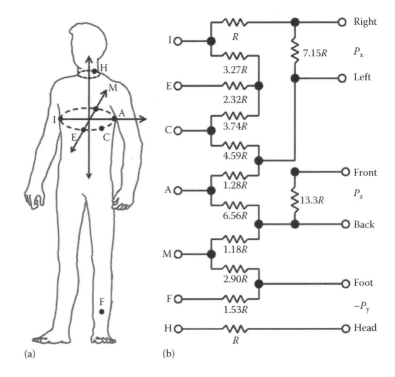

(a) (b)

FIGURE 3.7 (a) Frank lead electrode placement. (b) Resistor network for lead derivation.

- M: mid-spine

- H: junction of the neck and torso posteriorly

- F: left foot

From these electrodes, the leads X, Y, and Z are derived by the following weighted difference equations:

- $X = 0.610 \, (A - I) + 0.170 \, (C - I)$

- $Y = 0.345 \, (M - E) - 0.655 \, (H - F) + 0.345 \, (E - H)$

- $Z = 0.132 \, (A - I) + 0.372(M - E) + 0.365 \, (M - C) + 0.132 \, (C - I)$

The weighted differences can be achieved by using a resistor network as shown in Figure 3.7b or can be performed in software.

3.3.5 Lead Configurations for Bedside and Ambulatory Monitoring

Lead configurations for bedside and ambulatory monitoring have traditionally used reduced number of electrodes compared to the 12-lead ECG to save transmission bandwidth and recording space. Three-electrode and five-electrode systems are commonly used. The three-electrode system consists of a positive electrode, negative electrode, and a ground electrode to obtain one ECG channel. The electrodes can be configured to obtain the Mason–Likar equivalent of lead I, II, or III. Often a modified chest lead (MCL) is used (MCL1), which approximates a lead V1 signal by placing the positive electrode in the standard V1 location and the negative electrode on the left infraclavicular fossa. MCL1 will differ from the standard V1 due to the change in reference from the Wilson's central terminal to the left infraclavicular fossa. With a five-electrode system, the complete frontal plane leads can be simultaneously obtained (I, II, III, aVR, aVL, aVF) in addition to one chest lead (V1).

Philips Medical Systems (Eindhoven, Netherlands) also has a five-electrode system known as the EASI configuration that uses electrodes A, E, and I from the Frank lead system, an electrode on the sternal manubrium (S), and a ground electrode that can be placed anywhere on the torso (G). The advantage of this system is the ability to derive Mason–Likar 12-lead ECGs with only five electrodes instead of 10 electrodes facilitating setup and storage particularly for telemetry and bedside monitoring. Figure 3.8 shows the location of the EASI electrodes and the transformation coefficients for the derivation of the 12-lead signal.

3.3.6 Body Surface Mapping

Obtaining ECGs with high spatial coverage obtained by large numbers of electrodes is known as body surface mapping. The high density of recordings has been shown to allow the detection of the presence and location of myocardial infarction in cases where it was not possible with the 12-lead ECG. Body surface mapping has not gained wide

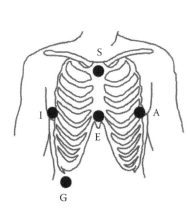

Transformation coefficients			
Lead	ES	AS	AI
I	0.026	−0.174	0.701
II	−0.002	1.098	−0.763
III	−0.028	1.272	−1.464
aVR	−0.012	−0.462	0.031
aVL	0.027	−0.723	1.082
aVF	−0.015	1.185	−1.114
V1	0.641	−0.391	0.080
V2	1.229	−1.050	1.021
V3	0.947	−0.539	0.987
V4	0.525	0.004	0.841
V5	0.179	0.278	0.630
V6	−0.043	0.431	0.213

FIGURE 3.8 EASI lead placement and transformation coefficients for approximation of the 12-lead ECG.

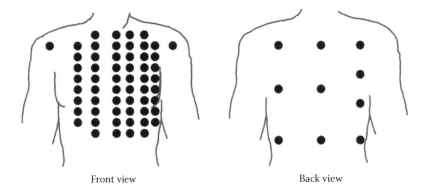

Front view   Back view

FIGURE 3.9 Electrode locations for the PRIME ECG body surface mapping system. Seven electrodes on the left side and six electrodes on the right side not shown.

acceptance. As such, there are no accepted standards regarding the number or locations of the electrodes. The PRIME ECG system by Heartscape Technologies (Bothell, WA) is one commercially available body surface mapping system that records from 80 electrodes placed in the front and back of the chest. The locations of the PRIME ECG electrodes are shown in Figure 3.9. Body surface mapping is also a topic of current research in which the surface potentials are backprojected onto either 3D MRI or CT images of the patient's heart to predict the propagation pattern of cardiac activation. This technique is known as electrocardiographic imaging.

3.4 AMPLIFIERS

The potentials recorded on the body surface have amplitudes on the order of 1 mV. A combination of amplification and filtering is needed to obtain diagnostic quality tracings. A block diagram of the typical components of an ECG amplifier circuit is shown in Figure 3.10.

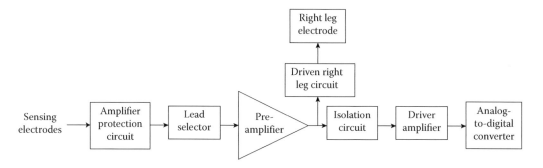

FIGURE 3.10   Block diagram of a basic ECG amplifier circuit.

The first component receiving the signals from the electrodes is the amplifier protection circuit. Diodes are often used to limit the effect of high voltage/current, such as in the cases of electrostatic discharge or defibrillation shock. Some ECG systems have lead failure detection circuits that operate at this stage as well. The next block is the lead selector circuit. This circuit is usually controlled by the microprocessor to select the combination of electrodes required to be amplified. Next is the preamplifier where initial amplification is performed. The preamplifier should have high common-mode rejection and input impedance. The next stage is the isolation circuit. For ECG machines running on AC power, the isolation circuit protects both the patient and the ECG machine from possible high voltages. Isolation circuitry is not needed for battery-powered recorders such as Holter monitors. The driven right leg circuit sets the right leg as the reference signal. Finally, the driver amplifier performs the final amplification as well as band-pass filtering before analog-to-digital conversion is performed. International guidelines recommend band-pass filtering with cutoff frequencies of 0.05–150 Hz and zero-phase distortion. Amplification is performed to achieve an amplitude resolution on the order of 10 μV or better prior to analog-to-digital conversion.

3.5 DISPLAYS

ECG recordings have traditionally been displayed with either oscilloscope-like tracings or printed directly to paper. Figure 3.11 shows an example of 12-lead ECG tracing in a typical clinically used format. The first three rows contain 2.5 s snapshots of each of the 12 leads. The fourth row consists of a 10 s rhythm strip of lead II. The ECGs are plotted on a grid pattern consisting of larger 5 by 5 mm boxes, which are in turn made up of smaller 1 by 1 mm boxes. The rectangular waveforms preceding each of the four rows are calibration pulses, which indicate the time scale (width of 200 ms) and amplitude scale (height of 1 mV). In this example, one big box (5 mm) equals 200 ms, and two big boxes (10 mm) equal 1 mV. It is important to examine these calibration pulses as ECG machines do allow alterations in both the time and amplitude scaling of the ECG.

Real-time displays present ECG waveform data as they are being collected. Real-time displays often display ECGs alongside heart rate measurements or other real-time physiologic measurement. The real-time waveforms can be either continuously scrolling or updating every few seconds at a determined sweep speed (typically 2–5 s).

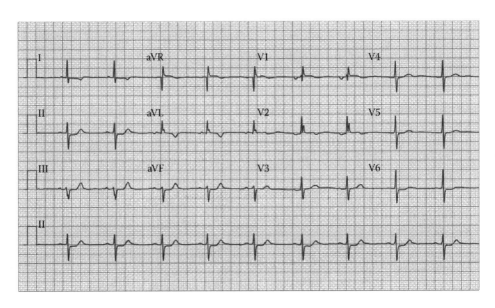

FIGURE 3.11 Example of 12-lead ECG tracing. The first three rows contain 2.5 s snapshots of each of the 12 leads. The fourth row consists of a 10 s rhythm strip of lead II. The ECGs are plotted on a grid pattern consisting of larger 5 by 5 mm boxes, which are in turn made up of smaller 1 by 1 mm boxes. The rectangular waveforms preceding each of the four rows are calibration pulses, which indicate the time scale (width of 200 ms) and amplitude scale (height of 1 mV).

3.6 ARRHYTHMIAS

Arrhythmias are abnormal heart rhythms with characteristics that differ from rhythms, which originate from the sinus node and conduct normally through the atria, atrioventricular node, and ventricles. Figure 3.12a shows an illustration of an ECG during normal sinus rhythm. The ECG is an important tool for the evaluation of many types of arrhythmias. Table 3.1 shows arrhythmias that can be detected via automatic algorithms processing the digital ECG waveform. Illustrations of ECG manifestations of different arrhythmias are shown in Figures 3.13 and 3.14.

3.7 AUTOMATIC DIAGNOSIS

Automatic algorithms utilizing digital signal processing are capable of providing a plethora of ECG measurements and diagnoses. These measurements can include interval and amplitude measurements that can be confirmed by a physician with calipers. Automated algorithms can also involve rhythm and conduction classifications, which are then confirmed by a trained physician. More sophisticated signal-processing algorithms can also be used to extract information from the ECG that cannot be easily determined or quantified by humans. The following are examples of ECG measurements that can be obtained automatically.

3.7.1 QRS Detection

Robust QRS detection is required not only for the measurement of heart rate but is the basis for almost all automatic ECG detection algorithms. QRS complexes are characterized

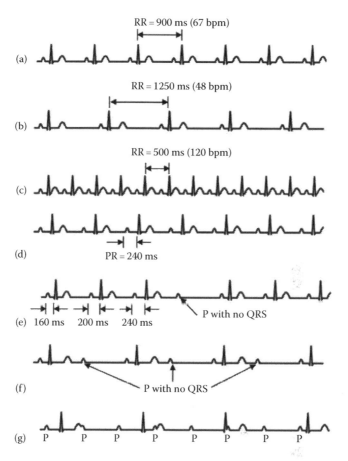

FIGURE 3.12   Illustrations of different heart rhythms. (a) Normal sinus rhythm. (b) Sinus brady-cardia. (c) Sinus tachycardia. (d) First degree AV block. (e) Type 1 second degree AV block. (f) Type 2 second degree AV block. (g) Third degree AV block.

by high amplitude and steep slopes. Both amplitude and slope can be used for detection, but slope usually offers better discrimination from the T-wave. Figure 3.14 compares amplitude versus slope-based QRS detection in an ECG where high-amplitude T-waves are present. QRS detection methods almost always include an assumption of a minimum refractory period to be used as a blanking period during which subsequent beats can-not be detected. Cross-correlation using a QRS template can be used to reduce timing jitter compared to relying on the maximum amplitude or slope alone as the fiducial point. Template comparisons can also be used in the detection and classification of premature ventricular complexes.

The time period between QRS complexes is known as the RR interval. Ventricular (heart) rate in beats per minute can be calculated by dividing 60,000 by the mean RR interval (in milliseconds) over a time period.

3.7.2 QRS Duration

The measurement of the duration of the QRS complex is used to evaluate intraventricular con-duction. Short QRS durations usually reflect normal and synchronized conduction through the

TABLE 3.1   Examples of Common Arrhythmias and Abnormalities That Can Be Determined from ECG Tracings

Arrhythmia/ Abnormality	Description	ECG Characteristics
Sinus bradycardia	Sinus rhythm with very slow heart rates (<60 beats/min).	Mean interval between QRS complexes is greater than 1 s (Figure 3.12b).
Sinus tachycardia	Sinus rhythm with very fast heart rates (>100 bpm).	Mean interval between QRS complexes is less than 600 ms (Figure 3.12c).
First degree AV block	Long conduction times through the atrioventricular node.	Long intervals between the P-wave onset and the QRS onset (>200 ms) (Figure 3.12d).
Type 1 second degree AV block (Wenckebach)	During a stable sinus rhythm, there is progressively longer conduction times through the AV node until conduction is blocked for one sinus beat, after which conduction times are reset and the pattern repeats.	The interval between the P-wave onset and the QRS onset progressively lengthens until there is a P-wave without a trailing QRS complex and T-wave, after which the pattern repeats (Figure 3.12e).
Type 2 second degree AV block	Sudden conduction block of a sinus beat without preceding lengthening of the AV nodal conduction time.	A P-wave is not followed by a QRS complex and T-wave, without preceding lengthening of the PR interval (Figure 3.12f).
Third degree AV block	No conduction from the atrium to the ventricle.	The P-waves and QRS complexes are completely dissociated, with an atrial rate that is faster than the ventricular rate; QRS complexes can result from escape beats initiated from AV junctional or ventricular sources (Figure 3.12g).
Premature atrial complexes	An early beat initiated from another atrial site other than the sinus node.	P-wave occurs earlier than sinus-node-initiated P-waves, often with a different morphology (Figure 3.13a).
Premature ventricular complexes	Early beat initiated from a ventricular site.	QRS complexes, often with different morphology, occur early without a preceding P-wave (Figure 3.13b).
Atrial flutter	Rapid atrial activity with continuous but repeated patterns due to a reentrant circuit around an anatomic obstacle.	P-waves are replaced with F waves, often with a continuous sawtooth or sinusoidal-like waveform, with consistent timing and morphology (Figure 3.13c).
Atrial fibrillation	Rapid and seemingly uncoordinated activation of the atria.	Discrete P-waves are replaced by an undulating baseline with changing timings and morphologies (Figure 3.13d).
Ventricular tachycardia	An abnormally fast rhythm (>100 bpm) originating from the ventricles.	Monomorphic ventricular tachycardia has a stable rate, rhythm, and QRS morphology; polymorphic ventricular tachycardia has changing rate, rhythm, and QRS morphology (Figure 3.13e).
Ventricular fibrillation	A type of polymorphic ventricular tachycardia with seemingly disorganized activity in the ventricles.	Undulating baseline with changing amplitudes and timing; no discrete QRS complexes or T-waves are present (Figure 3.13f).

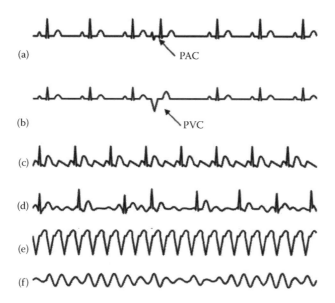

(a) PAC

(b) PVC

(c)

(d)

(e)

(f)

FIGURE 3.13 Illustrations of different arrhythmias. (a) Premature atrial contraction (PAC). (b) Premature ventricular contraction (PVC). (c) Atrial flutter. (d) Atrial fibrillation. (e) Ventricular tachycardia. (f) Ventricular fibrillation.

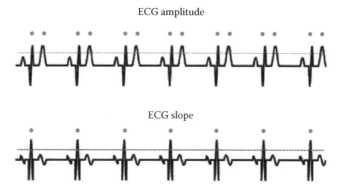

ECG amplitude

ECG slope

FIGURE 3.14 Comparison of ECG amplitude and slope for beat detection. The QRS and T-waves have similar amplitude and therefore are double counted when an amplitude threshold is used to detect the beats. The slope of the ECG accentuates the QRS complexes and allows beat detection without double counting.

ventricles, whereas long QRS durations may indicate conduction block somewhere along the normal conduction sequence of the ventricle. Normal QRS duration ranges from 60 to 110 ms.

The measurement of QRS duration requires the detection of the beginning (onset) and the end (offset) of the QRS complex (Figure 3.15). Because the QRS complex is typically characterized by sharp deflections, automatic algorithms used to detect QRS onsets and offsets rely on the slope of the signal. Starting with the fiducial point obtained by QRS detection, the QRS onset can be detected by the closest point preceding the fiducial point where the QRS slope falls below a threshold for a given amount of time. The QRS offset is

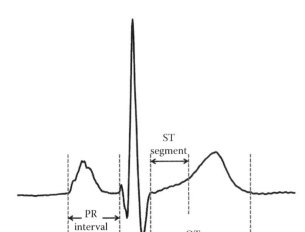

FIGURE 3.15   Illustration of the standard intervals and segments that are measured from the ECG.

similarly detected as the closest point following the fiducial point where the QRS slope falls below a threshold for a certain amount of time. The QRS offset is also known as the J-point.

3.7.3 QT Interval

The QT interval is a measure of the time period from the QRS onset to the end of the T-wave. QT intervals provide an estimate of the time required for the ventricles to repolarize following activation. Because the time to repolarize is dependent on the heart rate, a heart-rate-corrected QT interval (QTc) is often used. QTc is calculated by the following formula (Bazett's formula):

$$QTc = \frac{QT}{\sqrt{RR}}$$

where
   QT is in milliseconds
   RR is in seconds

A QTc interval of greater than 450 ms for males and 470 ms for females is an indicator for vulnerability to ventricular arrhythmias. Other rate correction formulae have also been used.

Automatic measurement of the QT interval requires detection of the QRS onset as previously discussed and the T-wave offset. The end of the T-wave is a challenge to accurately detect as the tail slowly returns to baseline, as shown in Figure 3.15. Small differences in amplitude or slope thresholds can drastically affect where the end of the T-wave is marked. A common alternative algorithm marks the end of the T-wave by projecting the tangent of the maximum descending slope of the T-wave to the baseline. Where the projection intersects with the baseline is considered the T-wave offset. The point of the maximum descending slope can be detected reliably, and using it provides a T-wave offset that is more stable than that obtained by fixed amplitude or slope thresholds.

3.7.4 PR Interval

The PR interval is the time period between the P-wave onset and the QRS onset (Figure 3.15). The PR interval is used in the evaluation of atrioventricular node conduction and to evaluate heart block. The P-wave onset is automatically detected by the presence of an isoelectric baseline for a certain amount of time preceding the P-wave. During fast heart rates, the P-wave may be masked by the end of the T-wave of the preceding beat.

3.7.5 ST Height

The portion of the ECG between the QRS offset and the beginning of the T-wave is known as the ST segment (Figure 3.15). The ST segment reflects the early portion of repolarization of the ventricle. ST-segment elevation can be used as an indicator of occurring or prior myocardial infarction (i.e., heart attack) where portions of the ventricle are damaged or distressed due to the lack of blood supply from a blocked blood artery. ST elevation of more than 1–2 mm (0.1–0.2 mV) in two or more contiguous leads is an indicator of acute myocardial infarction. ST height can be measured automatically by determining the amplitude of the ECG relative to the isoelectric line at points at or after the QRS offset.

3.7.6 QRS Axis

The polarity of the QRS complex in the leads that define the frontal plane can indicate whether the ventricular activation sequence is propagating in the expected direction. The direction of the QRS complex is known as the QRS axis. The normal QRS axis ranges from −30° to +90°. With a normal axis, lead I has a positive deflection, and lead aVF or II generally has positive deflections. QRS axis of −30° to −90° indicates left axis deviation (lead I is positive and aVF is negative), while right axis deviation is defined from +90° to +150° (lead I is negative and leads II and aVF are both positive). Left or right axis deviation could indicate an abnormality in ventricular conduction.

3.8 ADVANCED ECG SIGNAL PROCESSING

Most of the interval, amplitude, and axis measurements described earlier can be done not only with automatic algorithms but can also be performed manually with calipers. Some ECG signal-processing methods have been developed that offer diagnostic information that cannot be easily obtained through manual analysis.

3.8.1 Median Beat Analysis

Noise in the ECG due to electromyogram artifact, motion, or poor electrode contact can make any of the previously described measurement difficult, either manually or automatically. Noise is a particular issue when ECGs are recorded in an ambulatory setting or for the purposes of an ECG stress test. Filtering can be used to improve the signal-to-noise ratio of the ECG, but overfiltering can cause distortion of the waveform. Median

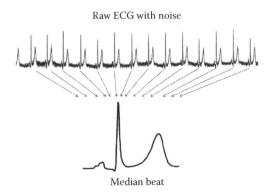

Raw ECG with noise

Median beat

FIGURE 3.16   Illustration of the creation of a median beat from a noisy ECG.

templates of ECG waveforms are a composite of several beats. With a normal rhythm and a stable heart rate, most waveforms of each beat should be very similar to each other. Noise, on the other hand, is generally uncorrelated with ECG activity. Therefore, a composite of several beats will give us a truer sense of what the waveform should look like without noise.

The process of creating a median beat is illustrated in Figure 3.16. First, QRS detection is performed as previously described. Next, fixed length windows, enough to encompass each beat from the beginning of the P-wave to the end of the T-wave, are obtained. All the windows are then aligned with each other with the maximum slope of the QRS used as the fiducial point. A median value for each sample corresponding to a point in the cardiac cycle is obtained and used as the representative amplitude for that time point. The median beat is obtained when all median values for each point in the cycle are calculated.

3.8.2 Pacemaker Detection

With interpretation of the ECG, it is important to know if a pacemaker is controlling the heart beats. Because the output pacing stimuli of pacemakers have very narrow pulse width (generally on the order of 0.5 ms), ECG recording devices with standard sampling rates of 1000 Hz and less are not adequate to reliably capture the pacing activity (sampling at 1000 Hz is every 1 ms; a typical ECG sampling rate is 250 Hz or every 4 ms). ECG systems with pacemaker detection features have specialized hardware and software that detect the pacemaker impulses during acquisition. The process usually includes band-pass filtering at a high-frequency band, an analog comparator circuit, and high sample rate analog-to-digital conversion. The times of the detected pulses are then saved as annotation on the lower sample rate ECG.

3.8.3 Heart Rate Variability

Heart rate variability reflects the modulation of beat-to-beat firing rate of the sinus node by the autonomic nervous system. The autonomic nervous system consists of two arms: the sympathetic nervous system that controls the body's "fight and flight" response and the parasympathetic nervous system that controls the body's "rest and recovery" response.

Increase in sympathetic neural activity results in an increase in heart rate, while increase in parasympathetic neural activity decreases heart rate. The magnitude of resting heart rate variability is largely reflective of parasympathetic neural activity and/or its respiratory modulation. However, the relative effects of sympathetic and parasympathetic neural activity will determine the spectral profiles of the heart rate variability signal. Low heart rate variability has been shown to have prognostic value in determining risk for sudden cardiac death.

All measures to quantify heart rate variability require QRS detection, calculation of the intervals between the R-waves (RR intervals), and the rejection of RR intervals due to nonsinus node activity. The resulting data are known as NN intervals. There are several common heart rate variability measures that are used. In the time domain, these include the standard deviation of NN intervals (SDNN), the root mean square of successive differences (RMSSD), and the percentage of NN intervals with greater than 50 ms change from the preceding NN interval (pNN50). Frequency domain measures that have also been proposed as different frequency bands are thought to reflect different autonomic processes. Frequency domain parameters are obtained by first interpolation of the NN intervals to obtain a signal with a constant sampling rate (usually 2 Hz). Linear detrending of the resampled signal is then performed. The power spectrum is then obtained. The power in a high-frequency band (0.15–0.4 Hz) has been shown to reflect parasympathetic modulation of heart rate. A low-frequency band (0.04–0.15 Hz) has been shown to reflect both sympathetic and parasympathetic modulations of the heart rate. An example of NN intervals in the frequency domain is shown in Figure 3.17.

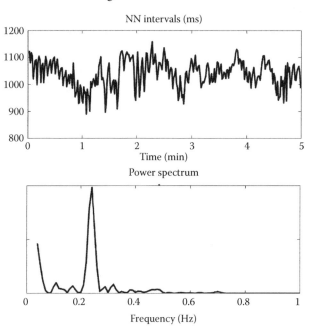

FIGURE 3.17   Example of a 5 min NN interval series at rest and the corresponding power spectrum. The peak at 0.24 Hz corresponds to the respiratory modulation of heart rate.

3.8.4 Signal-Averaged ECG

Signal-averaged ECG is a technique used to identify patients with "late potentials" that occur due to slow conduction through infarcted areas of the ventricle. These late potentials occur at the end of the QRS complex and may indicate vulnerability to ventricular arrhythmias. Because of the low amplitude of the late potentials, specific signal-processing operations consisting of a combination of signal averaging and filtering are used. Signal-averaged ECG requires several minutes of resting ECG recording using an orthogonal lead system with a sampling rate of 1 kHz. Signal-averaged (mean) waveforms of the X, Y, and Z leads are obtained in a similar process used in the median beat analysis described previously. After signal averaging, the noise levels are effectively reduced by an order of magnitude. High-pass filtering with cutoff frequency of 30 Hz is then performed. The vector magnitudes of the filtered X, Y, and Z signal-averaged beats are then calculated. This process is illustrated in Figure 3.18. Using the vector magnitude, the following measures are typically used to detect the presence of late potentials: filtered QRS duration (>114 ms), root-mean-square voltage in last 40 ms of the QRS (<20 μV), and signal duration in terminal QRS that is <40 μV (>38 ms). Patients meeting two or three of these criteria have been shown to be at increased risk for sudden cardiac death.

3.8.5 T-Wave Alternans

At high heart rates, the T-wave can exhibit alternating beat-to-beat patterns of A to B to A to B, where A and B each represents a specific amplitude, duration, or morphology of the T-wave. Detection of these "T-wave alternans" patterns at relatively low heart rates has been shown to be a risk predictor of cardiac arrhythmias. Because the beat-to-beat changes

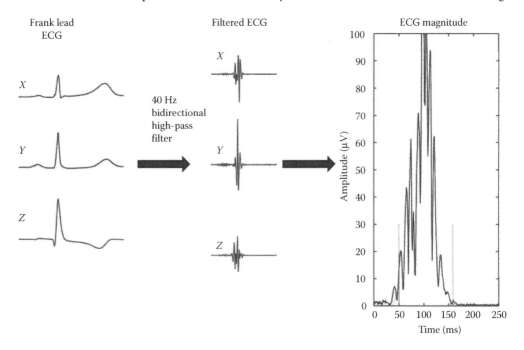

FIGURE 3.18 Illustration of the processing used in signal-averaged ECG analysis. Late potentials are measured in the filtered ECG magnitude.

often are in the order of microvolts, detection of alternans is not easily discerned visually and requires specialized signal processing.

Two prominent methodologies have emerged for the analysis of T-wave alternans that are used in the clinical setting. The first is a method developed by Cohen and colleagues that measures beat-to-beat alteration using the frequency domain. If a beat-to-beat sequence of T-wave amplitudes is transformed into the frequency domain, a prominent peak at the 0.5 cycles per beat frequency will be present in the power spectrum if alternans is present. If there is no T-wave variation or if the variation is not in an ABAB alternating pattern, the power at the 0.5 cycles per beat frequency will not be significantly different from the nearby frequencies. Figure 3.19 shows an example of a power spectrum with alternans present. A parameter called the "k-score" is used to determine if the alternans magnitude is significantly greater than the nearby "noise" frequencies. The k-score is calculated as the difference between alternans magnitude measured at the 0.5 cycles per beat frequency and the mean noise magnitude measured in the 0.43–0.47 cycles per beat band, which is then divided by the standard deviation of the noise magnitude. A k-score greater than 3 is considered significant alternans. Patients having significant alternans during a stress test at heart rates less than 110 beats/min have been shown to be a risk for ventricular arrhythmias. A commercial system performing this test is currently sold by Cambridge Heart (Bedford, MA).

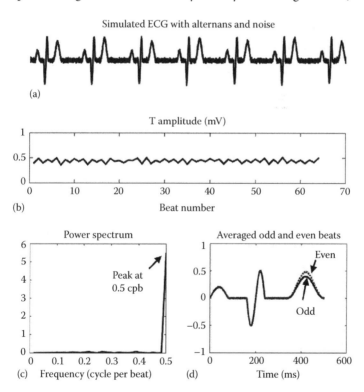

FIGURE 3.19   Simulation of alternans with added noise. (a) Raw ECG signal. (b) Measured T-wave amplitude. (c) Power spectrum of the amplitude values with a large peak at 0.5 cycles per beat indicating alternans. (d) Superimposed plots of the averaged even and odd beats. Separation occurs at the T-wave.

A second method to measure T-wave alternans is the "modified moving average" developed by Nearing and Verrier. This method is a time-domain approach that performs separate moving averages for even and odd beats. To account for noise and outliers, the algorithm limits the effect of any one beat on the calculations by using nonlinear update factors. If a certain amount of separation between the even and odd beats is detected, alternans is considered to be present. Patients with separation of more than 65 μV have been shown to be at high risk for sudden cardiac death. Figure 3.19 shows an example of an ECG tracing and the superimposed even and odd averaged beats. The test is available as part of the GE Healthcare's (Waukesha, WI) Holter monitoring analysis software.

3.9 AMBULATORY MONITORING

The evaluation of the ECG during resting or stationary conditions is limited because of its short duration and typical absence of symptoms at the time of the recording. Ambulatory monitors can be used to monitor or record ECG activities for long periods of time, while the patient follows a normal daily routine. The likelihood of capturing transient ECG events is thus improved when these monitors are used.

3.9.1 Holter Monitoring

Holter monitoring involves using a portable ECG recording device that can continuously record ECG over 24 h or longer. Holter monitors commonly record ECGs from three to five electrodes with leads connecting to the belt-worn monitor. Twelve-lead recording systems are also available. The leads are often taped to the body for extra security. Holter monitors also feature an "event" button that allows the patient to document the time of any symptoms, which can then be correlated with ECG to determine whether the symptom is associated with a cardiac arrhythmia.

Early versions of Holter monitors were battery-powered cassette tape recorders that stored amplified ECG waveforms on the tape in analog form. As solid-state ("flash") memory has become cheaper and more widespread; almost all Holter monitors currently store digital data. It is also common that data compression on the waveforms is performed prior to storage to save storage space. However, because the cost of flash memory is continuously decreasing, compression is becoming less necessary. Digital Holter recorders also have the advantage of not having any motorized parts, thus allowing greater battery life.

Although Holter monitors primarily function as a recording device, some monitors may have limited real-time capabilities such as lead checking and heart rate measurement. However, most processing occurs after the conclusion of the recording period when the signals are transferred to a workstation containing the analysis software. Some clinical information that is obtained in the analysis of Holter recordings includes QRS morphologies, heart rate, heart rate variability, intervals (PR, QRS duration, QT, etc.), and arrhythmia detection. Manual overreading is necessary with Holter monitoring because of the susceptibility of ambulatory recordings to noise. Motion and tremor artifacts are common sources of interference of the ECG signal. Thus, time segments with high levels of artifact must be excluded to prevent false reading of heart rate and other measures.

Some cardiac events occur too infrequently to be detected with even 24 or 48 h Holter monitoring. Recording longer continuous ECG for such situations may not be practical, not only from the standpoint of data storage but also for the amount of data that would need to be analyzed. Thus, an alternative to the Holter monitor is the event recorder.

3.9.2 Long-Term Ambulatory/Event Recorders

Event recorders are another form of ambulatory monitors. Event recorders differ from Holter monitors in that the ECG recordings are stored only if the patient starts the monitor when symptoms occur or if the device automatically detects an arrhythmia. Because the device is not recording continuously, event monitors are usually smaller in size than Holter monitors and can record for longer periods of time (30 days) because of reduced power requirements. The physician can then use the collected data to determine whether an arrhythmia was documented. Long-term continuous recorders are also available.

3.9.3 Implantable Loop Recorders

Holter and event monitors record and store ECG signals from surface electrodes. Implantable loop recorders are small devices that are implanted under the skin, usually beneath the collar bone. These devices record single-lead ECGs from electrodes that are on the case of the recorder. The loop recorder buffers a short amount of ECG at a time. The patient places a handheld activator over the area where the loop recorder is implanted when the patient senses symptoms. The recorder then transfers the ECG from the buffer to another memory location that can be accessed later. The loop recorders can continue to operate for up to 3 years after which they can be removed.

One application for these devices is to monitor patients for episodes of atrial fibrillation. These episodes can be both symptomatic and asymptomatic. Thus, a robust ECG signal processing is required to accurately detect these episodes. These recorders typically detect atrial fibrillation via the response of the ventricle, which is usually characterized by QRS complexes that are highly irregular in timing but similar in morphology. Medtronic (Minneapolis, MN) and St. Jude Medical (St. Paul, MN) both offer implantable loop recording devices. Loop recording is also performed in modern implantable pacemakers and defibrillators.

3.10 TELEMETRY

Telemetry is a system in which ECG recordings (and other physiologic measurements) are wirelessly transmitted to a separate monitor for real-time display and monitoring. The setup on the patient is similar to that of a Holter monitor. The ECG electrodes are connected to a portable wireless transmitter by leads. The transmitter contains amplifiers and modulators for radio-frequency transmission. In current systems, the transmitter performs analog-to-digital conversion before transmission using a digital wireless protocol.

Inpatient telemetry systems are used by hospitals for constant monitoring for serious cardiac events, such as arrhythmia or acute myocardial infarction. Data are transmitted to both the monitors in the patient's room and the central nursing station. Alarms alert the staff if a serious event is detected.

Outpatient telemetry systems are also available. However, these outpatient telemetry systems are more of an alternative to Holter or event monitoring than a replacement for inpatient telemetry. Patients with very transient and minimally symptomatic arrhythmic episodes (namely atrial fibrillation) are ideal candidates for outpatient telemetry. Similar to an inpatient monitoring system, the ECG is recorded by a transmitter worn by the patient that continuously transmits data to a monitoring station in the home. If the monitoring station detects an arrhythmia, the station will transmit the ECG by landline or cell phone to a center where a technician can overread and confirm the arrhythmia. Doctors can use this information to confirm the presence of arrhythmia and adjust medication on an outpatient basis if needed. Because the captured data are transmitted and not stored, monitoring for indefinite periods of time is possible with only battery changes/recharging needed.

3.11 CARDIAC STRESS TESTING

Cardiac stress testing involves recording an ECG while the patient follows an exercise protocol on a treadmill or stationary bicycle. Infusion of pharmacologic agents can also be used to simulate exercise conditions during a stress test. A stress test can indicate if there is insufficient blood flow to the heart during exertion, which is evident with depression of the ST segment of the ECG. There are special ECG systems designed to work in the stress test laboratory that interface with exercise equipment. Ultrasound or nuclear imaging may also be performed as part of a stress test to better assess heart function and blood flow to the heart muscle.

Specific exercise protocols can be programmed into the stress test systems to control the speed and slope of the treadmill or the resistance for the bicycle test. The Bruce protocol is the most commonly used exercise protocol that is used to screen for coronary artery disease. This protocol consists of seven stages that last 3 min each. In the first stage, the patient begins by walking at speed of 1.7 mph on a 10% incline. In the subsequent stages, the treadmill speed and incline grade are gradually increased until the speed is 6.0 mph and the incline is at 22%. In addition to ECG recording, blood pressure is measured at the end of every stage. A modified Bruce approach is used for patients who are more limited physically. In the modified approach, incline grade begins at 0% instead of 10%.

An example heart rate plot from a patient during a stress test is shown in Figure 3.20. The heart rate sharply increases at the start of exercise and continues to gradually increase

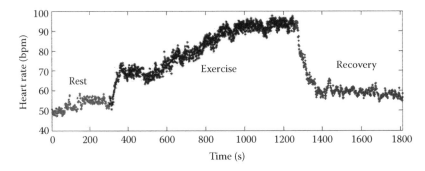

FIGURE 3.20 An example of the heart rate plot from a patient undergoing a stress test.

before reaching a plateau. When exercise is stopped, heart rate quickly recovers. Both the ability to reach the target heart rate during exercise (220 − age) and the speed of recovery once stopping exercise are indicators of cardiovascular health.

3.12 CARDIAC MAPPING

The surface ECG provides composite information of the heart's electrical activity in its entirety. However, this "global" view of the electrical activity may not offer sufficient enough detail of the activation sequences of the heart to pinpoint the locations responsible for certain arrhythmias. This detail is particularly needed for radio-frequency ablation procedures, where catheters are used to deliver radio-frequency energy to specific areas of the heart to destroy the culprit tissue. Local electrical activity can be recorded by electrodes that are in contact directly with the heart tissue. These contact electrical recordings are known as electrograms. In a cardiac electrophysiologic study, catheters with electrodes at the tip are guided through a vein or artery to the heart. X-ray machines that produce real-time images displayed on monitors are used to guide the catheters. The catheter used for radio-frequency ablation can also be used to record electrograms. A mapping system is used to amplify, filter, display, and record electrogram signals. Mapping systems also allow the selection of channels to perform pacing via an electrical stimulator.

3.12.1 Electrograms

Contact electrograms again differ from surface ECG recordings as they reflect local electrical activity at the site of recording. Electrograms are most commonly either unipolar or bipolar recordings. Unipolar electrograms are differential recordings made from a single electrode in contact with the heart referenced to a distant electrode not in contact with the heart. Figure 3.21 shows the typical waveform of a unipolar electrogram and how it relates to an activation wavefront. A sharp positive deflection occurs as the electrical propagation

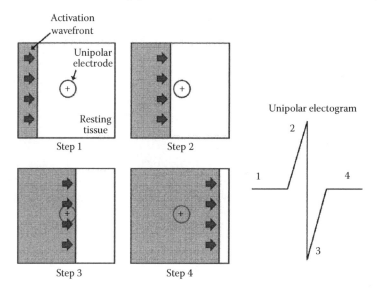

FIGURE 3.21   Illustration of the unipolar electrogram and how it relates to activation of tissue.

nears the electrode followed by a sharp negative deflection as the wavefront crosses the electrode and propagates away from the electrode. A slower deflection may follow that corresponds to the repolarization wavefront. The time of maximum negative slope is commonly regarded as the activation time of the recording site. With unipolar electrodes from multiple sites in the heart, the activation sequences can be determined. A greater number of electrodes will produce a more detailed activation map. However, if the cardiac rhythm consists of repetitive activation sequences, the detailed activation maps can be created with one roving catheter that sequentially determines activation times relative to the activation times of one stable electrode.

The morphology of the unipolar electrogram can also provide insight into the activation sequence. If the electrogram only has a negative deflection and not a positive deflection, this can be interpreted as the electrode site being the focal source of an activation sequence. Similarly, if the electrogram only has a positive deflection and no negative deflections, this would indicate that the recording site is at the end of the activation sequence.

A limitation of unipolar electrograms is that far-field activity may also be reflected in the signal, in addition to local activity. The additional of far-field activity may interfere with the ability to determine the activation time from the unipolar electrogram. For this reason, bipolar electrograms are commonly used. Bipolar electrograms are electrical recordings comprised of the difference between two closely spaced electrodes that are both in contact with the heart. Bipolar electrograms can also be thought of as the difference between two unipolar electrograms. The biggest difference between two closely spaced unipolar signals occurs at the time of activation of the two recording locations. Far-field activity, on the other hand, will have similar effects on the closely spaced electrodes and thus will cancel out with the bipolar recording.

The morphology of bipolar electrograms depends on a number of factors. First, unlike unipolar electrograms, bipolar electrograms are dependent on the direction of the activation wavefront. As illustrated in Figure 3.22, a wavefront that first crosses the negative electrode and then crosses the positive electrode will produce a bipolar electrogram that has the opposite polarity to the bipolar electrogram where the wavefront first crosses the positive electrode and then the negative electrode. A wavefront that crosses the positive and negative electrode perfectly perpendicular to the two electrodes will produce a bipolar electrogram with negligible potential, since the two electrodes will sense similar activity. Wider electrode spacing or slower conduction velocity will result in a bipolar deflection with longer duration. High-pass filtering (typically with a 30 Hz cutoff) is typically used to further accentuate the activation deflections during bipolar recordings. With atrial fibrillation, bipolar recording can also appear very fractionated, indicating complex conduction patterns in those areas. These fractionated electrograms have been studied extensively to understand their role in the maintenance of the arrhythmia.

The third kind of electrode recording is known as the monophasic action potential (MAP) recording. A MAP recording is obtained by an electrode that is in contact with the tissue referenced to the second electrode that is in close proximity but not in contact with the tissue. The result is a signal that approximates the shape of the transmembrane action potential. This recording technique is useful in the study of repolarization characteristics

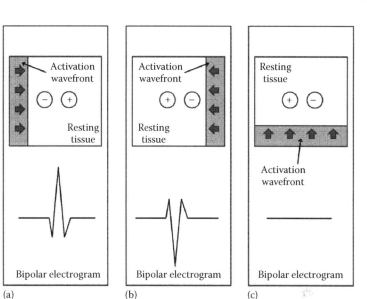

FIGURE 3.22 Illustration of bipolar electrogram morphology changes with changes in activation direction relative to the positive and negative recording electrodes. (a) Activation from left to right invokes an electrogram that is mostly positive in polarity. (b) Activation from right to left results in electrogram that is mostly negative in polarity. (c) Activation traveling perpendicular to the two electrodes results in a mostly isoelectric electrogram.

of the tissue, particularly repolarization dynamics. However, good-quality recordings require the contact electrode to have a constant pressure against the tissue. Suction or spring mechanisms are usually incorporated into the MAP catheter or probe to provide the constant pressure. Despite this, MAP signals will still degrade over a short time. Because of the technical difficulties in achieving good-quality signals, MAPs are rarely obtained during clinical procedures.

3.12.2 Effective Refractory Periods

Information of the refractory periods of cardiac tissue is not easily or reliably obtained with unipolar or bipolar recordings. The refractory period can instead be measured using programmed stimulation to pace the tissue. Because refractory periods are dependent on the heart rates, the stimulator is first set to pace at a stable heart rate. A common setting for this phase, known as the S1 interval, is 8 paced beats at a 400 ms interval. Following the S1 intervals, a stimulus with a shorter coupling interval (S2) is delivered. If the S2 stimulus elicits a response from the tissue, then the S1–S2 sequence is repeated with a decremented S2 until the tissue does respond to the S2. The longest S2 interval that fails to exhibit a response is the effective refractory period (ERP) for that particular S1 interval. An illustration of the process of determining the ERP with pacing is shown in Figure 3.23.

3.12.3 Electroanatomic Mapping

Determining the catheter locations with fluoroscopy requires a level of intuition by the cardiac electrophysiologist as the heart is not visible from the x-ray images. Furthermore,

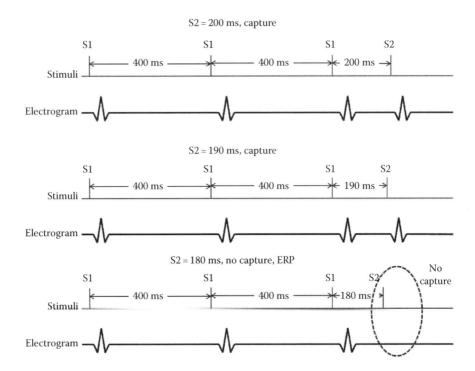

FIGURE 3.23 Illustration of pacing-based determination of the ERP. The basic cycle length (S1) is 400 ms. The extrastimulus (S2) is decremented from 200 ms, where an electrogram potential is invoked, to 180 ms, where no electrogram is invoked. The ERP is 180 ms.

the x-ray images are 2D, whereas the catheters are manipulated in 3D space. With complex ablation procedures such as atrial fibrillation ablation that require ablation of many sites, it is important to know what part of the atria has already been ablated. This is not easily determined with a fluoroscopy system alone. An electroanatomic mapping system is a tool utilized in the electrophysiology laboratory to enable tracking of the 3D location of the catheters.

Catheter tracking with electroanatomic mapping works similarly to satellite-based global positioning systems used to track a user's position on earth. The Carto electroanatomic mapping system by Biosense Webster (Diamond Bar, CA) determines the position of the catheters by triangulating low-level magnetic fields produced by a pad beneath the patient on the table. The Ensite NavX by St. Jude Medical (St. Paul, MN) uses high-frequency currents emitted by patches applied to the skin to triangulate catheter positions. Once the catheters are positioned within the heart chamber of interest, the electroanatomic mapping system will allow real-time tracking of the catheter and reduce the amount of x-ray exposure required for fluoroscopy. By collecting spatial points when the catheter tip is in contact with different parts of the chamber, a 3D geometry corresponding to the geometry of the endocardium can be created. This geometry can be merged with 3D data obtained from MRI or CT for more detailed anatomic information. Electroanatomic mapping can then be used to construct activation maps to determine activation sequences, create voltage

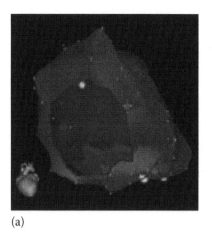
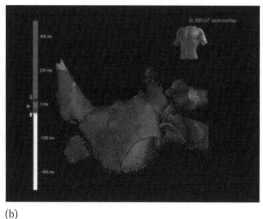

(a)   (b)

FIGURE 3.24 Examples of activation maps obtained with electroanatomic mapping systems. (a) Activation map of the right atrium created by the Carto XP system (http://www.biosensewebster.com). (b) Activation map of the left atrium created by the Ensite NavX system (http://www.sjmprofessional.com/).

maps to identify areas of scar, and document ablation sites. Examples of activation maps created by electroanatomic mapping systems are shown in Figure 3.24.

3.12.4 Noncontact Mapping

Contact mapping with catheters is limited by the number of catheters and electrodes available to perform electrogram recording. The St. Jude Medical EnSite system is a non-contact mapping system that records far-field potentials within a chamber of the heart with a multielectrode balloon catheter and uses these signals to predict what the potentials would be on the endocardial surface of the chamber. This process first requires the construction of the 3D geometry of the chamber with electroanatomic mapping and a roving catheter. Each of the 64 electrodes of the balloon catheter records far-field potentials from the entire chamber. By taking account of the distance of each electrode from a point on the wall, a virtual electrogram at that point can be calculated from the inverse solution to Laplace's law. An unlimited number of virtual electrograms can therefore be computed. A limitation of this system is that areas at greater distance from the balloon will have greater errors in the reconstruction, particularly if there are inaccuracies in the constructed geometry.

3.13 RADIO-FREQUENCY LESION MEASUREMENTS

3.13.1 Contact Sensing and Force Measurement

Electrogram recording and ablation require good contact of the catheter electrodes with the heart. Adequate contact force is a requirement for ability to make transmural ablation lesions. Lesions that are incomplete transmurally may not block activations as intended. Too much contact force comes with the risk of perforation. Thus, contact sensing is important to provide an electrophysiologist feedback regarding the amount of pressure the catheter has against the chamber wall, which is not easily determined from tactile feedback.

The two most common means to determine contact are electrogram signal quality and impedance measurement. Good-quality electrograms with high amplitude are possible only with good contact with the chamber wall. Impedance measurements drop significantly when the catheter is in contact with the heart. However, both these indications only indirectly measure contact force, and both have significant limitations. Electrogram amplitude by itself is not a good indicator of contact as both poor contact or scar tissue can be responsible for low amplitude. Neither electrogram amplitude nor impedance can provide good indication that too much force is being applied. Thus, more direct measures of contact force would be desirable.

There are a few technologies that are being developed and evaluated for contact force measurement. A catheter by Endosense (Geneva, Switzerland) uses optical fibers to sense catheter contact force. Bending of the optical fibers produces a change in wavelength that can be translated into a force measurement. A second technology by Biosense Webster (Diamond Bar, CA) uses a spring-loaded tip with a magnetic signal emitter and three magnetic sensors. Pressure applied to the tip will change the position of the sensors relative to the emitter and results in the change of the detected signal. The contact force is estimated based on the signal change. For both technologies, force is measured in grams. Because of the movement associated with the beating heart, contact force will vary during the time needed to apply the radio-frequency energy. To determine the total amount of contact "energy," a force–time integral can be calculated. The measured force–time integral has been shown to be a better predictor of lesion size than the peak force of the catheter.

3.13.2 Tissue Temperature Sensing

Radio-frequency ablation destroys tissue by heating the site of catheter contact. The radio-frequency current density is the highest at the point of contact; thus, most of the heat will be generated there, as current through a resistive medium generates heat. The temperature to which the tissue is heated is a factor in the formation of the lesion. Thus, temperature sensors on ablation catheters are useful in determining the power output needed to create adequate lesions to also prevent overheating. Thermistors and thermocouples are two common types of temperature sensors. Thermistors change resistance in response to changing temperature. By applying a constant current to the thermistor, a voltage is produced that decreases as temperature increases. Thermocouples are sensors that consist of two different conductors, which output a voltage proportional to the temperature. Unlike thermistors, thermocouples do not need to have current applied to them to output a voltage.

We previously discussed the importance of contact force and contact time for lesion generation. Convective heat transfer by the circulating blood pool is a counteracting factor against lesion formation. The greater the convective cooling, the greater amount of radio-frequency output that is needed to obtain the desired effect. Measurement of the convective heat transfer coefficient can be performed by measuring the temperature with two sensors: one at the tip and another nearby, which measures blood flow temperature. The difference between the temperatures divided by the product of the electric power consumed by heating the sensor and sensor area equals the convective heat transfer coefficient.

3.14 POSSIBLE FUTURE INSTRUMENTS

ECG and contact electrical recordings of the heart have stood the test of time for their utility in the diagnosis of arrhythmias and other heart conditions. The current techniques for ECG measurement may not significantly evolve from their present forms, although the hardware may improve and new algorithms used for automatic diagnosis or risk stratification may be developed. What we may see more of in the future is the integration of ECG with other imaging modalities to provide the context of heart function and substrate with the electrical activity. Three-dimensional MRI or CT can provide detailed characterization of the torso, the heart, and even scarring within the heart. How to use this information in conjunction with ECG body surface mapping to accurately characterize activation sequences in the heart is a subject of active investigation. If local conduction velocities and repolarization can be measured in this fashion, patient-specific computer simulation could be used to test arrhythmia vulnerability in the modeled heart and perhaps be used to plan ablation or other course of treatment. This would be a noninvasive and lower-risk alternative to a study performed in the cardiac electrophysiology laboratory involving catheter insertion and x-ray exposure.

The greatest number of sudden cardiac deaths occurs in the population with no known risk factors. Future devices that are low cost and minimally intrusive to monitor ECG could provide alerts to emergency services when arrhythmias are detected and could potentially save many lives. It would need to be determined whether this technology could be practically integrated into cell phones, watches, personal digital assistants, or made as a subcutaneous device.

BIBLIOGRAPHY

Goldberger J J and Ng J (eds.) 2010 *Practical Signal and Image Processing in Clinical Cardiology* (London: Springer)

Webster J G (ed.) 2009 *Medical Instrumentation: Application and Design*, 4th edition (Hoboken: John Wiley & Sons)

Zipes D J and Jalife J (eds.) 2009 *Cardiac Electrophysiology: From Cell to Bedside* (Philadelphia: Saunders)

CHAPTER 4

Cardiology

Hemodynamics

Sándor J. Kovács

CONTENTS

4.1 Overview of Hemodynamic Physiological Measurement in Cardiology 76
    4.1.1 Four-Chambered Heart 76
4.2 Tissue Perfusion 80
    4.2.1 Tissue Perfusion Measurement 81
    4.2.2 MRI 82
    4.2.3 CT Myocardial Perfusion Imaging 83
4.3 Ultrasonic Methods of Perfusion Determination 84
    4.3.1 Method of Myocardial Perfusion Analysis 85
4.4 Microspheres 86
    4.4.1 Flow in Capillaries 86
4.5 Venous Flow 87
    4.5.1 Function 87
4.6 Fick Principle 89
4.7 Arterial Flow Measurement 90
4.8 Cardiac Output 91
    4.8.1 Invasive Methods for Cardiac Output Determination 91
    4.8.2 Thermodilution 91
        4.8.2.1 Thermodilution Method: Technical Details 93
    4.8.3 Noninvasive Determination of Cardiac Output 94
4.9 Arterial Stiffness 94
4.10 Pulse Pressure 95
4.11 Pulse Wave Velocity 95
4.12 Ultrasound-Based Measurements 96
4.13 Pressure Waveform Analysis 97
4.14 Possible Future Instrumentation 98
References 98

4.1 OVERVIEW OF HEMODYNAMIC PHYSIOLOGICAL MEASUREMENT IN CARDIOLOGY

Cardiologists diagnose and treat diseases of the cardiovascular system. Diagnosis involves a detailed history, a physical examination, blood tests, and invasive and noninvasive measurement of cardiac function. Treatment depends on the diagnosis and may involve medical therapy, percutaneous procedures, or surgery (performed by surgeons—not cardiologists). This chapter will deal with hemodynamics from the organ system to the tissue perfusion level and describe the most important physiological measurements required that are important in medical diagnosis, the types of sensors used, and how the signal may be acquired and processed and then displayed. Because the author is a clinically active, academic cardiologist, the relative value and utility of the measurements in the *real world* will also be addressed. Although engineering and quantitative aspects of the physiology and how the heart works as a coordinated pump will be emphasized, it is important to recall that the entire system we are concerned with is composed of living biological tissue comprised of trillions of living cells. Therefore, it is subject to all the basic biological laws that govern genetics, embryology, cellular physiology, metabolism, inflammation, repair of injured tissue, and so forth, including all of the compensatory pathways that nature has developed through natural selection.

Elements of the material here have been covered to various levels of detail in Chapters 2 and 3. Selected aspects are reviewed to emphasize the context of the physiological measurements involved.

Blood circulation is achieved by the four-chambered heart, located in the pericardial sac in the midthorax. Total blood volume in a typical 70 kg male is 5.5 L, about 60% of which resides in the venous system at any given time. Assuming the average heart rate (HR) is 60 beats/min, over the lifetime of a healthy individual (90 years), the heart will deliver 3×10^9 cardiac cycles and will pump about 5 L/min at rest, or 2.6 million L/year or 250 million L/lifetime.

In simplest terms, the circulation can be approximated as a simple lumped circuit that obeys Ohm's law. The familiar $V = IR$ is expressed using physiologic variables as mean arterial pressure (MAP in mmHg) = cardiac output (L/min) × mean peripheral vascular resistance (mmHg·min/L). In actuality, because the system is pulsatile, and the flow occurs in elastic tubes (the vasculature), rigorous analysis of pressure and flow phenomena in the circulation requires measurement as a function of frequency. Accordingly, impedance, as a function of frequency and its higher harmonics, is the time-dependent measure of resistance to flow (see Figure 4.1; Milnor 1989).

4.1.1 Four-Chambered Heart

Fundamental physiologic principles govern key aspects of heart function. Although the four-chambered heart (see Figure 4.2) resides in one physical location, and its contraction (systole) and relaxation (diastole) are coordinated as a result of direct physical coupling by the conduction system (see Chapter 3 for details) and by virtue of the fact they reside in the same physical location (pericardial sac), circulation is actually achieved by two muscle-powered oscillatory pumps connected in series (see Figures 4.1 and 4.2).

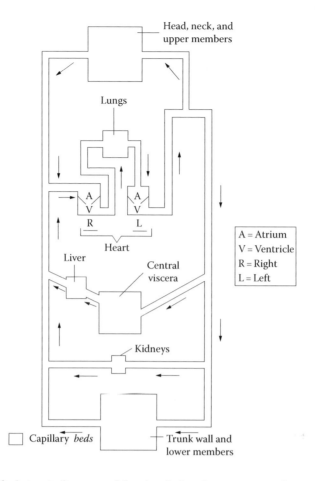

FIGURE 4.1 Simplified circuit diagram of the circulation. In computing the mean peripheral vascular resistance from the equivalent of Ohm's law, written as MAP = CO · PVR, all of the peripheral resistive components (head, neck, viscera, liver, kidneys, extremities, etc.) are lumped into a single resistance element. MAP, mean arterial pressure (mmHg); CO, cardiac output (L/min); PVR, peripheral vascular resistance (L/min)/mmHg. (From http://armymedical.tpub.com/MD0007/MD00070193.htm.)

The low-pressure (volume) pump consists of the right heart—comprised of the right atrium (RA) and ventricle (RV) and its associated valves. Its purpose is to advance deoxygenated venous blood that has returned to the heart at low pressure (<25 mmHg) into the pulmonary vasculature in order to eliminate CO_2 and absorb O_2. Accordingly, the right heart is best viewed functionally and physiologically as a *volume pump*, both in systole and in diastole. The left heart—consisting of left atrium (LA) and left ventricle (LV) and the associated valves—is primarily a volume (*suction*) pump in diastole (during filling) and a simultaneous volume–pressure pump in systole. All four cardiac valves are passive and are designed to open in response to suitably established pressure gradients and to close and seal—without leaking—when the sign of the pressure gradients governing forward flow is reversed.

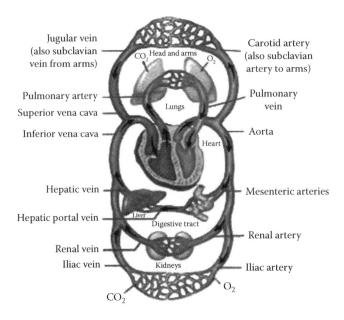

FIGURE 4.2 The circulation. Illustrated using anatomical terminology to emphasize the major organ systems. (From http://www.tutorvista.com/content/biology/biology-iv/circulation-animals/ blood-circulation-mammalian-heart.php.)

The four-chambered heart (Figures 4.2 and 4.3) is an almost perfect *constant-volume pump* (Bowman 2003). More specifically, the approximately 850 mL volumetric content of the pericardial sac varies only by about 5% between systole and diastole. The near constant-volume pumping feature is accomplished by spatially anchoring the apex behind the breastbone (sternum) and the back of the atria in the mediastinum, while there is simultaneous reciprocation of atrial and ventricular volumes, such that when the ventricles eject, the atria fill. This is accomplished by displacement of the plane of the A-V valves in the apex-to-base direction and wall thickening/thinning in systole/diastole. Importantly, the outer surface of the four-chambered heart, that is, the pericardial sac itself, has very slight inward motion during systole—primarily over the LV free wall. Therefore, the radial inward motion of the LV exterior surface during systole is *not* axially symmetric. When short-axis (transverse) end-systolic and end-diastolic magnetic resonance imaging (MRI) views of the epicardial (outer) surface are overlaid, the difference demonstrates the *crescent effect* (Waters 2005). In simple terms, the *crescent effect* is due to a slightly greater volume of blood being ejected through the aorta during systole than the volume of blood that simultaneously enters the LA from the lung through the four pulmonary veins. The blood entering the LA during systole corresponds to the pulmonary vein S-wave (S = systole) imaged during echocardiography. The slight difference in ejected versus atrial inflow volume during systole is made up in the next diastole and is imaged echocardiographically as the pulmonary vein D-wave (D = diastole) entering the atrium to fill the ventricle. Near *constant-volume* physiology has important consequences in mechanically coupling systole to diastole and causally explaining the volume conservation–imposed relationship

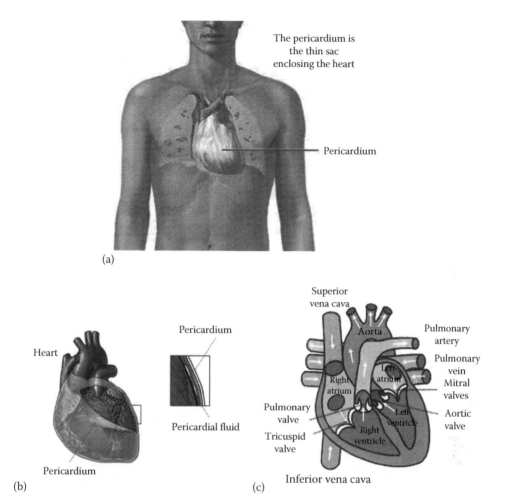

The pericardium is the thin sac enclosing the heart

Pericardium

(a)

Heart

Pericardium

Pericardial fluid

Pericardium

(b)

Superior vena cava

Aorta

Pulmonary artery

Pulmonary vein

Mitral valves

Aortic valve

Left atrium

Right atrium

Left ventricle

Right ventricle

Pulmonary valve

Tricuspid valve

Inferior vena cava

(c)

FIGURE 4.3 (a) The location of the four-chambered heart as it resides within the pericardial sack in the midthorax. (From http://www.mesothelioma911.org/.) (b) A small amount of pericardial fluid lubricates the surface of the beating heart. The (parietal) pericardium is fixed to the surrounding external tissues of the thorax, whereas the surface of the moving heart is the visceral pericardial surface that slides relative to the fixed parietal pericardium with each cardiac cycle. (From http://heartheavy.com/what-is-pericarditis/.) (c) Components of the four-chambered heart. Note that the pericardial sac includes the roots of the great vessels. (From http://quotations-com-loo.es.tc/images/circulatory-system-diagram-not-labeled.)

between certain echocardiographic indexes of diastolic function, such as the ratio of peak transmitral flow velocity (Doppler echocardiographic E-wave) to peak mitral annular motion velocity (Doppler echocardiographic E′-wave) that were previously viewed as being independent (Lisauskas 2001).

Simultaneous features of the cardiac cycle can be appreciated by considering the *Wiggers diagram* (Figure 4.4). The diagram is named after Carl J. Wiggers (1883–1963), an American physiologist/cardiologist who pioneered blood pressure measurement methods and the

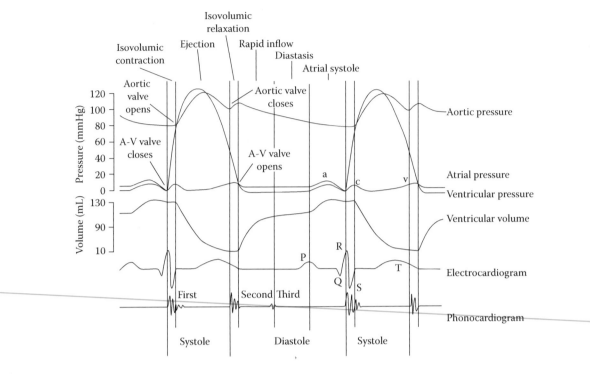

FIGURE 4.4 The Wiggers diagram showing the events of the cardiac cycle for the LV. Aortic pressure, LV pressure, LA pressure, LV volume ECG, and heart sounds on the phonocardiogram are shown. (From http://www.enotes.com/topic/Wiggers_diagram.)

effects of hypertension. The physiological measurements that are the gold standard of cardiac function assessment such as the ejection fraction (EF), peak rate of LV filling, atrial filling fraction, LV end-diastolic pressure, and peak systolic pressure can all be appreciated from the Wiggers diagram. The diagram conventionally illustrates the properties of the LV—although the timing of events and volumes is similar for the RV—with the proviso that the peak systolic pressures in the normal RV are typically <25 mmHg.

4.2 TISSUE PERFUSION

The perfusion of blood through the tissues is a regulated process. The control and feedback is organ and tissue specific. August Krogh was awarded the Nobel Prize in Physiology or Medicine in 1920 for his discovery of the mechanism of regulation of the capillaries in skeletal muscle. He was first to describe that blood perfusion in muscle and other organs responds to physiologic demand through opening (dilating) and closing (constricting) of the arterioles (typical diameter tens of micrometers) and capillaries (typical diameter 5–10 μm). Note that the control is achieved at the inlet (arteriolar) portion of the capillary bed, not at the outlet (venous) portion.

Tissue perfusion is usually measured using dimensions of milliliters of blood per gram of tissue per unit of time. It should not be confused with blood flow measured in milliliters per second. In dimensional terms, it is a measure of the volume of blood that flows through the capillaries in a particular tissue or organ in a specified interval of time. Measurements are often reported in milliliters of blood per 100 g of tissue per second. Insufficient tissue perfusion can affect any organ system or portion of the body. When blood perfusion to a particular region decreases, it results in reduced nutritional supply to the cells in this region and accumulation of metabolic products that can be problematic if continued over a prolonged period. The tolerance of various human organs and tissue compartments for variation of tissue perfusion is remarkable. A tourniquet on an extremity decreasing perfusion to zero can remain in place for hours, whereas a few minutes of significantly decreased or totally interrupted perfusion to the brain or other selected organs can cause irreversible tissue/organ damage.

4.2.1 Tissue Perfusion Measurement

The conceptual basis of all perfusion measurements relies on the ability to measure the concentration of an agent as a function of time either in 'absolute' or 'relative' terms.

By 'relative' measurement is meant signal (contrast) intensity as a function of space and time relative to a *control* anatomic area of known normal perfusion.

The perfusion of selected tissues can be readily measured in vivo using nuclear medicine methods. Nuclear medicine involves the administration and detection of short-lived radioactive compounds—usually injected intravenously and imaged using photomultiplier technology. The technology involved is inherently sophisticated and is proprietary depending on the manufacturer. What is consistent from case to case is the underlying physiology. Perfusion assessment methods include positron emission tomography (PET) and single-photon emission computed tomography (SPECT). Various radiopharmaceuticals targeted at specific organs have been developed. The most common of these are the following:

99mTc labeled HMPAO and ECD for brain perfusion (rCBF) studied with SPECT

99mTc labeled tetrofosmin and sestamibi for myocardial perfusion imaging with SPECT (see Figure 4.5)

133Xe-gas for absolute quantification of brain perfusion (rCBF) with SPECT

15O-labeled water for brain perfusion (rCBF) with PET (absolute quantification possible when measuring arterial radioactivity concentration)

82Rb-chloride for measuring myocardial perfusion with PET (absolute quantification is possible)

The approach uses each subject as its own control at rest and allows segmental comparison of perfusion in the various anatomical orientations of the LV shown under stress (usually pharmacologic increase in HR). The spatial resolution of the method is inherently limited by the nature of how the image is generated, that is, counting of photons via a collimated detector requiring data acquisition over many cardiac cycles (From http://en.wikipedia.org/wiki/Myocardial_perfusion_imaging).

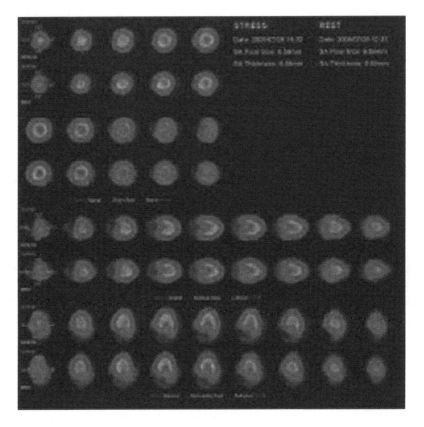

FIGURE 4.5 Nuclear myocardial perfusion image example showing myocardial perfusion scan using thallium-201 for the rest images (bottom rows) and Tc-sestamibi for the stress images (top rows).

4.2.2 MRI

It is assumed that the conceptual basis and physical principles that govern the mechanisms by which MRI machines operate and generate an image are familiar to the reader. These machines are inherently complex, and the programming and pulse sequences that determine the type of image displayed are also familiar—at least in conceptual and descriptive terms. As far as perfusion is concerned, two main types of MRI techniques are in current use.

Gadolinium (GAD)-based methods require the injection of the GAD contrast agent that changes the magnetic susceptibility of the blood. The MR signal is repeatedly measured during bolus passage through the organ or anatomic compartment of interest. Perfusion is computed from the change in signal intensity as a function of time. The method assumes that perfusion is normal in a specific control segment of the image.

In arterial spin labeling (ASL), the arterial blood is magnetically tagged before it enters into the tissue of interest. The amount of labeling is measured and compared to a control recording obtained without spin labeling.

A specific form of ASL is blood oxygen level–dependent (BOLD) imaging, which uses the change in magnetization between oxygen-rich and oxygen-poor blood as its basic signal-generating method. It is usually used in brain functional imaging, in part because the

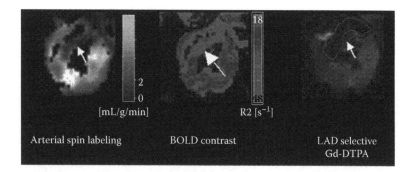

FIGURE 4.6   Myocardial ischemia detection using ASL and BOLD imaging at 3T: MRI detection of myocardial perfusion defect in a porcine model of coronary stenosis at 3 tesla using T_1-weighted-based perfusion imaging (*left image*) and T_2-weighted BOLD contrast imaging (*middle image*) during IV infusion of adenosine. Notice the close regional correspondence with the territory *at risk* delineated post-LAD selective injection of Gd-DTPA (*right image*). (From http://www.nmr.mgh. harvard.edu/cardiovasc/Perfusion-Metabolism.htm.)

brain is a static organ. Using it for cardiac perfusion assessment has been achieved, but it remains a research tool. See Figure 4.6.

4.2.3 CT Myocardial Perfusion Imaging

The assessment of myocardial perfusion by CT is based on the differential wash-in and washout of iodinated contrast in normal and ischemic or infarcted myocardium (Blankstein 2009). Images obtained at rest and during pharmacological stress are assessed for areas of relative hypoenhancement.

The potential for myocardial perfusion imaging with multidetector CT is established in animals (George *et al* 2006) and small studies of humans. See Figure 4.7. The sensitivity of CT perfusion ranges from 79% to 97% and specificity from 72% to 98%, depending on the scanner type, reference standard, and whether analysis is per patient, segment, or territory.

CT has a higher spatial resolution than PET and SPECT, and therefore, it can identify the transmural extent of myocardial perfusion abnormalities (George *et al* 2009). Myocardial ischemia is initially apparent in the subendocardium (Milner 1989, Nagel 2009) and MRI and CT imaging can detect nontransmural infarcts. However, CT can detect smaller areas of ischemia than other modalities and so can identify perfusion defects earlier (Hoffman 1987, Nagel *et al* 2009).

Artifacts limit the quantitative assessment of perfusion defects. Beam-hardening artifacts occur when x-rays pass through a dense object and low-energy photons are preferentially absorbed, thus locally increasing the x-ray beam mean energy and causing a false apparent reduction in attenuation in adjacent areas. Dense objects in the field of view include vertebrae, sternum, ribs, and the contrast-filled LV and aorta (Hundley *et al* 2010). Beam-hardening artifacts may be mistaken for perfusion defects (Hundley *et al* 2010) and are particularly common in the basal inferior wall (Blankstein *et al* 2009). Standard beam-hardening correction algorithms are designed to eliminate the effect of bone, and further improvements are required to allow for contrast-enhanced areas (George *et al* 2010).

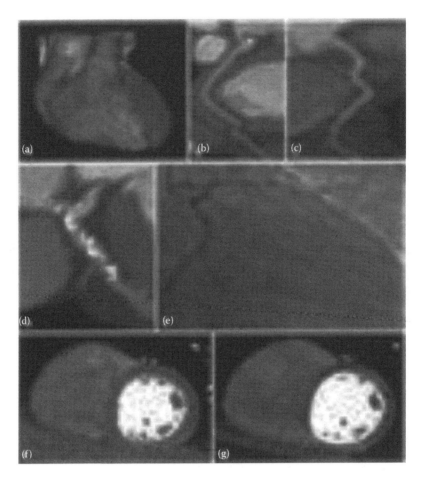

FIGURE 4.7   CT images showing the presence of coronary artery disease. (a) Three-dimensional reconstruction of the heart. (b) Curved planar reconstruction of the left circumflex artery with moderate proximal mixed plaque and moderate distal soft plaque. (c) Curved planar reconstruction of the right coronary artery with mild distal soft plaque. (d) Curved planar reconstruction of the left main stem and left anterior descending artery with a lesion of indeterminate significance in the left main stem. (e) Invasive coronary angiogram image highlighting the left main stem lesion. (f) Rest and (g) stress myocardial perfusion images showing perfusion defects in the mid-anteroseptal, mid-inferoseptal, and mid-inferolateral regions. (From http://www.medscape.com/viewarticle/746462_4.)

4.3 ULTRASONIC METHODS OF PERFUSION DETERMINATION

Substantial progress has been made in ultrasonic methods of perfusion imaging (Porter and Xie 2010). The invention of shells that can encapsulate microbubbles (Keller *et al* 1987, Unger *et al* 1992), their stabilization following venous injection by incorporating high-molecular-weight insoluble gas within the shell (Porter *et al* 1997), and the discovery that typical ultrasound imaging techniques at a high mechanical index (MI) cause collapse of microbubbles perfusing the myocardial microcirculation provide a means for assessing perfusion. As a result, the visibility of the myocardium on echocardiographic images was enhanced by a venous injection of microbubbles (Tsutsui *et al* 2005) or with very small intravenous (IV) bolus injections of perfluorocarbon-containing microbubbles (Porter *et al* 1997). Wei and Kaul discovered that these ultrasound techniques could quantify myocardial blood flow (Wei *et al* 1998a). Additional studies ultimately demonstrated that myocardial contrast echocardiography (MCE) can provide important clinical and physiologic information regarding myocardial blood flow at the bedside, during stress testing (Kaul *et al* 1997) and in acute and chronic assessment of myocardial viability (Balcells *et al* 2003) including rapid assessment in the emergency department.

4.3.1 Method of Myocardial Perfusion Analysis

Regardless of the route of microbubble injection, accurate myocardial contrast imaging requires a linear relationship between concentration and signal intensity. This is best achieved at low intramyocardial microbubble concentrations. Otherwise, the echocardiographic imaging system can saturate so that video intensity is no longer linearly proportional to microbubble concentration (Skyba *et al* 1994). After a bolus injection of contrast through an IV line, it is during the wash-out period that a linear relationship between concentration and signal intensity is achieved.

Instead of a bolus injection, a continuous peripheral venous infusion can also be used. This assumes that the input of microbubbles into the venous system and therefore the myocardium is constant. The practical advantage of a continuous infusion is that attenuation artifacts from the high contrast intensity in the left ventricular cavity that are always present as the initial result of a bolus injection can be reduced (Wei *et al* 1998), and the amount of contrast administered can be easily adjusted based on observed image quality (Skyba *et al* 1994).

Regardless of contrast administration technique, the plateau portion of background-subtracted myocardial contrast intensity of a selected myocardial region as a function of time is related to the capillary cross-sectional area. The initial slope of the intensity versus time plot at which this plateau stage is achieved is proportional to the blood flow velocity in that region. The slope times peak or plateau myocardial video intensity represents the measure of myocardial blood flow (Wei *et al* 1998b).

The method has limitations primarily due to attenuation of the basal segments of the myocardium. This is further detailed in Porter and Xie (1995).

Using appropriate ultrasonic instruments, blood flow volume rates, blood flow velocity profiles, pressure gradients, orifice areas, flow disturbances, jets, characteristics of blood vessels and the circulatory system, and tissue perfusion can all be investigated. Contemporary ultrasonic diagnosis employs ultrasound exposure levels that appear to be free from biological risk, but other factors need be taken into account in considering the prudent use of ultrasonic methods. Promising research is being carried out into the mechanism of ultrasonic scattering by blood, Doppler speckle, time-domain processing for blood flow imaging, methods for increasing the scanning speed, Doppler flow microscopy, and contrast agents.

The primary advantages of ultrasound are as follows: (1) very low risk to patients, (2) real-time imaging, (3) relatively low cost, and (4) bedside practicality of the technique.

The image resolution ranges from millimeters to tens of micrometers, depending on the frequency used and the penetration depth required.

To summarize, a relatively recent advance in ultrasound imaging is the use of micro-bubble contrast agents, having a gas core encapsulated by a shell, typically constructed of a lipid monolayer or cross-linked albumin. The most common type of ultrasound contrast agent is encapsulated microbubble, typically 1–5 μm in diameter. These microbubbles have acoustic scattering properties that are very different from the surrounding tissue. Even a single bubble can create significant and specific ultrasound echoes. In addition, the bubbles can be destroyed by the ultrasound itself, depending on the frequency and intensity setting used. In clinical applications, these microbubbles can serve as imaging agents and even as therapeutic agents in the diagnosis and treatment of a wide range of diseases.

4.4 MICROSPHERES

Regional organ perfusion can be estimated with hematogenously delivered microspheres (Heyman *et al* 1977). When appropriately sized microspheres are used, regional blood flow is proportional to the number of microspheres trapped in the region of interest (Bassingwaithe *et al* 1990). A number of excellent review articles describe and validate the use of micro-spheres for measurement of regional organ perfusion, but the classic review by Heyman *et al* (1977) contains many details for radioactive microsphere use that apply to fluorescent microspheres.

4.4.1 Flow in Capillaries

It is a physiologic fact that flow in the large arteries is pulsatile. The pulsatile nature of flow is eventually lost at the capillary level as a result of the inherent mechanical filtering effect due to the size of capillaries that are a few micrometers in diameter *and* the local control of capillary segment perfusion due to constriction or dilation of arterioles. When a viscous fluid such as blood flows through a tube of a certain length and internal radius r, a resistance to flow exists. As long as the Reynolds number is <2000 and the fluid can be considered Newtonian (meaning the stress vs. strain relation for the fluid is linear and its slope is the viscosity), well-established equations can be used to describe the flow. Under these conditions, the maximum speed is at the center of the tube, the velocity profile is approximately parabolic, and near the walls, the fluid remains nearly stationary with velocity = 0 as a result of the *no slip at the walls* condition. As a function of viscosity, the resistance to flow, R, for steady flow through a capillary of radius, r, is given by

$$R = \frac{8 \cdot \eta \cdot L}{\pi \cdot r^2}, \qquad (4.1)$$

where
  η is the coefficient of viscosity
  L is the length

The assumption of steady flow in a capillary is reasonable in light of the loss of pulsatility of pressure at that scale of vessel size. It is useful to appreciate that a given capillary may have zero flow for short periods as a result of upstream arteriolar control, but when the flow resumes, steady flow is a useful and realistic approximation. The unit for viscosity is poise, named after the French physician Jean Poiseuille (1797–1869).

As the relationship indicates, the resistance is inversely proportional to the square of the radius. For example, a change from 0.5 mm internal diameter capillary to 0.25 mm internal diameter results in an increase in resistance to flow by a factor of 4. By rearrangement of the previous equation, Poiseuille's equation is derived. It follows that the flow rate through a capillary can be expressed in terms of pressure drop. More importantly, the flow rate is expressed in terms of the capillary radius and is given by

$$F = \frac{\pi \cdot r^4}{8 \cdot \eta \cdot L} \cdot \Delta P. \tag{4.2}$$

The key result here is that volumetric flow rate, F (in mL/s), varies as the fourth power of the radius, that is, a factor of 2 change in radius creates a factor of 16 change in flow rate.

4.5 VENOUS FLOW

In the circulatory system (see simplified circuit diagram in Figure 4.1), *veins* are the blood vessels that return blood to the heart. See Figure 4.8. Most veins carry deoxygenated blood from the tissues back to the heart; an exception is the pulmonary vein that carries oxygenated blood from the lung to the heart. Veins differ from arteries in structure and function; for example, arteries are more muscular than veins, veins are often closer to the skin and contain valves that serve as rectifiers to help keep blood flowing toward the heart, while arteries carry blood away from the heart.

4.5.1 Function

Veins serve to return blood from organs to the heart. Veins are also called *capacitance vessels,* because 60% of the blood volume is contained within them and their walls are elastic and are easily distended. In systemic circulation, oxygenated blood is pumped by the LV through the arteries to the periphery (muscles and organs) of the body, where its nutrients and gases are exchanged at the capillary level (typical diameter of 5 μm), and the blood then enters venules and then the larger veins. The deoxygenated blood is directed by the veins to the RA, which transfers the blood to the RV, from where it is then pumped through the pulmonary arteries (PAs) to the lungs. In pulmonary circulation, the pulmonary veins return oxygenated blood from the lungs to the LA, which empties into the LV, completing the cycle of blood circulation.

The return of blood to the heart is assisted by the action of the skeletal-muscle pump and by the thoracic pump action of breathing itself. Standing or sitting for a prolonged period of time can cause low venous return as a result of venous pooling even to the point of low blood pressure and impairment of consciousness. Fainting can occur but usually baroreceptors in the wall of the atrium and the carotid arteries can initiate a baroreflex response by the action of angiotensin II and norepinephrine to stimulate vasoconstriction and a simultaneous increase in HR to increase cardiac output (CO) and MAP. Fainting due to

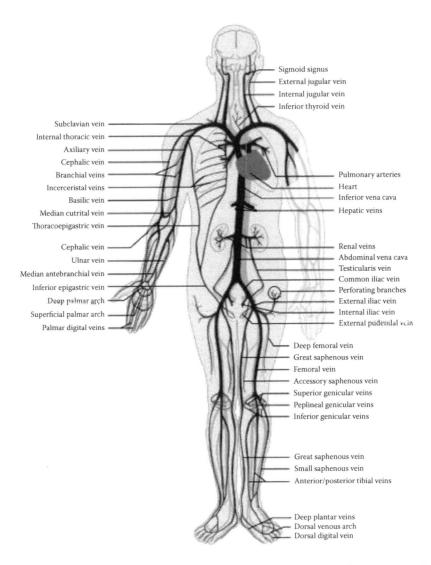

FIGURE 4.8 Schematic of the venous system. (From: http://images.google.com/imgres?q=venou
s+system+of+human+body&hl=en&biw=1215&bih=766&gbv=2&tbm=isch&tbnid=NzCBJv3yw
U6LqM:&imgrefurl=http://en.wikipedia.org/wiki/Vein&docid=xsbozzJ2hsibtM&imgurl=http://
upload.wikimedia.org/wikipedia/commons/thumb/c/c2/Venous_system_en.svg/250px-Venous_
system_en.svg.png&w=250&h=354&ei=4iOYT_XHN5LCgAeB04ThCA&zoom=1&iact=hc&vpx=
111&vpy=110&dur=112&hovh=267&hovw=189&tx=127&ty=162&sig=110324616525855842906&p
age=1&tbnh=129&tbnw=90&start=0&ndsp=35&ved=1t:429,r:0,s:0,i:69.)

decreased cerebral perfusion can be due to neurogenic or hypovolemic causes. Vasovagal
fainting is neurogenic in origin, since smooth muscles that govern venous tone relax and
become slack, causing the veins to be more distensible and fill with the majority of the
blood in the body. This maldistribution decreases arterial pressure then cerebral perfusion,
eventually causes unconsciousness. Jet pilots wear pressurized suits that act like support
hose for their entire lower and upper extremities to prevent venous pooling in the legs and
thereby help maintain venous return and arterial blood pressure.

The arteries carry oxygenated blood to the tissues, while veins carry deoxygenated blood back to the heart. This is true of the systemic circulation, the venous system is by far the larger of the two circuits of blood in the body, which transports oxygen from the heart to the tissues of the body. However, in the pulmonary circulation, the arteries carry deoxygenated blood from the heart to the lungs, and the pulmonary veins return oxygenated blood from the lungs to the heart. The difference between veins and arteries is their direction of flow (out of the heart by arteries, returning to the heart for veins), not their oxygen content. In addition, deoxygenated blood that is carried from the tissues back to the heart for reoxygenation in systemic circulation still carries some oxygen, though it is considerably less than that carried by the systemic arteries or pulmonary veins.

Although most veins take blood back to the heart, there is an exception. Portal veins carry blood between capillary beds. For example, the hepatic portal vein takes blood from the capillary beds in the digestive tract and transports it to the capillary beds in the liver. The blood is then drained in the gastrointestinal tract and spleen, where it is taken up by the hepatic veins, and blood is taken back into the heart. Since this is an important function in mammals, damage to the hepatic portal vein can be dangerous. Blood clotting in the hepatic portal vein can cause portal hypertension, which results in a decrease of blood flow to the liver.

4.6 FICK PRINCIPLE

The Fick principle is based on the concept of conservation of mass; more specifically, it is based on the observation that the total uptake of (or release of) a substance by tissues or a specific organ is equal to the product of the blood flow to the peripheral tissue or organ and the arterial–venous concentration gradient of the substance. The principle is most commonly employed for the determination of CO. The substance most commonly measured is the oxygen content of blood, thus giving the arteriovenous oxygen difference, and the flow calculated is the flow across the pulmonary system. This gives a simple way to calculate the CO:

$$CO\,(L/min) = \left\{ \frac{O_2 \text{ consumption}\,(mL \text{ of } O_2/min)}{\text{Arteriovenous difference}\,(mL\,O_2/dL)} \right\} \times 100.$$

Assuming there is no intracardiac shunt, the pulmonary blood flow equals the systemic blood flow. The measurement of the arterial and venous oxygen content of blood involves the sampling of blood from the PA (low oxygen content, the input side) and from the pulmonary vein (high oxygen content, the output side). In the lung, the oxygen content increases, so there is a step-up in O_2 concentration from PA to pulmonary vein. In practice, sampling of arterial blood anywhere in the peripheral arterial circulation is equivalent to sampling oxygenated blood from the pulmonary vein. Determination of the oxygen consumption of peripheral tissues, where there is a step-down in oxygen concentration between arterial input and venous drainage, is more complex because of variation in the input (arterial) and output (venous) locations. The measurement of the arterial and venous oxygen concentration of the blood is straightforward. Essentially all of the oxygen in the blood is bound to hemoglobin within red blood cells; a negligible amount is actually dissolved in the serum (the non-cellular component of blood). Measuring the amount of hemoglobin in the blood in g/dL and the percentage of saturation of hemoglobin (the oxygen saturation—see the following discussion on pulse oximetry) is a simple process. It is readily available in clinical practice. Since each gram of hemoglobin can carry 1.36 mL of O_2, the oxygen content of blood (either arterial or venous) can be estimated by the following formula:

$$O_2 \text{ content of blood} = [\text{Hb}](g/dL) \times 1.36 (\text{mL } O_2/g \text{ Hb}) \times O_2^{\text{saturation}} + 0.0032 P_{O_2} (\text{mmHg}).$$

Assuming a hemoglobin concentration of 15 g/dL and an oxygen saturation of 0.99 or 99% while breathing room air, the oxygen concentration of arterial blood is approximately 200 mL of O_2 per liter. Because Hb is essentially completely saturated with oxygen while breathing ambient room air, the administration of oxygen in healthy individuals results in an insignificant increase in oxygen delivery. If arterial O_2 saturation is below 90%, administration of supplemental oxygen is beneficial. The saturation of mixed venous blood measured in the RV or PA is approximately 75% in healthy individuals. The term *mixed* literally means mixing of venous blood from different parts of the body having different levels of saturation. Note that the O_2 saturation of venous blood is not at all uniform throughout the body and depends on where the venous effluent is being sampled. Certain organs, such as the heart, extract more O_2 and therefore have lower saturation levels (heart, 55%–65%; kidney, 77%; superior vena cava, 71%). Using the 75% *mixed* value in the equation, the oxygen concentration of mixed venous blood is approximately 150 mL of O_2 per liter. Oxygen saturation can be easily measured noninvasively using a fingertip pulse oximeter that costs about $30.00 such as the FDA-Approved, 100A Fingertip Pulse Oximeter unit. (http://www.devonsuperstore.com/100-A-Fingertip-Pulse-Oximeter-P264.aspx.)

The principle of O_2 saturation measurement relies on the difference in absorption spectra of oxygenated versus deoxygenated Hb. These devices are placed on the earlobe or fingertip that can be transilluminated, and typically, the sensor utilizes a pair of small light-emitting diodes (LEDs) facing a photodiode. One LED is red, emitting at a wavelength of 660 nm, and the other is infrared, emitting at 905, 910, or 940 nm. Absorption at these wavelengths differs significantly between oxyhemoglobin and deoxyhemoglobin; therefore, the oxyhemoglobin/deoxyhemoglobin ratio can be calculated from the ratio of the absorption of the red and infrared light. Invasive measurement requires obtaining an arterial blood sample—usually through the radial artery in the wrist. The (anticoagulated) sample is then inserted into an imager that uses the same optical principles as the fingertip oximeter.

4.7 ARTERIAL FLOW MEASUREMENT

The arterial system is the higher-pressure portion of the circulatory system. Arterial pressure varies between the peak pressure during heart contraction, the systolic pressure (typically about 120 mmHg), and the minimum, or diastolic, pressure between contractions (typically 80 mmHg), when the heart muscle relaxes and the ventricular chamber expands and refills. This pressure variation within the artery produces the peripheral pulse that can be felt in any artery; its intensity is a reflection of the heart's pumping activity as well as the distensibility of the artery itself. Arteries carry blood away from the heart, whereas veins direct blood toward the heart. PAs are the exception, because they carry venous blood away from the heart, into the lungs for oxygenation. Otherwise, all arteries carry oxygenated blood away from the heart to the peripheral tissues. The exceptions are the left and right coronary arteries, which carry oxygenated blood not to the periphery but to the heart muscle itself.

4.8 CARDIAC OUTPUT

CO (in mL/min) is the product of HR (beats/min) and stroke volume (SV) (mL), usually expressed in L/min. Table 4.1 shows that the average resting HR in a 70 kg male is 60–100 beats/min, and SV at rest is 55–100 mL, yielding an average CO of 4.0–8.0 L/min.

HR is usually measured by palpating the pulse rate in a peripheral artery (radial, ulnar, femoral) or by noting the R–R interval of the ECG on a monitor. SV can be determined by both invasive and noninvasive methods.

4.8.1 Invasive Methods for Cardiac Output Determination

- Quantitative ventriculography during cardiac catheterization and angiography (left ventriculography). Iodinated contrast is injected into the LV under direct visualization to opacify the LV. End-systolic and end-diastolic images are planimetered, and LVED and LVES volumes are computed. SV is the difference between the two volumes, that is, $SV = LVEDV - LVESV \cdot CO = SV \times HR$.

- Fick method of CO determination.

4.8.2 Thermodilution

The conceptual basis of the thermodilution method is the indicator-dilution method. CO is equal to the amount of an indicator (e.g., a dye) injected divided by its average concentration in the blood after a single circulation through the body. An indicator dye (indocyanine green) is injected, and the method assumes that the rate at which the indicator is diluted

TABLE 4.1   CO Parameters

Measure	Typical Value	Normal Range
LV EDV	120 mL	65–240 mL
LV ESV	50 mL	16–143 mL
LV SV	70 mL	55–100 mL
LV EF	58%	55%–70%
HR	75 bpm	60–100 bpm
CO	5.25 L/min	4.0–8.0 L/min
Cardiac index (CO/BSA)		

Source: Yang, S.S. et al. (eds.), *From Cardiac Catheterization Data to Hemodynamic Parameters*, 3rd edn., F.A. Davis, Philadelphia, PA, 1988.

reflects CO. The method measures the concentration of the dye at different points in the circulation, usually from an IV injection and then at a downstream sampling site, usually in a systemic artery. More specifically, the CO is equal to the quantity of indicator dye injected divided by the area under the dilution curve measured downstream (the Stewart [1897]–Hamilton [1932] equation): the trapezoid rule is often used as an approximation of this integral.

The thermodilution method uses the same dilution concept, but rather than measuring dye concentration, it measures temperature as a function of time after a specific volume of room temperature or cold saline is injected into the RA at a specified distance upstream from a temperature sensor. See Figure 4.9. The PA catheter-based thermodilution method involves injection of a small amount (10 mL) of cold glucose or saline at a known temperature into the RA and continuously measuring the temperature at the downstream temperature sensor as a function of time a known distance away (6–10 cm in the PA) in the same catheter.

The Swan–Ganz multilumen catheter has had an important impact on patient management in the intraoperative, post-op, and ICU setting. It allows reproducible calculation of cardiac output from a measured time/temperature curve (*thermodilution curve*) in addition to being able to assess the overall *volume status* of the circulation by assessment

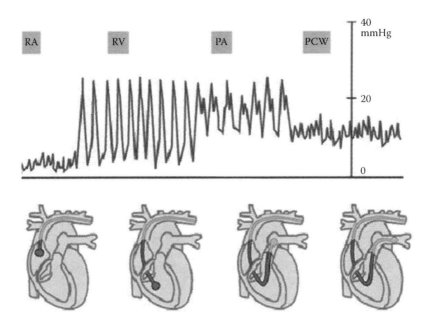

FIGURE 4.9 Advancement of balloon-tipped, flow-directed PA (Swan–Ganz) catheter from RA to RV to PA with associated pressure waveforms in each of the respective chambers shown earlier. When the balloon is inflated blocking flow in the branch of the PA where it resides, the catheter tip is exposed to the downstream pulmonary capillary wedge pressure (PCW)—approximating LA pressure. The catheter has a proximal lumen for cold saline injection and a distal thermistor for continuous temperature recording for thermodilution determination of CO. For CO determination, the balloon must be in the deflated state. (From Mathews 2007.)

of mean PA pressure and PCW pressure (LA pressure equivalent). Thermistor-based measurements revealed that low CO generates a slow decay in temperature change, and high CO inscribes a rapid decay in temperature change on the time versus temperature plot. The rate of change in temperature is directly proportional to the rate of flow, that is, CO. Based on these observations, three or four repeated thermodilution measurements or injections of the fluid are usually averaged to improve accuracy (Waters *et al* 2005, Reuter *et al* 2010). The most recent technological advancements have provided catheters having a heating filament that intermittently heats and measures the downstream thermodilution curve providing serial CO measurement. However, these average measurements require a 5–10 min recording range, depending on the stability of the circulation, and do not provide continuous monitoring.

4.8.2.1 Thermodilution Method: Technical Details

The thermodilution method adapts the indicator-dilution principle to a bolus injection of saline or dextrose that changes (decreases) blood temperature downstream. Classically, iced or room temperature saline 0.9% or dextrose 5% is injected.

Let T_0 (°C), σ_0 (J/kg°C), and ρ_0 (g/mL = 1) denote, respectively, the temperature, specific heat, and specific gravity of the injected fluid; let T_B, σ_B, and ρ_B denote the corresponding properties of the circulating whole blood. The injectate occupies a volume V_0 (mL), thus carrying an amount of cooling heat (cold indicator) $V_0 \sigma_0 \rho_0 (T_B - T_0)$ relative to blood temperature. Once introduced into the circulation, the cold injectate mixes with a volume V_1 of blood and cools it to temperature T_1. In accordance with the concept of conservation of thermal energy, $V_0 \sigma_0 \rho_0 (T_B - T_0) = V_1 \sigma_B \rho_B (T_B - T_1)$. The cooled blood is sensed by the thermistor downstream over a duration t(s). The CO F (mL/s) is then computed as

$$F = \frac{V_1}{t} = \frac{V_0}{t} \frac{\sigma_0 \rho_0 (T_B - T_0)}{\sigma_B \rho_B (T_B - T_1)},$$ (4.3)

which states that CO is inversely proportional to the temperature depression $T_B - T_1$ of cooled blood and the duration of its passage t (i.e., area under the curve). In reality, the nonuniform velocities caused by the injection itself and the blood and the continuous dilution of injectate result in a temperature depression $T_B(t) \approx T_B - T_1(t)$ at the thermistor site that initially increases, then gradually decreases. The thermodilution curve $\Delta T_B(t)$ is essentially the time–concentration curve. Modification of the constant temperature change $T_B - T_1$ in Equation 4.3 with the time-averaged temperature change yields the following thermodilution equation:

$$F = \frac{V_1}{t} = \frac{V_0 (T_B - T_0) K_1}{\int_t \Delta T_B dt},$$ (4.4)

where the density × heat capacity factor K_1 is given by

$$K_1 = \frac{\sigma_0 \rho_0}{\sigma_B \rho_B}. \tag{4.5}$$

$K_1 = 1.08$ for 5% dextrose. Hence, CO is inversely proportional to the mean blood temperature depression and the duration of transit of cooled blood (i.e., area under the $\Delta T_B(t)$ curve).

4.8.3 Noninvasive Determination of Cardiac Output

Echocardiography permits determination of SV (cm³) by integrating the contour of the Doppler echocardiographic aortic outflow waveform, that is, determining its time–velocity integral (in cm) and multiplying it by the aortic valve cross-sectional area (cm²) to yield the SV in cm³. If there are no shunts, the velocity–time integral across any of the four cardiac valves should yield the same (time-averaged) SV. The velocity time integrals can be computed in systole for the pulmonic or aortic valves and in diastole for the tricuspid or mitral valves. Because diastolic filling is biphasic, and consists of the suction-initiated early rapid filling wave (transmitral Doppler E-wave) followed by diastasis (no flow) followed by atrial systole (transmitral Doppler A-wave), velocity time integral calculations are most easily achieved by integrating systolic flow either through the aortic or pulmonic valves using Doppler echocardiography.

Cardiac MRI (CMR) has achieved rapid development in scanner technology and software, thereby generating an explosion in the indications for CMR procedures. CMR is the clinical gold standard for myocardial infarction detection and quantification owing to its excellent spatial and incomparable tissue characterization capability. CMR is also the most robust tool for ventricular function quantification of both the LV and RV, because it is not limited by acoustic windows required for echocardiography. CMR is the gold standard for LV chamber volume, wall thickness, and mass assessment, and as such, SV is easily determined by semiautomated edge-detection methods by computing the difference between end-systolic and end-diastolic volumes (ESV and EDV) (see Chapter 18 Section 18.9).

4.9 ARTERIAL STIFFNESS

Arterial stiffness is essentially assessment of the rigidity of the wall of the artery and can be measured using various techniques (Mackenzie *et al* 2002). At present, most measurements are made for experimental and physiological studies rather than for clinical diagnosis or assessment of therapy. Pulse pressure and pulse wave velocity (PWV) are surrogate measures of arterial stiffness, which is a key component of the development of systolic hypertension, a disease of the elderly. Many terms have been used interchangeably when referring to arterial stiffness and are shown in Table 4.2.

TABLE 4.2   Indices of Arterial Stiffness

Term	Definition	Methods of Measurement
Elastic modulus[b]	The pressure change required for theoretical 100% stretch from resting diameter	Ultrasound[a]
	$(\Delta P \times D)/\Delta D$ (mmHg)	MRI
Young's modulus[b]	Elastic modulus per unit area	Ultrasound[a]
	$(\Delta P \times D)/(\Delta D \times h)$ (mmHg/cm)	MRI
Arterial distensibility[b]	Relative change in diameter (or area) for a given pressure change; inverse of elastic modulus	Ultrasound[a]
	$\Delta D/(\Delta P \times D)$ (mmHg^{-1})	MRI
Arterial compliance[b]	Absolute diameter (or area) change for a given pressure step	Ultrasound[a]
	$\Delta D/\Delta P$ (cm/mmHg) (or cm^2/mmHg)	MRI
PWV	Velocity of travel of the pulse along a length of artery	Pressure waveform[a]
	Distance/Δt (cm/s)	Volume waveform
		Ultrasound
		MRI
Augmentation index	The difference between the second and first systolic peaks as a percentage of pulse pressure	Pressure waveform[a]
Stiffness index (β)[b]	Ratio of ln(systolic/diastolic pressures) to (relative change in diameter) $\beta = \ln(P_s/P_d) (D_s - D_d)/D_d$	Ultrasound[a]
Capacitive compliance	Relationship between pressure change and volume change in the arteries during the exponential component of diastolic pressure decay $\Delta V/\Delta P$ (cm^3/mmHg)	Pressure waveform[a]
Oscillatory compliance	Relationship between oscillating pressure change and oscillating volume change around the exponential pressure decay during diastole $\Delta V/\Delta P$ (cm^3/mmHg)	Pressure waveform

Source: Downloaded from http://qjmed.oxfordjournals.org/.

P, pressure; *D*, diameter; *V*, volume; *h*, wall thickness; *t*, time; *v*, velocity; *s*, systolic; *d*, diastolic.

[a] Most common method of measurement.

[b] Also requires pressure measurements.

4.10 PULSE PRESSURE

The pulse pressure is the difference between systolic and diastolic pressures. It depends on CO, stiffness of large arteries (where it is measured), and wave reflection. The pulse pressure is increased in the setting of aortic insufficiency—hence, valvular competence needs to be assessed echocardiographically in concert with arterial stiffness determination. It is a surrogate for arterial stiffness, as noted by Bramwell and Hill (1922): "Hence the difference between systolic and diastolic pressure, that is the pulse pressure, other things being equal will vary directly as the rigidity of the arterial walls."

4.11 PULSE WAVE VELOCITY

The PWV is the speed at which the forward-moving pressure wave travels from the aorta distally. PWV is calculated by measuring the time interval for the arterial waveform to pass between two points separated by a known distance. It requires simultaneous readings at the two sites or gating separate recordings to the R wave of the ECG. Both invasive and noninvasive methods can be used to record either flow or pressure waves. The key important point is the reference point on the waveforms themselves. Usually, the foot-to-foot methodology (meaning the initial upstroke of the pressure wave) is used, because it avoids the confounding influence of reflected waves. Usually, larger superficial arteries are used and precision is limited if the recording points are close together. Clearly, the result applies only to the arterial segment being measured, and not to the overall (lumped) arterial stiffness.

The Moens–Korteweg equation is used to calculate PWV:

$$PWV = \sqrt{\frac{E_{inc} \cdot h}{2r\rho}}, \tag{4.6}$$

where
  E is the incremental Young's modulus of elasticity of wall material
  h is the wall thickness of vessel
  r is the inside radius of vessel
  ρ is the density of blood

Another version is the Bramwell and Hill equation, relating PWV to distensibility:

$$PWV = \sqrt{\frac{\Delta PV}{\Delta V\rho}} \cdot \sqrt{\frac{1}{\rho D}}, \tag{4.7}$$

where
  $\Delta PV/\Delta V$ is the relative volume elasticity of vessel segment
  ρ is the density of blood
  D is the distensibility

4.12 ULTRASOUND-BASED MEASUREMENTS

Ultrasound can measure arterial stiffness (distensibility and compliance), in the larger and more accessible arteries, such as brachial, femoral, and carotid arteries and the abdominal aorta. Recording of multiple vessel wall images throughout the cardiac cycle provides maximum and minimum areas usually calculated by wall tracking and computerized edge-detection software. Once blood pressure has been recorded, distensibility and compliance (inverse of stiffness) are calculated as

$$\text{Distensibility} = \frac{\Delta V}{\Delta P \cdot V} \quad \text{and} \quad \text{Compliance} = \frac{\Delta V}{\Delta P}, \tag{4.8}$$

where
  ΔV is the change in volume
  ΔP is the change in pressure
  V is volume

Note that MRI methods can also be used for these measurements, particularly in the large central artery such as the aorta, (Resnick *et al* 1997) although spatiotemporal resolution is inherently limited by the method itself.

4.13 PRESSURE WAVEFORM ANALYSIS

The arterial pressure waveform is a composite of the forward-traveling pressure wave generated by systolic contraction and backward-traveling reflected wave(s). Waves are reflected from branches in the arterial tree such as the aortoiliac bifurcation. It therefore follows that the pressure waveform itself is a function of the site of recording and depends on local vessel wall properties. In general, the stiffer the vessel, the higher is the velocity. The phenomenon of *peripheral amplification of pressure* (Hashimoto *et al* 2010) is well established and is a manifestation of the change in wave velocities as a function of vessel diameter and wall stiffness.

A pulse wave analysis (PWA) system has been developed by O'Rourke and Gallagher (1996). It employs applanation tonometry (Figure 4.10) to record pressure waveforms at the radial or carotid artery and utilizes a validated generalized transfer function to derive the corresponding central waveform. The augmentation index, the difference between the first and second systolic peaks, is expressed as a percentage of the pulse pressure, and a measure of systemic stiffness is derived. The tonometer is about the size of a pen, and it is easily portable and, therefore, is clinically useful. The piezoelectric sensor is placed over the radial artery and held in place by a wrist strap. The system can automatically readjust to detect the best possible pressure waveform. The advantages include less operator dependency than with a handheld tonometer and continuous measurements.

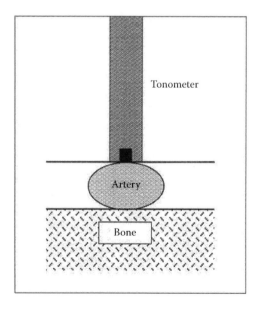

FIGURE 4.10 The basis of applanation tonometry. The radial artery is gently compressed against the underlying bone, thus flattening it and equalizing circumferential pressures, allowing a high-fidelity pressure waveform to be electronically recorded with the tonometer. (From Mackenzie, I. et al., *QJM*, 95, 67, 2002.)

Another technique for noninvasive arterial stiffness determination that involves tonometry is pressure pulse contour analysis of the radial artery (Cohn *et al* 1995, McVeigh *et al* 1999). However, compliance is derived using a modified Windkessel model of the circulation and assessment of diastolic pressure decay. This method calculates large-vessel and peripheral compliance. Augmentation index and peripheral compliance, calculated from pulse contour analysis, are related, but O'Rourke's augmentation index characterizes drug-induced hemodynamic changes more accurately (Segers *et al* 2001).

4.14 POSSIBLE FUTURE INSTRUMENTATION

Fundamental conceptual issues related to measurement theory play a role in anticipating any type of future instrumentation suited for hemodynamic assessment in cardiology.

One must keep in mind the difference between *absolute* and *relative* measurements of any hemodynamic parameter, index or variable. The importance of *absolute measurement* versus *relative measurement* depends on your application. By analogy, consider a straight line given by $y = mx + b$, where m is the slope and b is the y intercept. If we can only determine m by our method of measurement, we have measured a relative index, because an infinite set of straight lines may have slope m. If we can also determine b, we have made an absolute measurement, since we know where in the x–y plane the line resides. In cardiology, absolute indexes include pressure, volume, temperature, and LV EF. Relative indexes include dP/dt, dV/dt = flow rate, stiffness dP/dV, compliance dV/dP, and the time constant of isovolumic relaxation tau. Note that absolute metrics often require invasive (cardiac catheterization or insertion of a tube into the arterial system) measurement, such as pressure, whereas a pressure gradient, such as occurring when blood flows across a valve of the heart, is a relative measure (obtainable noninvasively via Doppler echocardiography), since it merely requires the measurement of velocity differences across the valve and estimation of the pressure gradient using the simplified Bernoulli equation.

The most significant instrumentation advances in recent years pertain to invasive and noninvasive imaging. The elucidation of biologic and genetic pathways in concert with molecular imaging and nanotechnology-derived methods holds enormous future promise for more complete characterization of the physiology and pathophysiology of the cardiovascular system. The time when, as in Star Trek, Dr. McCoy is able to diagnose presence and severity of cardiovascular illness by using a handheld sensor device is consistent with the fundamental laws of nature and therefore not beyond the realm of future possibility.

REFERENCES

Balcells E, Powers E R, Lepper W *et al* 2003 Detection of myocardial viability by contrast echocardiography in acute infarction predicts recovery of resting function and contractile reserve *J. Am. Coll. Cardiol.* **41** 827–33

Bassingthwaighte J B, Malone M A, Moffett T C, King R B, Chan I S, Link J M and Krohn K A 1990 Molecular and particulate depositions for regional myocardial flows in sheep *Circ. Res.* **66** 1328–44

Blankstein R, Shturman L D, Rogers I S *et al* 2009 Adenosine-induced stress myocardial perfusion imaging using dual-source cardiac computed tomography *J. Am. Coll. Cardiol.* **54** 1072–84

Bowman A W and Kovács S J 2003 Assessment and consequences of the constant-volume attribute of the four-chambered heart *Am. J. Physiol. Heart Circ. Physiol.* **285** H2027–33

Bramwell J C and Hill A V 1922 Velocity of transmission of the pulse-wave and elasticity of the arteries. *Lancet* **1** 891–2

Cohn J N, Finkelstein S, McVeigh G *et al* 1995 Noninvasive pulse wave analysis for the early detection of vascular disease *Hypertension* **26** 503–8

George R T, Arbab-Zadeh A, Miller J M *et al* 2009 Adenosine stress 64- and 256-row detector computed tomography angiography and perfusion imaging: a pilot study evaluating the transmural extent of perfusion abnormalities to predict atherosclerosis causing myocardial ischemia *Circ. Cardiovasc. Imaging*: **2** 174–82

George R T, Bengel F M and Lardo A C 2010 Coronary flow reserve by CT perfusion *J. Nucl. Cardiol.* **17** 540–3

George R T, Silva C, Cordeiro M A *et al* 2006 Multidetector computed tomography myocardial perfusion imaging during adenosine stress *J. Am. Coll. Cardiol.* **48** 153–60

Hashimoto J and Ito S 2010 Pulse pressure amplification, arterial stiffness, and peripheral wave reflection determine pulsatile flow waveform of the femoral artery *Hypertension* **56** 926–33

Heyman M A, Payne B D, Hoffman J I and Rudolf A M 1977 Blood flow measurements with radionuclide-labeled particles *Prog. Cardiovasc. Dis.* **20** 55–79

Hoffman J I 1987 Transmural myocardial perfusion *Prog. Cardiovasc. Dis.* **29** 429–64

Hundley W G, Bluemke D A, Finn J P *et al* 2010 ACCF/ACR/AHA/NASCI/SCMR 2010 expert consensus document on cardiovascular magnetic resonance: a report of the American College of Cardiology Foundation Task Force on Expert Consensus Documents *J. Am. Coll. Cardiol.* **55** 2614–62

Kaul S, Senior R, Dittrich H, Raval U, Khattar R and Lahiri A 1997 Detection of coronary artery disease with myocardial contrast echocardiography: comparison with 99mTc-sestamibi single-photon emission computed tomography *Circulation* **96** 785–92

Keller M W, Feinstein S B and Watson D D 1987 Successful left ventricular opacification following peripheral venous injection of sonicated contrast agent: an experimental evaluation *Am. Heart J.* **114** 570–5

Lindner J R 2004 Microbubbles in medical imaging: current applications and future directions *Nat. Rev. Drug Discov.* **3** 527–33

Lindner J R, Villanueva F S, Dent J M, Wei K, Sklenar J and Kaul S 2000 Assessment of resting perfusion with myocardial contrast echocardiography: theoretical and practical considerations *Am. Heart J.* **139** 231–40

Lisauskas J B, Singh J, Courtois M R and Kovács S J, Jr. 2001 The relation of the peak Doppler E-wave to peak mitral annulus velocity ratio to diastolic function *Ultrasound Med. Biol.* **27** 499–507

Mackenzie I S, Wilkinson I B and Cockcroft J R 2002 Assessment of arterial stiffness in clinical practice *QJM* **95** 67–74

Mathews L 2007 Paradigm shift in hemodynamic monitoring *Internet J. Anesthesiol.* **11**(2)

McVeigh G E, Bratteli C W, Morgan D J, Alinder C M, Glasser S P, Finkelstein S M and Cohn J N 1999 Age-related abnormalities in arterial compliance identified by pressure pulse contour analysis: aging and arterial compliance *Hypertension* **33** 1392–8

Milnor W R 1989 *Hemodynamics* (Baltimore: Williams & Wilkins)

Nagel E, Lima J A, George R T *et al* 2009 Newer methods for noninvasive assessment of myocardial perfusion: cardiac magnetic resonance or cardiac computed tomography? *JACC Cardiovasc. Imaging* **2** 656–60

O'Rourke M F and Gallagher D E 1996 Pulse wave analysis *J. Hypertens.* **14** 147–57

Porter T R, Li S, Kricsfeld D and Armbruster R W 1997 Detection of myocardial perfusion in multiple echocardiographic windows with one intravenous injection of microbubbles using transient response second harmonic imaging *J. Am. Coll. Cardiol.* **29** 791–9

Porter T R and Xie F 1995 Visually discernible myocardial echocardiographic contrast following intravenous injection of sonicated dextrose albumin microbubbles containing high molecular weight, less soluble gases *J. Am. Coll. Cardiol.* **25** 509–615

Porter T R and Xie F 2010 Myocardial perfusion imaging with contrast ultrasound *J. Am. Coll. Cardiol. Imaging* **3** 176–87

Resnick L M, Militianu D, Cunnings A J, Pipe J G, Evelhoch J L and Soulen R L 1997 Direct magnetic resonance determination of aortic distensibility in essential hypertension: relation to age, abdominal visceral fat and in situ intracellular free magnesium *Hypertension* **30** 654–9

Segers P, Qasem A, De Backer T, Carlier S, Verdonck P and Avolio A. 2001 Peripheral 'oscillatory' compliance is associated with aortic augmentation index *Hypertension* **37** 1434–9

Skyba D M, Jayaweera A R, Goodman N C, Ismail S, Camarano G and Kaul S 1994 Quantification of myocardial perfusion with myocardial contrast echocardiography during left atrial injection of contrast: implication for venous injection *Circulation* **90** 1513–21

Tsutsui J M, Elhendy A, Anderson J R, Xie F, McGrain A C and Porter T R 2005 Prognostic value of dobutamine stress myocardial contrast perfusion echocardiography *Circulation* **112** 1444–50

Unger E C, Lund P J, Shen D K, Fritz T A, Yellowhair D and New T E 1992 Nitrogen-filled liposomes as a vascular US contrast agent *Radiology* **185** 453–6

Waters E A, Bowman A W and Kovács S J 2005 MRI-determined left ventricular "crescent effect": a consequence of the slight deviation of contents of the pericardial sack from the constant-volume state *Am. J. Physiol. Heart Circ. Physiol.* **288** H848–53

Wei K, Jayaweera A R, Firoozan S, Linka A, Skyba D M and Kaul S. 1998a Basis for detection of stenosis using venous administration of microbubbles during myocardial contrast echocardiography: bolus or continuous infusion? *J. Am. Coll. Cardiol.* **32** 252–60

Wei K, Jayaweera A R, Firoozan S, Linka A, Skyba D M and Kaul S 1998b Quantification of myocardial blood flow with ultrasound-induced destruction of microbubbles administered as a constant venous infusion *Circulation* **97** 473–83

Yang S S, Bentivoglio L G, Maranhao V and Goldberg H (eds.) 1988 *From Cardiac Catheterization Data to Hemodynamic Parameters*, 3rd edition (Philadelphia: F.A. Davis)

CHAPTER 5

Phonocardiography

Recording of Heart Sounds and Murmurs

Homer Nazeran

CONTENTS

5.1	Introduction	101
5.2	Origin (Genesis) of Heart Sounds	103
5.3	Summary of the Origin of Heart Sounds	108
5.4	Auscultation and Characteristics of Normal Heart Sounds	109
5.5	Pathological Heart Sounds	111
	5.5.1 Pathological First Heart Sound (S1)	113
	5.5.2 Pathological Second Heart Sound (S2)	113
	5.5.3 Pathological Third Heart Sound (S3)	113
	5.5.4 Pathological Fourth Heart Sound (S4)	113
5.6	Stethoscopes and Phonocardiographs	114
	5.6.1 Acoustical Stethoscopes	114
	5.6.2 Electronic Stethoscopes	117
5.7	Phonocardiographs	118
5.8	Digital Signal Processing Techniques in Phonocardiography	119
5.9	Digital Phonocardiography and Computer-Aided Classifications of Heart Sounds	119
References		120

5.1 INTRODUCTION

The heart like other mechanical pumps generates characteristic sounds with each beat. Listening to the sounds generated by the pumping action of the heart and blood flow through its valvular system provides valuable information for the clinician to assess the functional integrity of the heart and the cardiovascular system. The cardiovascular system could be considered as a fluid-filled balloon, which when stimulated or triggered by pressure gradients (i.e., due to ventricular contraction and movement of blood through the valves) *vibrates* as a whole and generates audible sounds on the body surface. The generated sounds, called the *heart sounds*, are best heard externally on the chest at specific locations known as mitral, aortic, pulmonary, and tricuspid areas.

The dynamic interplay between the contraction and the relaxation of the atria and ventricles, valve movements, and blood flow generates the heart sounds (Durand and Pibarot 1995). It is important to distinguish between heart sounds and murmurs. Even though there are many theories attempting to describe the origin of the heart sounds and murmurs, it is now widely believed that heart sounds are generated by the rate of change in the velocity (*acceleration* and *deceleration*) of blood flow and murmurs are vibrations generated by the *nature* of blood flow, namely, turbulence (Peura 2010).

Heart sounds and murmurs can signal many pathological conditions before they manifest themselves as changes in other physiological signals such as the electrocardiogram (ECG) or blood pressure (BP); therefore, *auscultation* (listening to heart sounds and murmurs) has become a primary test performed by the physician to assess the functional integrity of the heart. As such, auscultation is a fundamental component in cardiac diagnosis. It is the most commonly used technique for screening and diagnosis in primary health care. In some circumstances, particularly in remote areas or some developing countries, auscultation may be the only means available to the clinician for this purpose. However, it should be recognized that due to the subjectivity of the auscultation method, namely, the limitations of the human ear in listening to the heart sounds to detect relevant symptoms, forming a diagnosis based on sounds heard through a stethoscope is a clinical skill that can take years to acquire and refine. Additionally, due to the lack of a solid theory on the genesis of heart sounds, the objectivity of any cardiac diagnosis based solely on auscultation may be questionable. For a reliable and accurate diagnosis, physicians require more information to correlate the temporal relationships between the heart sounds, the atrial and ventricular BP waveforms (mechanical events), and the ECG signal (electrical events) during the cardiac cycle. *Phonocardiography* (PCG), which is the method of *graphical recording of heart sounds and murmurs*, constitutes an integral component of this cardiovascular profile.

The basic aim of PCG is to provide clinicians with a complementary tool to *record* the heart sounds and murmurs heard (perceived) during cardiac auscultation. This information facilitates the recognition of pathological heart sounds and murmurs and their relationships to underlying cardiac structures. Since PCG is noninvasive and provides valuable information concerning the functional integrity of the heart, it has a high potential for detecting various heart diseases (Durand and Pibarot 1995, Rangayyan *et al* 1987).

This chapter first discusses the origin (genesis) of heart sounds by giving a relatively detailed description of the events taking place during the cardiac cycle and their temporal relationships with each other. Second, it summarizes the origin of each heart sound and presents information on the signal characteristics of normal and abnormal heart sounds. Third, it highlights auscultation of normal heart sounds and presents an overview of stethoscopes and phonocardiographs. Finally, it highlights the general class of digital signal processing (DSP) techniques used in PCG and concludes with computer-aided analysis of heart sounds and development of a personal digital assistant or PDA-based intelligent phonocardiograph as an example.

5.2 ORIGIN (GENESIS) OF HEART SOUNDS

The heart is basically a dual, four-chamber, well-synchronized electrophysiological pump that propels blood through the circulatory system for a lifetime. It is configured as two serial pumps: one on the right side and the other on the left side. It is composed of four heart valves: the right atrioventricular (AV) (tricuspid) valve, which separates the right atrium from the right ventricle; the pulmonary valve (tricuspid), which connects the right ventricle to the pulmonary circulation; the mitral or the left AV valve (bicuspid), which connects the left atrium to the left ventricle; and the aortic valve (tricuspid), which connects the left ventricle to the systemic circulation. The aortic and pulmonary valves are known as *semilunar valves*, whereas the right AV and the mitral valves are known as the *AV valves*. Figure 5.1 shows the four heart valves.

The generation of heart sounds involves a dynamic and well-coordinated interplay between atria, ventricles, blood flow, and the heart valves during the cardiac cycle. To understand the genesis (origin) of the heart sounds and their graphic recordings, it is essential to develop a spatiotemporal perspective of the processes involved. This facilitates focusing on physiological changes (events) that occur in different structures of the heart during the cardiac cycle with reference to the temporal relationships between electrical and mechanical events and their correlations with the heart sounds. Figure 5.2, known as the Wiggers diagram of the cardiac cycle, shows the temporal relationships between different BP waveforms, ventricular blood volume, the ECG, and the heart sounds (for a review of cardiovascular physiology and the origin of the ECG signals, please refer to Chapters 2 through 4 in this book).

In clinical PCG, carotid pulse and apex cardiogram are recorded as well. A very informative recording would include the logic states of heart valves along with these tracings. A sample clinical phonocardiogram is shown in Figure 5.3 (Vermarien 1998).

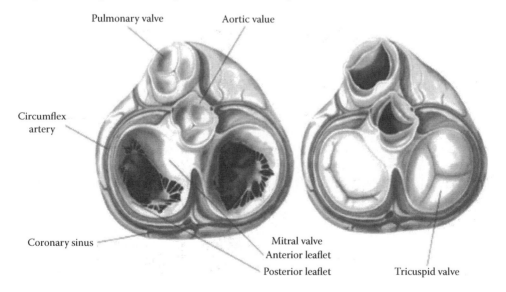

FIGURE 5.1   The four heart valves. (Carpentier A, Adams D H, and Filsoufi F, Reconstructive Valve Surgery: From Valve Analysis to Valve Reconstruction, Saunders/Elsevier, Philadelphia, PA, 2010.)

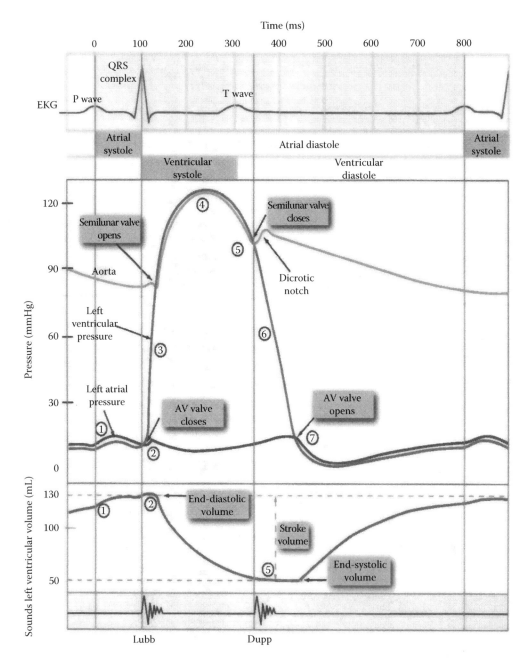

FIGURE 5.2 Wiggers diagram of the cardiac cycle (left heart). (Courtesy of http://fau.pearlashes. com/anatomy/chapter%2033a/images/Chapter%2033A_img_2.jpg).

The *cardiac cycle* (from the beginning of one heartbeat to the next) consists of two periods: *systole* and *diastole*. Diastole is a period of *relaxation*, during which the heart fills with blood. Systole is a period of *contraction*, during which the heart squeezes closed and pumps the blood into systemic and pulmonary circulations.

Each normal cardiac cycle is initiated by spontaneous generation of an action potential (rhythmic cardiac impulse) in the sinoatrial (SA) node, which is a specialized collection of

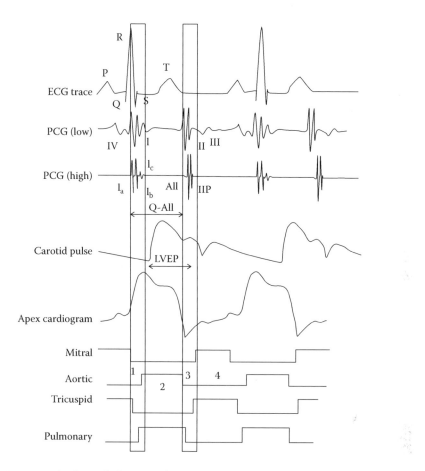

FIGURE 5.3   An example clinical phonocardiogram. (Courtesy of Vermarien, 1998.)

fibrous tissue (3 mm wide, 15 mm long, 1 mm thick) located below and lateral to the superior vena cava. The SA node (also known as the natural *pacemaker* of the heart) is capable of self-excitation and autorhythmicity. As the SA node fibers are directly connected with the atrial muscle fibers, the cardiac impulse immediately excites the atria (Guyton and Hall 2011).

The self-excitation of the SA node causes depolarization of the atrial muscles. The spread of depolarization through the atria causes the occurrence of the P wave in the ECG. This is followed by the atrial contraction (systole). The atrial contraction (shown as the encircled 1 wave on the left atrial pressure recording in Figure 5.2) causes a small rise in pressure in the atria (about 4–6 mmHg in the right atrium and about 7–8 mmHg in the left atrium). Atrial filling (diastole) accounts for 75% of blood flow into the atria, which flows on directly into the ventricles through the tricuspid valve on the right side and the mitral valve on the left side. Atrial contraction causes an additional 25% filling of the ventricles. As such, the atria serve as primer pumps for the ventricles, thereby improving their pumping effectiveness (Guyton and Hall 2011).

With the arrival of the atrial wave of depolarization at the AV node, there is about a 160 ms time delay, allowing enough time for blood to flow from the atria into the ventricles and complete the ventricular filling. This period corresponds to the PQ interval on the ECG. During this period, there is a slight rise in the ventricular volume and pressure. After this period, the ventricles depolarize and the QRS complex appears on the ECG. This corresponds with the closure of the AV valves. The QRS complex triggers the ventricles and initiates their contraction, which causes the ventricular pressure to begin to rise sharply. This corresponds to the encircled 2 wave on the atrial pressure recording, which signals the beginning of ventricular contraction. The occurrence of this wave is partly due to the slight backflow of blood from contracting ventricles into the atria and is mainly caused by the bulging of the tricuspid and mitral (AV) valves toward the atria as a consequence of ventricular pressure rise (Guyton and Hall 2011).

The peak of the R wave in the ECG signals the closure of the AV valves, the initiation of the ventricular systole, and the beginning of *isovolumetric ventricular contraction*. The ventricular volume recording clearly shows that during this time, the volume of blood in the ventricles remains constant.

Figure 5.2 shows that the first heart sound (shown as first in Figure 5.2 and abbreviated as S1 or I in Figure 5.3), audible as *lub*, occurs during the QRS complex of the ECG and is coincident with the onset of ventricular contraction (systole). It should be pointed out that the exact mechanisms of transmission of the heart sounds and murmurs within the heart and the chest are not well known (Durand and Pibarot 1995). According to the valvular theory of the genesis of heart sounds, S1 is composed of two major high-frequency components, M1 and T1, which coincide with the closure of the mitral and tricuspid valves, respectively. As a result of asynchronous closure of these valves, there is a delay (splitting) of 20–30 ms in S1. According to Rushmer's (cardiohemic) theory of the genesis of the heart sounds, S1 has four components. The initial component in S1 occurs during ventricular systole when myocardial contractions shift blood toward the atria, thus sealing the tricuspid and mitral valves. The second component in S1 starts with development of abrupt tension in the closed AV valves, slowing down the blood flow into the ventricles. Following the closure of AV valves, the pulmonic and aortic valves open and the blood is ejected out of the ventricles. The third component of S1 is due to oscillations of blood between the root of the aorta and ventricular walls. The fourth component of S1 is associated with oscillations caused by turbulent blood flow through the ascending aorta and the pulmonary artery (Durand and Pibarot 1995). (For our purposes here, we adopt the conceptual framework that heart sounds originate from the acceleration or deceleration of blood flow and murmurs are caused by blood turbulence.) Within this framework, S1 is caused by the movement of blood during ventricular systole. With contraction of the ventricles, blood shifts toward the atria, closes the AV valves, and causes oscillation of blood—*the closure of AV valves is a landmark event with substantial contributions to genesis of S1*. Oscillations of blood between the descending root of the aorta and ventricles and vibrations due to blood turbulence at the aortic and pulmonic valves also contribute to the genesis of S1 (Peura 2010).

During the ventricular isovolumetric (isometric) contraction, which lasts about 20–30 ms, the blood in the ventricles builds up sufficient pressure to open the semilunar (aortic and pulmonic) valves against pressures in the large (aorta and pulmonary) arteries. The opening of these valves is coincident with the end of the QRS complex in the ECG.

At the end of the ventricular isovolumetric contraction when the BP in the ventricles rises slightly above pressures in the large arteries (about 80 mmHg in the aorta and about 8 mmHg in the pulmonary artery), the ventricular ejection period begins. During the first one-third of the ejection period, known as the *rapid ejection period*, approximately 70% of ventricular emptying takes place. The remaining 30% of ventricular emptying occurs during the next two-thirds of the ejection period. This period is known as the *slow ejection period*. The rapid ejection period correlates with the ST interval, and the slow ejection period coincides with the T wave on the ECG. The end of the rapid ejection period also marks the end of *ventricular systole* (Guyton and Hall 2011).

At the end of the ventricular ejection period, *isovolumetric ventricular relaxation* (diastole) suddenly begins and lasts about 60 ms. This causes the intraventricular BPs to drop below pressures in the large arteries, snapping the *aortic* and *pulmonary* valves closed. This closure occurs as a consequence of elevated pressures in the distended large arteries, pushing blood back toward the ventricles. During ventricular isovolumetric relaxation, the intraventricular BPs drop rapidly back toward their low diastolic values. This marks the end of the T wave on the ECG. At this stage, the AV valves open and a new cycle of ventricular pumping begins (Guyton and Hall 2011).

The second heart sound (S2 or II), audible as *dub* or *dup*, occurs toward the completion of the T wave on the ECG. Its initial phase consists of two high-frequency vibrations, one due to the closure of the aortic valve (A2) and the other due to the closure of the pulmonary (P2) valve, with the aortic component leading and louder than the pulmonary component. At the end of ventricular diastole, the systolic ejection into the aorta and the pulmonary artery drops and the rising pressures in these arteries exceed intraventricular BPs, which reverse the flow of blood and cause the closure of the semilunar valves. The second heart sound usually has higher-frequency components as compared with the first heart sound. As a result of the higher pressures in the aorta compared with the pulmonary artery, the aortic valve tends to close before the pulmonary valve, so the S2 may have an audible split. In normal individuals, respiratory variations exist in the splitting of S2. (During the expiration phase, the interval between the two components is small: <30 ms. However, during inspiration, the splitting of the two components is evident. Clinical evaluation of S2 is a useful bedside technique and is a valuable screening test for heart disease.) (Abbas and Rahsa 2009)

With the closure of the AV valves and occurrence of the S2, the ventricular diastolic phase begins. During this phase, the ventricular volume increases from 110 to 120 mL (*end-diastolic volume*). With ventricular emptying during systole, this volume decreases about 70 mL, which corresponds to the stroke volume output, with about 50 mL remaining in the ventricle at this time, which is called *end-systolic volume* (Guyton and Hall 2011).

As the AV valves open, there is a rapid inflow of blood from the atria into the ventricles. The termination of the rapid ventricular filling phase from the atria causes the ventricular muscle walls, which are relaxed, to vibrate. This generates a low-amplitude, low-frequency vibration that is called the third heart sound (S3 or III) and is audible in children and in some adults.

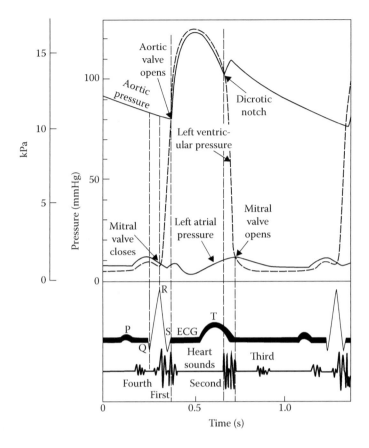

FIGURE 5.4 Electrical and mechanical events during the cardiac cycle. Please note the timing relationship of heart sounds, especially the fourth (atrial) heart sounds with these events. (From Peura, R.E., Blood pressure and sound, In: Webster, J.G. (ed.), *Medical Instrumentation: Application and Design*, 4th edn., John Wiley & Sons, Hoboken, NJ, 2010.)

After the rapid inflow of blood from the atria into the ventricles, there is a period of *diastasis*. This is coincident with the middle stage of diastole. During this phase, the initial passive ventricular filling has slowed down. The ventricular blood volume during this phase rises slightly. Diastasis leads into the initiation of the P wave on the ECG. After diastasis, there is atrial systole, during which the atria contract to complete the ventricles filling (the *a wave*). This period marks the end of diastole in the cardiac cycle.

The fourth heart sound (S4 or IV) is known as the atrial heart sound. It is not audible but can be recorded in a phonocardiogram. S4 occurs during atrial systole, which is marked by the PQ segment in the ECG. Figure 5.4 shows the occurrence of S4 marked as fourth in the lowest trace.

5.3 SUMMARY OF THE ORIGIN OF HEART SOUNDS

In a normal cardiac cycle, there are two major heart sounds: the first heart sound (S1 or I) heard as *lub* and the second heart sound (S2 or II) heard as *dub or dup*. S1 is associated with the onset of ventricular contraction and occurs during the QRS complex of the ECG signal.

S1 has four components. The initial component in S1 occurs during ventricular systole when myocardial contractions shift blood toward the atria, sealing the AV valves. The second component in S1 starts with development of abrupt tension in the closed AV valves, which slows down the blood flow into the ventricles. Following the closure of the AV valves, the pulmonary (P) and aortic (A) valves open and the blood is ejected out of the ventricles. The third component of S1 is due to oscillations of blood between the root of the aorta and the ventricular walls. The fourth component of S1 is associated with oscillations caused by turbulent blood flow through the ascending aorta and the pulmonary artery.

Following the systolic pause in the normal cardiac cycle, the second heart sound (S2) occurs. It is coincident with the completion of the T wave in the ECG signal. S2 is mainly associated with the slowing down and reversal of blood flow in the aorta and pulmonary artery as well as the closure of these valves. While the primary oscillations occur in the arteries due to the deceleration of blood, the ventricles and atria also vibrate due to the transmission of oscillations through the blood, valves, and the valve rings. S2 has two components. The first component of S2 (A2) is due to the aortic valve closure, and the second component of S2 (P2) is associated with pulmonary valve closure. The aortic valve normally closes before the pulmonary valve, and there are a few milliseconds time difference between A2 and P2. Pathological conditions could affect this timing relationship with a prolongation of the time delay and even a reversal of the time of occurrence of A2 and P2.

In some cases, a third heart sound (S3 or III) may be heard. S3 occurs after the completion of the T wave and before the initiation of the P wave in the ECG signal. S3 is mainly attributed to the sudden termination of the rapid ventricular filling phase from the atria. As the ventricles are filled with blood and their walls are relaxed, these oscillations are very low frequency ones. S3 is audible in children and in some adults.

In late diastole, during the PQ interval of the ECG signal, a fourth (atrial) heart sound (S4 or IV), not audible but recorded in PCG, may sometimes occur. This is associated with atrial contraction displacing blood into the relaxed and distended ventricles. In addition to the four heart sounds, valvular clicks and snaps are occasionally heard.

5.4 AUSCULTATION AND CHARACTERISTICS OF NORMAL HEART SOUNDS

The heart as an electrophysiologically controlled mechanical pump generates characteristic acoustical vibrations with each beat. These vibrations can be classified as normal heart sounds and murmurs. Heart sounds are short in duration and have musical character, whereas murmurs generally are longer in duration and have a noisy character (Vermarien 1988). As was explained in detail in the previous section, there are four normal heart sounds: S1, S2, S3, and S4. The first and second heart sounds, audible as *lub–dub*, have the largest intensities, respectively, and are mainly related to the closing of heart valves. The *lub* sound signals the closure of AV valves at the beginning of ventricular contraction, and the *dub* or *dup* sound signals the closure of the semilunar valves at the end of the ventricular contraction (Guyton and Hall 2011). The third heart sound is a weak (low-amplitude) and dull (low-frequency) vibration and is audible in children and in some adults. It is attributed to the vibrations of the relaxed ventricular muscle walls due to the sudden termination of

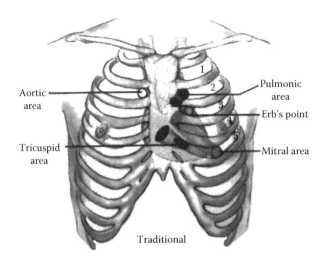

Traditional

FIGURE 5.5 The anatomical sites on the chest wall for auscultation of heart sounds. (Courtesy of http://acousticheart.com/auscultation_techniques.html).

the rapid ventricular filling phase. The fourth heart sound is not audible but is observable in phonocardiographic recordings and is attributed to atrial contraction and the movement of blood into the ventricles.

To be heard by the aid of a stethoscope, the heart sounds must travel from their source of generation (the heart) to the chest surface through major blood vessels. The acoustical properties of this sound transmission path attenuate the heart sounds. The most compressible tissues, such as the lungs and the fat layers, produce the largest attenuation of these acoustical waves.

The best recording or auscultation sites are those that provide the maximum intensity of heart sounds at the chest surface. Heart sound transmission through solid tissues or through paths with minimal thickness of inflated lung provides the highest sound intensity (Peura 2010). Figure 5.5 shows the four areas on the chest wall from which different heart sounds can best be distinguished.

Figure 5.5 shows the anatomical positions for the cardiac auscultation sites as follows: The aortic (A) area corresponds to the right second intercostal space; the pulmonic (P) area corresponds to the left second intercostal space; the mitral (M) area corresponds to the apex of the heart, located in the midclavicular line at the fifth intercostal space; and the tricuspid (T) area corresponds to the right fifth intercostal space.

The heart sounds generated by all the valves travel in all radial directions and can be heard from all these areas. However, the clinician distinguishes the sounds by a process of elimination. By moving the stethoscope from one auscultation area to the next, the clinician notes the intensity of the sound components from each valve (Guyton and Hall 2011).

Note that the optimal auscultation sites are not necessarily directly over the anatomical positions of the heart valves. As observed in Figure 5.5, the aortic area is upward along the aorta as sound transmission is in an upward direction in the aorta. The pulmonic area is also upward along the pulmonary artery. The auscultation site for the tricuspid valve is over the apex of the heart due to the fact that the apex is the portion of the left ventricle, which is closest to the chest surface as the heart is rotated so that most of the left ventricle is hidden behind the right ventricle (Guyton and Hall 2011).

Generally speaking, heart sounds are very-small-amplitude signals with a frequency spectrum spanning 0.1–2000 Hz (Peura 2010). Practically speaking, the lower limit (20 Hz) is set by the physiological limitations of human hearing. As a result, two difficulties may arise in their auscultation and recording. (1) At frequencies below 20 Hz, the amplitude of heart sounds is below the threshold of audibility of the human ear and is not heard. (2) Even though the high-frequency components of heart sounds are quite perceptible and occur in the region of maximum sensitivity of the human ear, however, recording them on a phonocardiograph poses a limitation due to the high-frequency characteristics of the signal. To address this limitation, a light beam inkjet, or digital array recorder, should be used (Peura 2010).

The normal heart sounds are low-amplitude (100 μPa range), low-frequency acoustical vibrations with major frequency components approximately 10–400 Hz. Murmurs have higher frequency than normal heart sounds (up to 600 Hz). The frequency spectrum of the first heart sound (S1) contains peaks in a low-frequency range (10–50 Hz) and a medium-frequency range (50–140 Hz). The frequency spectrum of the second heart sound (S2) contains peaks in a low-frequency range (10–80 Hz), a medium-frequency range (80–220 Hz), and a high-frequency range (220–400 Hz). In PCG, a frequency range of 5–2000 Hz, a dynamic range of 80 dB, and a threshold of 100 μPa are used (Rangayyan 1988). Features of heart sounds and murmurs, such as intensity, frequency spectrum, and timing, are affected considerably by a number of physical and physiological factors such as the recording site, intervening thoracic structures, left ventricular contractility, the degree of defect, the blood flow velocity, and the heart rate.

5.5 PATHOLOGICAL HEART SOUNDS

The diseases of the cardiohemic system impact the intensity, timing, frequency content, and splitting of the normal heart sounds. The origin of murmurs, caused mainly as a result of rapidly moving blood, is known. The occurrence of murmurs in children is common. Murmurs are normally heard in adults after exercise. The narrowing (stenosis) and leakage (regurgitation) of the mitral, pulmonary, and aortic valves may cause abnormal murmurs. The time of occurrence in the cardiac cycle and their location at the time of measurement are important factors in their detection (Peura 2010).

Aortic and mitral regurgitation and stenosis are two common causes of heart murmurs. In aortic regurgitation, blood flows backward from the aorta into the left ventricle during diastole. This causes a *blowing* murmur of relatively high pitch with a swishing quality, which is heard best over the left ventricle. This murmur results from the turbulence of blood jetting backward from the aorta into the blood in the left ventricle. No sound is heard during systole. In aortic stenosis, blood is ejected from the left ventricle through a narrowed aortic valve. A nozzle effect is created during systole, with blood jetting at a very high velocity. This causes severe turbulence in the root of the aorta. This turbulence generates an intensive vibration, and a loud sound is heard in the upper chest and the lower neck. No sound is heard during diastole.

In mitral regurgitation, blood flows through the mitral valve during systole. This causes a high-pitch *blowing* sound of swishing quality, which is heard best at the apex of the heart. This murmur results from the turbulence of blood ejecting backward from the left ventricle through the valve into the blood already in the left atrium or against the atrial walls. No sound is heard during diastole.

In mitral stenosis, the blood encounters a strong resistance while flowing from the left atrium into the left ventricle. This causes a low pitch murmur, which is difficult to hear. In mild stenosis, the murmur lasts only during the first half of the middle third of diastole, but in severe stenosis, the murmur can start early in diastole and may persist for the entire diastolic period. Figure 5.6 shows a phonocardiogram of normal and abnormal heart sounds caused by valvular lesions. Figure 5.6 also shows a continuous murmur, which is generated as a consequence of a connection between pulmonary artery and aorta (patent ductus arteriosus).

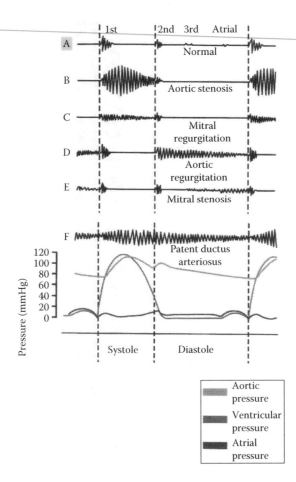

FIGURE 5.6 Phonocardiograms of normal and abnormal heart sounds. (A) normal (B) aortic stenosis, (C) mitral regurgitation, (D) aortic regurgitation, (E) mitral stenosis, and (F) patent ductus arteriosus. (Courtesy of http://www.textbookofcardiology.org/images/e/e4/Phonocardiograms_from_normal_and_abnormal_heart_sounds.svg).

5.5.1 Pathological First Heart Sound (S1)

Different cardiovascular pathologies impact the timing and intensities of heart sounds as recorded on phonocardiograms. The right bundle branch (RBB) block, the narrowing of the tricuspid (T) valve, and the atrial septal defect due to a delayed tricuspid component cause a wide splitting in S1. In the left bundle branch (LBB) block, the splitting between the two observable components of S1 disappears and this heart sound appears as a single component. The reduced contractility (due to myocardial infarction, cardiomyopathy, and heart failure), LBB block, mitral (M) valve leakage, and aortic (A) valve narrowing reduce the intensity of S1. Mitral stenosis and atrial septal defect generate an intensified S1 (Vermarien 1988).

5.5.2 Pathological Second Heart Sound (S2)

S2 manifests the closure of the aortic (A) valve followed by the closure of the pulmonary (P) valve. This causes a splitting of S2 into an aortic (A2) and a pulmonary (P2) component. Inspiration causes an increased splitting between these components. This is due to an increased difference in the duration of left and right ventricular systole caused by an increased right and decreased left ventricular filling. At the end of expiration, A2 and P2 may fuse together. Paradoxical splitting manifested as P2 preceding A2 is pathological. A2 normally has a larger intensity than P2. An increased intensity in P2 compared to A2 is generally pathological. A reduction in the intensity of S2 may reflect the stiffening of valve leaflets. An increased S2 may be indicative of a higher valve radius or a lowered blood viscosity. Delayed pulmonary or advanced aortic valve closure causes a wide splitting in S2. Pathologies like RBB block, the narrowing of the pulmonary valve, pulmonary hypertension, and atrial septal defect cause a delayed pulmonary valve closure. Mitral regurgitation and ventricular septal defect cause an advanced aortic valve closure. Paradoxical splitting of S2 could be indicative of a delayed aortic valve closure (caused by LBB block, aortic narrowing, and arteriosclerotic heart disease) or advanced pulmonary valve closure (caused by tricuspid regurgitation and advanced right ventricular activation). The aortic component (A2) of S2 decreases in aortic regurgitation and is pathologically decreased in left ventricular performance (Vermarien 1988).

5.5.3 Pathological Third Heart Sound (S3)

As mentioned earlier, S3 is mainly attributed to the rapid ventricular filling phase. It is often audible in children and adolescents and may be observable in the phonocardiograms of some adults. It may disappear as a consequence of aging due to an increased myocardial mass. The intensity of S3 may increase due to a high ventricular filling rate or altered physical properties of the ventricles. The reappearance of S3 after 40 years of age is often indicative of pathology. Mitral regurgitation, aortic stenosis, and ischemic heart disease generate a pathological S3 (Vermarien 1988).

5.5.4 Pathological Fourth Heart Sound (S4)

S4 is attributed to atrial contraction. It may be audible in older people but rarely in normal individuals. Augmented ventricular filling or diminished ventricular distensibility results in

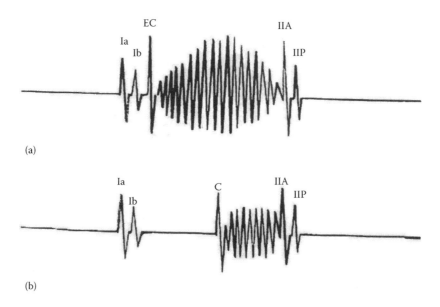

FIGURE 5.7 Examples of (a) pathological sounds and (b) murmurs in phonocardiographic recordings. (Courtesy of Vermarien, 1998.)

an increased S4. Mitral regurgitation, aortic stenosis, hypertensive cardiovascular disease, and ischemic heart disease lead to the generation of a pathological S4 (Vermarien 1988).

Other pathological heart sounds may arise during systole as the ejection sound and the nonejection click. Congenital aortic or pulmonary valvular stenosis with restrictions in the opening of the cusps may produce an ejection systolic sound. A sudden mitral valve prolapse into the left atrium produces a nonejection systolic click. In cases of mitral valve stenosis, an opening diastolic snap sound may occur. Figure 5.7 (a) and 5.7 (b) show some examples of pathological sounds and murmurs, respectively.

5.6 STETHOSCOPES AND PHONOCARDIOGRAPHS

Effective surgical treatment of cardiac disease requires exact and reliable diagnosis. Cardiac catheterization and medical imaging techniques offer a great degree of accuracy in the diagnosis of heart disease. However, careful auscultation of the heart and PCG play an essential role in the *bedside diagnosis* of cardiac disease. To that end, stethoscopes and phonocardiographs serve as important diagnostic devices in cardiology.

5.6.1 Acoustical Stethoscopes

The acoustic stethoscope was invented by Rene Laennec in 1916. It was a rigid monaural medical device made from a wooden tube. As such, it looked like a common ear trumpet. Figure 5.8 shows the Laennec stethoscope and its parts.

It was used for auscultation of sounds generated in the animal and human body. The main function of the acoustic stethoscope is to transmit sounds (acoustical vibrations) from the chest wall (by means of the *chest piece*) to the human ear (by means of the *ear piece*) through a sound transmission tube. Modern stethoscopes use binaural ear pieces,

FIGURE 5.8 Laennec's stethoscope and its parts (chest piece, transmission tube, and ear piece). (Courtesy of http://www.antiquescientifica.com/web.stethoscope.monaural.laennec.htm).

a flexible sound transmission tube, and a thin but rigid vibrating membrane (*diaphragm*) over the chest piece opening to pick up sound. The diaphragm provides differentiation between low-frequency and high-frequency sounds, in the diaphragm and bell modes, respectively. Figure 5.9 shows the details of a modern stethoscope.

The most important function of an acoustic stethoscope is airtight (leakage-free) acoustical coupling from the patient chest wall to the examiner's ears. There must be good contact between the patient's skin and the stethoscope diaphragm. Loose-fitting ear pieces compromise the quality of coupling between the chest wall and the ears. This causes a consequent decrease in the listener's perception of heart sounds and murmurs. The ear pieces must be comfortable and fit perfectly well into the listener's ears. Soft rubber ear pieces are comfortable and fit well into the ears and provide good acoustic coupling between the ear and the transmission tube. Important operational parameters of acoustic stethoscope are size, flexibility, loudness of sounds (perceived sound amplitude), clarity or crispness of sounds (the ability to distinguish between different cardiac sounds), ergonomic design,

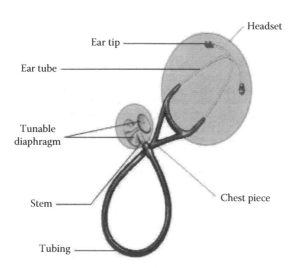

FIGURE 5.9 A modern acoustical stethoscope. (Courtesy of http://www.mountainside-medical.com/products/ADC-Adscope-602-Cardiology-Stethoscope.html).

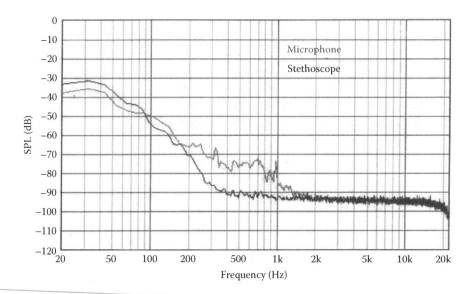

FIGURE 5.10 Frequency response of a typical stethoscope and a microphone. (Courtesy of http://www.forusdocs.com/reviews/Acoustic_Stethoscope_Review.htm).

immunity (sound proofing) against external noise, and the ability to respond equally well when listening to sounds from thin and obese individuals and patients with murmurs, chronic obstructive pulmonary disease (COPD), congestive heart failure, and vascular problems. Objective performance testing of acoustic stethoscopes involves loudness testing, clarity testing, frequency response (transfer function), and measurements under different diaphragmatic pressure conditions (mild and firm pressures). The stethoscope housing, which is in the shape of a bell, makes contact with the skin and serves as the diaphragm at the rim. By applying pressure on the diaphragm, it becomes taut and this causes an attenuation at low frequencies. For measuring the frequency response of a typical acoustic stethoscope, a known audiofrequency signal is applied to the bell of the stethoscope by means of a headphone coupler, and the audio output of the ear pieces is measured by using a coupler microphone system. A critical performance measure of the acoustical stethoscope is the amount of sound attenuation near the listener's threshold of hearing. In this region, even a small attenuation as little as 3 dB could make a difference between being able to hear the sound or miss it (Peura 2010). Figure 5.10 shows a typical frequency response for an acoustic stethoscope. This shows that the acoustic stethoscope has an uneven frequency response.

The materials and construction of modern stethoscopes have undergone tremendous improvements over the last several decades, and tens of patents have been filed in different countries in the search for the ideal stethoscope. In spite of all these efforts, the acoustical stethoscope still enjoys more popularity and acceptance over its electronic versions in the clinical arena. This is mainly due to their ergonomic design (easy to wear around the neck or be placed in the white coat pocket), robustness, being able to withstand abuse (dropped on the floor without damage), a durability of 15–20 years, and a low price.

5.6.2 Electronic Stethoscopes

Recently, many electronic stethoscopes have been developed. Basically, an electronic stetho-scope (also known as an e-stethoscope or a stethophone) converts the acoustical vibrations generated inside the body into electronic signals and processes (amplifies, shapes, and filters), displays, records, and, in some cases, even analyzes these signals. These devices offer a range of frequency response characteristics, from totally flat response to typical acoustical stetho-scope characteristics with a variety of frequency selective band-pass responses in between.

Electronic stethoscopes are composed of a heart sound *sensor* (transducer, a more gen-eral term) that converts sound (acoustical vibrations) energy into electronic (voltage) sig-nals. There are generally two types of sensors: the absolute pickup and the relative pickup. An absolute pickup sensor (contact type: piezoelectric/accelerometer, rigidly applied to the chest wall) detects the vibration averaged over the area of the measurement site. The rela-tive pickup (air-coupled) sensors (i.e., microphones) detect the sound averaged over the measurement area with respect to a reference area. As such, they are considered differential pickup sensors. Air-coupled sensors generally consist of a circular air filled cavity, with the edge of the cavity rigidly coupled to the chest wall in an airtight manner. This arrangement provides a differential measurement of the displacement under the cavity with respect to displacement under the edge as explained in Vermarien (1988).

Essentially, the electronic stethoscope could be considered as an acoustic stethoscope with a built-in microphone (a relative pickup device). As such, relative displacements in the chest wall produce air pressure changes in the cavity of the sensor, which in turn move the membrane of the microphone. The movements of the microphone are then transformed into an electrical signal by the moving coil (dynamic type) or the variable capacitance (electret type) sensing action. After detection of sounds by the sensor, electrical signals are fed into a preamplifier with desirable characteristics including input impedance, gain, common mode rejection ratio, bandwidth, and output impedance, the details of which are outside the scope of this chapter. The combined capabilities of the preamplifier and the sensor deter-mine the noise level of the electronic stethoscope. A variety of vibrations could deteriorate the quality of the combined sensor–preamplifier performance. These include physiological vibrations generated in organs other than the heart (such as respiration sounds, especially in patients with lung disease) and environmental vibrations (such as air-transmitted noises, especially at high frequencies where cardiac signals are weak). Recording (or listening to) heart sounds in an expired resting state could reduce the disturbance from lung sounds and applying the air-coupled sensor to the chest wall in an airtight manner could diminish envi-ronmental noise pick up. However, it is not possible to fully remove these disturbances, and therefore, they impose a limitation on the threshold at which heart sounds and murmurs can be recognized (Vermarien 1988).

The type and frequency response of the preamplifier connected to the sound sensor is also of great importance. For example, piezoelectric sensors present a capacitive output to the input resistance of a voltage preamplifier. This combination produces a first-order high-pass filtering effect that will distort the amplified signals. To address this problem, charge or cur-rent amplifiers with appropriate frequency response must be considered (Vermarien 1988).

After preamplification, high-pass or band-pass filtering must be deployed to compensate for the fact that the recorded (visualized) heart sound does not produce the same audio perception that an acoustical stethoscope provides during auscultation. To address this limitation, further amplification and filtering must be used to compensate for decreasing amplitudes of sounds at increasing frequencies (Vermarein 1988).

Modern electronic stethoscopes are compact, convenient to use, and immune to noise. Even though they are not yet widely accepted in the clinical arena (mainly because clinicians are not fully familiar with the sounds heard by using these devices), it seems that their combined advantages and capabilities with computing technologies (e.g., portable personal computers equipped with appropriate data acquisition hardware and software for virtual instrumentation and signal processing, visualization, archiving, and hard copy generation) are beginning to attract interest in their clinical deployment. Some of these advantages are ease of data acquisition and playback, high-quality visualization, sophisticated signal processing and feature extraction, pattern recognition, patient databank-based diagnosis, and telemedicine applications.

Wireless electronic stethoscopes equipped with noise reduction, signal enhancement, and audio and visual capabilities are now in existence. Computer-aided phonocardiographic analysis (algorithms) software programs are now available that can effectively analyze the recorded heart sounds and murmurs and distinguish between innocent and pathological murmurs. These capabilities facilitate and enable remote diagnosis (telemedicine) and teaching.

5.7 PHONOCARDIOGRAPHS

Basically, a phonocardiograph is a medical device that provides a graphical recording of heart sounds and murmurs (phonocardiogram—PCG signal). These devices were intended to address the limitations and subjectivity of acoustical (medical or mechanical) stethoscopes. However, by extension and in clinical evaluation of a patient, the term phonocardiogram includes a simultaneous recording of the ECG, carotid arterial pulse, jugular venous pulse, and apex cardiogram. A phonocardiogram eliminates the subjectivity associated with the interpretation of heart sounds and murmurs as it enables the evaluation of these acoustic phenomena in relation to the electric and mechanical events during the cardiac cycle. Figure 5.2 shows these events, and Figure 5.3 shows an extended phonocardiogram (including carotid arterial pulse and apex cardiogram as well as logic states of the heart valves).

A phonocardiograph is composed of a microphone, signal conditioning (processing) circuitry, and a display. A contact or air-coupled microphone is held against the patient's chest. A variety of microphones are used. However, most use piezoelectric crystal or dynamic type sensing. The crystal microphone is more rugged, more sensitive (produces a larger output signal for a given input), and is less expensive. The dynamic microphone uses a moving coil acoustically coupled to its diaphragm and encircles a permanent magnet loudspeaker. The dynamic microphone produces a frequency response similar to the acoustical stethoscope. Appropriate preamplifiers, amplifiers, and filtering (high-pass or band-pass) circuitry are used to amplify and process the PCG signal. (The detailed characteristics of these electronic circuits are outside the scope of this chapter.) For recording of the carotid, jugular, and apex cardiogram, an air-coupled microphone with a frequency response of 0.1–100 Hz is used. For display, oscilloscopes (or graphical user interfaces in PC-based data acquisition and analysis systems) are implemented to visualize the PCG signals while they are being recorded. For hard copies, strip chart recorders are used. As the direct recording of the phonocardiogram requires a frequency response that extends to 1 kHz or more, an optical or a high-velocity inkjet recorder is used for this purpose. Clinicians use morphological and temporal characteristics (shape, timing, duration, and intensity) of the PCG signals to carry out their clinical evaluations. Extended phonocardiographic information enables clinicians to evaluate the heart sounds and murmurs with respect to the electrical and mechanical events of the cardiac cycle and provides them with a more comprehensive profile of the patient's cardiac health.

5.8 DIGITAL SIGNAL PROCESSING TECHNIQUES IN PHONOCARDIOGRAPHY

The advent of modern electronic stethoscopes and improvements in sophisticated DSP algorithms have greatly facilitated the digital recording and computer-aided analysis of heart sounds and murmurs. These combined capabilities have added a lot of diagnostic value and importance to phonocardiograms. Such algorithms allow the extraction of characteristic signal components (features) that form the basis for the reliable detection of various cardiac diseases.

Recently, a tremendous amount of research effort has been devoted to the development of robust DSP algorithms in the detection, characterization, and analysis of different heart sounds. These algorithms are divided into three general groups: (1) spectrum estimation, (2) time–frequency analysis, and (3) nonlinear analysis.

As the current book is mainly about physiological measurements, a detailed description of these DSP algorithms and their performance characteristics is outside the scope of this chapter. For a comprehensive review of the PCG signal processing framework, addressing topics such as PCG signal presentation, PCG signal denoising and filtering, PCG signal presentation, cardiac sound modeling and identification, modeling abnormal heart sounds, and model-based acoustic PCG acoustic signal processing topics, among others, please refer to Abbas and Rahsa (2009).

5.9 DIGITAL PHONOCARDIOGRAPHY AND COMPUTER-AIDED CLASSIFICATIONS OF HEART SOUNDS

The advancement of intracardiac PCG combined with modern DSP techniques has strongly renewed researchers' interest in quantitative studies of the heart sounds and murmurs. In recent years, new research areas in digital PCG have emerged and opened up opportunities for many applications.

Previous research has shown that time–frequency/scale methods applied to the characterization of the heart sounds are suitable for heart sound signal analysis (Durand and Pibarot 1995, Rangayyan *et al* 1987), feature extraction, and classification (Liang *et al* 1997a,b, Say *et al* 2002). Researchers have obtained promising results by applying many

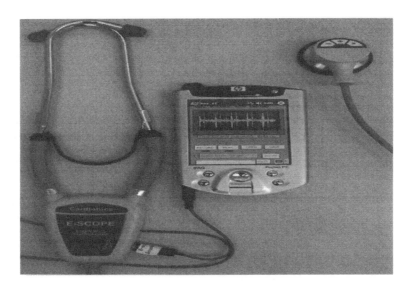

FIGURE 5.11 An intelligent PDA-based phonocardiograph. (Reprinted from Nazeran, H., *Methods Inf. Med.*, 46, 135, 2007. With permission from Schattauer GmbH Verlag für Medizin und Naturwissenschaften.)

modern DSP techniques including frequency content analysis, bilinear time–frequency representation techniques, cardiac acoustic mapping, spectrogram analysis, and wavelet transform to analyze and interpret heart sounds (Buthan and Yetkin 2003, Doyle *et al* 2003, El Asir *et al* 1996, Lee *et al* 1999, Obaidat and Matalgah 1992).

The digital analysis of phonocardiograms allows extracting features from the heart sounds that could not be detected by the human ear, such as muffled components of heart sounds, musical murmurs, rumble, or whiffs. As heart sounds are nonstationary signals, it is important to study their temporal information as well as their frequency content.

The main characteristics of heart sounds such as their timing relationships and components, frequency content, time of occurrence in the cardiac cycle, and envelope shape of murmurs can be quantified using advanced DSP techniques (Durand *et al* 1990, Obaidat and Matalgah 1992).

As an example integrating some of the aforementioned topics, the different stages involved in implementation of an intelligent PDA-based phonocardiograph are presented (Nazeran 2007). In this example, first, different applicable DSP methods to analyze heart sounds are compared, and then a wavelet-based segmentation method using Shannon energy is deployed to accurately distinguish between the first (S1) and second (S2) heart sounds. Finally, wavelet transform is used to extract informative features from heart sounds to classify them. Figure 5.11 shows an intelligent PDA-based phonocardiograph.

REFERENCES.

Abbas K A and Rahsa B 2009 *Phonocardiography Signal Processing*. J. Enderle (Series Editor). Morgan & Claypool

Buthan E and Yetkin T 2003 Optimal continuous wavelet analysis of phonocardiogram signals *Proc. Int. Conf. Signal Process.* **1**(2), 188–92

Carpentier A, Adams D H and Filsoufi F 2010 Reconstructive Valve Surgery: From Valve Analysis to Valve Reconstruction, Saunders/Elsevier, Philadelphia, PA

Doyle J D, Gopakumaran B and Smith W A 2003 Color spectrographic analysis of the phonocardiogram: A future noninvasive cardiac monitoring technology? *Proceedings of the 25th Annual International Conference of the IEEE Engineering in Medicine and Biology Society*, Cancun, Mexico, Vol 3, pp 2251–4

Durand L G, Blanchard M, Cloutier G, Sabbah H N and Stein P D 1990 Comparison of pattern recognition methods for computer–assisted classification of spectra of heart sounds in patients with a porcine bioprosthetic valve implanted in the mitral position *IEEE Trans. Biomed. Eng.* 37(12) 1121–9

Durand L G and Pibarot P 1995 Digital signal processing of the phonocardiogram: Review of the most recent advancements *Crit. Rev. Biomed. Eng.* 23 116–219

El-Asir B, Kahadra L, Al-Abbasi A H and Mohamed M M 1996 Time-frequency analysis of heart sounds *IEEE TENCON—Digital Signal Proc. Appl.* Vol. 2, 553–8

Guyton A C and Hall J E 2011 *Guyton and Hall Textbook of Medical Physiology*, 12th edition (Philadelphia: Saunders)

Lee J J, Lee S M, Young K, Min H K, and Hong S H 1999 Comparison between short time Fourier and wavelet transform for feature extraction of heart sounds. *IEEE TENCON*, Vol. 2, 1547–50

Liang H, Lukkarinen S and Hartimo I 1997a Heart sound segmentation algorithm based on heart sound envelogram *Comput. Cardiol.* 24 105–8

Liang H, Lukkarinen S and Hartimo I 1997b A heart sound segmentation algorithm using wavelet decomposition and reconstruction *Proceedings of the 19th Annual International Conference of the IEEE Engineering in Medicine and Biology Society*, Chicago IL, pp 1630–3

Nazeran H. 2007 Wavelet-based segmentation and feature extraction of heart sounds for intelligent PDA-based phonocardiography *Methods Inf. Med.* 46 135–41

Obaidat M S and Matalgah M M 1992 Performance of the short time Fourier transform and wavelet transform to phonocardiogram signal analysis *Symposium on Applied Computing: Technological Challenges 1990's* (New York: ACM) pp 856–62

Peura R E 2010 Blood pressure and sound in J. G. Webster (ed.), *Medical Instrumentation: Application and Design*, 4th edition (Hoboken: John Wiley & Sons) pp 293–337

Rangayyan R M and Lehner R J 1987 Phonocardiogram signal analysis: A review *Crit. Rev. Biomed. Eng.* 15(3) 211–36

Say O, Dokur Z and Ölmez T 2002 Classification of heart sounds by using wavelet transforms *Proceedings of the Second Joint EMBS/BMES Conference* (Houston, TX) pp 128–9

Vermarien H 1988 Phonocardiography in J. G. Webster (ed.), *Encyclopedia of Medical Devices and Instrumentation* (New York: Wiley) pp 2265–77

CHAPTER 6

Dermatology

John G. Webster

CONTENTS

6.1	Skin Color	123
6.2	Skin Thickness	125
6.3	Skin Water Loss	125
6.4	Skin Hydration	126
6.5	Skin Elasticity	126
6.6	Skin Stretch	126
6.7	Skin Peel Force	127
6.8	Skin Foreign Bodies	127
6.9	Skin Wound Healing	127
6.10	Skin Blood Flow	128
6.11	Skin Cancer Detection	128
6.12	Skin Electrical Measurements	129
References		129

6.1 SKIN COLOR

Colors may be described by their hue (color position in the color wheel), lightness (called value), and saturation (called chroma). Any color is expressed as a combination of hue, value, and chroma (H/V/C) (Berardesca et al 1995). Most measurements in dermatology are done visually, where the dermatologist uses optical magnification to carefully examine the entire human skin for possible lesions. However, there are instruments that can be used for the examination. Figure 6.1 shows a Cortex Technology (Hadsund, Denmark) DSM II Colorimeter. An optional LabVIEW-based application software package permits connection to a PC via the built-in USB port in order to import data directly into Excel spreadsheets.

Sommers et al (2013) investigated analysis of skin color, which is useful to understand skin phenotype, compare injured and noninjured skin, describe pathology, and monitor the outcomes from therapeutic interventions. They determined the intra- and interrater reliability of data from digital image color analysis between an expert and novice analyst. Both analysts used Adobe Photoshop lasso or color sampler tools based on the type of

FIGURE 6.1   The Cortex DSM II Colorimeter illuminates with red, green, and blue LEDs to display skin color on a handheld display for CIE L*a*b* and erythema/melanin skin applications. (From Cortex Technology 2014.)

image file. After color correction with PictoColor in camera software, they recorded L*a*b* (L* = light/dark; a* = red/green; b* = yellow/blue) color values for all skin sites. They found that data from novice analysts can achieve high levels of agreement with data from expert analysts with training and the use of a detailed, standard protocol.

Kikuchi *et al* (2013) used a spectral camera to develop a noncontact image-processing system that was capable of capturing a wide area of the face to visualize the distribution of the relative concentrations of skin chromophores in the face. The distribution of skin chromophores in the face, including melanin and oxy- and deoxyhemoglobin, was calculated from the reflectance data for each pixel of the spectral images. Melanin content increased, and hemoglobin oxygen saturation ratio decreased locally in the infraorbital areas of women with dark circles compared with those of women without dark circles.

Choi *et al* (2014) used computer-aided image analysis to test and compare the three global objective erythema index (OEI) candidates. In the aspect of clinical correlation, the red-blue difference index (RBI) was unsatisfactory despite its simple algorithm. The a* color values measured by colorimeter or erythema index (EI) by Mexameter and erythema dose (ED) were positively correlated with each other, and both showed a good correlation with subjective erythema index (SEI).

Dolotov *et al* (2004) developed a three-wavelength technique for determining pigmentation index (PI) and the EI of human skin as well as its particular implementation using LEDs operating at wavelengths of 560, 650, and 710 nm and a large-area photodiode. They evaluated the instrument both in vitro and in vivo. In vitro, they observed good correlation between the measured indices and results obtained with commercially available techniques. In addition, they established linearity of the PI with melanin concentration in the phantom medium up to 7×10^{-3} nm^{-1} (defined as a slope of the optical density spectrum). In vivo, they demonstrated feasibility of using the technique for predicting the minimal erythema dose (MED), minimal phototoxic dose (MPD), and the threshold of epidermal damage in a photothermal treatment.

6.2 SKIN THICKNESS

Polanska *et al* (2013) used high-frequency skin ultrasonography (USG) (Derma Scan Cortex Technology, version 3, Hadsund, Germany) to diagnose atopic dermatitis. USG examination produces a cross-sectional image of human skin with an axial resolution of 80 μm and lateral resolution of 200 μm with a sharp-focusing transducer. The ultrasonic wave is partially reflected at the boundary between adjacent structures and produces echoes of different amplitudes. The intensity of the reflection echoes is evaluated by the microprocessor and is visualized as a color-coded 2D B-mode image (standardized code of 255 levels). The average amplitude of the echoes in a defined area of the image is known as echogenicity, which may be objectively measured with the computer-assisted image analysis. Hypoechogenic is defined as an intensity <30 pixels. The A-mode interfaces are shown as well-defined peaks. A typical gain curve was 25 to 70 dB. The velocity of ultrasound in the skin was set at 1580 m/s. In all AD patients, skin USG was performed with the use of 20 MHz probe in B-mode scan within the skin region. Two skin parameters were assessed: skin echogenicity (%) and the thickness of hypoechoic band. The average hypoechoic band thickness was evaluated in millimeters by measuring the vertical distance between the lower edge of the entry echo and the posterior margin of the hypoechoic zone (A-mode scans).

Böhling *et al* (2014) compared stratum corneum (SC) thickness calculated from confocal Raman spectroscopy (CRS) data with results of SC thickness based on confocal laser scanning microscopy (CLSM) measurements. Raman spectra are obtained by focusing low-power laser light on the skin and measuring the Raman scattered light from the laser focus. A small part of the scattered light is found at wavelengths higher than the incident laser light. A River Diagnostics Model 3510 Skin Composition Analyzer (RiverD International B.V., Rotterdam, the Netherlands) with a fiber-coupled diode laser (wavelength: 671 nm) was used to measure the high wave number range of the Raman spectrum (2500 to 4000 cm^{-1}). A laser spot of 1 μm diameter was focused to the different depths of skin (substitute vertical for spatial resolution of about 5 μm) by means of a specific inverted microscope objective. A confocal laser scanning microscope (CLSM, Vivascope 1500; Lucid Inc., Rochester, New York) with a depth resolution of approximately 5 μm was used. Both methods, CRS and CLSM, were found to be suitable to measure SC thickness correctly.

6.3 SKIN WATER LOSS

Transepidermal water loss (TWL) refers to the rate at which water migrates from the viable dermal tissues through the layers of the SC to the external environment. In the absence of profuse sweating, the TWL is predominantly controlled by the diffusion of water vapor in the SC caused by the difference in vapor concentration between the inside and outside surfaces.

TWL measurements have been used in studying the restoration of barrier function in wounded skin, as well as in evaluating occlusive properties of topical preparations. TWL can be measured using the direct or indirect measurement techniques. Flow hygrometry is a commonly used technique for direct measurement. The indirect measurement technique relies on establishing a boundary air layer over skin of known geometry. The two commonly used techniques of indirect measurement are the closed cup and the open cup methods (Marks and Payne 1981).

In open cup flow hygrometry, the flux of water vapor out of a fixed area of skin is determined by measuring the increase in water vapor concentration in the flowing gas stream. The increase in humidity is read at a sensor output once steady-state conditions have been reached.

In this method, transfer of water vapor to the sensor is by convection, which normally requires gas tanks or pumps, valves, and tubing, in addition to the sensor and the skin chamber. In contrast, the closed cup method requires little auxiliary equipment (Smallwood and Thomas 1985).

In the closed cup method, the humidity sensor is sealed into one end of a cylinder of known length. The cylinder is placed on the skin, trapping a volume of unstirred air between the skin and the detector. Within a few seconds, the sensor voltage begins to rise steadily followed by a gradual decrease in rate of change. The application of diffusion principles to the water vapor in the volume of trapped air predicts this behavior and shows that the TWL is directly proportional to the slope of the transient linear portion of the detector output curve.

6.4 SKIN HYDRATION

Ohno *et al* (2013) examined biophysical parameters, including skin conductance and TWL, and biomechanical parameters of skin distension/retraction before and after suction at the forehead, lateral canthus, and cheek, with or without mist, in a testing environment (24 °C, 35% relative humidity) for 120 min. Skin deformation of the face was improved by mist, suggesting hydration of the SC by mist.

6.5 SKIN ELASTICITY

For measuring skin elasticity and firmness, Neto *et al* (2013) tested three different probes.

The Frictiometer FR 700 consists of a 16 mm diameter friction Teflon indented sensor head with eight groves, which is rotated by a motor on the skin surface. The resistance that the skin offers to torsion is related to its elasticity and plasticity characteristics.

The Reviscometer principle is based on a resonance running time (RRT) measurement. Two needle sensors are placed on the skin: one emits an acoustical shockwave, and the other receives it. The shockwave propagates through the skin fibers differently according to moisture and elasticity, and the time the wave needs to go from transmitter to receiver is the measured parameter.

The Cutometer probe generates negative pressure, which can be varied between 20 and 500 mbar. The skin area to be measured is drawn into the aperture of the probe as a result of this negative pressure, causing a vertical deformation. The penetration depth of the skin into the aperture is determined without contact by an optical measuring system.

6.6 SKIN STRETCH

Leveque *et al* (2014) examined living epidermis of human adults by means of in vivo reflectance confocal microscopy. Distances between skin surface and papillae apex and pegs of the dermal–epidermal junction (DEJ) were, respectively, recorded in both relaxed and stretched skin situation. The number of papillae present within a single image (field of view, 500 μm × 500 μm) was also measured. An in vivo reflectance confocal microscope (RCM; VIVASCOPE 1500 SYSTEM) was used. Skin was manually stretched in its plane to about a 30% extension. They found that skin extension has no effect upon the distance between skin surface and the apex of papillae. In contrast, the distance between skin surface and the pegs of papillae decreases. On the other hand, skin extension leads to a significant decrease in the number of papillae within a single image. Epidermal atrophy and structural changes observed in the DEJ with aging may be, by some extent, related to daily and repetitive skin deformations all along the life span.

6.7 SKIN PEEL FORCE

Krueger *et al* (2013) measured the force required to peel away a standard width of medical adhesive using an MTS Insight™ and a cyberDERM Inc. Mini Peel Tester (CMPT). They found that the CMPT provides the capability to obtain reliable peel force measurements in a variety of situations, including the more challenging conditions of obtaining measurements from a living human.

6.8 SKIN FOREIGN BODIES

Cinotti *et al* (2013) describe Raman spectroscopy, a technique that analyzes materials by detecting their molecular vibrations. It studies the scattered radiation arising from the interaction between a monochromatic light source and the molecules of the system to be studied. When a radiation passes through a medium, a fraction of its light is scattered. The energy exchange between the incident light beam and the medium results in a change of wavelength of the scattered light (Raman scattering) that is characteristic of the medium. Raman spectroscopy is getting more and more applications in dermatology, having already been used to identify cancerous lesions such as basal cell carcinoma and melanoma, to assess the amount of water in the dermis, to differentiate photoaging from chronoaging, and to follow the cutaneous penetration and metabolism of some drugs. It has also recently been employed for the chemical characterization of cutaneous foreign bodies to complete the histopathological examination of birefringent material in two skin biopsies, identifying two plastic materials used in previous surgical procedures in one case and an antibiotic (carbenicillin) in the other case.

6.9 SKIN WOUND HEALING

Kuck *et al* (2014) investigated optical coherence tomography (OCT), an imaging technique for in vivo evaluation of skin diseases with a resolution close to histopathology. The OCT system Callisto (Thorlabs GmbH, Lubeck, Germany) was used in this study for in vivo evaluation of wound healing. This method is a spectral domain OCT based on low-coherence interferometry to detect backscattered light from biological tissues by measuring the optical path length. The system utilizes broadband, lower-coherence light source and contains a 930 nm superluminescent diode light source and a linear CCD array-based spectrometer. Analogous to ultrasound, using light instead of sound, the skin is probed with a beam of light and the reflected light interferes with a reference beam that stems from the same light source. After recombining and forming the interference pattern, their spectrum was disturbed with a diffraction grating and determined by a CCD line camera. After conversion, an optical imaging is presented. This 1D depth scan is called the A-scan, and by recording many adjacent A-scans, 2D or 3D images of the specimen are obtained. The Callisto operates at a maximum A-scan rate of 1.2 kHz, which results in two frames per second. It contains a bandwidth of 100 nm (3 dB) and a sensitivity of 105 dB. The lateral resolution is up to 8 μm, and the axial resolution averages a value close to 5.0 μm by the refraction index of the skin. The maximum penetration depth of the Callisto is up to 1.7 mm in air. This value has to be divided by the refractive index of the skin, and for that reason, the maximum penetration depth is around 1.2 to 1.3 mm, dependent on the sample of the skin.

Comparable results between histological findings and OCT were achieved. OCT allowed the detection of partial loss of the epidermis, vasoconstriction, vasodilatation, and epithelialization. For monitoring purposes, OCT is sufficient and can reduce the number of invasive skin biopsies.

6.10 SKIN BLOOD FLOW

Igaki *et al* (2014) measured the increase in skin blood flow when using moist heat conditions. Air chambers adjusted to a 40 °C temperature and 80%–100% relative humidity and 40 °C temperature and 10% to 20% relative humidity were prepared as moist and dry heat conditions. They used a confocal microscopic Raman spectroscope (Nanofinder 30, Tokyo Instruments Inc., Tokyo, Japan; excitation laser wavelength, 632.8 nm; objective, Nikon S Fluor NA 1.3, Nikon Instruments Co., Tokyo, Japan). The moisture level was calculated from the Raman spectrum of the skin. The skin surface temperature was measured, employing a chromel–alumel thermocouple. Skin surface heat flux (energy level flowing in a unit area) was measured using a heat flux measurement device developed by the authors. Skin blood flow was measured using a laser Doppler flow meter (ALF21R, Advance Co., Tokyo, Japan). A sweating meter (SKD-2000, Skinos Co., Nagoya, Japan) was attached to investigate sweating during measurements. SC moisture levels and skin surface heat conductivity were higher in the moist heat condition, and skin blood flow was significantly greater than that in skin exposed to dry heat. Therefore, moist heat is more efficient at warming the body than dry heat.

6.11 SKIN CANCER DETECTION

Dancila *et al* (2014) used a micromachined silicon waveguide probe with an inspection area of 600 μm × 200 μm. The high planar resolution is required for histology and early-state skin-cancer detection. Three phantoms with different water contents, that is, 50%, 75%, and 95%, mimicking dielectric properties of human skin were characterized in the frequency range of 95 to 105 GHz. The complex permittivity values of the skin were obtained from the variation in frequency and amplitude of the reflection coefficient (S11), measured with a vector network analyzer (VNA). The resonance frequency lowers, from the idle situation when it is probing air, respectively, by 0.7, 1.2, and 4.26 GHz when a phantom material of 50%, 75%, and 95% water content is measured. The probe has the potential to discriminate between normal and pathological skin tissues. Further, improved information compared to the optical histological inspection can be obtained; that is, the complex permittivity characterization is obtained with a high resolution, due to the highly reduced measurement area of the probe tip.

6.12 SKIN ELECTRICAL MEASUREMENTS

Tam and Webster (1977) showed that the skin has a voltage across it that when the skin stretches, the voltage changes, and this causes motion artifacts during electrocardiography. Abrading the skin with sandpaper short-circuits the voltage and prevents motion artifact.

Burbank and Webster (1978) simultaneously measured skin potential and skin impedance and verified a simple electrical model of the skin. They recommended a needle-puncture method for reducing motion artifacts during electrocardiography as less traumatic than sandpaper abrasion.

Rosell *et al* (1988) measured skin impedance versus frequency and showed it was about 100 Ω at high frequencies and varied from 10 kΩ to 1 MΩ at low frequencies.

de Talhouet and Webster (1996) successively stripped 12 layers of the skin using Scotch Tape and found skin motion artifact decreased with each stripping.

Bahr *et al* (2014) developed a skin conductance monitoring system to reliably acquire and record hot flash events in both supervised laboratory and unsupervised ambulatory conditions. The 7.2 cm × 3.8 cm × 1.2 cm monitor consists of a disposable adhesive patch supporting two hydrogel electrodes and a reusable, miniaturized, enclosed electronic circuit board that snaps onto the electrodes. A good reference on skin is Wilhelm *et al* (1996).

REFERENCES

Bahr D E, Webster J G, Grady D, Kronenberg F, Creasman J, Macer J, Shults M, Tyler M and Zhou X 2014 Miniature ambulatory skin conductance monitor and algorithm for investigating hot flash events *Physiol. Meas.* **35** 95–110

Berardesca E, Elsner P, Wilhem K-P and Maibach H I 1995 *Bioengineering of the Skin: Methods and Instrumentation* (Boca Raton: CRC Press)

Böhling A, Bielfeldt S, Himmelmann A, Keskin M and Wilhelm K-P 2014 Comparison of the stratum corneum thickness measured in vivo with confocal Raman spectroscopy and confocal reflectance microscopy. *Skin Res. Technol.* **20** 50–7

Burbank D P and Webster J G 1978 Reducing skin potential motion artefact by skin abrasion *Med. Biol. Eng. Comput.* **16** 31–8

Choi J W, Kwon S H, Jo S M and Youn S W 2014 Erythema dose—A novel global objective index for facial erythema by computer-aided image analysis *Skin Res. Technol.* **20** 8–13

Cinotti E, Labeille B, Perrot J L, Boukenter A, Ouerdane Y and Cambazard F 2013 Characterization of cutaneous foreign bodies by Raman spectroscopy *Skin Res. Technol.* **19** 508–9

Cortex Technology 2014 http://www.cortex.dk/

Dancila D, Augustine R, Topfer F, Dudorov S, Hu X, Emtestam L, Tenerz L, Oberhammer J and Rydberg A 2014 Millimeter wave silicon micromachined waveguide probe as an aid for skin diagnosis—Results of measurements on phantom material with varied water content *Skin Res. Technol.* **20** 116–23

de Talhouet H and Webster J G 1996 The origin of skin-stretch-caused motion artifacts under electrodes *Physiol. Meas.* **17** 81–93

Dolotov LE, Sinichkin YP, Tuchin VV, Utz SR, Altshuler GB and Yaroslavsky IV 2004 Design and evaluation of a novel portable erythema-melanin-meter *Lasers Surg. Med.* **34** 127–35

Igaki M, Higashi T, Hamamoto S, Kodama S, Naito S and Tokuhara S 2014 A study of the behavior and mechanism of thermal conduction in the skin under moist and dry heat conditions *Skin Res. Technol.* **20** 43–9

Kikuchi K, Masuda Y and Hirao T 2013 Imaging of hemoglobin oxygen saturation ratio in the face by spectral camera and its application to evaluate dark circles *Skin Res. Technol.* **19** 499–507

Krueger E M, Cullum M E, Nichols T R, Taylor M G, Sexton W L and Murahata R I 2013 Novel instrumentation to determine peel force in vivo and preliminary studies with adhesive skin barriers *Skin Res. Technol.* **19** 398–404

Kuck M, Strese H, Alawi S A, Meinke M C, Fluhr J W, Burbach G J, Krah M, Sterry W and Lademann J 2014 Evaluation of optical coherence tomography as a non-invasive diagnostic tool in cutaneous wound healing *Skin Res. Technol.* **20** 1–7

Leveque J-L, Fanian F and Humbert P 2014 Influence of skin extension upon the epidermal morphometry, an in vivo study *Skin Res. Technol.* **20** 58–61

Marks R and Payne R A (eds.) 1981 *Bioengineering and the Skin* (Boston: MTP Press Ltd)

Neto P, Ferreira M, Bahia F and Costa P 2013 Improvement of the methods for skin mechanical properties evaluation through correlation between different techniques and factor analysis *Skin Res. Technol.* **19** 405–16

Ohno H, Nishimura N, Yamada K, Shimizu Y, Iwase S, Sugenoya J and Sato M 2013 Effects of water nanodroplets on skin moisture and viscoelasticity during air-conditioning *Skin Res. Technol.* **19** 375–83

Polanska A, Danczak-Pazdrowska A, Silny W, Wozniak A, Maksin K, Jenerowicz D and Janicka-Jedynska M 2013 Comparison between high-frequency ultrasonography (Dermascan C, version 3) and histopathology in atopic dermatitis *Skin Res. Technol.* **19** 432–7

Rosell J, Colominas J, Riu P, Pallas-Areny R and Webster J G 1988 Skin impedance from 1 Hz to 1 MHz *IEEE Trans. Biomed. Eng.* **35** 649–51

Smallwood R H and Thomas S E 1985 An inexpensive portable monitor for measuring evaporative water loss *Clin. Phys. Physiol. Meas.* **6** 147–54

Sommers M, Beacham B, Baker R and Fargo J 2013 Intra- and inter-rater reliability of digital image analysis for skin color measurement *Skin Res. Technol.* **19** 484–91

Tam H W and Webster J G 1977 Minimizing electrode motion artifact by skin abrasion *IEEE Trans. Biomed. Eng.* **BME-24** 134–9

Wilhelm K-P, Elsner P, Berardesca E and Maibach H I 1996 *Bioengineering of the Skin*, 2nd edition (Boca Raton: CRC Press)

CHAPTER 7

Gastroenterology

John William Arkwright and Phil Dinning

CONTENTS

7.1	Introduction	132
7.2	Gastrointestinal Endoscopy	133
7.3	Manometry: Measurement of Pressure and Contractile Force	134
7.4	Manometry Displays	139
7.5	Manometry within the Gastrointestinal Tract	140
	7.5.1 Esophageal Manometry	140
	7.5.2 Antroduodenal Manometry	140
	7.5.3 Sphincter of Oddi Manometry	141
	7.5.4 Colonic Manometry	141
	7.5.5 Anorectal Manometry	142
	7.5.6 Catheter Dimensions and Recording Characteristics	144
7.6	Impedance Monitoring of Bolus Presence	144
	7.6.1 Impedance Manometry	145
7.7	pH Monitoring of Acidity Levels	146
7.8	Ingested Pill Techniques	146
	7.8.1 SmartPill	146
	7.8.2 PillCam	147
7.9	X-Ray Fluoroscopy and Scintigraphy Imaging	148
	7.9.1 Barium Meals and Enemas	148
	7.9.2 Radiopaque Markers	148
	7.9.3 Scintigraphy	149
	7.9.4 Barostat Studies	151
7.10	Electrogastrography	152
7.11	Emerging Technologies	154
	7.11.1 Motility Tracking System	154
	7.11.2 Small Bowel HRM	155
	7.11.3 3D-HRM	155
	7.11.4 Multimodal Analyses	156
	7.11.5 Predictive Algorithms (Combined Esophageal HRM and Impedance)	156

7.11.6 Functional Luminal Imaging Probe   157
7.11.7 Intraluminal Ultrasound   157
7.11.8 Optical Coherence Tomography   158
7.12 Concluding Remarks   159
References   159

7.1 INTRODUCTION

Gastroenterology can be described as the study of diseases and motor (motility) abnormalities of the human gastrointestinal (GI) tract (gut) where motility refers to the contractions and relaxations of the longitudinal and circular smooth muscle layers of the gut wall. Motility of the GI tract plays a central role in normal digestive health, mixing contents with secretions, exposing them to the surface for absorption of nutrients, and propelling them in a controlled fashion. The human gut is 5 to 7 m in length, extending from the mouth through to the esophagus, stomach, small intestine, colon, rectum, and anus, and each of these segments has distinctive motility patterns adapted to their particular functions. The key feature of normal motility is that it involves *coordinated patterns* of muscular contraction and relaxation, distributed over considerable lengths of gut and all hidden within the body. A primary aim of the modern gastroenterologist therefore is to infer this activity in the least invasive manner possible in order to determine the normal physiology and importantly the pathophysiology of their patients.

The study of the human digestive tract has roots back to the ancient Egyptians who demonstrated an understanding of digestive processes, the use of laxatives, and rectal procedures on medical papyri as early as 2125 BC. The modern age of GI measurement in humans, however, is usually considered to have had its genesis in the mid-1880s with Kronecker and Meltzer who studied intraluminal motility events using air- and water-filled balloons inserted into the esophagus.

Just over 20 years later, following the discovery of x-rays by Wilhelm Rontgen, radiologists and gastroenterologists were able to directly study the passage of content through the digestive tract. The ability to follow the dynamics of the movement of content through the gut quickly followed with the advent of long-duration x-ray cinematography. These studies were performed before the dangers of exposure to x-rays were known and their prolonged nature allowed for detailed descriptions of transit through the esophagus, stomach, small intestine, and colon. While this type of investigation is no longer permissible due to the appropriate application of ethical constraints on excessive x-ray exposure, many lessons can be learned from these early data. Close study of this material shows that the gut undergoes many types of contraction along its length, some of which are purely propulsive, as seen in the esophagus, and other forms of which provide mixing and arresting functions that are assumed to ensure that, in health, the imbibed material traverses the gut at the correct speed and that the correct amount of nutrition is extracted from it at each location.

Over the next 100 years, no single technique has been able to describe the relationship that exists between wall motion and flow in all regions of the human gut with such clarity. As a result, much of the diagnostic apparatus currently available and under development seeks to provide analogous details of the function of the gut without resorting to the direct imagery available in the past. To achieve this, the modern gastroenterologist has been provided with a veritable arsenal of devices for testing, monitoring, and diagnosing a wide range of functional GI disorders. It is now possible to measure and record such physical parameters as pressure, temperature, pH, muscular tone or elasticity, the presence of luminal content, luminal shape, and makeup of the luminal wall. We can effectively view the entire length of the gut by means of endoscopes and ingested PillCams. We can also record details of transit times through different sections of the gut using radiopaque markers, SmartPills with onboard electronics, and remotely tracked magnetic capsules.

The wide range of parameters that can be observed, the functionality of the hardware designed to do so, and, importantly, the means of displaying these parameters in a clear and intuitive manner can be daunting. This chapter provides details of some of the techniques available for measurement and monitoring of GI motor function, and we attempt to clarify the physical basis for each measurement and provide the reader with sufficient information to interpret and understand the array of data that can be accessed using each technique.

The first part of the chapter is organized with respect to the basic types of measurement available to the clinician and covers the principles of measurement, the data obtained, and, where pertinent, a description of typical displays provided by commercially available instrumentation.

The second part of the chapter covers emerging techniques that we consider to be particularly exciting for future GI diagnostics.

7.2 GASTROINTESTINAL ENDOSCOPY

No review of GI measurement would be complete without covering endoscopy in its various forms. While not really a *measurement* tool, endoscopes and their multitudinous variants (e.g., gastroscopes, colonoscopes, sigmoidoscopes) provide vitally important information about the condition of the gut via direct, real-time imagery and hence often form part of primary diagnostics for GI disorders. They consist of either a flexible fiber-optic bundle that transfers images from the tip of the instrument to a camera located at the proximal end or a miniature video camera located at the tip of the instrument with the image transferred via electrical wiring running down the snakelike body of the device. The scopes also contain a means of steering the tip of the scope, via control knobs on the handle, to facilitate placement into the body lumen, a light source for the camera, ports for flushing and extracting fluid around the tip of the instrument, a pressure/vacuum port to allow the lumen to be insufflated (inflated) to facilitate forward movement of the scope and collapsed again on removal, and one or more instrument ports through which various devices such as biopsy forceps, snares, and endoclips can be passed to allow the operator to inspect, remove, or directly treat any features seen within the lumen. Figure 7.1 shows a labeled image of a typical endoscope tip.

Endoscopes provide the gastroenterologist with images of the accessible regions of the GI tract, including the esophagus, stomach, and proximal duodenum, the anorectum, and

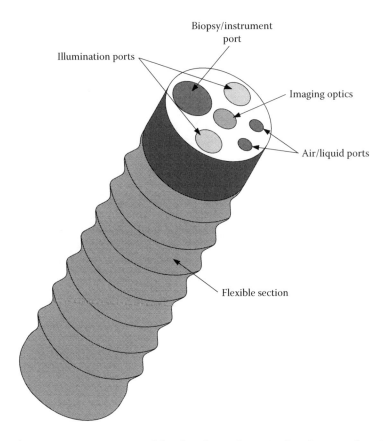

FIGURE 7.1   Schematic representation of the distal tip of a typical endoscope showing the imaging and illumination optics, the instrument port, the air port used to inflate the lumen to facilitate forward motion and to allow improved viewing of the luminal wall, and a liquid jet for removing intraluminal material from the field of view. The distal section of an endoscope is flexible and can be steered by means of controls on the scope handle to facilitate advancement and to correctly position any instruments used during the procedure.

all regions of the colon—their use in the distal small bowel is limited due to the long length of this organ. The direct visualization of the luminal surface provides a perfect primary means of diagnosis for disorders that have an identifiable morphology, such as Barrett's esophagus or ulcerative colitis, as well as the identification of polyps and regions of cancerous tissue. However, scoping is not effective as a direct diagnostic tool for functional GI disorders (which typically have no identifiable surface features) such as nonobstructive dysphagia or slow-transit constipation, and hence, more quantitative methods are required in these cases.

7.3 MANOMETRY: MEASUREMENT OF PRESSURE AND CONTRACTILE FORCE

The word *manometry* refers specifically to the measurement of pressure; however, in GI circles, it is now used synonymously for the measurement of pressure and/or contact force between the luminal wall and the manometry catheter generated by muscular contractions in the lumen of the GI tract. Manometric catheters consist of long flexible tubes containing one or more pressure sensors that can be placed within the GI tract to record either a pressure within a lumen filled with semisolid, liquid, or gas content during a nonoccluding contraction (Sinnott *et al* 2012) or direct contractile force during a lumen occluding event. It is important to note that these sensors will record a signal in both instances; however, the amplitude and dynamic behavior of the recording will depend on the nature of the luminal content, the degree of closure of distant sections of the lumen, and the rate and extent of contraction of the luminal wall (Arkwright *et al* 2013).

The trend in GI manometry is now increasingly toward *high-resolution manometry* (HRM) with catheters having a large number of sensors spaced at intervals of ≤10 mm; however, there are still many low-resolution sensors in use (with four or five sensors spaced at 5 cm intervals) servicing both installed infrastructure and the low-cost disposable market. HRM allows a pseudocontinuous recording of muscular activity along significant lengths of the GI tract. This HRM approach has led to a dramatic simplification in the interpretation of manometric signals in the esophagus and is now providing information on subtyping and preferred means of therapy for sufferers of dysphagia. HRM catheters are now commonly available with up to 36 sensing elements, and emerging fiber-optic technologies are increasing the sensor counts out to 120 sensors and beyond (Dinning *et al* 2013).

There are, broadly speaking, four types of manometry catheters currently available: air-charged, water-perfused, solid-state, and, a recent new entrant to the field, fiber-optic-based devices.

Air-charged devices are arguably the simplest possible design and consist of a number of air-filled balloons spaced along a length of flexible tube, each being inflated via a separate lumen within the catheter tube once it is placed in the body. The balloons are partially inflated, to avoid any confounding effects of the elasticity of the balloon itself, and then monitoring the pressure inside each balloon allows a direct measure of the luminal activity inside the body. Figure 7.2 shows a 5 cm spaced four-channel air-charged device supplied by Sandhill Scientific Instruments. The air-charged design records the averaged force applied from all angles around the sensing balloon and is hence referred to as *circumferential* sensors. These devices provide a low-cost, potentially disposable option in situations where low sensor counts are all that is required.

Water-perfused devices consist of a length of extruded silicone or polyvinyl chloride (PVC) containing multiple lumens, each of which is vented to the outer surface of the catheter at the point where sensing occurs, often referred to as a *side hole*. Water is pumped or *perfused* into each lumen at a slow rate so that the liquid slowly seeps out of the side hole. The pressure in the region of the side hole can be inferred from the flow rate of water being fed into the proximal end of each lumen in much the same way that the flow rate through a hose pipe is affected by placing your thumb over the end of the pipe. These devices require the patient to maintain their position relative to the recording equipment to avoid background drift in the signal due to changes in water head pressure (the difference in height between the flow meter and the output vent of the catheter). As water is perfused into the lumen, these tests can present problems when recording in patients

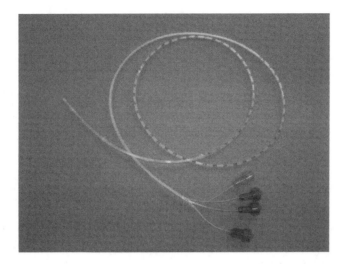

FIGURE 7.2   Four-channel air-charged manometry catheter. (Courtesy of Sandhill Scientific, Highlands Ranch, CO.)

with severe swallowing or aspiration disorders or for investigations that run for multiple days such as with slow-transit constipation. The assembly of water-perfused devices, and the hardware needed to monitor each channel, gets increasingly complex as more sensing sites are added, and so they are currently limited to approximately 32 sensors per catheter. Due to the nature of the water-perfused catheter design, the signal is usually directional in nature (i.e., the recording site is one side of the catheter), and the direction of peak sensitivity varies along the length of the catheter. This needs to be considered when recording in an asymmetric region of the GI tract that may shift longitudinally during measurement such as the esophageal sphincters. A variant of the water-perfused design, known as a *sleeve sensor* or *Dent sleeve* after its originator Professor John Dent, overcomes the directionality and also the wide sensor spacings typical of some water-perfused catheters.

Sleeves work by directing the flow of water from a vent underneath an outer tube. Compression anywhere along or around the sleeve further restricts the flow and is therefore indicative of the mean peak squeeze felt at any point along or around the length of the tube. These devices are commonly used to detect peak pressures within sphincters that can be both radially asymmetric and rapidly varying in the axial direction. Figure 7.3 shows details of the sensing region of a water-perfused device and indicates the varying directional nature of the sensing ports. These devices tend to be cheap, at least for small sensor counts, and the catheters are very durable and can be reused many times as long as cleaning procedures, including a thorough flushing of each internal lumen, are scrupulously maintained.

Solid-state sensors use one of a number of different types of electrically operated pressure sensors (e.g., microelectromechanical systems [MEMS], resistive strain gage bridges, or piezoresistive devices) to record pressure at one or more locations along the catheter. Solid-state devices respond to forces either from a specific direction (directional)

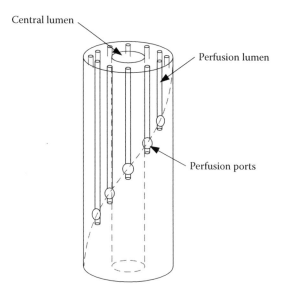

Central lumen

Perfusion lumen

Perfusion ports

FIGURE 7.3   A schematic representation of a water-perfused HRM catheter showing the arrangement of the perfusion lumen and the sensor ports. The arrangement of the sensing ports around the outer edge of the catheter results in the direction of maximum sensitivity varying for each successive sensor.

or uniformly in all directions (circumferential). Each mode of operation has its supporters, for example, in the esophageal sphincters where the measured profile varies in different directions around the catheter; some prefer an average measured value, while others prefer to know the value in a known orientation such as anterior–posterior. There appears to be no consensus opinion regarding the diagnostic capability of these two options at present.

Solid-state sensors provide excellent resolution and can be calibrated to work at body temperature but can be prone to baseline drift and sensor dropout if poorly handled. Since solid-state pressure sensors need wired electrical connections, these devices are typically limited to around 36 sensors because the increasing sensor count, and hence the increasing wire count, tends to reduce the flexibility of the catheter. These devices are available from a number of suppliers and have revolutionized the measurement of peristalsis in the esophagus due to their ease of use and ability to reliably record pressures at closely spaced intervals along their length. Solid-state devices have a significant advantage over water-perfused devices in that only one connector is necessary to connect the catheter to the data acquisition hardware. Figure 7.4 shows a solid-state HRM catheter as supplied by Unisensor AG of Switzerland.

Fiber-optic sensors use either Fabry–Perot sensors (for single location) or fiber Bragg grating (fbg) sensors (for multiple locations) to record pressure based on observed variations in the optical spectrum. The Fabry–Perot sensors have a pressure-sensitive microcavity located at the tip of the fiber and can be extremely small in diameter (<0.5 mm). They are finding primary applications in urology, cardiology, and intracranial pressure monitoring and will not be considered further in this chapter. The fbg

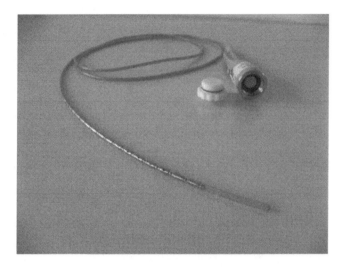

FIGURE 7.4   A 10 French solid-state HRM catheter with circumferential pressure sensors. (Courtesy of Unisensor AG, Wiesendangen, Switzerland.)

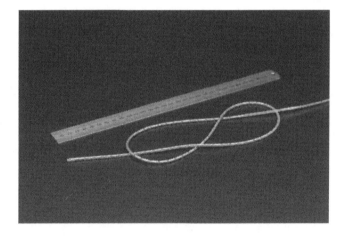

FIGURE 7.5   A 72-element fiber-optic HRM catheter indicating the flexibility of the sensing region.

devices can be formed into extremely flexible catheters with relatively small outer diameters (≤3 mm) having smooth and continuous outer sleeves. High–sensor count fiber-optic catheters (Arkwright *et al* 2012) are just emerging onto the market and can provide extremely long HRM sensing regions without impacting on the small diameter or the overall flexibility (Figure 7.5). These devices are typically unidirectional, with the direction of sensitivity pointing in opposite directions for successive sensors along the catheter. Their primary application is for measurement in long or hard to reach regions of the GI tract such as the colon and small intestine. Early results from the human colon are providing the same dramatic improvements in signal interpretation for the lower gut that solid-state sensors have provided for the esophagus. Fiber-optic sensors have the advantage of being inherently nonmagnetic and so can be used in high–magnetic field environments such as MRI suites.

Due to fluctuations in ambient temperature and pressure, all manometry sensors must be regularly calibrated, and this should form part of the setup prior to each and every recording session.

7.4 MANOMETRY DISPLAYS

At this point, it is worth taking time to describe the data that result from a manometric study. Figure 7.6 shows some typical output data from a manometric recording of a swallow in the esophagus of a healthy volunteer. Although the data all came from a fiber-optic HRM device, in Figure 7.6a, only four channels have been shown spaced at 5 cm intervals to simulate the output of a typical air-charged device. The relaxation of the upper esophageal sphincter and the peristaltic wave traveling down the esophagus are clear, and in experienced hands with *pull-through* techniques that reposition the catheter during the study, these four-channel devices provide sufficient information of the sections of the esophagus for clinical diagnostics. Figure 7.6b shows the data from every sensor element along the catheter and indicates the increase in information and overall span that can be achieved by switching to high–sensor count HRM recording. Figure 7.6c shows the same HRM data as an intensity plot (also referred to as a spatiotemporal map) that is the currently preferred manner for displaying such data. These maps are generated by interpolating the data from each sensor along the catheter for every time increment and then displaying the variations in pressure as changes in color or gray scale. The x-scale is again time, the y-axis indicates position along the catheter or organ, and the z-axis represents variations in pressure and/or contact force. The combination of HRM and spatiotemporal mapping provides a very convenient means of interpreting manometric data and also of accurately positioning catheters with respect to distinctive morphological features such as the esophageal sphincters and the pylorus (see later).

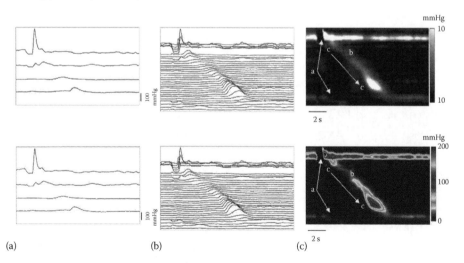

(a)   (b)   (c)

FIGURE 7.6 Manometric data from a 10 mL water swallow in a human volunteer. (a) Simulated data from a manometry catheter with four sensors at 5 cm intervals, the first one being in the upper esophageal sphincter, (b) full HRM data set, and (c) a spatiotemporal map of the same data with pressure indicated by changes in intensity with key features marked (see text for further details).

7.5 MANOMETRY WITHIN THE GASTROINTESTINAL TRACT

7.5.1 Esophageal Manometry

Esophageal manometry is the most common form of upper GI manometry currently performed. It usually involves intubation of the catheter via the nose, although oral intubation is also possible. Using the spatiotemporal representation, the catheter can be readily located in the esophagus by identifying the upper and lower sphincters characterized by the bands of increased pressure. This provides a simple and fast technique for correct positioning of the device during intubation, hence speeding up the overall procedure for esophageal function testing. Once the catheter is positioned in the region to be studied, data acquisition is started, and the patient is given a series of controlled boluses to swallow, typically, 5 or 10 mL of water injected into the mouth. The patient is then asked to relax, holding the water in their mouth for a few seconds, and then to swallow. The exact time of the swallow is recorded to allow the data to be correctly interpreted after the study. A full study will consist of a number of such swallows (typically 10) and then repeats of the procedure using either different volumes of water or different types of bolus such as yoghurt or bread depending on the patient's specific disorder.

The spatiotemporal maps that are thus generated allow features of the swallow such as the relaxation of the upper and lower sphincters (marked "a" in Figure 7.6c), the characteristic dropout as the wave passes the transition zone between striated and smooth muscle (marked "b"), and the coordinated contractions running toward the stomach (marked "c–c") to be readily identified. The corollary of this is that any breakdown in the correct function of the esophagus can be similarly recognized.

7.5.2 Antroduodenal Manometry

Antroduodenal manometry is typically used in cases of suspected dyspepsia, gastroparesis, or chronic intestinal pseudo-obstruction. A manometry catheter is passed through the nose and into the stomach and then guided through the pylorus into the duodenum. Pyloric placement can be achieved in a number of different ways including the endoscopic placement of a guidewire into the duodenum; the catheter is then positioned by passing it over the top of the guidewire until it is correctly located in the duodenum. Obviously, the catheter needs an open central lumen of sufficient diameter to accommodate the guidewire for this approach to work. Alternatively, a catheter with a small balloon at the distal tip can be used. The catheter is again passed via the nose into the stomach, the patient is then positioned in the left supine position, and the balloon is inflated with water to approximately 1 cm in diameter. This has two purposes: first, the additional weight of the water-filled balloon causes the catheter to drop toward the antrum and pylorus, and second, as the balloon engages with the antral contractions, it draws the catheter through the pylorus and into the duodenum. Natural peristalsis then continues to draw the catheter further into the duodenum until the correct location is achieved. Confirmation of duodenal placement is achieved through fluoroscopic imaging.

Antroduodenal measurements can span between 5 and 24 h and should always include a number of hours of recording in the fasted state as this shows the most characteristic small bowel activity.

To date, the majority of antroduodenal manometry has been performed with widely spaced sensors simply because the technology has not been available to allow high-resolution data to be recorded. More recently, there have been some reports of antroduodenal HRM studies that again show significant complexity of activity that is hard to pick up with traditional equipment, and it is likely that HRM will again become the preferred option as suitable technologies emerge.

7.5.3 Sphincter of Oddi Manometry

Sphincter of Oddi manometry (SOM) is a specialist branch of GI motility dealing with dysfunction of the sphincter that controls the release of bile into the GI tract. The sphincter is located in the proximal section of the duodenum and is difficult to see and access as it is located in the wall of the lumen. A small diameter manometry catheter, typically five French water perfused with three sensor ports 2 mm apart, is introduced into the sphincter using a side-view endoscope. The side-view endoscope has the option to position its optics facing at right angles to the axis of the scope allowing the wall of the duodenum to be viewed. This allows the catheter to be inserted into the sphincter by means of the endoscope elevator. SOM involves locating the sphincter endoscopically and introducing a thin guidewire through the sphincter into the bile ducts. The catheter is then fed over the guidewire and the basal pressure recorded in the duodenum prior to being passed through the sphincter and into the duct. The catheter is then slowly withdrawn in 1 to 2 mm steps with a pause of 15 to 20 s at each step until the sensors are fully in the duodenum once again. The outcome of the procedure is given by the pressure difference between the basal duodenal value and the average plateau pressures recorded on each channel during the pull-through. A typical abnormal value is considered to be >35 to 40 mmHg above the basal level measured in the duodenum.

SOM is a difficult procedure that needs specialist equipment and high levels of experience to perform well. There is considered to be a finite risk of pancreatitis following the procedure, which may be exacerbated by the perfusion of liquid into the ducts, and so the procedure should only be performed in instances where successful therapeutic sphincterotomy is indicated.

7.5.4 Colonic Manometry

Colonic manometry is a standardized diagnostic test in pediatrics in the United States. Indeed, colonic manometry is one of those rare instances where a technology has established its value in pediatrics before it has in adults, where the technique is still largely considered a research only tool. It was listed among the tests to be performed in children unresponsive to medical and behavioral management in a position paper of the North American Society for Pediatric Gastroenterology, Hepatology, and Nutrition. Most pediatric centers use water-perfused catheters that are placed either radiologically or with aid of a colonoscope. When using water-perfused manometry, care must be given to the rate of infusion, because prolonged studies may result in a potentially dangerous amount of water being infused in the colon of young infants and toddlers. In adults, catheters are commonly placed under fluoroscopic guidance with the tip of the catheter clipped to

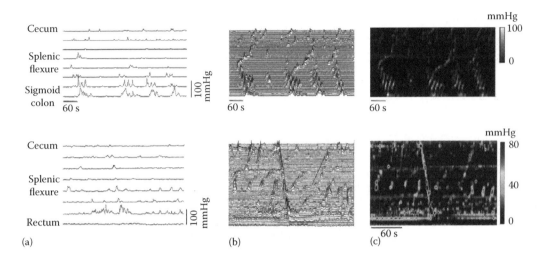

FIGURE 7.7   A colonic high-resolution fiber-optic manometry pressure trace with recording sights spanning from the cecum to the rectum. (a) Display of data from every 10th sensor to match the standard low-resolution colonic traces. (b) Display of the entire data set. (c) A spatiotemporal intensity plot of the data shown in (b). The arrows in (b) represent episodes of retrograde activity that only become clearly apparent in the high-resolution trace. This activity is largely lost with the low-resolution recording at (a).

the mucosa, preferable in the cecum, although due to technical constraints limiting the length of the catheters, the majority of the studies can only place catheters in the transverse colon. In both pediatric and adult colonic manometry, the recordings continue long enough to pick up a variety of colonic responses suitable for subsequent analyses. In children, recordings tend to be limited to <6 h, and in adults, recording can continue for up to 24 h. The physiological responses looked for include a colonic response to a meal, chemical stimulation, and evidence of high-amplitude propagating pressure waves, all of which should be present in healthy subjects.

The typical sensor spacing on these catheters is 7–15 cm; however, it is now clear from recent fiber-optic HRM studies that these low-resolution catheters can potentially miss a portion of the propagating activity that exists in the colon. With the advent of fiber-optic HRM catheters, it is now possible to record colonic motility with the same precision that is common in the esophagus (Dinning *et al* 2012). Figure 7.7 shows the output from a colonic study of a patient with previously diagnosed slow-transit constipation. As with the esophageal data shown, Figures 7.6 and 7.7a show data from widely spaced sensors, in this instance 10 cm spaced to indicate the output from a traditional colonic manometry catheter, Figure 7.7b shows the full set of HRM data, and Figure 7.10c shows the same data on a spatiotemporal plot. In Figure 7.7b and c, the full extent of colonic activity is shown with both antegrade and retrograde propagation that are clearly evident.

7.5.5 Anorectal Manometry

Traditional anorectal manometry is used to test both internal and external anal sphincter (EAS) function by recording resting anal sphincter pressure, the response to a voluntary squeeze, and the effect of rectal distension. Devices consist of a number of sensors spanning the anal sphincters and one additional sensor positioned in the rectum. A balloon is located at the tip of the device and often includes a further sensor to record the balloon inflation pressure. In operation, the patient is placed in the left lateral position, and the probe is inserted into the rectum. The resting anal pressure is first measured, then the increase in sphincter pressure in response to the patient voluntarily tightening their EAS and the involuntary contraction during a cough (this can be of use with patients who have difficulty carrying out voluntary contraction of the anal sphincter), and finally the response of the anal sphincter and rectal wall during local distension of the rectum by inflating the balloon at the tip of the device.

Normal values for sphincter resting pressures are from 60 to 100 cm H_2O (~45 to 75 mmHg), and the increase in pressure during a voluntary squeeze or a cough is considered to be >60 cm H_2O. It must be noted though that these values can vary considerably depending on the instrumentation used and the precise orientation of patient and measurement setup. The expected healthy response to balloon inflation should be a simultaneous increase in rectal pressure and a drop in sphincter pressure in the upper anal canal immediately followed by recovery of pressure. Figure 7.8 indicates idealized responses to squeeze pressure and balloon inflation.

The balloon can also be used to detect rectal sensory capacitance by asking the patient to indicate the balloon volumes that produce a first sensation (15 to 30 mL), a first urge (50 to 100 mL), and the maximum tolerable volume (>100 mL). This is analogous to the barostat studies detailed in Section 7.9.4.

As with other forms of clinical manometry, high-resolution devices are now beginning to emerge that give far greater spatial resolution of the anorectal region. Although not yet fully integrated into clinical practice, these devices are likely to show a significant improvement in the evaluation and assessment of sphincteric morphology and function and the anorectal coordination in defecatory disorders (fecal incontinence and obstructed defecation).

Anorectal manometry is often followed by a balloon expulsion test in which a balloon is inserted into the rectum and filled with water. The patient is then asked to visit the

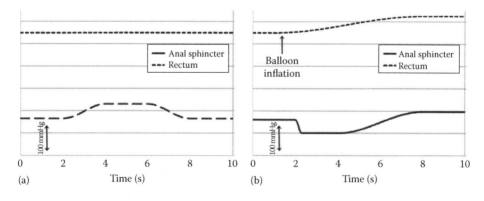

FIGURE 7.8 An idealized representation of (a) the expected response to anorectal squeeze pressure and (b) the EAS response to balloon inflation in the rectum.

bathroom and attempt to bear down to expel the balloon. An excessive time taken for expulsion suggests possible anorectal dysfunction.

7.5.6 Catheter Dimensions and Recording Characteristics

The ability of small diameter manometry catheters to record variations in pressure in a large open lumen such as the colon is sometimes questioned, specifically the belief that as the luminal diameter increases, manometry catheters are not able to pick up contractions that do not fully occlude the luminal cavity or provide direct contact with the catheter itself. This is not the case and can be simply discounted by the fact that calibration of manometry catheters is usually carried out by increasing the pressure applied to a rigid vessel containing either air or water. Because these misconceptions prevail however, it is worth spending a bit of time describing the situation with respect to the colon given the variable nature of its content. Manometry will successfully record luminal activity in all situations with the following caveats applied. When the colon is empty of content, it is closed, and the luminal walls will be in direct contact with the catheter, so it is clear that the effect of muscular contractions will be recorded. In this instance, the manometric signals will be a direct response to the muscular contractions within the luminal wall and should more correctly be referred to as a measurement of contractile force. When the colon is locally filled with gas or liquid, then contractions of the wall will be rapidly and uniformly transferred to the catheter, and so again, the contractions will be recorded, but in this instance, all sensors within the localized filled region (e.g., within a single haustral pouch) will record the same pressure. This is often referred to as a *common cavity* event. In the third, and most common, situation, where the colon is filled with semisolid viscous material, the contractions will again be transferred to the catheter, but in this instance, the pressures will vary from location to location as the material is locally squeezed and moved from one region to another as a result of either peristalsis or haustration. The response measured by the catheter in this instance will depend on the viscosity of the luminal content, the speed of contractions, and the geometry of the lumen in the vicinity of the measurement, specifically, if haustra are formed that partially occlude the lumen at a distance from the measuring site. This can be likened to a tube of toothpaste in which a squeeze on the outer surface results in the toothpaste being ejected through the small opening in the end of the tube. This would not happen if there was no transfer of force from the outer surface of the tube to the content inside the tube.

7.6 IMPEDANCE MONITORING OF BOLUS PRESENCE

Impedance monitoring is a relatively new technology for determining the presence of luminal content in the esophagus. It works by detecting the characteristic electrical impedance between pairs of electrodes on the outer surface of a catheter due to the contact of the luminal wall and/or the luminal content with the catheter. The measured signal is derived by applying an alternating current between each successive pair of electrodes on the catheter and recording the impedance between them. As the passage of a bolus pushes the luminal wall away from the impedance electrodes, the measured impedances drop because the bolus tends to have a higher electrical conductivity compared to the mucosal wall. To date, impedance measurements have been generally restricted to esophageal use because of the ability to control the imbibed bolus.

7.6.1 Impedance Manometry

When combined with manometry, the impedance monitoring technique provides a very powerful means of detecting the muscular activity in the esophagus and also the effectiveness of the resulting contractions. The clear advantages of this dual approach are that it can provide an indication of bolus movement without the need for x-ray irradiation and it can also detect nonacidic reflux events that cannot be detected from pH recordings alone.

Reading impedance data can be challenging, but once again, spatiotemporal mapping techniques have simplified interpretation. Figure 7.9 shows the representation initially developed by Dr. Michal Szczesniak et al. (2009) to intuitively combine manometric and impedance data. The plot shows the isobaric contours (lines of equal pressure) of a typical swallowing event in the lower esophagus with the impedance data overlayed in gray scale to indicate bolus presence. The combined plot shows the bolus rapidly filling the lower esophagus, as indicated by the grayscale overlay in the lower left-hand quadrant of the image and then being pushed into the stomach by the descending peristaltic wave. After the swallow, the lower sphincter closes again completely sealing off the esophagus against potential reflux events. The effective clearance of the bolus from the esophagus after the swallow is indicated by the absence of the grayscale shading in the upper right-hand quadrant of

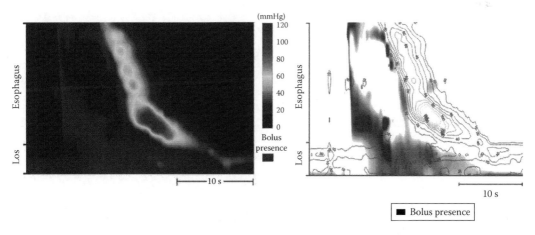

FIGURE 7.9 Composite spatiotemporal plot of pressure and impedance recorded during a single swallow (10 mL saline) through the esophagus and the lower esophageal sphincter (LES) in a healthy human. The impedance-defined bolus presence is shown by the grayscale shading. The diagonal-contoured region represents the peristaltic contraction moving down the esophagus moving the swallowed bolus into the stomach. The grayscale shading has largely disappeared from the top right-hand side of the figure indicating that the bolus is being cleared from the esophagus by the peristaltic pressure wave.

the image. This means of display readily indicates a correct swallow, and deviations from this can indicate disorders such as incomplete esophageal clearance and acid or nonacid reflux (Bredenoord *et al* 2007).

7.7 PH MONITORING OF ACIDITY LEVELS

pH monitoring can be achieved in a number of ways: using catheter-based sensors, a sensor temporarily clipped to the mucosa (the Bravo™ capsule), or by means of an ingestible capsule. The first two methods provide a means of detecting acid reflux in the esophagus. The ingestible capsule uses pH as a means of identifying the location of the capsule in the body and will be dealt with in Section 7.8. Once placed, the Bravo provides a simple means of gathering ambulatory data without the need for the presence of a nasal catheter. The capsule is clipped to the mucosa of the esophagus and stays in place for the duration of the study. It eventually sloughs off with the epithelial mucosal layers and is passed naturally from the body. The data from a catheter-based pH probe are similar but require the user to wear a nasal catheter linked to an external data logger for the duration of the measurement. Both approaches are similarly tolerated by patients, and diagnostic yields appear to be equivalent.

The primary need for pH monitoring is to detect the presence of acid reflux, and this measurement is often associated with manometry. The data from a pH sensor show the change in acidity of the local environment, for example, a reflux event in the lower esophagus due to stomach acid passing the gastroesophageal junction would be typified by a drop in pH to below a value of 4.

Since the identification of nonacidic reflux, it may transpire that pH sensing alone is insufficient, and a move to combined pH–impedance–manometry may gain popularity.

7.8 INGESTED PILL TECHNIQUES

There are a number of different types of ingestible data logging capsules now on the market. The common features are that the capsule is ingested by the patient in a clinical setting and then allowed to resume normal activities as the capsule passes through the different regions of the body recording and transmitting data to a data recorder external to the body. All capsule types provide a moving snapshot of the measured parameters as the capsule traverses the different regions of the body rather than an overall picture of motile activity or of propagating events within the gut. The two common ingestible capsule types are described in more details in the sections later, and a third emerging technology is described in Section 7.11 (McCaffrey *et al* 2008).

7.8.1 SmartPill

The SmartPill provides radiotelemetry of a number of in vivo parameters as it traverses the gut, specifically temperature, pressure, and pH. These devices are also used to determine segmental and total transit time through the gut. The onboard batteries power the device for a number of days and acquire data at various rates throughout this period. Table 7.1 shows the data acquisition rates for the first 24 h.

The changes in temperature as the capsule is swallowed and then again as it exits the body provide a simple means of assessing total transit times, and the characteristic

TABLE 7.1 Data Acquisition Rates for the SmartPill

Parameter	0–24 h	24 h Onwards
Temperature	Once every 20 s	Once every 40 s
Pressure	Twice per second	Once per second
pH	Once every 5 s	Once every 10 s (24 to 48 h), once every 5 min thereafter

changes in pH seen when the capsule enters and leaves the stomach and again as it passes from the small intestine to the colon provide details on the segmental transit times through the key regions of the gut. The transit times provide a ready means of identifying stomach-emptying disorders such as gastroparesis and for subtyping slow and normal transit constipation. The SmartPill has been validated for measurement of the colonic transit and in the clinical discrimination of normal from slow colonic transit (Camilleri *et al* 2010).

The SmartPill is approved for use in the United States and in Europe.

7.8.2 PillCam

PillCam offers a unique view of the upper GI tract and the small bowel by means of a microcamera located in an ingestible capsule. There are currently PillCam variants for the esophagus (PillCam ESO), the small bowel (PillCam SB), and the colon (PillCam COLON), each with different imaging rates suitable for their intended purpose. The PillCam ESO and PillCam SB have a single camera mounted at the front of the device and capture images at 14 and 2 frames/s, respectively (Figure 7.10). The PillCam COLON has a camera at each end, optimized for image capture in the larger colonic lumen, each at 2 frames/s. The image data are transmitted wirelessly to a data recorder worn on the patient's belt. At the end of the trial, the data recorder is returned to the supervising clinician for the analysis of the images.

At present, the PillCam SB represents the only viable way of imaging deep into the small bowel as this region of the body is effectively inaccessible using normal endoscopic

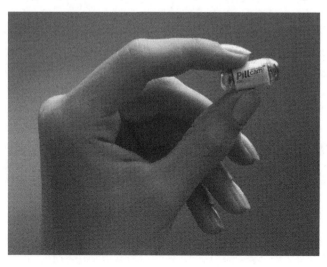

FIGURE 7.10 Photo of a given imaging PillCam. (Courtesy of Given Imaging, Duluth, GA.)

procedures. The device produces vast numbers of images and can be very effective for identifying disorders such as Crohn's disease, small bowel tumors, malabsorption disorders (such as celiac disease), and suspected small bowel bleeding. The recently introduced PillCam COLON has a camera at both ends of the capsule that is intended for use specifically within the larger luminal diameters of the colon. At present, the battery life of these devices is limited to approximately 8 h, so imaging of the whole bowel is not always achieved; however, improvements in battery power will doubtlessly extend this lifetime in the near future giving access to the most distal parts of the GI tract.

At the time of writing, the PillCam ESO and PillCam SB have been approved for use in Europe and the United States, and the PillCam COLON has been approved in Europe and has been submitted for 510(k) approval in the United States (PillCam Capsule Endoscopy 2014).

7.9 X-RAY FLUOROSCOPY AND SCINTIGRAPHY IMAGING

Three standard techniques exist for the routine assessment of transit through the digestive tract, all involve irradiation of the subjects: barium isotope, radiopaque markers, and radionuclide scintigraphy (Kolodny 1991).

7.9.1 Barium Meals and Enemas

Barium meal studies require the patient to eat a standard meal including a barium-based isotope that provides a radiopaque image on x-rays. This allows video imaging (often referred to as x-ray videofluoroscopy) of a specific event such as a swallow and/or snapshots of longer-duration events such as gastric emptying. In the esophagus, videofluoroscopy is often combined with esophageal manometry to provide a detailed view of both muscular activity and bolus transport (Figure 7.11). Because of the highly controlled nature of esophageal studies, the ability to initiate a swallow at a specific time, and the relatively short duration of a swallowing event (typically <10 s), x-ray exposure can be kept to a minimum. The analysis of videofluoroscopy is often qualitative in nature and requires expert interpretation to determine bolus clearance.

Barium enemas are used for video defecography or the study of patients with expected outlet disorders (the inability to expel stool from the rectum). In these studies, a paste mixed with a barium isotope and having the approximate consistency of feces is injected into the rectum, and an x-ray video is taken while the patient is asked to expel the paste. Again, the controlled nature of the event allows significant details to be gathered in a short time. Video defecography is analyzed by visual inspection of the x-ray images and is useful for identifying structural anomalies that impede stool expulsion such as intussusception (part of the intestine has invaginated into another section of intestine), rectal prolapse, or a rectocele (the protrusion of the rectum into the vagina). The barium enema can help to determine if surgical intervention is required to fix structural defects in the rectum.

7.9.2 Radiopaque Markers

These tests involve the ingestion of radiopaque (e.g., barium-impregnated PVC) markers and assessing their movement through the GI tract by use of plain abdominal x-rays. There is no single predominant technique with at least 15 different methods currently

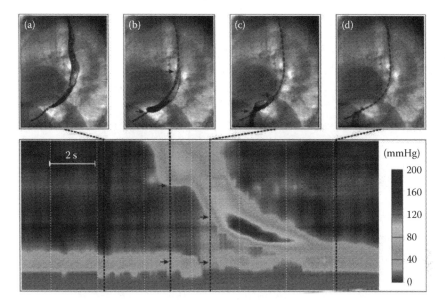

FIGURE 7.11 A video-esophageal manometry display of a subject swallowing 10 mL bolus containing a barium contrast agent. (a–d) represent a series of fluoroscopic images of the bolus moving down the esophagus. The individual sensors of the manometry catheter can be seen in each of these images as a series of closely spaced black rectangles. The bottom image is a spatiotemporal plot of the pressure in the final regions of the esophagus and LES. The hatched black lines indicate the equivalent times in images (a–d) and the spatiotemporal pressure plot. (a) The barium bolus fills the esophagus; (b) propagating peristaltic wave travels down the body of the esophagus, and the LES remains contracted; (c) LES relaxes and barium is emptied into the stomach; and (d) radiograph shows complete clearance of barium.

published (Dinning *et al* 2009). The capsules containing a number of radiopaque markers (Sitz markers) are taken orally and have an outer coating that dissolves either slowly or rapidly in response to a given in vivo chemistry such as the marked increase in pH found at the extreme distal regions of the ileum. This ensures that the package of markers is delivered to the region of the body of interest, typically the colon, in close proximity to each other. Recording static x-ray images of the abdomen at given times after ingestion (e.g., 1, 5, 7 days) can provide details of transit times and the spread of digesta along the colon. A variant on the aforementioned technique that reduces overall x-ray exposure involves the patient swallowing a number of these pills on consecutive days, each containing different shaped markers that can be individually identified on an x-ray film. In this way, a single film can provide greater temporal details without the need for additional hospital visits. Figure 7.12 shows an abdominal x-ray image 100 h after ingestion of a Sitz marker capsule. In this instance, the majority of the markers are still located in the descending and sigmoid colon, and the patient was identified with slow-transit constipation.

7.9.3 Scintigraphy

Scintigraphy refers to the process of monitoring the progress of a radioisotope through the GI tract by means of a *scintillation* or *gamma* camera (*scintillation* refers to the brief flash

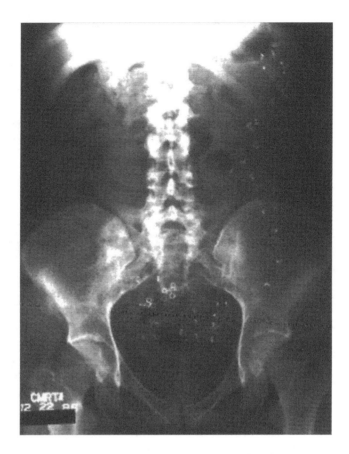

FIGURE 7.12 Abdominal x-ray of Sitz markers taken 100 h after the ingestion. The Sitz markers can be seen as shown as small white shapes, and they are located in the descending and sigmoid colon. The upper limit of normal retention at 100 h is <20% (approximately 10 markers); hence, this patient demonstrates slow colonic transit.

of light that occurs when a γ-ray photon hits the outer crystalline layer of the camera's detector). These cameras pick up the gamma radiation emitted by a very weakly radioactive isotope that is administered orally to the patient. Images of the patients' abdomen are then recorded periodically (at 4, 24, 48, 72, 96 h) to show the movement of the radioisotope through the GI tract. There are two techniques used. First, as with the radiopaque marker capsules, the scintigraphy capsules can be coated with a pH-sensitive layer that breaks down in the mildly alkaline environment of the distal ileum. Again, this ensures a close-packed bolus of material is released into the colon. Intraluminal movement is usually expressed by calculating the geometric center of the isotope mass from the gamma camera images for each of a number of discretely defined regions of the colon. The second technique is an oral administration of a liquid form of 111 indium, bound to diethylenetriaminepentaacetic acid (DTPA). In addition to colonic transit, the latter technique can also be used to measure gastric and small bowel transit.

Scintigraphy is considered by many to be the *gold standard* for assessing colonic transit times, which is of critical importance for subtyping the three most common forms of constipation: slow-transit, normal-transit, and outlet obstruction constipation. It should be noted however that there is a considerable overlap in the definitions of each subtype and the field is in real need of finding a better way of identifying subtypes and, of course, finding viable treatments for each.

7.9.4 Barostat Studies

A barostat is an instrument that can provide a range of constant pressures that, when connected to a balloon placed intraluminally, will inflate the balloon against the walls of the organ being studied. Figure 7.13 shows the functional components of a barostat coupled to a balloon catheter for controlled rectal distension.

The primary use of a barostat is the investigation of sensory anorectal disorders (Whitehead and Delvaux 1997). In this instance, the balloon is slowly inflated, usually in a stepwise fashion, until the patient is first aware of the intraluminal distension, and then on further inflation, when either the urge to defecate is felt or a pain threshold is reached. Depending upon the patient's response to these distensions, hypersensitivity or hyposensitivity can be defined, both of which may play a role in functional GI disorders. These tests can provide information on the local mechanical sensors in the gut; however, care needs to be taken with these tests as the response to pain generated by the balloon distension is always subjective and may vary widely between subjects.

Barostats are also used to measure tone and elasticity of the luminal wall. In this instance, the means of operation is to place the barostat balloon into the section of gut under investigation and then slowly raise the barostat pressure in discrete steps to its working value. The amount of air injected into the balloon provides a direct measure of the volume of the balloon and hence, from a knowledge of the dimensions of the balloon, the degree of luminal distension caused. A greater volume of air for a given pressure indicates that the section of lumen expands more readily in response to the distending pressure within

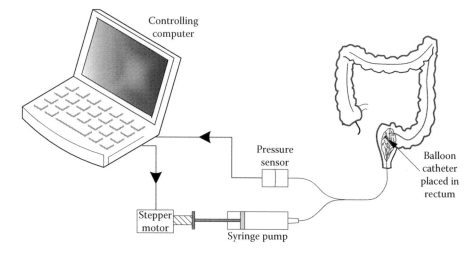

FIGURE 7.13 Schematic of a barostat and balloon catheter system. A feedback loop is used to operate the syringe pump in order to maintain a preset pressure within the barostat and the balloon based on input signals from the pressure sensor.

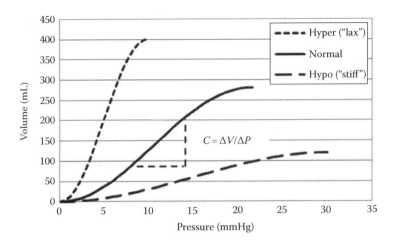

FIGURE 7.14   Plot of volume against pressure recorded during a barostat compliance measurement. The slope of the steep linear section of the graph is used as a measure of rectal compliance (C). Hypercompliance results in a larger rectal volume for a given applied pressure, whereas hypocompliance results in lower volumes.

the balloon; hence, the luminal wall has a lower elasticity or *tone*. It can also be described in terms of the compliance of the organ (e.g., the ability of the luminal wall to stretch or *comply* with the local balloon distension) in which case, the greater the volume achieved for a given pressure, the higher the compliance of the organ. The output of a barostat test in this instance is a record of the change of volume as a function of the change in pressure, that is, the slope of a graph of volume against pressure as shown in Figure 7.14.

It is important to note that the use of a barostat and inflating it to its *operating pressure or volume* must not be confused with a measurement of the effective diameter of the organ. The volume measured by a barostat (and hence the diameter inferred by the dimensions of the barostat balloon) is only caused by the distension of the balloon itself. Without this distending influence, or the presence of intraluminal content such as gas, liquid, or semisolid material, organs within the GI tract will remain closed. This is in direct contrast to the bronchial lumens that are kept open by cartilaginous rings. The operating pressure or volume of a barostat balloon is therefore determined by the maximum volume of *infinite compliance* of the bag. This is the volume to which the balloon can be inflated before it offers any appreciable resistance to the inflating forces and is usually achieved by using a floppy bag of a plastic such as polyethylene inflated to less than its maximum capacity.

7.10 ELECTROGASTROGRAPHY

Electrogastrography (EGG) is a technique for recording the myoelectric activity emanating from the stomach (Parkman *et al* 2003). It is well known that, in health, myoelectric slow waves are always present and propagate from a pacemaker area in the corpus, or major curvature, of the stomach and propagate toward the antrum at a frequency of approximately 3 cycles/min. These slow waves are made up of fronts of electrical depolarization and repolarization of cellular potential, and at any one time, there are typically two to three propagation fronts present between the major curvature and the antrum. These slow waves are not of themselves responsible for the generation of muscular contractions within the stomach wall, but they are responsible for controlling the frequency and direction of contractions. Contractions are initiated when spike potentials are superimposed on the slow waves, typically in response to ingestion of food and/or distension of the stomach by the presence of imbibed material.

A major advantage of EGG is that it is noninvasive and does not require any significant preparation of the patient, other than an overnight fast and possible cessation of drugs known to affect gastric motility or emptying. Recording of EGG signals is achieved by placing adhesive electrodes on the abdomen of the patient. The location for each electrode is not rigorously controlled; however, a common configuration uses two active and one reference electrode. One active electrode is placed on the midline halfway between the xiphoid and umbilicus. A second active electrode is placed 5 cm to the left of this electrode, at least 2 cm below the rib cage. The reference electrode is located distant from the active electrodes such as the left midaxillary line or right midclavicular line. Figure 7.15 shows one common placement of EGG electrodes.

Although EGG can provide useful data on gastric emptying disorders, there are some important issues to be aware of when analyzing the data. The most pertinent aspect of the output data is the frequency of the gastric slow waves, and information on the frequencies present in the data is usually determined by carrying out a fast Fourier transform (FFT) of the raw data. An FFT is a mathematical method of extracting periodic frequencies from recorded data in much the same way that the human ear can pick out and identify many musical notes from the background noise in a crowded music venue. Figure 7.16 shows typical raw data recorded during an EGG session and also the FFT frequency analysis.

There is evidence that cutaneous measurement using EGG can correlate well with data taken from in vivo electrodes placed on the serosal layer in animals; however, difficulties arise from the complexity of the data recorded using cutaneous EGG electrodes. Because

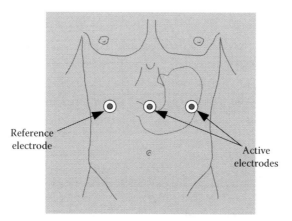

FIGURE 7.15　One common arrangement of electrodes for an EGG measurement. One active electrode is placed on the midline halfway between the xiphoid and umbilicus, and the second active electrode is placed 5 cm to the left of this electrode, at least 2 cm below the rib cage. The reference electrode is located at a distance from the active electrodes such as on the left midaxillary line or right midclavicular line (shown).

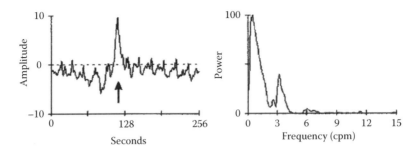

FIGURE 7.16 Raw data and FFT frequency analysis of an EGG measurement. In this instance, the artifact indicated by the arrow in (a) image results in a significant low-frequency (bradygastria) component in the frequency domain (b). (Courtesy of the American Gastroenterological Association, Bethesda, MD.)

the electrodes are positioned further away from the tissue generating the electrical activity and there are typically two to three slow wave events sweeping across the wall of the stomach at any one time, the recorded EGG data tend to be a composite of these multiple events. In addition, the signals can be very prone to the movement of the subject and electromagnetic interference (EMI) from external sources, both of which can cause measurement artifacts giving rise to anomalous results in the frequency analysis of the recorded data. Because of this, it is important to closely study the raw data and to remove any obvious movement or EMI artifacts before proceeding to the frequency analysis. This can take considerable skill to do effectively and if not done correctly can lead to the presence of low- and/or high-frequency components in the data analysis that can be mistaken for bradygastria or tachygastria events, respectively.

7.11 EMERGING TECHNOLOGIES

This section provides an incomplete overview of some of the recently reported and emerging technologies that the authors consider will develop into useful diagnostic techniques for GI professionals in the near future. The review should be considered as subjective as it only covers technologies that we have had direct or close indirect experience with. The field of emerging diagnostic technologies is fascinating and constantly changing, so this section should always be read as a *work in progress* providing a somewhat blinkered snapshot of technologies being discussed in journals and conferences at the time of writing.

7.11.1 Motility Tracking System

The motility tracking system (MTS) (Motilis Medica SA, Lausanne, Switzerland) provides a means of tracking the movement of an ingestible silicone-coated pill (18 × 5.5 mm) that contains a magnet (Stathopoulos *et al* 2005). The location of the pill is monitored using a grid of 5 cm spaced magnetic sensors placed above the abdomen. The sensor array can accurately locate the position of the pill within the digestive tract using an algorithm of five coordinates (positions x, y, and z and inclination angles θ and φ). The MTS essentially measures transit of the pill through the gut; however, unlike scintigraphy and x-ray marker studies, the MTS can provide far greater temporal details of intraluminal movement and also provides a 3D map of the pill's progress as it traverses the GI tract. In addition, the system can provide details on the velocity, direction of movement, and regional intestinal length, and by monitoring the subtle oscillatory movement, the MTS may even detect the underlying contractile frequencies responsible for the motility within the region.

A drawback of the current system is that the subject must sit or lie as still as possible under the detection grid during data acquisition that makes prolonged studies difficult. However, the MTS has recently been used in a clinical setting to measure gastric and small bowel transit in patients with spinal cord injury and patients with sclerosis, and a new ambulatory system is under development that may elevate this technique to being a viable clinical tool.

The diagnostic benefits of these devices are yet to be proven, but the data beginning to flow from early studies provide a fascinating insight into colonic motility that matches the pioneering work of the early GI pioneers using x-ray imagery.

7.11.2 Small Bowel HRM

Measurement of small bowel motility is currently restricted to specialized centers and is used when patients present with upper GI obstructions or idiopathic nausea and vomiting (Quigley *et al* 1997). There are very few published studies utilizing HRM in the small bowel, and those that have been performed report the relative advantages of HRM in identifying pyloric sphincter length and tone and the ease with which motor patterns such as the migrating motor complex (a classic fasted motility pattern of the small bowel) can be detected. At present, this technique is restricted to the proximal section of the small bowel due to the relatively short length of available HRM catheters, and the potential clinical worth of small bowel HRM is yet to be established. This may change as viable long-length HRM catheters become more widely available.

7.11.3 3D-HRM

3D-HRM (sometimes referred to as high-definition manometry [HDM]) is a recent addition to the clinical armory (Dinning *et al* 2010). As opposed to HRM, 3D-HRM consists of a high number of sensors at a spatial resolution of <1 cm that can provide data at each axial location *and* from a number of discrete radial directions. The 3D-HRM sensors provide detailed 3D maps of asymmetric regions such as the sphincteric complexes where neither circumferential averaging nor directional measurements provide sufficient measurement details.

Current designs consist of 12 rings divided in 4 equal banks. Each ring is spaced at 3 mm, and the banks of rings are separated by 4 mm. Each of the rings contains 8 radial sensors, thus providing 96 pressure recording sites (Kahrilas *et al* 2008). The remaining probe utilizes 32 circumferential HRM sensors on 1 cm spacing for a total sensing length of approximately 40 cm. This technology allows for a sufficient number of high-resolution sensors to span the whole esophagus and a high-definition recording length of 9 cm to span the esophagogastric junction (EGJ). In the recent past, prototype devices for mapping the anorectum have become available and are undergoing clinical evaluation; early results show some dramatic imagery of the 3D pressure profiles in this region. These probes utilize

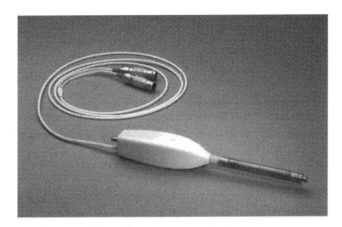

FIGURE 7.17   Sierra 3D-HRM anorectal device. (Courtesy of Given Imaging, Duluth, GA.)

16 levels at 4 mm spacing with 16 sensors at each level for a total of 256 distinct pressure measurements. In this instance, maintaining a small overall diameter and high flexibility is not so critical, and the devices seen at trade shows are <11 mm in diameter and rigid, making them viable for anorectal investigation, but not for intubation further into the gut (Figure 7.17). While the increase in information seen to date is highly suggestive of improved diagnostic ability, the full worth of anorectal 3D-HRM still needs to be validated.

7.11.4 Multimodal Analyses

Many diseased states involve a number of different root causes that combine to give rise to the symptoms experienced by the patient, and different subtypes of a common disorder may involve the activation of a range of different receptors, sensory pathways, and mechanisms. Therefore, it is often required to run a series of diagnostic tests before the clinician can gain insight into the correct course of treatment. To avoid the need for excessive invasive tests, a Danish group led by Hans Gregersen has developed a multimodal probe with the capability to provide chemical, mechanical (impedance planimetric bag distension), electrical, and cold and warmth stimuli to the esophagus (Drewes *et al* 2003). This multimodal probe allows stimulation of receptors in both superficial and deep layers of the esophagus during a single procedure, and hence, greater differentiation and diagnostic insight can be gained in a shorter timescale and with less inconvenience to the patient. Multimodal probes have been used in clinical studies to differentiate between patients with noncardiac chest pain and gastroesophageal reflux disease and have also been applied successfully to the study of the effect of drugs in pharmaceutical trials.

7.11.5 Predictive Algorithms (Combined Esophageal HRM and Impedance)

In a similar vein to the multimodal stimulation approach discussed earlier, Omari et al. have recently developed an automated algorithm to predict the likelihood of reaspiration in patients with swallowing disorders (Omari *et al* 2011). This group has demonstrated the strength of combining multiple diagnostic measurements with advanced computer algorithms to predict patient risk factors.

The technique uses four variables measured simultaneously from esophageal HRM and impedance catheters during the pharyngeal phase of a liquid bolus swallow (peak pressure, pressure at minimum impedance, time interval from minimum impedance to peak pressure, and the interval of impedance drop in the distal pharynx) from which a *swallow risk index* (SRI) can be calculated. The study indicates that SRI provides a robust prediction of the risk of aspiration and highlights the advantages of combining multiple techniques to improve the diagnostic potential of currently available discrete techniques.

Approaches such as this and the multimodal approach using multiple parameters to improve diagnostic and therapeutic outcomes are sure to play an increasing role in future GI diagnostics as it is certain that many disorders of the GI tract will have multiple root causes and are therefore unlikely to ever be conducive to simple single-parameter analyses. The increased effectiveness of multiple diagnostic techniques is sure to be a cornerstone for personalized medicine and improved matching of therapies to correctly identified receptive patient groups.

7.11.6 Functional Luminal Imaging Probe

While HRM can provide excellent details of the pressures exerted along the GI tract, it cannot provide details of local function or geometry of areas such as the EGJ. Gregersen et al. have developed the functional luminal imaging probe (FLIP) in an attempt to provide this information (McMahon *et al* 2007). The technique is based on the recording of a cross-sectional area (CSA) through the measurement of impedance planimetry. This technique is similar to esophageal impedance measurements except that it uses saline solution of known electrical characteristics contained in a flexible bag to calculate local variations in CSA. The FLIP consists of a saline solution in a bag fitted on the distal end of a catheter and covering a number of electrodes along its length. By measuring the impedance between the electrodes, multiple CSAs can be determined at known locations, hence enabling a 3D image of the region under investigation to be built up. The technique can display accurate images of sphincter characteristics during distension with the saline-filled bag that represent the EGJ when it is physiologically active (McMahon *et al* 2007, 2011). The FLIP technique has been used to differentiate the functionality of the EGJ between patients with gastroesophageal reflux disease and healthy controls (e.g., a large CSA during distension represents an incompetent EGJ, which is associated with excessive reflux [McMahon *et al* 2007]). The FLIP has been approved for use in the United States and Europe and is now available from Crospon, Galway, Ireland, under the name EndoFLIP.

7.11.7 Intraluminal Ultrasound

Intraluminal ultrasound provides a window into the subsurface structure of the GI lumen. The technology is a miniaturized version of standard medical imaging ultrasound in which very short ultrasonic pulses are generated and are then projected at right angles to the device axis by a rotating scanner at the device tip. The pulses are reflected back from the interfaces between the subsurface layers, and the distance between the device and each reflective interface can then be determined using the speed of sound in the body and the time taken for the reflected pulse to return to the transmitter. The measured time delays

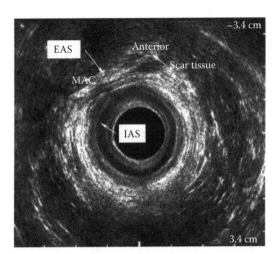

FIGURE 7.18 An ultrasonic image of the midanal canal. The rings from the center outward are as follows: transducer, mucosa, internal anal sphincter (IAS), EAS, and perianal fat. *Note:* The anterior attenuation of the external sphincter and evidence of scar tissue are common following perineal tear and repair.

between reflections from interfaces between regions of the lumen such as the luminal content, mucosa, and longitudinal and circular muscle layers are used to form an image of the subsurface structure. The rotating scanner forms an image that is similar to the rotating radar image used in aircraft and ships and consists of contrasting bands that indicate the interfaces between each sublayer. Larger side scan ultrasound devices are already in clinical use and can detect structural defects in the anal canal. Figure 7.18 shows an ultrasound image of the midanal canal showing subsurface details of the region.

Smaller devices with diameters less than 3 mm are now becoming available for measurement of regions deeper within the GI tract such as the esophagus. These devices are in the form of a flexible catheter that is introduced into the lumen via the port of an endoscope to ensure correct location relative to anatomical features. These devices have been used successfully for research applications such as investigating the timing between contractions in the longitudinal and circumferential muscle layers and are finding clinical applications for the identification of defects in the luminal musculature. The basic device will generate a 2D cross-sectional image of the luminal wall; however, 3D images can be built up by recording successive images as the catheter is slowly withdrawn either manually or using an automated stepper motor to relocate the scanner tip during each successive sets of 2D scans. Depending on the extent of the data acquisition, it is often necessary to time the recordings so that they coincide with cardiac or respiratory cycles. This process is known as *gating* and provides 3D images built up of measurements taken at similar phases of the cycles.

7.11.8 Optical Coherence Tomography

Optical coherence tomography (OCT) is similar in concept to intraluminal ultrasound in that it provides 2D images of subsurface detail, but differs in that it uses light waves instead of sound waves. With OCT however, light travels too fast to allow the simple time of flight analysis used for ultrasound, so the light from a single source (typically a high-powered light-emitting diode [LED]) is split into two paths, one of which is directed at right angles to the probe and into the surrounding tissue and the other is reflected from an external reflector to provide a known reference for a subsequent interferometric comparison. During the measurement, the reference reflector is scanned back and forth to vary the optical path length back to the detector. Recombining the two light paths and looking for optical interference effects that occur when the reference and measurement arms are precisely the same optical distance apart allow very accurate measurements of the depth of reflection (~1 to 10 µm) from the tissue substructure. However, the overall depth that can be probed is limited by the distance that the light beam can penetrate into the tissue (typically 1 to 2 mm). Hence, OCT provides a much greater detail of the subsurface structure but at a cost of reduced penetration depth.

Because of its increased resolution in the upper tissue layers, the technique is better for detecting early-stage predictors of conditions such as Barrett's esophagus and colorectal dysplasia and to evaluate the depth of penetration of early-stage neoplastic lesions (Drexler and Fujimoto 2008).

7.12 CONCLUDING REMARKS

This chapter has provided a brief overview of the common methods used by gastroenterologists to measure, record, and infer the physiological and pathophysiological functions of the human digestive tract. The field is much larger than these few pages can cover, and so we direct the interested reader to the excellent review given in Feldman et al. (2010) for further details of GI measurement techniques.

REFERENCES

Arkwright J W, Anneliese Dickson S A, Maunder N G et al 2013 The effect of luminal content and rate of occlusion on the interpretation of colonic manometry *Neurogastroenterol. Motil.* **25**, e52–9

Arkwright J W, Blenman N G, Underhill I D et al 2012 Measurement of muscular activity associated with peristalsis in the human gut using fiber Bragg grating arrays *IEEE Sens. J.* **12** 113–7

Bredenoord A J, Tutuian R, Smout A J and Castell D O 2007 Technology review: Esophageal impedance monitoring *Am. J. Gastroenterol.* **102** 187–94

Camilleri M, Thorne N K, Ringel Y et al 2010 Wireless pH-motility capsule for colonic transit: Prospective comparison with radiopaque markers in chronic constipation *Neurogastroenterol. Motil.* **22** 874–82, e233.

Dinning P G, Arkwright J W, Gregersen H, O'Grady G and Scott S M 2010 Technical advances in monitoring human motility patterns *Neurogastroenterol. Motil.* **22** 366–80

Dinning P G, Hunt L, Arkwright J W et al 2012 Pancolonic response to sub-and suprasensory sacral nerve stimulation in patients with slow transit constipation *Br. J. Surg.* **99** 1002–10

Dinning P G, Smith T K, Scott S M 2009 Pathophysiology of colonic causes of chronic constipation. *Neurogastroenterol. Motil.* **21** 20–30

Dinning P G, Wiklendt L, Gibbins I, Patton V, Bampton P, Lubowski D Z, Cook I J and Arkwright J W 2013 Low-resolution colonic manometry leads to a gross misinterpretation of the frequency and polarity of propagating sequences: Initial results from fiber-optic high-resolution manometry studies *Neurogastroenterol. Motil.* **10** e640–9

Drewes A M, Gregersen H and Arendt-Nielsen L 2003 Experimental pain in gastroenterology: A reappraisal of human studies. *Scand. J. Gastroenterol.* **38** 1115–30

Drexler W and Fujimoto J G 2008 *Optical Coherence Tomography: Technology and Applications* (Berlin: Springer)

Feldman M, Friedman L S and Brandt L J (eds.) 2010 *Sleisenger and Fordtran's: Gastrointestinal and Liver Disease*, 9th edition (Philadelphia: Saunders Elsevier)

Kahrilas P J, Ghosh S K and Pandolfino J E 2008 Challenging the limits of esophageal manometry. *Gastroenterology* **134** 16–8

Kolodny G M 1991 *Nuclear Medicine in Gastroenterology* (Dordrecht: Kluwer Academic Publishers Group)

McCaffrey C, Chevalerias O, O'Mathuna C and Twomey K 2008 Swallowable-capsule technology *Pervasive Comput.* **7** 23–9

McMahon B P, Frokjaer J B, Kunwald P *et al* 2007 The functional lumen imaging probe (FLIP) for evaluation of the esophagogastric junction *Am. J. Physiol. Gastrointest. Liver Physiol.* **292** G377–84

McMahon B P, Rao S S, Gregersen H *et al* 2011 Distensibility testing of the esophagus *Ann. N. Y. Acad. Sci.* **1232** 331–40

Omari T, Dejaeger E, Van Beckevoort D *et al* 2011 A method to objectively assess swallow function in adults with suspected aspiration *Gastroenterology* **140** 1454–63

Parkman H P, Hasler W L, Barnett J L and Eaker E Y 2003 Electrogastrography: A document prepared by the gastric section of the American Motility Society Clinical GI Motility Testing Task Force *Neurogastroenterol. Motil.* **15** 89–102

PillCam Capsule Endoscopy 2014 http://www.givenimaging.com/en-us/Innovative-Solutions/Capsule-Endoscopy/Pages/default.aspx

Quigley E M, Deprez P H, Hellstrom P M *et al* 1997 Ambulatory intestinal manometry: A consensus report on its clinical role *Dig. Dis. Sci.* **42** 2395–400

Sinnott M D, Cleary P W, Arkwright J W, Wang C and Dinning P G 2012 Investigating the relationships between peristaltic contraction and fluid transport in the human colon using smoothed particle hydrodynamics *Comput. Biol. Med.* **42**(4) 492–503

Stathopoulos E, Schlageter V, Meyrat B, Ribaupierre Y and Kucera P 2005 Magnetic pill tracking: A novel non-invasive tool for investigation of human digestive motility *Neurogastroenterol. Motil.* **17** 148–54

Szczesniak M M, Rommel N, Dinning P G, Fuentealba S E, Cook I J and Omari T I 2009 Intraluminal impedance detects failure of pharyngeal bolus clearance during swallowing: A validation study in adults with dysphagia. *Neurogastroenterol. Motil.* **21** 244–52

Whitehead W E and Delvaux M 1997 Standardization of barostat procedures for testing smooth muscle tone and sensory thresholds in the gastrointestinal tract. The working team *Dig. Dis. Sci.* **42** 223–41

CHAPTER 8

Measuring Glomerular Filtration Rate in Clinical Practice

A. Michael Peters

CONTENTS

8.1 Introduction   161
8.2 Theory   162
8.3 Scaling GFR for Body Size   168
8.4 Normal Values   168
8.5 Age and Gender Differences   168
8.6 Clinical Indications for Measuring GFR   169
8.7 Conclusion   169
References   170

8.1 INTRODUCTION

Glomerular filtration rate (GFR) is the single most useful test of renal function. It is measured using tracers or indicators (filtration markers) that display no binding to proteins in plasma, accumulate exclusively in the kidneys as a result of glomerular filtration, and are excreted in the urine with no tubular secretion or reabsorption. Filtration markers are small hydrophilic solutes with molecular weights in the range ~350 to 6000 Da that exchange by diffusion between plasma and the extravascular extracellular space. They are excluded from cells so their distribution volumes approximate the extracellular fluid volume (ECV).

The plasma, renal, and urinary clearances of an ideal filtration marker are identical to GFR. The classical method of measurement is from the urinary clearance of inulin, which has a molecular weight of ~6000 Da. Inulin is infused i.v. at a constant rate until a steady state is reached, when the infusion rate is equal to the urinary excretion rate, which is equal to the product of urinary concentration and urinary flow rate, which in turn is equal to the product of GFR and plasma concentration. GFR can therefore be measured from a timed urine collection and indicator concentrations in urine and plasma. This is a difficult and cumbersome method, compounded by the technical demands of inulin assay, and has essentially been replaced by methods that measure plasma or renal clearance at a nonsteady state and use radioactive isotopes that can be easily assayed in a well counter or imaged with a gamma camera.

The role of gamma camera imaging is to measure the renal clearance of Tc-99m-DTPA for each kidney separately (single kidney GFR). The clinical value of this, however, is limited as Tc-99m-DTPA has been largely replaced for routine isotope renography by Tc-99m-MAG3, which is not a glomerular agent. So plasma clearance following bolus injection is the preferred routine clinical method for measuring GFR.

The filtration markers in common use for measuring GFR by non-steady-state plasma clearance are Cr-51-EDTA (mostly in Europe, it does not have a license in the United States); Tc-99m-DTPA and I-125-iothalamate, all of which are radioactive; and iohexol, the concentration of which can be measured accurately using HPLC. As a result of minimal degrees of protein binding, extrarenal clearance, and tubular handling, none of these agents, unlike inulin, completely meets the ideal requirements of a filtration marker, but they are nevertheless satisfactory for routine clinical use, except at very low levels of GFR (<10 mL/min).

8.2 THEORY

The blood (or plasma) clearance of any substance following bolus i.v. injection is equal to the ratio of the injected amount to the total area under curve (AUC) enclosed under the subsequent blood (or plasma) time–concentration curve. The sense of this can be appreciated by considering the units (which is a useful general approach to any equation describing a pharmacokinetic model). The units of AUC are equal to the product of the y-axis and x-axis units of the curve, which, for a radioactive tracer, are MBq/mL and min, respectively. Therefore,

$$\frac{\text{MBq}\,(\text{administered activity})}{[\text{MBq/mL}]\times[\text{min}](\text{AUC})} = \frac{\text{mL}}{\text{min}} \tag{8.1}$$

It can be seen that clearance has the same units as blood flow (mL/min). This is because clearance is the virtual incoming blood (or plasma) flow that can be imagined to be completely cleared of the substance in a single pass. The proportion of blood flow that is cleared is the extraction fraction, which is an important concept in renal physiology. In the context of glomerular filtration, the renal extraction fraction of a filtration marker is the ratio of GFR to renal plasma flow (filtration fraction).

Various mathematical approaches have been suggested for measuring the AUC. The most widely used approach is compartmental analysis, which requires the resolution of the curve into its component exponentials. The plasma clearance curve of a filtration marker is multiexponential, especially if sampling is continued beyond 5 h. On venous sampling up to 4 to 5 h, two exponentials dominate the curve. The faster, which disappears by about 90 to 120 min, represents mixing of tracer throughout the ECV, while the slower (the *terminal* exponential) represents the fraction of ECV cleared of tracer by GFR in unit time. The terminal exponential (the rate constant of which will be termed α_2) is therefore close to the ratio of GFR to ECV ($[mL/min]/mL = min^{-1}$). The AUC is obtained by summing the areas under the two exponentials (the area under an exponential is its zero-time intercept divided by its rate constant).

With arterial instead of venous sampling, a substantial additional fast exponential can also be identified that disappears by about 20 min (Figure 8.1). Thus, up to 4 to 5 h, the curve is, in fact, a triple exponential. This can be explained by the heterogeneous nature of the extravascular extracellular fluid space, which has two major functional compartments in series with each other (Figure 8.2): first the fluid phase and second the gel phase, wherein

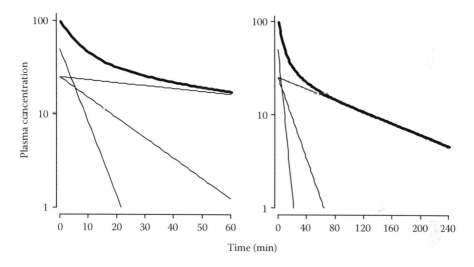

FIGURE 8.1   Representative arterial plasma clearance curves (bold lines) obtained following i.v. injection of a filtration marker. The straight fine lines are the fitted exponentials. Left panel, sampling up to 60 min; right panel, sampling up to 240 min. Note the logarithmic vertical axes.

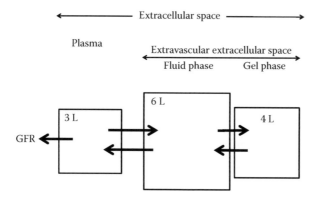

FIGURE 8.2   Simple compartmental model that forms a useful basis for the kinetics of a filtration marker following intravascular injection. (Data based on Tc-99m-DTPA; From Cousins, C. et al., *Eur. J. Nucl. Med.*, 29, 655, 2002.)

lie vast molecules, called the glycosaminoglycans (GAGs). Mixing of tracers between plasma and the fluid phase is rapid, approaches completion at about 20 min postinjection, and is reflected by the fast exponential. Penetration of the gel phase, however, takes longer. Scaled for average body size, the combined volume of plasma and fluid phase, when measured with Tc-99m-DTPA, is about 9 L and that of the gel phase is about 4 L (Cousins *et al* 2002). The fast exponential is not clearly seen on antecubital venous sampling because of, first, extraction of Tc-99m-DTPA from plasma into the extracellular space of the forearm (~0.5 at resting blood flow; Cousins *et al* 2002) and, second, vascular delay across the forearm. The venous curve, therefore, appears essentially biexponential.

Multiple venous samplings are generally considered too labor intensive for routine clinical measurements of GFR, so a widespread approach is to start sampling at 120 min and record only the terminal exponential from 3 or 4 plasma samples. This will obviously overestimate GFR as a result of underestimation of AUC. The area under the faster exponential recorded from venous sampling is relatively constant and independent of GFR (Brochner-Mortensen 1972). The magnitude of the underestimation of AUC therefore depends on GFR, since this determines the area under the terminal exponential (Figure 8.3). Thus, AUC under the terminal exponential in a patient with GFR almost equal to zero would be infinitely large, so that the contribution to total AUC made by the faster exponential

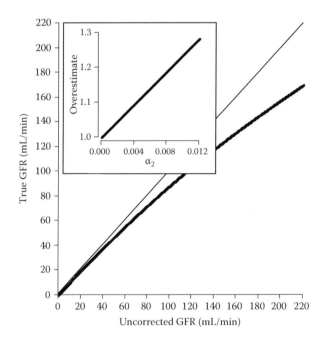

FIGURE 8.3 The bold line shows the relation between uncorrected (one-compartment) GFR measured from samples obtained from 120 min and *true* GFR based on biexponential analysis that includes sample concentrations obtained from soon after injection. The fine line is identity. The overestimation of true GFR is a function of GFR. The inset shows that the overestimation, expressed as the quotient uncorrected GFR/true GFR, is a linear function of α_2.

would be negligible. A correction equation that is based on the raw GFR value can therefore be applied to account for the underestimation in AUC (the one-compartment correction). The most widely used equation was described by Brochner-Mortensen (1972), as follows:

$$\text{Corrected GFR} = \left(0.991 \times \text{uncorrected GFR}\right) - \left(0.00122 \times \left[\text{uncorrected GFR}^2\right]\right) \text{mL/min}$$

(8.2)

Note that uncorrected GFR needs to be scaled for body size, such as body surface area (BSA), prior to correction (see the following texts for different scaling methods).

This technique is known as the *slope–intercept* technique, where the intercept is the value of the terminal exponential after extrapolation back to zero time. The ratio of administered activity (MBq) to the intercept (MBq/mL) is called the volume of distribution (V_d), but it should not be confused with ECV, which is the *true* volume of indicator distribution and which V_d overestimates by an amount that depends on GFR (40% with normal GFR; Peters [2004]).

Then,

$$\text{Uncorrected GFR} = \alpha_2 \times V_d \left(\text{mL/min} = \text{min}^{-1} \times \text{mL}\right) \qquad (8.3)$$

Equation 8.2 is an example of a second-order polynomial. Such an equation approaches a maximum value for corrected GFR, which then declines as the uncorrected GFR further increases. This is an undesirable feature of the equation, even though the maximum is reached at only extremely high levels of the uncorrected GFR. Nevertheless, Jodal and Brochner Mortensen (2009) developed an alternative correction equation, to avoid this problem, as follows:

$$\text{Corrected GFR} = \frac{\text{Uncorrected GFR}}{\left(1 + \left[f \cdot \text{uncorrected GFR}\right]\right)} \qquad (8.4)$$

where

$$f = 0.0032 \cdot \text{BSA}^{-1.3} \qquad (8.5)$$

BSA is estimated from height and weight.

As already mentioned, α_2 in Equation 8.3 is close to the ratio of GFR to ECV. This ratio is quite constant across species and subject size and represents the rate at which the ECV is *turned over* by glomerular filtration. Analogous to the one-compartment correction for slope–intercept GFR, α_2 has to be corrected to make it equal to GFR/ECV using an equation described by Bird *et al* (2009) in which

$$\text{Corrected GFR/ECV} = \alpha_2 + \left(\alpha_2^{\,2} \times 15.4\right)(\text{mL/min})/\text{mL} \tag{8.6}$$

Although not immediately obvious, this correction is necessary because after mixing throughout the extravascular space, extravascular tracer concentration exceeds plasma concentration (otherwise, the extravascular space would never be cleared of tracers; Figure 8.4) and both concentrations fall with the same rate constant, whereas before mixing, plasma concentration is higher as a result of net entry of tracer into the extravascular space.

GFR/ECV, expressed as (mL/min)/mL, is often not understood by referring clinicians. However, because ECV of a *standard man* is 13 L, GFR/ECV can be multiplied by 13 to give a value that is numerically comparable with GFR scaled to a BSA of 1.73 m². Expression of GFR exclusively in terms of corrected α_2 has been termed *slope-only* GFR.

A significant advantage of slope-only GFR is its simplicity—there is, for example, no need to measure administered activity—but its main disadvantage is its sensitivity to error in α_2. Slope–intercept GFR, in contrast, has an intrinsic resistance to error in α_2 because any error tends to be counterbalanced by an opposing error in V_d. It is, however, sensitive to error in administered activity (a silent error, moreover).

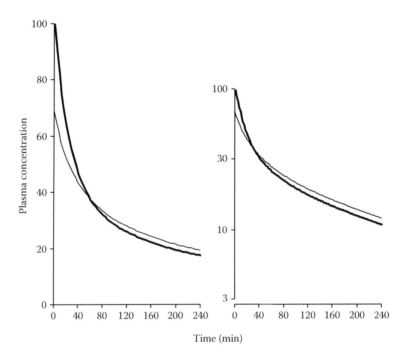

FIGURE 8.4 Plasma arterial (bold lines) and venous (fine lines) concentration curves following intravascular injection of a filtration marker: left panel, linear vertical axis and right panel, logarithmic vertical axis. Note that, initially, arterial concentration exceeds venous concentration, but this gradient is reversed after mixing of filtration markers throughout its distribution volume.

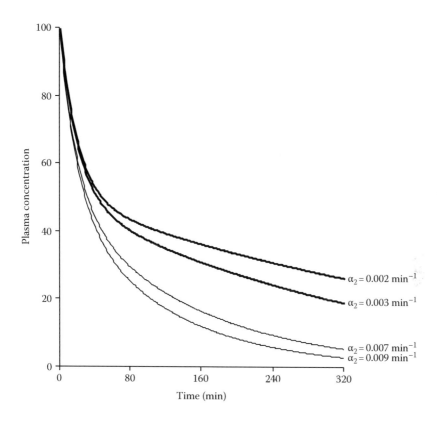

FIGURE 8.5 Theoretical blood concentration–time curves following intravascular injection of a filtration marker for different values of α_2, illustrating the basis of single-sample GFR. A biexponential curve was assumed, with the area under the faster exponential assumed to be constant. Impaired function is illustrated by bold lines and normal function by fine lines.

Some departments consider that even 3 or 4 samples is too labor intensive and take only one sample at 180 or 240 min—the single-sample technique. This method works because if GFR is impaired, a higher proportion of the injected activity is still present in plasma at late time points (Figure 8.5). It performs surprisingly well (Bird *et al* 2009, Gaspari *et al* 1996) and, for example, in a recent study was shown to outperform independently measured 6-sample GFR (Bird *et al* 2009). As can be seen from Figure 8.5, it is better able to separate low values of GFR from each other than higher values, because by the time mixing is complete, plasma concentrations are becoming very low at high values of GFR.

A disadvantage of single-sample GFR is that it is sensitive to deviations in ECV. A further major drawback is that it cannot be subjected to quality control (QC). QC is possible with both slope–intercept and slope-only GFR. For example, the British Nuclear Medicine Society guidelines suggest that the correlation coefficient of the exponential fit to the plasma isotope concentrations between 120 and 240 to 300 min should be >0.985. This tests the accuracy of sample timing and accuracy with which the plasma concentrations are measured. The correlation coefficient, however, is insensitive to errors in the measurement of the administered activity or inadvertent partially *tissued* injection. A useful QC procedure is to compare the slope–intercept value ([mL/min]/1.73 m^2) with the slope-only value ([mL/min]/13 L). Substantial disagreement indicates an error at some stage in the measurement or less likely a significant deviation of ECV from normal (which is usually clinically evident).

8.3 SCALING GFR FOR BODY SIZE

A mention has already been made of scaling GFR to BSA, which in a *standard man* is considered to be 1.73 m^2. Scaling GFR to body size is essential, and BSA is the established method. It has, however, been criticized by several authorities (Turner and Reilly 1995) and alternatives suggested, including ECV (White and Strydom 1991), lean body mass (LBM) (Peters *et al* 2010), and total body water (TBW) (Eriksen *et al* 2011). Slope-only GFR represents GFR that is already scaled to ECV, but ECV can be estimated from height and weight and used for scaling corrected slope–intercept GFR (Bird *et al* 2003).

An obvious problem with BSA is that it is 2D, whereas GFR is 3D. This means that BSA is disproportionately high in small individuals (e.g., the volume to surface area ratio of a golf ball is lower than that of a soccer ball) and results in the regressions of GFR and ECV on BSA having significant negative intercepts (Boer 1984, Turner and Reilly 1995). Analogous plots against LBM (Peters *et al* 2010) and TBW (Eriksen *et al* 2011), in contrast, pass through the origin. The attraction of the new equation described by Jodal and Brochner-Mortensen (2009; see previous text) is that raising BSA to the index of 1.3 addresses the nonlinear relations of BSA with GFR and ECV. Thus, the plot of ECV on BSA$^{1.3}$ does pass through the origin (Jodal and Brochner-Mortensen 2009).

8.4 NORMAL VALUES

The normal value for GFR scaled to BSA in young adults is often given as 120 (mL/min)/1.73 m^2, but it is probably lower than this and closer to 100 (mL/min)/1.73 m^2. Chronic kidney disease is categorized according to GFR as follows: grade 1, >90; grade 2, 60 to 90; grade 3, 30 to 60; grade 4, 15 to 30; and grade 5, <15 (mL/min)/1.73 m^2.

GFR may also be abnormally increased, and this is well recognized in diabetic patients. It may also be increased in children with certain forms of cancer, although this is less well recognized and the mechanism is unknown. It may also be abnormally increased in the morbidly obese, but this is controversial (see the following texts).

8.5 AGE AND GENDER DIFFERENCES

Scaled to BSA, GFR in children is lower than in adults and increases with age from 6 months to about 12 years. This, however, is simply the result of growth (as from a golf ball to a soccer ball) and the changing impact of BSA. GFR/ECV, in contrast, is higher in children than adults (Bird *et al* 2003).

In adults, scaled GFR declines with advancing age from about age 40 at a rate of 0.5 to 1 (mL/min)/year (Grewal and Blake 2005, Peters *et al* 2012). It declines more rapidly in women (Peters *et al* 2012, Xu *et al* 2010). Women have a lower value of GFR/BSA than men but a higher value of GFR/ECV (Peters *et al* 2012), again probably because of the effects of BSA as a scaling variable. By age >60 years, however, men have a higher GFR/ECV. When scaled to LBM or TBW, ECV is higher in women. It was recently shown in large populations of healthy men and women (Peters *et al* 2012) that while ECV/BSA was higher in men, there was little difference between the genders with respect to ECV/weight, even though women have a higher percentage of body fat. This is another example of how scaling to BSA can be misleading. A likely reason why BSA lacks validity for scaling is that the equation for estimating it from height and weight is the same for men and women, unlike the equations for estimating TBW and LBM.

As in diabetics, GFR appears to be elevated in obese subjects (Chagnac *et al* 2008). However, this may be related to the difficulties of scaling GFR in the obese, since so much of their weight is in the form of adipose tissue, and indeed scaling to LBM appears to normalize GFR (Janmahasatian *et al* 2008). Critically, GFR/ECV is not elevated in the obese but tends to be reduced in the very obese, especially in women and especially in older women (Peters *et al* 2012).

Differences between men and women, not only with respect to GFR but also to ECV, are important and should be recognized. There has been a tendency when assessing scaling variables to consider a variable to be valid if it equalizes GFR between the genders (Daugirdas *et al* 2009, Ericksen *et al* 2011). This however is misguided if there are real gender differences.

8.6 CLINICAL INDICATIONS FOR MEASURING GFR

The two main groups of patients that frequently undergo routine clinical measurement of GFR are cancer patients and patients with nephrourological disease. Scaling in the latter group is important in order to quantify the degree of renal impairment. Cancer patients need GFR measurement because many chemotherapeutic drugs that are toxic to the bone marrow undergo renal excretion. The drug dose therefore needs to be *titrated* against GFR; otherwise, in the presence of a low GFR, for example, the bone marrow may be poisoned. Scaling is less important in this group as drug dosage is titrated against absolute GFR. The test is also used in a small proportion of patients to assess the nephrotoxic effects of certain drugs, for example, cyclosporin.

8.7 CONCLUSION

In spite of the increasing use of equations to estimate GFR (eGFR) from the patient's serum creatinine, age, gender, and ethnicity (MDRD 4-variable equation), formal measurement of GFR from the plasma clearance of a filtration marker will remain an important clinical procedure. Two issues, however, demand an urgent concern. The first is the choice of the best scaling variable. The second is the development of QC. So while in general GFR is accurately measured from plasma clearance, the test is complex, demands high levels of technical expertise, and probably gives an erroneous result more often than generally recognized. Departments performing the test need to identify robust QC endpoints that enable them first to audit their performance at a departmental level and second identify unreliable results at the level of the individual patient.

REFERENCES

Bird N J, Henderson B L, Lui D, Ballinger J R and Peters A M 2003 Indexing glomerular filtration rate to suit children *J. Nucl. Med.* **44** 1037–43

Bird N J, Peters C, Michell A R and Peters A M 2009 Assessment of single sample GFR from repeatability, response to food, and comparison with simultaneous and independent multi-sample reference technique *Am. J. Kidney Dis.* **54** 278–88

Boer P 1984 Estimated lean body mass as an index for normalization of body fluid volumes in man *Am. J. Physiol.* **247** F632–5

Brochner-Mortensen J 1972 A simple method for the determination of glomerular filtration rate. *Scand. J. Clin. Lab. Invest.* **30** 271–4

Chagnac A, Herman M, Zingerman B, Erman A, Rozen-Zbi B, Hirsh J and Gafter U 2008 Obesity-induced glomerular hyperfiltration: its involvement in the pathogenesis of tubular sodium reabsorption *Nephrol. Dial. Transplant* **23** 3946–52

Cousins C, Skehan S J, Rolph S, Flaxman M, Ballinger J R, Bird N J, Barber R W and Peters A M 2002 Comparative microvascular exchange kinetics of [^{77}Br]bromide and ^{99m}Tc-DTPA in humans *Eur. J. Nucl. Med.* **29** 655–62

Daugirdas J T, Meyer K, Greene T, Butler E S and Poggio E D 2009 Scaling of measured glomerular filtration rate in kidney donor candidates by anthropometric estimates of body surface area, body water, metabolic rate, or liver size *Clin. J. Am. Soc. Nephrol.* **4** 1575–83

Eriksen B O, Melsom T, Mathisen U D, Jenssen T G, Solbu M D and Toft I 2011 GFR Normalized to total body water allows comparisons across genders and body sizes *J. Am. Soc. Nephrol.* **22** 1517–25

Gaspari F, Guerini E, Perico N, Mosconi L, Ruggenenti P and Remuzzi G 1996 Glomerular filtration rate determined from a single plasma sample after intravenous iohexol injection: is it reliable? *J. Am. Soc. Nephrol.* **7** 2689–93

Grewal G S and Blake G M 2005 Reference data for ^{51}Cr-EDTA measurements of the glomerular filtration rate derived from live kidney donors *Nucl. Med. Commun.* **26** 61–5

Janmahasatian S, Duffull S B, Chagnac A, Kirkpatrick C M J and Green B 2008 Lean body mass normalizes the effect of obesity on renal function *Br. J. Clin. Pharmacol.* **65** 964–5

Jodal L and Brochner-Mortensen J 2009 Reassessment of a classical single injection ^{51}Cr-EDTA clearance method for determination of renal function in children and adults. Part I: analytically correct relationship between total and one-pool clearance *Scand. J. Clin. Lab. Invest.* **69** 305–13

Peters A M 2004 The kinetic basis of glomerular filtration rate measurement and new concepts of indexation to body size *Eur. J. Nucl. Med. Mol. Imaging* **31** 137–49

Peters A M, Glass D M, Love S and Bird N J 2010 Estimated lean body mass is more appropriate than body surface area for scaling glomerular filtration rate and extracellular fluid volume *Nephron Clin. Pract.* **116** 75–80

Peters A M, Perry L, Hooker C A, Howard B, Neilly M D J, Seshadri N *et al* 2012 Extracellular fluid volume and glomerular filtration rate in 1,878 healthy potential renal transplant donors: effects of age, gender, obesity and scaling *Nephrol. Dial. Transplant* **27** 1429–37

Turner S T and Reilly S L 1995 Fallacy of indexing renal and systemic hemodynamic measurements for body surface area *Am. J. Physiol.* **268** R978–88

White A J and Strydom W J 1991 Normalization of glomerular-filtration rate measurements *Eur. J. Nucl. Med.* **18** 385–90

Xu R, Zhang L-X, Zhang P-H, Wang F, Zuo L and Wang H-Y 2010 Gender differences in age-related decline in glomerular filtration rates in healthy people and chronic kidney disease patients. *BMC Nephrol.* **11** 20–6

CHAPTER 9

Neurology

Central Nervous System

Richard B. Reilly

CONTENTS

9.1   Historical Context for the Measurement of Neural Activity   172
9.2   Neurophysiology of the Human Brain   173
9.3   Neuron   176
9.4   Resting Potential   177
9.5   Cell Membrane Electrical Equivalent Circuit   179
9.6   Action Potential Generation   179
9.7   Cortical Organization   180
9.8   Implanted Electrophysiological Recordings   183
9.9   Implanted Electrode Design   183
9.10   Extracellular Signal Acquisition   185
9.11   Analysis of Extracellular Signals   185
       9.11.1 Spike Analysis   186
9.12   Electroencephalography   189
       9.12.1 Brain Imaging   189
9.13   Acquisition Methods   190
       9.13.1 EEG Recording Equipment   193
       9.13.2 Electrodes and Electrode Placement   193
       9.13.3 Basic EEG Amplifier   197
              9.13.3.1 Inverting and Noninverting Amplifier   197
              9.13.3.2 Input Impedance   198
              9.13.3.3 Instrumentation Amplifier   199
       9.13.4 EEG Recording Circuitry   199
       9.13.5 Referencing and Bipolar Recordings   200
       9.13.6 Artifacts   200
9.14   EEG Signal Properties   204
       9.14.1 Rhythmic Brain Activity   206
       9.14.2 Event-Related Potentials   208
       9.14.3 Event-Related Synchronization/Desynchronization   209

9.15 Applications   210
    9.15.1 Biofeedback   211
9.16 Summary   211
9.17 CNS Measurement Equipment   212
References   212

9.1 HISTORICAL CONTEXT FOR THE MEASUREMENT OF NEURAL ACTIVITY

The great Greek philosophers such as Aristotle and Plato spawned an inquisitive era for hypothesizing the anatomical and biological makeup of the human body. Herophilus (335 to 280 BC), often referred to as the *father of anatomy*, was the first to begin meticulously cataloging his anatomical findings. Through human insatiable curiosity, this quest developed into modern-day medicine.

Throughout these periods, the human brain was the great medical and philosophical fascination of the human body. It began with human autopsies to explore the anatomical structure and then developed to live situ in brain experiments to explore its behavior and functionality. These were first performed on animals and then later on humans with severe cranial fractures or terminal illnesses. Slowly, on a trial-by-trial basis, an understanding of the neurophysiology of the human brain was developed. It generated much interest for philosophers, psychologists, and surgeons alike. Human kind was no longer content to simply know the anatomical makeup of the brain, but now wanted to know how this organ performed so many complex functions.

From the mid-nineteenth century, a multidisciplinary collaboration began between neurosurgery and engineering. Its goal was to explore and understand the physiological and psychological behavior of the brain, spinal cord, and peripheral nerves that make up the body's complex integrated information processing and control system.

Among those at the forefront of neurosurgery included Sir Victor Horsley (1857 to 1916) who experimented with electrical stimulation to understand the complexity of sensory processing in the central nervous system (CNS). He is best known for developing the Horsley–Clarke apparatus, together with Robert H. Clarke, for performing stereotactic neurosurgery, whereby a set of precise numerical coordinates were used to locate each brain structure. Another leading neurosurgeon and a pioneer in this domain was Harvey Williams Cushing (1869 to 1939) who is often referred to as the *father of modern neurosurgery*. Cushing also electrically stimulated and recorded electrical activity from within the CNS to inform his diagnoses.

While stimulation was advancing as a means to probe brain function at a macro level, the end of the nineteenth and start of the twentieth century saw investigations into the microscopic structure of the brain. Santiago Ramón y Cajal (1852 to 1934) researching in Barcelona learned to use Golgi's silver nitrate preparation and focused his research on the CNS making extensive studies of neural material covering the major regions of the brain. Ramón y Cajal is credited with discovering the mechanism of axonal growth, developing detailed descriptions of cell types associated with neural structures, and produced excellent depictions of structures and their connectivity. He is also associated with providing definitive evidence for what is known as *neuron theory*, experimentally demonstrating that nerve cells or neurons are not continuous structures but link in a contiguity relationship. Neuron theory is a foundation of modern neuroscience and biomedical engineering, providing a means to understand the neurophysiological structure of the CNS.

9.2 NEUROPHYSIOLOGY OF THE HUMAN BRAIN

The average adult human brain weighs around 1.4 kg and is surrounded by cerebrospinal fluid that suspends it within the skull and protects it by acting as a motion dampener. The CNS, composed of the brain and the spinal cord, receives sensory information from the nervous system and controls the body's responses. The CNS is the coordinating system of the body and is differentiated from the peripheral nervous system, which connects the CNS to sensory organs (such as the eye and ear) and other organs of the body, muscles, blood vessels, and glands.

The CNS itself is a complex collection of tissues that coordinate and control the functions of the body. These tissues are groups of neurons that function together to perform specialized functions. Anatomically, the brain can be divided into the three largest structures, brain stem (hindbrain), cerebrum, and cerebellum (forebrain), and into a number of distinct anatomical areas (Figure 9.1). The functions of these areas are summarized as follows:

- The *brain stem* controls the reflexes and autonomic nerve functions (respiration, heart rate, blood pressure).

- The *cerebellum* integrates information from the vestibular system that indicates position and movement and uses this information to coordinate limb movements and maintain balance.

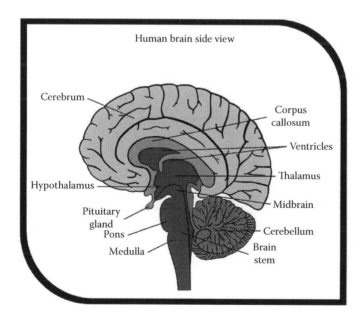

FIGURE 9.1   Anatomical areas of brain.

- The *hypothalamus* and pituitary gland control visceral functions, body temperature, and behavioral responses such as feeding, drinking, sexual response, aggression, and pleasure.

- The *thalamus* or specifically the thalamic sensory nuclei input is crucial to the generation and modulation of rhythmic cortical activity.

- The *cerebrum* consists of the cortex, large fiber tracts (corpus callosum), and some deeper structures (basal ganglia, amygdala, hippocampus). It integrates information from all of the sense organs, initiates motor functions, controls emotions, and holds memory and higher thought processes.

The cerebrum can be spatially subdivided into two hemispheres, left and right, connected to each other via the corpus callosum. The right hemisphere senses information from the left side of the body and controls movement on the left side. Similarly, the left hemisphere is connected to the right side of the body. Each hemisphere can be further subdivided into four lobes: the frontal, parietal, occipital, and temporal lobes. The cerebral cortex is the outer layer of the cerebrum and is formed as a thin convoluted layer of gray matter, 2 to 4 mm thick, composed of folds or ridges known as gyri, along with grooves known as sulci (see Figure 9.2).

The cortex is commonly referred to as gray matter from its color, which contrasts it with white matter, which dominates the layer below in the cerebrum. The highest centers of the brain are located in the cerebral cortex. Due to its surface position, the activity of the cerebral cortex has the greatest influence on recordings made on the scalp surface, such as the electroencephalogram. Higher-order cognitive tasks such as problem solving, language comprehension, and processing of complex visual information all take place in the cortex.

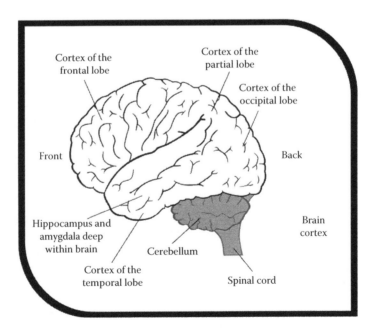

FIGURE 9.2   Anatomical areas of the cortex, with sulci and gyri illustrated.

The functional activity of the brain is highly localized. This facilitates the cerebral cortex to be divided into several areas responsible for different brain functions. The areas are depicted in Figure 9.3 and the related functions are described in Table 9.1.

Given that the CNS is the complex collection of tissues that coordinate and control the functions of the body and that these tissues are made up of groups of neurons that function

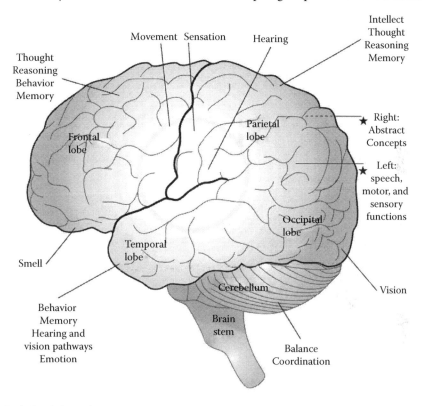

★ Right-handed people
(for left-handed people, these are the other way around)

FIGURE 9.3   Functional areas of the brain.

TABLE 9.1   Cortical Areas of the Brain and Their Function

Cortical Area	Function
Auditory association area	Complex processing of auditory information
Auditory cortex	Detection of sound quality (loudness, tone)
Broca's area (speech center)	Speech production and articulation
Prefrontal cortex	Problem solving, emotion, complex thought
Premotor cortex	Coordination of complex movement
Primary motor cortex	Initiation of voluntary movement
Primary somatosensory cortex	Receiving of tactile information from the body
Sensory association area	Processing of multisensory information
Gustatory area	Processing of taste information
Wernicke's area	Language comprehension
Primary visual cortex	Complex processing of visual information

together, it is important to consider the fundamental properties of the basic element of the CNS: the neuron. Along with neurons, multiple other cell types make up the CNS, including astrocytes, microglia, oligodendrocytes, Schwann cells, and neural precursor cells.

9.3 NEURON

The human brain at birth consists of approximately 100 billion (10^{11}) neurons at an average density of 10^4 neurons per cubic mm. The number of neurons decreases with age. Neurons share the same characteristics and have the same parts as other biological cells, but the electrochemical aspect lets them transmit electrical signals and thus pass information to each other over long distances. Neurons have three basic components (see Figure 9.4):

- *Soma*—The soma or cell body is the main part of the neuron and contains all of the necessary components of the cell, such as the nucleus (contains DNA), endoplasmic reticulum and ribosomes (for building proteins), and mitochondria (for supplying energy). If the cell body dies, the neuron dies.

- *Axon*—This long, cable-like projection of the cell carries electrochemical information along the length of the cell. The connection between the soma and the axon is known as the axon hillock.

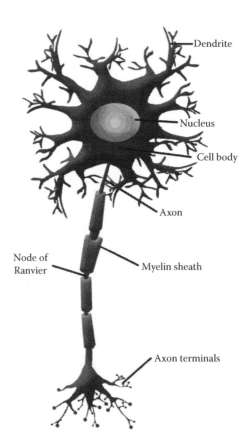

FIGURE 9.4   The structure of the neuron.

Cells of the central nervous system

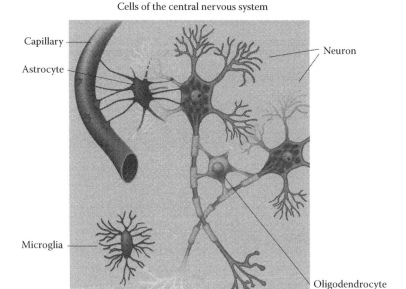

FIGURE 9.5   Cells of the CNS, demonstrating their structure and interconnectivity.

- *Dendrites*—These small, branch-like projections of the cell make connections to other cells and allow communication with other neurons. Dendrites can be located on one or both ends of the cell (see Figure 9.5).

The most active element of the neuron is the membrane, which consists of lipid bilayer separated by an insulator. This membrane contains ion channels, which allow the passage of ions in and out of the cell. These channels serve the function of controlling the current flow in and out of the cell's membrane. Interneuron communication and thus resulting brain electrical activity are based on the movement of ions, Na^+, K^+, Ca^{++}, and Cl^- ions, which are pumped through the channels in the neuron membrane in the direction governed by the membrane potential. Any biometric potential observed on the skin is due to the flow of ion-based electrical currents within the body.

The ion channels are classified according to their role into resting (passive) and gated ion channels (active). Resting ion channels are not controlled by external factors and are always open, while gated ion channels are controlled by several factors including the membrane potential. At rest, the cell has excess of positive charges outside of the membrane and excess negative charges on the inside. This charge separation across the membrane results in a voltage difference that is referred to as the resting membrane potential.

9.4 RESTING POTENTIAL

The membrane potential is defined as the voltage difference between inside and outside the cell. By convention, the potential outside the cell is defined as zero. Therefore, the resting membrane potential is equivalent to the voltage inside the cell. This voltage is usually in the range of −60 to −90 mV depending on the cell type. Electrical signaling of the cell involves changes to this resting potential. These changes occur as a result of selective opening and closing of ion channels within the membrane. This leads to the passage of ions through the membrane that causes current flow and membrane voltage changes from its resting state.

There are four main ions (K^+, organic anions A^-, Na^+, and Cl^-) that contribute to the voltage difference observed across the membrane. At rest, there is a higher concentration of Na^+ and Cl^- outside the cell, while K^+ and organic anions A^- are mainly concentrated in the fluid inside the cell. Neuron potential is determined by the gradient concentration of these ions and the permeability of the membrane. Nerve cells are permeable to K^+, Na^+, and Cl^-.

There are two opposing forces acting upon the ions: The first force is due to diffusion and depends on the concentration gradient of the ions, whereby ions move from areas of higher concentration to lower concentration. The second force depends on the potential difference and is called the electrical force. These forces oppose each other and cause ions to move in and out of the cell until equilibrium is reached where both forces are equal.

Assuming the membrane is permeable to only one ion, this equilibrium potential can be calculated using the Nernst equation. For example, given the concentration of K^+ outside $[K]_o$ and inside $[K]_i$ the cell, the Nernst equation can be used to estimate the potassium equilibrium potential E_K:

$$E_K = \frac{RT}{nF} \ln \frac{[K]_o}{[K]_i} \tag{9.1}$$

where
  R is the gas constant
  T is the temperature in Kelvin
  n is the valence of potassium
  F is the Faraday constant

Since there is more than one ion contributing to the resting potential of the membrane and taking into account the permeability coefficient of each ion, the resting potential (V_m) can be estimated using the Goldman equation:

$$V_m = \frac{RT}{F} \ln \frac{P_K[K]_o + P_{Na}[Na]_o + P_{Cl}[Cl]_i}{P_K[K]_i + P_{Na}[Na]_i + P_{Cl}[Cl]_o} \tag{9.2}$$

When the relative permeability P of a single ion is much higher than other ions, the resting potential can be approximated by the Nernst equation for that particular ion.

Depolarization is the process where ions flow in the direction such that the resting potential becomes less negative. Hyperpolarization on the other hand is the process whereby the membrane potential becomes more negative. When depolarization reaches a critical threshold, the cell responds by opening voltage-gated ion channels; this causes the cell to produce an action potential (AP) that lasts approximately 1 to 5 ms. These APs can travel along the axons of the cell for large distances without the loss of amplitude and facilitate transmission of information between neurons.

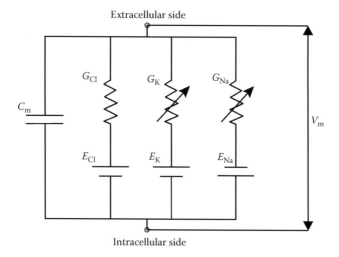

Extracellular side

Intracellular side

FIGURE 9.6   Equivalent circuit model of the cell.

9.5 CELL MEMBRANE ELECTRICAL EQUIVALENT CIRCUIT

Basic circuit elements can be used to model properties of the neuron and can be used to describe time-dependent generation of the neurons electrical signals. In the equivalent circuit, ion channels are represented by resistance (R) or conductance (G) ($G = 1/R$). This represents the ability of the ion channels to control (resist) movement of ions. The charge separations of ions are modeled by voltage sources with their potential estimated using the Nernst equation. Capacitance arises when two conductors are separated by an insulating material; hence, the lipid bilayer of the membrane (insulator) along with the intra- and extracellular fluids (conductors) is represented by a capacitor C (Figure 9.6).

9.6 ACTION POTENTIAL GENERATION

Alan Hodgkin and Andrew Huxley introduced a model describing the generation of APs based on voltage clamp experiments in squid giant axons. The voltage clamp was a new experimental tool that allowed Hodgkin and Huxley to measure the time varying conductance of Na^+ and K^+ channels for several values of membrane potentials. These observations were then analyzed to model the AP using nonlinear empirical equations. According to the measurements by Hodgkin and Huxley, depolarization of the membrane beyond the critical threshold causes the opening of the Na^+ channels, or in other words, it causes an increase in G_{Na} that results in an inward current. This current causes further depolarization of the membrane, and hence more Na^+ ion channels open. This process continues and drives the membrane potential toward E_{Na}. This process reaches a stage where the Na^+ channels gradually close. At a certain time, K^+ voltage-gated channels open allowing outward current flow but at a slower rate (increase in G_K), which causes the cell to repolarize and reach E_K. During this state, the cell briefly becomes hyperpolarized until all K^+ voltage-gated channels return to their closed state. See Figure 9.7 for a summary of the generation process of APs, a process that is an all-or-nothing process. That is to say, once

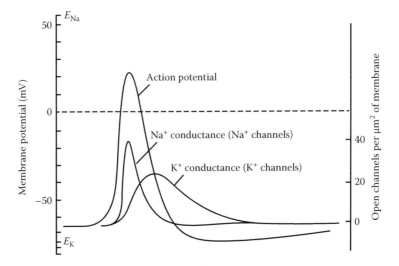

FIGURE 9.7   The process of AP and corresponding change in G_K and G_{Na}.

the critical threshold is reached, the complete process of AP generation will occur. After each AP, a period of time exists where the cell is unable to generate another AP; this period is referred to as the refractory period and lasts for approximately 2 ms.

As we have seen, when neurons are activated by means of an electrochemical concentration gradient, local current flows are produced. The electrical activity of neurons can be divided into two subsets: APs and postsynaptic potentials (PSPs). PSPs are categorized into two forms: excitatory and inhibitory. An excitatory PSP (EPSP) is a temporary depolarization of the postsynaptic membrane potential caused by the flow of positively charged ions into the postsynaptic cell. A PSP is defined as excitatory if it makes it easier for the neuron to fire and generate an AP.

They are the opposite of inhibitory PSPs (IPSPs), which usually result from the flow of negative ions into the cell. An IPSP is the change in membrane voltage of a post-synaptic neuron, which results from synaptic activation of inhibitory neurotransmitter receptors. A PSP is considered inhibitory when the resulting change in membrane voltage makes it more difficult for the cell to fire an AP, thus lowering the firing rate of the neuron.

9.7 CORTICAL ORGANIZATION

The cortex contains the vast majority of neurons of the brain (10^{10} of the total 10^{11}) and is organized into both columns and layers. The greater part of the cortex is organized into six horizontal layers. This laminar architecture extends from Layer I, the molecular layer at the cortical surface; to Layer II, the external granular layer; to Layer III, the external pyramidal layer; to Layer IV, the internal granular layer; to Layer V, the large or giant pyramidal layer (ganglionic layer); and finally to Layer VI, the fusiform layer. The cortex is also formed of columns oriented orthogonally to the brain surface. These can be minicolumns, which are the width of only one neuron (50 μm), or macrocolumns, which can be up to 3 mm in diameter and accommodating 10^5 neurons (see Figure 9.8).

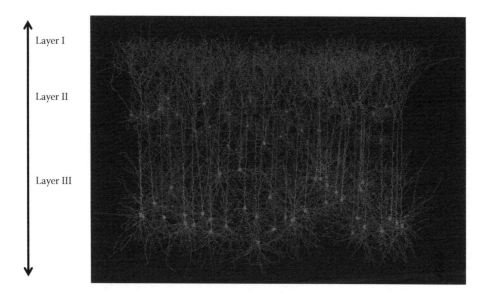

Layer I

Layer II

Layer III

FIGURE 9.8 This image (Courtesy of Professor M. Hausser, UCL, Wellcome Images.) shows synthetic neurons that represent the optimized size, shape, and connectivity of pyramidal neurons analogous to those found in the cortex of the brain. Pyramidal neurons are so called because they have a pyramid-shaped cell body (soma); they are also characterized by long branching dendrites. They are found in the forebrain (cortex and hippocampus) of mammals and are thought to be involved in cognitive function. The structure of the cerebral cortex consists of six layers. Layer III contains external pyramidal neurons, while Layer V contains large or giant pyramidal layer.

Cortical neurons can also be divided into two major groups: spiny and smooth. The spiny neurons are so called because they contain spines at the end of their dendrites onto which synapses are formed. Spiny neurons are mostly excitatory and include pyramidal neurons.

The pyramidal neuron cell is the most prevalent neuron cell in the cerebral cortex, particularly in the cortical folds and grooves (gyri and sulci, respectively) that are parallel to the scalp. Pyramidal neurons are typically seen in Layer II and Layer V of the cortex. These neurons have a pyramid-shaped soma and two distinct dendritic trees. The basal dendrites emerge from the base and the apical dendrites from the apex of the pyramidal soma that extends up perpendicularly toward the surface of the brain. Hence, most neurons in the cerebral cortex due to its structure have dendrites, which align in parallel. This has the effect that if potentials are generated in pyramidal neurons in the same macrocolumn, then this causes a summation of potentials in one direction.

The ability of pyramidal neurons to integrate information depends on the number and distribution of the synaptic inputs they receive. A single pyramidal cell receives about 30,000 excitatory inputs and 1,700 inhibitory inputs. EPSP inputs terminate exclusively on the dendritic spines, while IPSP inputs terminate on dendritic shafts, the soma, and even the axon. Pyramidal neurons use glutamate as their excitatory neurotransmitter and GABA as their inhibitory neurotransmitter.

Smooth neurons tend to be inhibitory and include cortical basket cells, which form baskets around the cell bodies of the target neurons, which typically are pyramidal neurons. There exist different types of smooth neurons that can have varying structures depending on their functionality, including chandelier cells and double bouquet cells.

The electrical potentials recordable on the scalp surface are generated by low-frequency summed IPSPs and EPSPs from pyramidal neuron cells that create electrical dipoles between the soma and their apical dendrites (see Figure 9.9). These PSPs summate in the cortex and extend to the scalp surface where they are recorded as the electroencephalogram. Neuronal APs have a much smaller potential field distribution and are much shorter in duration than PSPs. Only large populations of active neurons can generate electrical activity recordable on the scalp.

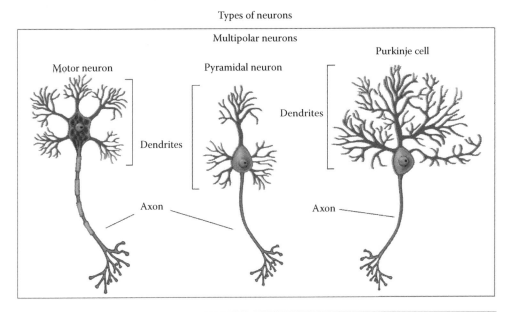

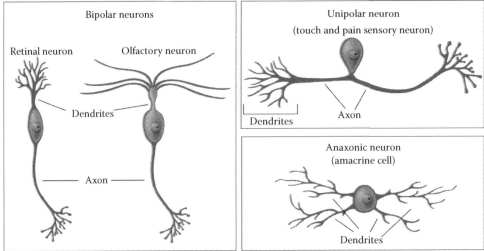

FIGURE 9.9   Types of neurons in the CNS.

Typically, four prerequisites must be met for the activity of any network of neurons to be visible in surface scalp recordings:

1. The neurons must generate most of their electrical signals along a specific axis oriented perpendicular to the scalp.

2. The neuronal dendrites must be aligned in parallel so that their field potentials summate to create a signal that is detectable at a distance.

3. The neurons should fire in near synchrony.

4. The electrical activity produced by each neuron needs to have the same electrical sign.

9.8 IMPLANTED ELECTROPHYSIOLOGICAL RECORDINGS

The electrical activity of neurons can be measured by inserting a small electrode (1 μm diameter) with a large resistance into the cell and it measures the voltage fluctuations with respect to an electrode in the extracellular space. This method allows recordings from single neurons with fine detail. However, due to the fragility of the electrode and the contact requirement of the method, this technique is only feasible in a controlled environment where the electrode is secured to the skull to prevent movement. A different approach is to place electrodes in the extracellular fluid to measure the electrical activity of neurons. When an electrode is placed close to the membrane of the cell, it enables the detection of voltage changes arising from in- and outflow of ions across the cell membrane. Thus, this method allows the recordings of voltage changes arising from the cell's APs. Electrodes placed in the extracellular space record neurons' electrical activities such as APs. These neural activities occur over the duration of a few milliseconds. In addition, these extracellular recordings contain lower-frequency activity which commonly referred to as the local field potential (LFP). It is thought that the LFP reflects the synaptic potentials and measures the average electrical activity around the electrode.

Implanted recordings involve surgical placement of electrodes in the brain, which may limit their usability in human subjects and clinical trials.

9.9 IMPLANTED ELECTRODE DESIGN

The first challenge is extracellular recordings in the tissue/electrode interface. After electrode implantation, the body reacts to this new foreign object and an inflammation response is generated. In addition, the insertion of the electrode causes damage to the surrounding tissue along the electrode insertion path. This triggers the wound healing response. Therefore, electrode design must take in account biocompatibility factors in order to minimize these while also maintaining mechanical and electrical stability.

Microwire electrodes are one of the older electrode types for implanted recording yet remain commonly used in many experiments. These wires are composed of conducting material and are coated with insulating noncytotoxic material (polyimide, parylene, or

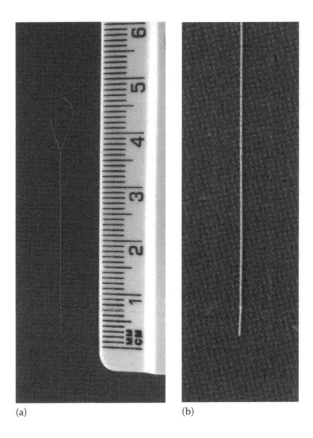

(a)   (b)

FIGURE 9.10 Microwire electrodes. In (a), electrode is composed of four electrodes together. A close-up of the tip is shown in (b).

tetrafluoroethylene) with the exception of the tip. The diameter of microwire varies from 25 to 125 μm. These electrodes are manufactured from conducting materials such as platinum, gold, tungsten, iridium, and stainless steel. The fabrication is relatively straightforward and as a consequence, the manufacturing costs are low. These electrodes can be arranged into several configurations during manufacture. For example, tetrodes are formed from four wires closely spaced, and thus such an array cannot allow recording over a larger tissue area. Knowledge of the dimensions of the tetrode can aid in the interpretation and analysis of the recorded data (Figure 9.10).

The second type of electrode design is silicon micromachined electrodes. These electrodes allow more flexibility in their design, and hence more complex designs can be achieved using this type of electrodes. Silicon electrodes offer fine control over electrode area, contact size, and spacing between recording sites, which is the main advantage of these electrodes compared to microwire electrodes. These electrodes allow recording from a larger number of larger samplings of neurons with less damage to the tissue. However, further development is required for silicon micromachined electrodes to improve their biocompatibility in chronic (recording over long duration) experiments.

Other electrode types include probes that contain microfluidic channels in addition to recording electrodes. These electrodes allow simultaneous drug delivery at the cellular

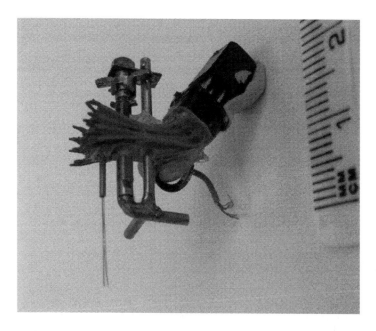

FIGURE 9.11   Headstage with electrode visible as well as a connector for connection to signal amplifiers.

level as well as recording neural activity. These probes allow detailed investigations into neural responses to chemical stimuli delivered at cellular level.

9.10 EXTRACELLULAR SIGNAL ACQUISITION

Typically, the electrodes mentioned in the preceding section are connected to a preamplifier close to the implantation area; commonly, these are referred to as the headstage (see Figure 9.11). The headstage is then connected to an acquisition box with filters, variable gain amplifier, and analog-to-digital (A/D) converter where the signals are sampled and subsequently stored for further analysis. Typically, the LFP signal is sampled at low frequencies (~200 Hz), while APs are sampled at higher frequencies (>20 kHz). Filtering parameters are chosen according to the type of signal required; LFPs are normally low-pass filtered, while APs are typically band-pass filtered. Neural signals are in the order of few microvolts (50 to 500 µV); hence, signal amplification with high gains is required. The A/D conversion bit depth can vary from 8- to 12-bit resolution. Several factors affect the design of these acquisition systems such as power consumption, size of the instruments, and manufacturing costs (see Figure 9.12).

9.11 ANALYSIS OF EXTRACELLULAR SIGNALS

LFPs are continuous signal and traditional time and frequency signal processing methods can be applied in their analysis. The EEG and LFP share similar frequency characteristics; hence, methods applied in EEG analysis can be also employed to analyze LFPs.

APs on the other hand are discrete events commonly referred to as point processes and require a different approach in their analysis. APs acquired from implanted electrodes are commonly referred to as spikes. Hence, analysis of APs is often termed spike analysis.

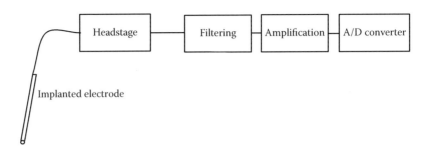

FIGURE 9.12   Extracellular recording experimental setup.

9.11.1 Spike Analysis

The first step in analyzing neural spike activity is referred to as spike sorting. When an electrode is placed in the extracellular space, it may record spike activity from more than one neuron in the vicinity of the electrode. The sorting role is therefore to separate the spikes generated by each neuron so that their function can be examined in detail.

Spike sorting algorithms are typically composed of four steps in total (illustrated in Figure 9.13). The first step involves detecting spike segments. The second step consists of extracting features that best discriminate the spikes produced by the different neurons. In the third step, the number of neurons is estimated, and often, this step is carried in conjunction with the final step. In the final step, each spike is assigned to the neuron that generated it.

The first step in spike sorting aims at separating the spikes from the LFP; typically, this step is carried out by band-pass filtering recorded data to remove low frequency interference, with the spikes subsequently detected by observing recorded segments that have amplitudes that are larger than a preset threshold. Other methods such as nonlinear energy operators and wavelet transforms can be also applied to achieve this task.

Subsequent steps aim at separating spikes generated by different neurons. Typically, feature extraction methods that discriminate the spikes of different neurons are extracted; commonly used include principal component analysis, wavelet transform, or temporal features of the spikes such as spike amplitude and width. Then, unsupervised clustering methods are employed to find the number of neurons contributing to the recording and assign each spike to the neuron that produced it. For further details on these methods, refer to Lewicki (1998).

Subsequent analysis to the spike sorting stage is to examine the roles of individual neurons. One of the most important features in assessing the roles of individual neurons is to examine the firing rate of each neuron. For example, there is a subset of cells reported in animal studies where the neuron's firing rate is dependent on the animal's head position. These cells are commonly referred to as head-directional cells, where the cell fires maximally when the animal's head is in specific direction and the firing decreases as the animal moves its head away from this specific direction. A simple analysis to evaluate the relationship between cell spikes and behavioral state (e.g., head direction) is to

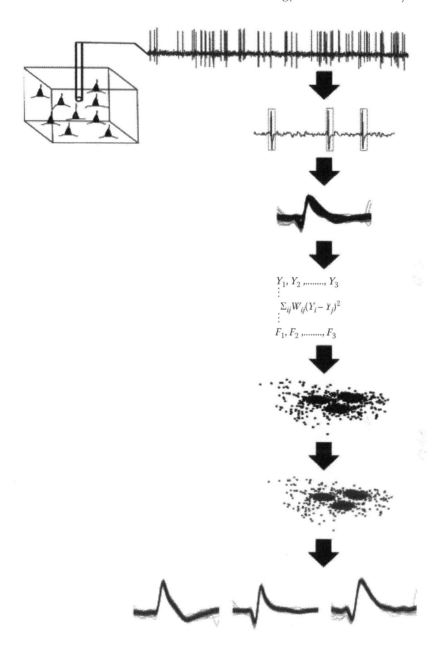

FIGURE 9.13 Spike sorting process, spike detection, feature extraction, identification of the number of neurons, and clustering.

graphically plot the firing rate against the head direction (Figure 9.14). These plots are known as tuning curves for the cell type.

Information theory can be also applied to assess how much information the neurons convey about the subject's behavioral state. Information theory is extensively employed for this purpose. As in the head-directional cell example, information theory can be used to calculate the amount of information conveyed by each spike regarding head direction.

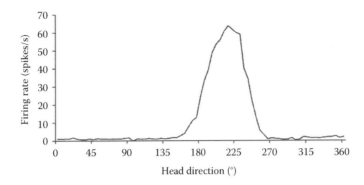

FIGURE 9.14 An example of firing rate–based analysis where the firing rate is graphically represented against behavioral state (head direction).

If the head direction (0° to 360°) is divided into j bins representing degrees (e.g., 5° or 10°), then the information content I can be calculated using the formula

$$I = \sum p_j \frac{\lambda_j}{\lambda} \log_2 \left(\frac{\lambda_j}{\lambda} \right) \tag{9.3}$$

where
  p_j is the probability of occupying bin j
  λ_j is the average firing rate in bin j
  λ is the average firing rate of the cell

Decoding algorithms are also applied to decipher the neural information encoded in spike data. These methods attempt to predict the behavioral state from the recorded neural response. Bayesian decoding is one such method. Such methods typically require training data sets with the decoding parameters obtained from the training set and the decoding performance assessed on a data set that was not included in the training process.

It has to be noted that these analysis methods can also be applied to study the effect of specific stimuli rather than behavioral activity. For example, if a picture stimulus is presented, then information theory can be applied to study the spiking activity of neurons in response to the visual stimulus presented. For a detailed review of spike analysis based on information theory, refer to Quian and Panzeri (2009).

The trend in neural implanted recordings is moving toward simultaneous recording of hundreds of neurons. Hence, the challenge is to apply automated analysis, which will also factor the interaction between these recorded neurons to understand activity in the CNS. It has been shown from analysis that oscillations observed in the LFP encode information that complement information conveyed by spikes. Hence, taking into consideration the information carried between different modalities can yield a better understanding of the underlying process in the CNS. While analysis methods such as cross correlogram and coherence can be used to study the relation between different neurons, these methods tend to be limited in examining the relation between a few neurons.

9.12 ELECTROENCEPHALOGRAPHY

The generation of the electrochemical currents and the laminar architecture of the cortex with its parallel macrocolumns creates the situation whereby the summation of these currents can be measured by scalp electrodes. The existence of electrical currents in the brain was reported in 1875 by a Liverpool surgeon named Richard Caton (1842 to 1926). He studied APs from the exposed brains of rabbits and monkeys. Hans Berger (1873 to 1941), a German neuropsychiatrist working on cerebral localization and intracranial blood circulation, followed on from Caton's work. In 1924, he used his ordinary radio equipment to amplify the brain's electrical activity measured on the human scalp. This was the first electroencephalogram (EEG) recording of humans. He showed that weak electrical currents generated in the brain can be recorded without opening the skull and be depicted graphically on a strip of paper:

> We see in the electroencephalogram a concomitant phenomenon of the continuous nerve processes which take place in the brain.
>
> BERGER (1929)

The activity that he observed changed according to the functional status of the brain, such as in sleep, anesthesia, lack of oxygen, and certain neural diseases such as in epilepsy. He was correct in his assertion that brain activity changes in a consistent and recognizable way when the general status of the subject changes, as from relaxation to alertness. Berger was the first to use the word *electroencephalogram* to describe the brain electrical potentials in humans. He laid the foundations for many of the present applications for EEG and as a result earned himself the title as the *father of EEG*.

The modern-day electroencephalogram (EEG) is defined as the electrical activity of an alternating type generated by brain structures and recorded from the scalp surface by metal electrodes and conductive media. Many variations of the scalp-recorded EEG exist. EEG measured directly from the cortical surface using subdural electrodes is called the electrocorticogram (ECoG), while when using a depth probe, it is called electrogram. In this chapter, we will refer only to EEG recorded from the surface of the scalp. In the following subsection, electroencephalography is compared with other brain imaging technologies, highlighting its advantages for the purposes of measurement of CNS activity.

9.12.1 Brain Imaging

Modern medicine applies a variety of imaging methodologies to analyze the functioning of the human body. The group of electrobiological measurements consists of electrocardiography (ECG, heart), electromyography (EMG, muscle contraction), electrogastrography (EGG, stomach), electrooculography (EOG, eye dipole field), and electroencephalography

(EEG, brain). EEG involves the recording of scalp electrical activity generated by brain structures. It is just one of the many brain imaging methodologies that have been developed in the pursuit of the ability to visualize brain condition and understand brain function. Based on various physical properties, other brain imaging methods include x-ray computed (axial) tomography (CT), positron emission tomography (PET), single-photon emission computed tomography (SPECT), rapid-rate transcranial magnetic stimulation (rTMS), event-related optical signal (EROS), magnetoencephalography (MEG), and (functional) magnetic resonance imaging (fMRI). Table 9.2 provides a summary of each brain imaging methodology, its physical measurement property, and highlights its applications and presents its relative advantages and disadvantages.

Brain scans are subject to a phenomenon analogous to an uncertainty principle: we can detect either the localization or the timing of neural activation, but not both. Figure 9.15 depicts the relative spatiotemporal resolutions for some brain imaging methodologies and the progression toward a method with both good temporal and spatial resolution. Multimodal fusion of imaging methods such as EEG and fMRI is becoming more prevalent and may facilitate further improvements in imaging resolution of real-time activity.

Despite EEG's poor spatial resolution, it has excellent temporal resolution of less than 1 ms. It is also relatively inexpensive and simple to acquire making it the only practical noninvasive brain imaging modality for repeated real-time brain behavioral analysis. For this reason, the remainder of this chapter will focus on EEG as the input brain imaging modality for activity in the CNS.

With the development of the MRI scanner, neuroscience has become a hugely popular research area spanning disciplines such as neurophysiology, psychology, engineering, mathematics, and clinical rehabilitation. It has inspired many to explore every aspect of clinical and cognitive brain imaging. fMRI has enabled the location with high spatial precision of regions within the brain associated with specific function, for example, the response to a sensory or cognitive stimulus. Similarly, it has inspired the use of EEG in parallel to observe with high precision the dynamics of such brain activity. Since the architecture of the brain is nonuniform and the cortex is functionally organized, interpreting EEG waveforms is very dependent on the acquisition methods employed.

9.13 ACQUISITION METHODS

In scalp-recorded EEG, the neuronal electrical activity is recorded noninvasively, typically using small metal plate electrodes. Recordings can be made using either reference electrode(s) or bipolar linkages. While the number of the electrodes employed varies from study to study, they are typically placed at specific scalp locations. The voltages, of the order of microvolts (μV), must be carefully recorded to avoid interference and digitized so that they can be stored and analyzed.

The amplitude of the recorded potentials depends on the intensity of the electrical source, on its distance from the recording electrodes, on its spatial orientation, and on the electrical properties of the structures between the source and the recording electrode. The greatest contributions to the scalp-recorded signals result from potential changes that (a) occur near the recording electrodes, (b) are generated by cortical dipole layers that are

TABLE 9.2   Comparison of Brain Imaging Methodologies

Brain Imaging Technology	Physical Measurement Property	Advantages	Disadvantages
EEG	Macroscopic brain electrophysiology.	Inexpense and ease of acquisition High temporal resolution Noninvasive	Poor spatial resolution Trial-to-trial variability of ERPs Volumetric smearing effects of skull
ECoG	Electrophysiology of extracellular currents.	High temporal resolution Good spatial resolution Equipment inexpensive	Highly invasive Requires surgery, craniotomy required Procedurally expensive
MEG	Cortical magnetic fields associated with the electrical activity.	Completely noninvasive Spatial resolution up to 3 mm Good temporal resolution	Extremely expensive equipment Difficult to acquire, requires magnetic isolation room Not practical for real-time analysis
CT	X-ray brightness intensity maps in relation to brain tissue density. Superseded by MRI technology	Excellent spatial resolution	Only anatomical information, none on cognitive function Extremely expensive X-ray radiation hazard
SPECT	Tracking of radioactive tracers in blood stream. The measurement of blood flow, oxygen, and glucose reflects the amount of brain activity.	Do not require on-site cyclotron to produce SPECT tracers, unlike PET Less technical and medical staff required than PET	Measures blood flow instead of electrophysiology Poor time and spatial resolution Ionizing radiation hazard Procedurally expensive More limited than PET tracers
PET	Tracking of gamma radiation from decaying radioactive tracers in the blood stream. Measures the regional cerebral metabolism and blood flow that reflect the amount of brain activity.	More versatile than SPECT More spatial resolution than SPECT, particularly for deeper brain structures able to identify which brain receptors are being activated by neurotransmitters, abused drugs, and potential treatment compounds	Measures metabolism of oxygen and sugar rather than electrophysiology Ionizing radiation hazard Poor time resolution (~2 min) With measurements that cannot be repeated, annual maximum dosage is one examination Equipment extremely expensive (cyclotron required) Highly qualified staff required
MRI	Radio waves pass through a large magnetic field (~1.5 T). A computer monitors the variations in the radio waves due to the electromagnetic activity in the brain to generate a picture.	High anatomical detail (spatial resolution) Noninvasive	Only anatomical information, none on cognitive functionality Poor temporal resolution Expensive equipment and procedure

(Continued)

TABLE 9.2 (*Continued*)   Comparison of Brain Imaging Methodologies

Brain Imaging Technology	Physical Measurement Property	Advantages	Disadvantages
fMRI	Magnetic fields and radio waves exploit the magnetic properties of blood to track its flow. It involves monitoring the blood oxygenation-level dependence (BOLD) in response to a function or stimulus.	Good spatial resolution Noninvasive	Depends on hemodynamic response of blood that introduces an inherent lag Trades off some spatial resolution from MRI to improve temporal resolution Expensive equipment and procedure Temporal resolution (~1 s) not as good as EEG
(r)TMS	Induced electrical activity of neuronal regions by pulsed magnetic fields to stimulate a brain region.	Diagnostic aid to check that the nerve pathways are intact Able to influence many brain functions, including movement, visual perception, memory, reaction time, speech, and mood	Stimulator of brain function rather than an imaging technique Unknown health risks
EROS	Changes occur in optical parameters (scattering and absorption) of cortical tissue when active.	Good temporal resolution Noninvasive Good spatial resolution Nononizing radiation Relatively low cost	Penetration of only several centimeters Studying the cortical activity rather than the subcortical Infancy of its development

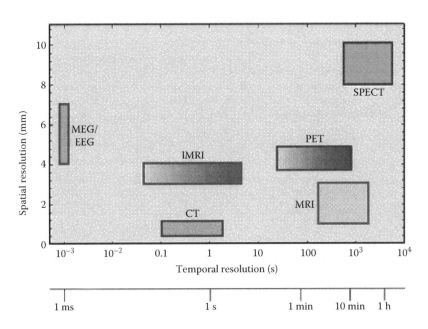

FIGURE 9.15   Scale of spatiotemporal resolution for various brain imaging technologies.

oriented toward the recording electrode at a 90° angle to the scalp surface, (c) are generated in a large area of tissue, and (d) rise and fall at rapid rate.

This section will briefly highlight the equipment, methods, and standards involved in acquiring a scalp-recorded EEG. It serves as a review of the key issues related to acquisition.

9.13.1 EEG Recording Equipment

The basic EEG recording system consists of electrodes with conductive media, amplifiers with filters, an A/D converter, and finally a recording device to store the data. Electrodes, in conjunction with the electrode gel, sense the signal from the scalp surface; amplifiers bring the microvolt and often nanovolt signals into a range where they can be digitized accurately; and the A/D converter changes signals from analog to digital form that can be finally stored or viewed on a computer. Table 9.3 provides a summary of the necessary EEG acquisition equipment and the typical specifications or products. The equipment typically employed for EEG acquisition in a neurophysiological laboratory is shown in Figure 9.16.

9.13.2 Electrodes and Electrode Placement

An electrode is a small conductive plate that picks up the electrical activity of the medium that it is in contact with. In the case of EEG, electrodes provide the interface between the skin and the recording apparatus by transforming the ionic current on the skin to the electrical current in the electrode. Conductive electrolyte media ensure a good electrical contact by lowering the contact impedance at the electrode–skin interface.

The following types of electrodes are available:

- Disposable (gel-free and pregelled types)

- Reusable cup electrodes (gold, silver, stainless steel, or tin)

- Electrode caps

- Needle electrodes

- Nasopharyngeal and sphenoidal electrodes

TABLE 9.3   Necessary EEG Acquisition Equipment Specifications

EEG Acquisition Component	Typical Specifications or Products
Electrodes	Electrode cap with conductive jelly Ag–AgCl or Au disc electrodes with conductive paste
Amplifiers (with filters)	Amp gain between 100 and 100k Input impedance >100 MΩ Common-mode rejection ratio >100 dB High-pass filter with corner frequency in the range 0.1–0.7 Hz Low-pass filter with corner frequency below half the sampling rate Notch filter at power line frequency (50/60 Hz)
A/D converter	At least a 12 = bit A/D converter with accuracy lower than overall noise (0.3–2 µV pp) and sampling frequency typically between 128 and 1024 Hz per channel
Storage/visualization unit	Sufficiently fast PC for presentation, processing, and storage

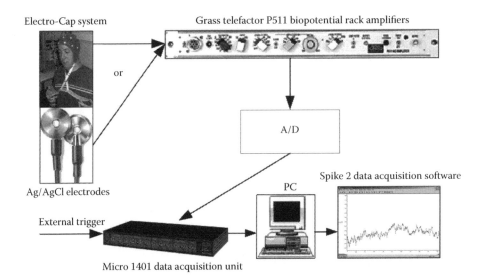

FIGURE 9.16   EEG acquisition system in a typical laboratory. Equipment suppliers include electrode caps, low-noise amplifiers for neurophysiological signals, data acquisition system, and computer for storage and analysis.

For large multichannel montages comprising of up to 256 or 512 active electrodes, electrode caps such as those shown in Figure 9.17 are preferred to facilitate quicker setup of high-density recordings. Commonly, Ag–AgCl cup or disc electrodes of approximately 1 cm diameter are used for low-density or variable placement recordings as shown in Figure 9.18. The space between the electrode disc and the skin is filled with conductive paste, which also helps them to bond to the scalp.

To improve the stability of the signal, the outer layer of the skin (stratum corneum) should be removed under the electrodes by light abrasion. In the case of the electrode caps, the blunt syringe for conductive gel insertion is also used for skin scraping. The

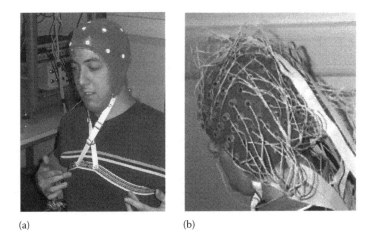

(a)   (b)

FIGURE 9.17   (a) Electro–Cap System consisting of 21 electrodes according to 10–20 system, (b) 128-channel electrode cap based on a modified 10–20 system.

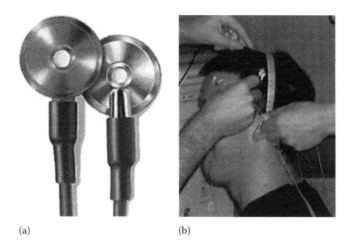

(a)   (b)

FIGURE 9.18   (a) Gold (Au) and silver–silver chloride (Ag–AgCl) disc electrodes, (b) electrode placement over C3 and C4. The interelectrode and skin–electrode impedance is checked.

cleaning of the skin surface of sweat, oil, hair products, and dried skin is also highly recommended. Sterile alcoholic medical wipes are useful here to prepare the skin and maintain hygiene.

In order to standardize the methodology of applying the electrodes on the skull, in 1949, the International Federation of Societies for Electroencephalography and Clinical Neurophysiology (IFSECN) adopted a system that has been adopted worldwide and is referred to as the 10–20 electrode placement international standard. This system, consisting of 21 electrodes, standardized physical placement and nomenclature of electrodes on the scalp. This allowed researchers to compare their findings in a more consistent manner. In the system, the head is divided into proportional distances from prominent skull landmarks (nasion, inion, mastoid, and preauricular points). The 10–20 label in the system title designates the proportional distances in percentage between the nasion and inion in the anterior–posterior plane and between the mastoids in the dorsal–ventral plane (see Figure 9.19).

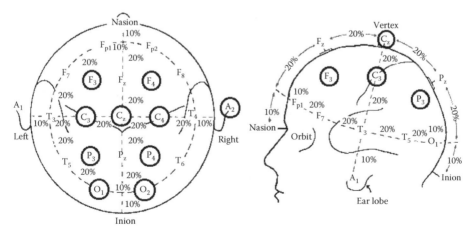

FIGURE 9.19   The international 10–20 electrode system: top and side views.

Electrode placements are labeled according to adjacent brain regions: F (frontal), C (central), P (parietal), T (temporal), and O (occipital). The letters are accompanied by odd numbers for electrodes on the ventral (left) side and even numbers for those on the dorsal (right) side. The letter *z* instead of a number denotes the midline electrodes. The left and right sides are considered by convention from the point of view of the subject. Based on the principles of the 10–20 system, a 10–10 system and a 10–5 system have been introduced as extensions to further promote standardization in high-resolution EEG studies (see Figures 9.20 and 9.21).

The high-density EEG electrode placement can help pinpoint more accurately the brain region contributing to the recording at a given electrode. This is known as source localization. There exist signal processing methods to generate inverse solution models to estimate the origins of the main components of the EEG recordings. However, the scalp electrodes may not reflect the activity of particular areas of the cortex, as the exact location of the active

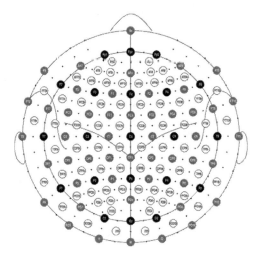

FIGURE 9.20   The international 10–20 electrode system: top view. Electrode positions and labels for various systems. Black circles indicate positions of the original 10–20 system, gray circles indicate additional positions in the 10–10 extension, and small dots indicate additional positions in the 10–5 extension.

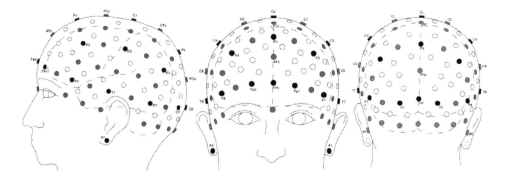

FIGURE 9.21   Selection of 10–10 electrode positions in a realistic display: lateral, frontal, and posterior views. The head and brain contours based on typical models. Black circles indicate positions of the original 10–20 system; gray circles indicate additional positions in the 10–10 extension.

sources is still unknown due to the limitations caused by the nonhomogenous properties of the skull, different orientation of the cortex sources, and coherences between the sources.

9.13.3 Basic EEG Amplifier

The amplifier that is often used in conjunction with EEG electrodes is based on a configuration of operational amplifiers or op-amps. An op-amp is a differential amplifier that amplifies the difference between two input signals. One input has a positive effect on the output signal; the other input has a negative effect on the output. The perfect op-amp theoretically has an infinite voltage gain, being able to amplify a signal to any level; theoretically has infinite bandwidth, being able to amplifier any frequency without distortion; and theoretically has infinite input impedance, being able to sense an input voltage level without distorting that voltage in any way. The ideal op-amp also has $0\ \Omega$ output impedance, allowing it to be connected to any other electronic device without distortion. In reality, such characteristics are not met, but real op-amps do achieve great performance when correctly configured.

9.13.3.1 Inverting and Noninverting Amplifier

The inverting amplifier is one such configuration. The op-amp is connected using two resistors R_A and R_B such that the input signal is applied in series with R_A and the output is connected as feedback to the inverting input through R_B. The noninverting input is connected to a reference voltage, typically electrical ground or 0 V. This can be achieved by connecting to the center tap of the dual-polarity power supply (Figure 9.22a).

During operation, as the input signal becomes positive, the output will become negative. The opposite is true as the input signal becomes negative. The amount of voltage change at the output relative to the input depends on the ratio of the two resistors R_A and R_B. As the

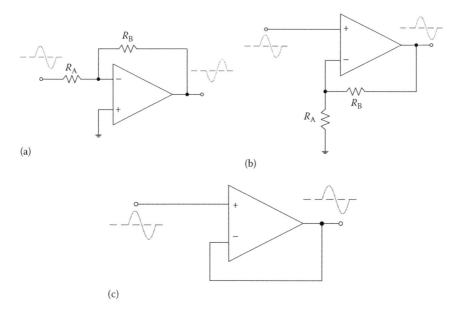

(a)

(b)

(c)

FIGURE 9.22   (a) Inverting amplifier, (b) noninverting amplifier, and (c) voltage follower.

input changes either positively or negatively, the output will change in the opposite direction, so that the voltage at the inverting input remains constant or 0 V in this case. If R_A is 1 kΩ and R_B is 10 kΩ and if the input is +1 V, then there will be 1 mA of current flowing through R_A, and the output will have to move to −10 V to supply the same current through R_B and keep the voltage at the inverting input at zero. The voltage gain in this case would be R_B/R_A or 10 kΩ/1 kΩ = 10. Since the voltage at the inverting input is always zero, the input signal will see input impedance equal to R_A, or 1 kΩ in this case. For higher input impedances, both resistor values can be increased.

Another configuration is the noninverting amplifier, where the input signal is connected to the noninverting input of the op-amp with resistor R_A connected between the inverting input and electrical ground (Figure 9.22b). As the input signal changes either positively or negatively, the output will follow in phase to maintain the inverting input at the same voltage as the input. The voltage gain is always greater than 1 and can be calculated as $V_{gain} = 1 + R_B/R_A$.

Another configuration known as the voltage follower, also called an electrical buffer, provides high input impedance, low output impedance and unity gain. As the input voltage changes, the output and inverting input will change by an equal amount (Figure 9.22c). The voltage follower is often used to electrically isolate and impedance balance the electrode and the data acquisition computer, to allow subsequent analysis and signal processing to take place.

9.13.3.2 Input Impedance

Impedance (symbol Z) is a measure of the overall opposition of a circuit to current flow. It may be a simple resistance, but it may also take into account the effects of capacitance and inductance. Impedance is more complex than resistance because the effects of capacitance and inductance vary with the frequency of the current passing through the circuit, thus implying that impedance varies with frequency. Input impedance (Z_{IN}) is the impedance resulting from the connection of any input to a circuit or device (such as an amplifier). It is the combined effect of all the resistance, capacitance, and inductance connected to the input of the amplifier. The effects of capacitance and inductance are generally most significant at high frequencies. It is a critical parameter for the amplification of quality EEG signals.

For good signal amplification, the input impedances should be high, at least 10 times the output impedance of the sensor supplying a signal to the input. This ensures that the amplifier input will not overload the EEG signal and will thus reduce the voltage of the EEG signal by a substantial amount.

Human skin presents a large impedance if connected via an electrode to an amplifier. The impedance is greatly reduced if the surface layer of the skin is removed. This layer is known as the stratum corneum. In practice, as mentioned earlier, for EEG recording, the stratum corneum is removed by lightly abrading the skin. The skin contact with the electrolyte gel and silver–silver chloride electrode now presents significantly lower impedance. For EEG, input impedance at each electrode site is recommended to be kept below 5 kΩ.

9.13.3.3 Instrumentation Amplifier

An instrumentation amplifier is a specific type of differential amplifier that has been specifically designed to have characteristics suitable for use in physiological measurement. Its characteristics include very low DC offset, low electrical drift, low noise, very high gain, very high common-mode rejection ratio, and very high input impedance. Instrumentation amplifiers are used in recording EEG, where great accuracy and stability of the circuit, both short and long terms, are required. The ideal common-mode gain of an instrumentation amplifier is zero. While instrumentation amplifiers can be built with individual op-amps and precision resistors, they are most useful when available in integrated circuit form. EEG instrumentation amplifiers typically have very stringent characteristics. The peak-to-peak output noise is typically less than 1 μV in the EEG range of interest 1 to 100 Hz.

9.13.4 EEG Recording Circuitry

A typical EEG channel amplification circuit is shown in Figure 9.23. The EEG signal is first amplified by a high-quality instrumentation preamplifier, which measures the voltage difference between two locations on the scalp. As mentioned earlier given its differential configuration, this ensures that similar electrical interference present at one or both electrodes never enters the system, as the level of the electrical interference on those two scalp locations is essentially the same. The instrumentation preamplifier typically applies an amplification factor of 20.

Following this preamplification, the signal is amplified by a factor of 40 in a second amplifier stage. There are two main reasons to divide the amplification into two stages. First, we can site the preamplifier close to the electrode site, which improves signal quality. Second, the two-stage process allows for the insertion of a high-pass filter, which removes any DC voltage offsets generated by the preamp stage. Materials employed for electrodes, such as gold or steel, are polarizable. This means that electric charge can accumulate on the surface of the electrode, building up a large DC offset voltage, sometimes several hundreds of millivolts.

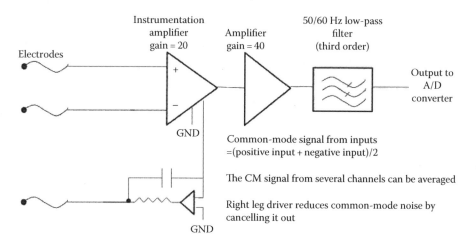

FIGURE 9.23   A typical EEG channel amplification circuit.

Finally, the signal is amplified with a high amplification factor and low-pass filtered. Such filtering is implemented to prevent aliasing effects when the signal is digitized.

To improve the quality of recorded signal, a third amplifier circuit is used. This is the driven-right-leg (DRL) circuit, a name derived historically from ECG circuits. The purpose of the DRL is to reduce common-mode signals, such as electrical interference (50 or 60 Hz), by cancelling them out. The EEG signal can very easily be swamped by electrical interference, which is transmitted capacitively from power line electrical wiring. With the DRL circuit, the system can attenuate electrical interference many times more efficiently than the instrumentation amplifier working in isolation.

9.13.5 Referencing and Bipolar Recordings

EEG recordings can be divided into two major categories: reference recordings and scalp-to-scalp bipolar linkage recordings. In the reference recording, each electrode is referred either to a distant reference electrode, one common electrode on each side of the head, or to the combined activity of two or more electrodes. The reference electrode(s) must be placed on the parts of the body where the electrical potential remains fairly constant. Several different recording reference electrode placements are mentioned in the literature. These include references such as the vertex (Cz), linked ears, linked mastoids (bones behind the ear), ipsilateral ear, contralateral ear, FPz, AFz, and the tip of the nose. In addition to one single reference electrode, two reference electrodes shorted together can be used. The choice of reference may produce topographic distortion if relatively electrically inactive locations such as the ear and mastoid are not chosen. Linking reference electrodes from both earlobes or mastoids reduces the likelihood of artificially biasing activity in one hemisphere.

The most preferred reference is linked ears due to the relative electrical inactivity and ease of use and symmetry, which prevents a hemispheric bias being introduced. The vertex (Cz) is also predominant due to its central location. Reference-free methods are represented by common average reference, weighted average reference, and source derivation. These methods use the average or weighted average of the activity at all electrodes as the reference. This can be carried out by means of a resistor network or digitally as part of a postprocessing method. These do not suffer the same artifact problems associated with an actual physical reference.

Bipolar recordings are differential measurements that are made between successive pairs of electrodes. In the literature, C3–C3′ refers to a bipolar link from one channel at C3 and the other in close proximity (typically around 2 to 3 cm from electrode in direction away from Cz). Closely linked bipolar recordings are affected less by some artifacts, particularly ECG, due to the differential cancelling out of signals similarly picked up at the pair of electrodes. Bipolar referencing is not commonly used due to placement issues and a lack of spatial resolution.

9.13.6 Artifacts

Artifacts are undesirable potentials of noncerebral origin that contaminate the EEG signals. As they can potentially masquerade and be misinterpreted as originating from the

TABLE 9.4   Groups of Physiological Artifacts and Their Origins during EEG Recordings

Physiological Artifact Type	Possible Sources
Movements	Movements of the head, body, scalp or skin stretch
Bioelectrical potentials	Moving electrical potentials within the body, such as those produced by eye, tongue, and pharyngeal muscle movement. Also those generated by the scalp muscles, heart (ECG), or sweat glands
Skin resistance fluctuations	Sweat gland activity, perspiration, and vasomotor activity

brain, there is a need to avoid, minimize, or ideally remove them from the EEG recording to facilitate accurate interpretation.

Typical EEG artifacts originate from two sources, technical and physiological. Technical artifacts are mainly due to line interference and equipment malfunction or result from poor electrode contact. Incorrect gain, offset, or filter settings for the amplifier will cause clipping, saturation, or distortion of the recorded signals. Technical artifacts can be avoided through proper apparatus setup, meticulous inspection of equipment, and consistent monitoring. Physiological artifacts arise from a variety of body activities that are due to either movements such as skin stretch, other bioelectrical potentials, or skin resistance fluctuations as summarized in Table 9.4.

The predominant physiological artifacts include electrooculographic activity (EOG, eye), scalp-recorded electromyographic activity (EMG, muscle), electrocardiographic activity (ECG, heart), ballistocardiographic activity (heart-related pulsatile motion), and respiration. These artifacts are always present to some extent and are typically much more prominent on the scalp than the macroscopic cerebral potentials. This results in an undesirable low signal-to-noise ratio in the EEG. Physiological artifacts are often involuntary and hence cannot be controlled or *turned off* during acquisition. They pose a much greater challenge than technical artifacts to avoid or remove them. Traditionally, noncerebral recordings such as EOG, ECG, or EMG are also performed to aid in the discrimination and potentially the removal of the relevant artifacts from the EEG signals. Table 9.5 presents a summary of the most common technical and physiological artifacts, their probable causes, and potential steps to avoid/remove them.

Vertical eye movements such as eye closure or blinking, which are recorded as vertical electrooculographic (vEOG) spikes, spread to a certain degree to all scalp locations, particularly the frontal sites. Although subjects are typically asked not to move or blink during acquisition, blinking is for the most part an involuntary response to dry eyes or prolonged visual focus.

Eye movement artifacts in EEG can be identified by their frontal distribution, their symmetry, and their characteristic shape. The amplitude of vertical eye movement decreases in successive channels from anterior to posterior. In addition, typical eye blinks are 100 ms in duration. Therefore, repetitive eye movements may mimic cerebral rhythms at around 10 Hz (alpha band), which is an important rhythm in most EEG studies. EOG artifacts can sometimes be eliminated or at least minimized by signal processing methods.

TABLE 9.5 Common Technical and Physiological Artifacts and Their Possible Causes and Solutions

Origin	Artifact Examples	Potential Sources	Possible solutions
Technical	Line interference (50/60 Hz)	Power supply interference—the surrounding electrical equipment may induce a 50 Hz (Europe) or 60 Hz (USA) component in the signal especially in the case of high electrode impedance at contact.	Use shorter electrode wires. Reduce electrode impedance. Scalp–electrode impedance <5 kΩ. Electrode–electrode impedance ~1 MΩ. Perform recordings within a shielded room. Use an analog or digital notch filter.
	Electrode impedance fluctuations	Loose or damaged wire contacts. Inconsistencies in electrode gel application (contaminants or dried pieces). Cable movements.	Check impedances with digital multimeter and ensure good electrode contact.
Physiological	Electrooculographic (EOG), eye activity	Eye movements, blinking. Saccadic activity related to focus variation.	Instruct subject to minimize eye movement. Do not stare, as this will force eye blinking. Allow brief intervals between stimuli for subject to blink.
	Electromyographic (EMG), muscle activity	Muscle movement or tension particularly the masticatory (jaw and tongue), neck, and forehead causes EMG artifacts.	Ensure no movement particularly the neck, mouth, tongue, or face.
	Electrocardiographic (ECG), heart activity	Heart beating. Ballistocardiographic (pulsatile motion of heart). Respiration. Pacer.	Close bipolar referencing. Reference electrode having similar ECG activity.
	Skin conduction variation	Sweating offers an epidermal layer of lower impedance for current conduction. It produces a slow baseline drift in EEG recording.	Cool environment.

There are three main approaches to combat the problem of artifacts in EEG recordings:

1. Artifact minimization or avoidance

2. Artifact rejection

3. Artifact removal

Although these approaches may appear similar, they have very different methodologies and resulting EEG data.

Artifact minimization is a prerequisite to most EEG recordings, whether or not arti-fact rejection or artifact removal is performed in conjunction. It uses a thorough under-standing of the artifacts' origins to first identify them and then reduce or eliminate their impact on the EEG by some appropriate steps. This is predominantly effective for technical artifacts but can also be applied to physiological artifacts. For example, potential changes generated by the heart are picked up mainly in EEG recordings with wide interelectrode distances, especially in interhemispheric linkages across the head or referencing to the left ear and particularly in subjects who are overweight. This understanding can then be used to avoid or at least limit the impact the ECG artifact may have on the EEG signals by employing linked-ear referencing or a grounding collar. Similarly, for other artifacts, an understanding of their origins can be exploited to minimize their influence.

Artifact rejection involves the identification and exclusion of the artifact segments from the EEG trace. In the past, it was conventionally performed by trained experts who visually scored the EEG data or the artifact activity itself (such as EOG, EMG, and ECG), rejecting any periods of EEG with unacceptable levels of artifact activity. This is an extremely laborious and inconsistent approach but is typical in clinical studies. More popular today are automated signal processing approaches (e.g., threshold-based rejection methods) that make it much less labor intensive to implement but at the cost of under- or overrejection of EEG data. The goal in artifact rejection is to produce EEG data that are as clean as possible, ignoring the cost of dis-carding valuable periods of EEG recordings. This is a huge drawback and would be unaccept-able for the purposes of processing EEG signals on a real-time or on a single-trial basis where every segment of EEG data is required. This requires the artifact to be removed from the con-tinuous EEG and can only be performed with the aid of digital signal processing algorithms.

Artifact removal methods utilize such algorithms to isolate and remove, as best as pos-sible, certain artifacts from the recorded EEG activity most importantly, unlike the other approaches retaining the period of EEG. These methods are a very topical research area and may be divided into two different approaches: filtering and higher-order statistical separation.

Filtering involves the development of a filter model to emulate the artifact activity and use it to remove the artifact from the EEG recorded signals. The filter coefficients can be established either from definitive artifact properties such as line interference at 50 Hz or from empirical processing of the noncerebral bioelectrical artifact recordings such as ECG, EMG, and EOG. This may result in a conventional low-pass, high-pass, band-pass, or notch filter or a more complex filter model. Typically, continuous adaptive regressive filtering is used in this approach. Regressive filtering methods in the time or frequency domain can, despite their computational efficiency, overcompensate for the artifact contribution result-ing in the loss of EEG information or the introduction of new artifacts.

Ideally, a method is required for removing artifacts on a single-trial basis from EEG recordings without corrupting or smearing the underlying EEG data, thus removing the need to reject corrupted periods or trials of EEG data. However, due to the uncertainty and variability in the desired EEG signals, the large number of potential artifact sources, and the significant background neuronal (EEG) noise, this is difficult to achieve. Due to the large number of unknown sources of neuronal and nonneuronal origins contributing to the recorded signal, it becomes a blind source separation (BSS) challenge to identify and remove artifacts on a single-trial basis. The BSS method exploits the higher-order statistical differences between the contributory signals to discriminate possible artifact and cerebral components. This is a very difficult problem if one is uncertain of the properties of the artifacts one wishes to separate from the EEG. Artifacts such as EOG, ECG, and EMG activity can be monitored and used as inputs in such algorithms to identify the isolated components and subsequently filter out their contributions to the EEG recordings without smearing the underlying EEG activity. Makeig et al. (1997) successfully employed the independent component analysis (ICA), a BSS method, to remove blink (EOG), EMG, and line interference artifacts and highlight its superiority over adaptive regressive methods.

The ICA and regressive filter approaches have their own relative advantages and disadvantages in terms of effectiveness, computational complexity, and implementation practicality. Despite this research topic receiving much attention, a de facto standard has not yet been set for EEG artifact removal due to the underlying uncertainty of the contributory neuronal and nonneuronal signals.

9.14 EEG SIGNAL PROPERTIES

This section serves as an introduction to the analysis and characteristics of scalp-recorded EEG signals from both clinical and signal processing points of view. The focus here is to characterize EEG activity but more importantly the generated EEG-related features that are associated with the brain's response to a specific function or specific stimulus.

The recorded EEG originates from a multitude of different neural communities from various regions of the brain as mentioned earlier. These neural communities produce electrical contributions or components that can differ by a number of characteristics such as topographic location, firing rate (frequency), amplitude, and latency. An assumption that is often made in these methods is that these components are independent of one another, thereby inferring that certain neural populations are acting in isolation of one another, even if it is in response to the same stimulus. Due to the complex interconnectivity of neuronal cells from one brain region to another, it would be difficult to justify such an assumption physiologically. The volumetric effect of the cerebrospinal fluid, skull, and scalp results in a smearing of these many electrical components that result in the scalp-recorded EEG macropotential. Similar coherent electrical activity can be picked up in nearby electrodes.

Through the use of high-density EEG recordings, one can hope to perform an inverse solution to undo, as best as possible, the volumetric smearing effects of the medium through which the neuroelectric signals permeate. BSS methods attempt to isolate dominant (independent) components and localize them to particular brain regions to aid in the understanding of the cognitive processes that resulted in their creation or to identify abnormalities. The results are often used to create 2D or 3D color topographic brain maps to enhance visualization. Whether it is the macropotential EEG signal or its contributory components, there is a need to characterize the EEG waveforms using standard terminology to help characterize the functional behavior of the brain.

This section highlights the many descriptors that are used with EEG recorded signals or its decomposed components to help in the categorization and description of complex brain activity. A brief summary of these is listed in Table 9.6 with examples. Clinical

TABLE 9.6   Common Descriptors of EEG Activity: Explanation, Examples and Comments

Descriptor of EEG Signals	Explanation	Characterization Examples	Comments
Morphology	Shape of the wave	Rhythmical (regular) Arrhythmical (irregular) Sinusoidal Spindles Complexes Spikes Polyspikes Sharp waves	Morphology is the primary EEG descriptor in epileptic studies. Brain patterns form wave shapes that are commonly sinusoidal. A complex can be made up of a combination of sharp and slow waves and tend to last longer than 0.25 s, therefore not repeating over rates of 4 Hz.
Repetition	Defines the type of waveform occurrence	Rhythmic Semirhythmic Irregular	Spindles are groups of rhythmical repetitive waves that gradually increase and then decrease in amplitude.
Frequency	How often a repetitive wave recurs	Frequency Bands Delta Theta Alpha Beta	A symmetry seen in one bipolar montage should be verified using another montage oriented at 90° to the first or reference montage. Differences of amplitude are sometimes caused by factors outside the brain, especially by unequal spacing and impedance at recording electrodes.
Amplitude	Measured in microvolts (µV) peak to peak or from the calibrated zero reference	Clinical reference: Low (<20 µV) Medium (20–50 µV) High (>50 µV) Amplitude asymmetry	Typically between 10 and 100 µV (in adults more commonly between 10 and 50 µV). Longer interelectrode distances produce increasing amplitudes up to an interelectrode distance of about 8 cm.
Distribution	The occurrence of electrical activity recorded by electrodes positioned over different parts of the head	Widespread, diffuse (generalized) Lateralized Localized (focal)	Lateralized activity is suggestive of a cerebral abnormality. In describing the location of local patterns, electrode names should be used, not head regions or brain areas.
Phase relation	The relative timing and polarity of components of waves in one or more channels, e.g, do the troughs and peaks line up?	In phase Out of phase Phase angle	In a single channel, phase refers to the time relationship between different components of a rhythm.
Timing	Relative occurrence of activity in time at different parts of the brain recorded by different channels	Simultaneous (synchronous) Independent (asynchronous) Bilaterally synchronous	The resolution of time relations deteriorates if more distant channels are compared. Gives an idea of possible triggering mechanism.

(Continued)

TABLE 9.6 (*Continued*)   Common Descriptors of EEG Activity: Explanation, Examples and Comments

Descriptor of EEG Signals	Explanation	Characterization Examples	Comments
Persistence	How often a wave or pattern occurs during a recording session	Index percentage (proportion of time for which these waves appear in the recording) Poorly/well sustained High, moderate, and low persistence	Very common in polysomnography (sleep staging studies). The persistence and amplitude are often described together in terms of quantity, amount, and prominence.
Reactivity	Refers to changes that can be produced in some normal and abnormal patterns by various maneuvers or functions	EEG alteration in response to closing the eyes, hyperventilation, visual or sensory stimulation, changes in levels of alertness, and movements	Useful in abnormality or drug addiction studies. Useful to prepare or evaluate the subject's condition prior to recording.

electroencephalography uses a large number of these descriptors, particularly in the study of epilepsy, to facilitate accurate analysis. In relation to cognitive function research, the most important aspects of EEG activity are distribution, frequency, amplitude, morphology, periodicity, and importantly the behavioral and functional correlates. In summary, EEG requires a considerable level of experience to accurately identify and characterize the signals.

For the purposes of CNS measurement and analysis, there exist various EEG signal properties that discriminate brain function and hence can be employed as investigative tools in research studies or in specific EEG-based applications. EEG signal properties can be categorized into one of the following groups:

1. Rhythmic brain activity

2. Event-related potentials (ERPs)

3. Event-related desynchronization (ERD) and event-related synchronization (ERS)

9.14.1 Rhythmic Brain Activity

Frequency is one of the most important criteria for assessing abnormality in clinical EEG and for understanding functional behavior in cognitive research. With billions of oscillating communities of neurons as its source, the human EEG potentials are manifested as aperiodic unpredictable oscillations with intermittent bursts of oscillations having spectral peaks in certain observed bands: 0.1–3.5 Hz (delta, δ), 4–7.5 Hz (theta, θ), 8–13 Hz (alpha, α), 14–30 Hz (beta, β), and >30 Hz (gamma, γ). The band range limits associated with the brain rhythms, particularly beta and gamma, can be subject to debate in the literature and are often further subdivided into subbands that can further distinguish brain processes. Table 9.7 summarizes some common brain rhythms and their resulting EEG characteristics such as typical amplitude, frequency, location, and reactivity.

TABLE 9.7 Normal EEG Rhythms Characteristics

Brain Rhythm	Typical Frequency Range (Hz)	Normal Amp. (μV)	Where It Can Be Found	Reactivity	Comments
Delta	0.1–4	<100	Dominant in infants, during deep stages of adult sleep and serious organic brain disease Central cerebrum and parietal lobes	Focal in pathologies. Occur after transactions of the upper brain stem separating the reticular activating system from the cerebral cortex.	Polymorphic delta, severe, acute, or ongoing injury to cortical neurons. Rhythmic discharge, psychophysiologic dysfunction.
Theta	4–8	<100	In drowsy normal adult, in frontal, parietal, and temporal regions In children when awake	Rare in EEG of awake adults. During emotional stress in some adults. Sudden removal of something causing pleasure will cause about 20 s of theta waves.	Niedermeyer lists some studies in which the theta activity of 6–7 Hz over frontal midline region had been correlated with mental activity. Focal or lateralized theta indicates focal pathology. Diffuse theta more generalized neurologic syndrome.
Alpha	8–13	20–60	The most prominent rhythm in the normal alert adult brain Most prominent at occipital and parietal regions About 25% stronger over the right hemisphere	Fully present when a subject is mentally inactive, alert, and with eyes closed. Blocked or attenuated by deep sleep, attention, especially visual, and mental effort. When a person is alert and their attention is directed to a specific activity, the alpha waves are replaced by asynchronous waves of higher frequency and lower amplitude. Eye opening/ closure offers the most effective manipulation.	Location of central alpha peak declines with age and in dementia. Alpha waves are usually attributed to summated dendrite potentials. Slowing is considered a nonspecific abnormality in metabolic, toxic, and infectious conditions. Asymmetries—unilateral lesions. Loss of reactivity—a lesion in the temporal lobe. Loss of alpha—brainstem lesion.

(Continued)

TABLE 9.7 (*Continued*)   Normal EEG Rhythms Characteristics

Brain Rhythm	Typical Frequency Range (Hz)	Normal Amp. (µV)	Where It Can be Found	Reactivity	Comments
Mu	10–12	<50	Central electrodes, over motor and somatosensory cortex	Does not react to opening of eyes like alpha rhythm. Blocking before movement of the contralateral hand. Blocking for light tactile stimuli. Blocking for readiness or imagination to move limbs.	Physiologically and topographically different to alpha rhythm. Not clinically useful. Useful as a brain–computer interface input in relation to actual and imagined limb movement. Typically appears as part of a background rhythm.
Beta	14–30	<20	Three basic types: frontal beta, widespread, and posterior	Frontal beta blocked by movement. Widespread beta often unreactive. Posterior beta shows reactivity to eye opening.	Beta I waves, lower frequencies, which disappear during mental activity. Beta II waves, higher frequencies, which appear during tension and intense mental activity. Underintense mental activity beta can extend up as far as 50 Hz.
Gamma	>30	<2	Widespread	Found when the subject is paying attention or is having some other sensory stimulations.	Due to its really low amplitude, it is very difficult to isolate without band-pass filtering.

Activity that is either less than 0.5 Hz or greater than 20 Hz is often considered to be of limited clinical utility. However, some published studies in the literature report the existence of a cognitive brain process with activity reflected in the beta and gamma bands.

Human EEG rhythms are affected by different actions, thoughts, and stimuli. For example, the planning of a movement can block or attenuate the mu rhythm (as described in Table 9.7). The fact that mere thoughts affect the rhythmical activity of the brain can be used as the basis for a clinical diagnosis.

9.14.2 Event-Related Potentials

ERP is a common title for potential changes in the EEG that occur in response to a particular *event* or a stimulus. ERPs provide a suitable methodology for studying the aspects of cognitive processes of both normal and abnormal nature, such as neurological or psychiatric disorders. Mental operations such as those involved in perception, selective attention, language processing, and memory proceed over time ranges in the order of tens of milliseconds. Whereas PET and MRI can localize regions of activation during a given mental task, ERPs can help in defining the time course of these activations.

Amplitudes of ERP components are often much smaller than spontaneous EEG components, typically by a factor of ~10. They are subsequently unrecognizable from the raw EEG trace. They can be elicited by ensemble averaging EEG epochs time locked to repeated sensory, cognitive, or motor events. The assumption is that the event-related activity, or signal of interest, has a more or less fixed time delay to the stimulus, while the spontaneous background EEG fluctuations are random relative to the time when the stimulus occurred. Averaging across the time-locked epochs highlights the underlying ERP by averaging out the random background EEG activity (similar to additive white noise), thus improving the signal-to-noise ratio. These electrical signals reflect only the activity that is consistently associated with the stimulus processing in a time-locked manner. The ERP thus reflects, with high temporal resolution, the patterns of neuronal activity evoked by a stimulus. This approximation to the real ERP, however, offers no phasic information. ERPs can be divided into exogenous and endogenous. Exogenous ERPs occur up to about 100 ms after the stimulus onset. They depend on the properties of the physical stimulus. The potentials prestimulus and 100 ms after the stimulus are termed endogenous. DC shifts or slow cortical potentials are endogenous ERPs and are used by Birbaumer and Rodden (2007) in their *thought translation device* brain–computer interface (BCI) system.

Evoked potentials (EPs) are a subset of the ERPs that occur in response to or during attention to certain physical stimuli (auditory, visual, somatosensory, etc.). They can be considered to result from a reorganization of the phases of the ongoing EEG signals. The EPs can have distinguishable properties related to different stimuli properties. For example, the visual evoked potential (VEP) over the visual cortex is elicited by a time-varying visual stimuli.

The readiness potential (RP) or movement-related potential (MRP) is an ERP of the order of 1 µV generated in response to a cognitive desire to perform or to imagine of limb movement and is prevalent at highly localized areas of the brain related to that function (primary motor cortex).

9.14.3 Event-Related Synchronization/Desynchronization

Pfurtscheller and Aranibar (1977) first quantified ERD. Pfurtscheller developed a BCI known as the Graz BCI in the 1990s that was based on detecting ERD and ERS of the different mu and beta rhythm bands during the imagination of left- and right-hand movements. ERD is associated with amplitude attenuation and ERS with amplitude enhancement of a certain EEG rhythm. In order to measure ERD or ERS, the power of a chosen frequency band is calculated before and after a specific event. The average power across a number of trials is then measured relative to the power of the reference interval. The reference interval can be an arbitrary period prior to the event and represent a period of inactivity or rest. The ERS is the power increase (in percent) and the ERD is the power decrease relative to the reference interval that is defined as 100%. ERD/ERS measurements selected over specific frequency ranges are typically used to produce a spatiotemporal map to visualize the functional behavior across the brain.

9.15 APPLICATIONS

This section highlights the many clinical and cognitive research applications that have been established over the years for EEG. Hans Berger in his original research laid the foundations for most of today's clinical uses for EEG. The greatest advantage of EEG over other brain imaging methodologies is its temporal resolution coupled with this speed of acquisition and low cost. Complex patterns of neural activity can be recorded with millisecond resolution. As mentioned earlier, EEG however provides poor spatial resolution (~1 cm) compared to MRI or PET (<1 mm). Thus, for improved insight into the origins of neural activity, EEG inverse solution images are often compared with MRI scans for a similarly executed experiment. Research and clinical applications of EEG in humans and animals include

- Monitoring alertness, coma, and brain death
- Locating areas of damage following head injury, stroke, tumor etc.
- Testing afferent pathways (by EPs)
- Monitoring cognitive engagement (alpha rhythm)
- Controlling anesthesia depth (*servo anesthesia*)
- Investigating epilepsy and locating seizure origin
- Testing pharmacological effects on brain function
- Assisting in experimental cortical excision of epileptic focus
- Monitoring human and animal brain development
- Investigating sleep disorder and physiology

One important diagnostic analysis method often employed is assessing the symmetry of alpha activity within hemispheres. In cases of restricted lesions such as tumor, hemorrhage, and thrombosis, it is usual for the cortex to generate lower EEG frequencies. EEG signal distortion can be manifested by reduction in amplitude, decrease of dominant frequencies beyond the normal limit, and production of spikes or special patterns. Epileptic conditions produce low-frequency, rhythmic activity in certain areas of the cortex and the appearance of high-voltage waves (up to 1000 μV) referred to as spikes or spike waves. EEG waveforms have been shown to be modified by a wide range of variables, including biochemical, metabolic, circulatory, hormonal, neuroelectric, and behavioral factors. By tracking changes of electrical activity during drug-induced-related phenomena, as euphoria and craving, brain areas and patterns of activity that mark these phenomena can be determined. As the EEG recording procedure is noninvasive and painless, EEG- and ERP-related activities are widely used in the study of brain organization of cognitive processes such as perception, memory, attention, language comprehension, and emotion in normal adults and children.

Taking the functional and cognitive insight that EEG offers a step further would allow humans to modulate their brainwave activity in order to communicate. This gives rise to the rapidly developing area of neuroscience and engineering that is BCI design. Its development is in its infancy with respect to other communication technologies but offers great potential within the area of rehabilitation. A common goal in BCI research is to detect small ERPs of approximately 1 μV from the background EEG (10 to 50 μV) and extract distinguishable features and classify them for subsequent use as inputs to a computer interface. The greatest proportion of BCI systems is based on using ERPs. The ERP pattern recognition approach to BCI design produces a more natural brain response directly related to function. Most relevant for the BCI use is the fact that these patterns do not require actual physical movement to elicit them and can be generated by imagined movements. Most present-day BCIs based on imagined movements use a synchronous cued approach to time lock the event. This restricts the user to fixed time periods but dramatically simplifies the ERP pattern classification process. The system knows the exact timing of when a brain pattern can occur and consequently only has to classify the event rather than both the occurrence of an event and then the event itself. In contrast, an asynchronous paradigm has the difficult task of detecting an event from continuous EEG amidst the widely varying baseline or background EEG activity.

9.15.1 Biofeedback

Biofeedback is the process in which a subject receives feedback information about his physiological state. It creates an external loop by which a subject can monitor one or more of their physiological states to aid in training or performance of a task. The most popular biofeedback methods include galvanic skin response (GSR), body temperature, EEG, ECG, heart rate variability (HRV), EMG, and respiration. These various methods are employed for different clinical purposes to report on the physiological state of the human body. EEG biofeedback or neurofeedback can be dated back to the early 1970s when the self-regulation of the alpha rhythm was used to aid in relaxation and meditation. Today, it proposes to offer a greater insight into the physiological and mental state of a subject compared to other biofeedback modalities.

While many researchers are skeptical about neurofeedback, some contend that one can improve their mental performance, normalize behavior, and stabilize mood through its use. There exists a wide range of systems, which incorporate the use of EEG biofeedback. It has been employed in the study of conditions such as attention deficit hyperactivity disorder (ADHD), depression, anxiety, epilepsy, sleep disorders, and alcoholism. EEG data as neurofeedback input are typically employed to decipher the physiological or mental state of the subject. This allows the feedback to be presented to the subject in a more stimulating and meaningful manner using a combination of sensory feedback methods such as visual, auditory, and tactile. Biofeedback can be a powerful therapeutic tool, which can help individuals in learning strategies to self-regulate certain behaviors.

9.16 SUMMARY

EEG is limited by the vast number of electrically active neuronal elements; the complex electrical and spatial geometry of the CNS, brain and head; and the disconcerting trial-to-trial variability of brain function. The layers of cerebrospinal fluid, skull, and scalp between the electrodes and the brain itself act as an electrical shield severely attenuating and volume smearing the electrical contributions of neuron communities to EEG recordings. This presents a difficult challenge to localize and decipher the contributions of neuron communities to the EEG recorded macropotentials. Despite EEG's poor spatial resolution, its excellent temporal resolution of less than 1 ms and ease of acquisition make it the only practical brain imaging modality for real-time analysis of CNS and brain function systems. Despite the challenges for utilizing EEG as a means to measure CNS activity, recent advances in signal processing methods make it more feasible than ever before.

9.17 CNS MEASUREMENT EQUIPMENT

NeuroNexus: NeuroNexus is a neurotechnology company that develops and commercializes neural interface technology, components, and systems for neuroscience and clinical studies (NeuroNexus 2014).

BioSemi: BioSemi's goal is to provide the scientific community with state-of-the-art instrumentation for electrophysiology research (BioSemi 2014).

Neuroscan: Neuroscan provides complete solutions for a wide range of neuroscience applications (Compumedics NeuroScan 2014).

REFERENCES

BioSemi 2014 http://www.biosemi.com/

Birbaumer N and Rodden F A 2007 The thought-translation-device in H. Cohen and B. Stemmer (eds.), *Consciousness and Cognition. Fragments of Mind and Brain* (Atlanta GA: Academic Press-Elsevier) pp. 69–81

Buzsaki G 2006 *Rhythms of the Brain* (New York: Oxford University Press)

Compumedics NeuroScan 2014 http://compumedicsneuroscan.com/

Greenfield Jr L J, Geyer J D and Carney P R 2010 *Reading EEGs—A Practical Approach* (Philadelphia: Wolters Kluwer)

Kandel E, Schwartz J H and Jessell T 2000 *Principles of Neural Science* (New York: McGraw-Hill Medical)

Lewicki M S 1998 A review of methods for spike sorting: The detection and classification of neural action potentials *Network* **9** R53–78

Makeig S, Jung T-P, Bell A J and Sejnowski T J 1997 Blind separation of auditory event-related brain responses into independent components *Proc. Natl. Acad. Sci. USA* **94** 10979–84

NeuroNexus 2014 http://www.neuronexus.com/

Pfurtscheller G and Aranibar A 1977 Event-related cortical desynchronization detected by power measurements of scalp EEG *Electroencephalog. Clin. Neurophysiol.* **42** 817–26

Quian Quiroga R and Panzeri S 2009 Extracting information from neuronal populations: Information theory and decoding approaches *Nat. Rev. Neurosci.* **10** 173–85

Reilly J P 1998 *Applied Bioelectricity: From Electrical Stimulation to Electropathology* (New York: Springer)

CHAPTER 10

Neurology

Peripheral Nerves and Muscles

Ernest Nlandu Kamavuako and Dario Farina

CONTENTS

10.1 Introduction   214
10.2 Basic Physiology   214
    10.2.1 Nerve   214
    10.2.2 Muscle   215
    10.2.3 Action Potentials   216
    10.2.4 Recording of the Extracellular Potential   216
    10.2.5 Amplifier Block   217
    10.2.6 Analog-to-Digital Convertor Block   217
10.3 Electroneurography   217
    10.3.1 Neural Interfaces   218
        10.3.1.1 Extraneural Electrodes   218
        10.3.1.2 Intraneural Electrodes   219
        10.3.1.3 Regenerative Electrodes   220
10.4 Recording of Nerve Activity   220
    10.4.1 Recording Configuration   220
    10.4.2 Interference   221
    10.4.3 Signal Characteristics   221
10.5 Processing and Applications of Nerve Recordings   222
    10.5.1 Signal Processing   222
    10.5.2 Applications Based on Peripheral Nerve Recordings   223
10.6 Electromyography   224
    10.6.1 Intramuscular EMG Electrodes   224
    10.6.2 Surface EMG Electrodes   226
    10.6.3 Recording of Muscle Activity   226
    10.6.4 Recording Configuration   226
    10.6.5 Characteristics of Muscle Recordings   227
    10.6.6 Processing and Applications of Muscle Recordings   229

10.6.7 Surface EMG   229
10.6.8 Systems for Recording Nerve and Muscle Activities   230
10.7 Evoked Potentials and Conduction Velocity   230
10.7.1 Evoked Potentials   230
10.7.2 Nerve Conduction Velocity   231
10.7.3 Collision Techniques   232
10.7.4 Muscle Fiber Conduction Velocity   234
10.8 Summary   235
Acknowledgments   235
References   235

10.1 INTRODUCTION

Neurology is the diagnosis and treatment of all categories of diseases involving the central, peripheral, and autonomic nervous systems. The first step in the treatment of these disorders is the assessment or measurement of the functionality of the system. In this chapter, techniques for assessing the peripheral nervous system and muscles are described. The chapter focuses on four main topics. The first relates to the electrical signals generated from peripheral nerves, referred to as electroneurogram (ENG). The second deals with the electrical potentials of muscles or electromyogram (EMG), which is measured either on the surface of the skin (surface EMG [sEMG]) or from inside the muscle tissue (intramuscular). For both nerve and muscle recordings, we present basic information on anatomy and physiology, physiological measurements required, and instrumentations, such as sensors, amplifiers, and signal processing. The third topic is the measurement of conduction velocity of nerve and muscle impulses, where assessment techniques and applications are described. Finally, this chapter describes the instrumentations and means of recording evoked potentials.

10.2 BASIC PHYSIOLOGY

The ability to perform movement is a result of complex mechanisms that involve the central nervous system (CNS), the spinal cord, peripheral nerves, and muscles. The command signals (or efferent signals) are transmitted via the spinal cord and further to the muscles through peripheral nerves. All information about a given movement is encoded into electrical impulses that are referred to as action potentials. These impulses propagate along the peripheral nerve down to the muscle that is to contract. In the following, we will review the basic nerve and muscle physiology in order to clarify the mechanisms of signal generation.

10.2.1 Nerve

A nerve is a bundle of conducting cells that transmits impulses from the brain or spinal cord to the muscle and glands (motor nerves) or inward from the sense organs to the spinal cord and brain (sensory nerves). A nerve is part of the nervous system and its function is to coordinate, with the endocrine system, the activities of other body systems. The nervous system is composed of two principal categories of cells: neurons and neuroglia (glial cells). Neurons are the basic units and neuroglia cells are supportive cells aiding the function of neurons. All neurons have three principal components: a cell body, dendrites, and an axon. The cell body contains the nucleus and the machinery for protein synthesis. The dendrites respond to specific stimuli and conduct impulse to the cell body. The axon or nerve fiber is relatively long and conducts impulses away from the cell body. The main axon has branches, called collaterals, and each branch ends in an axon terminal responsible for releasing neurotransmitters. Thus, peripheral nerve refers to axons that run from the spinal cord to different muscles. The functional classification of neurons is based on the direction of conduction. Sensory or afferent neurons convey information from the tissues into the CNS, and motor or efferent neurons transmit impulses from the CNS to the effector cells, for example, muscles. Association neurons or interneurons connect neurons within the CNS. However, most nerves are composed of both motor and sensory neurons and are thus called mixed nerves.

Anatomically, each nerve is covered externally by a dense sheath of connective tissue called the epineurium. The perineurium lies under the epineurium forming a complete sleeve around a bundle of axons (nerve fascicle). Surrounding each fiber is the endoneurium, which constitutes an unbroken tube that extends from the surface of the spinal cord to the level at which the axon synapses with its muscle fibers or ends in sensory receptors.

10.2.2 Muscle

The human body is composed of mainly three types of muscles (skeletal, smooth, and cardiac), of which the skeletal muscle will be the focus of this chapter. The skeletal muscle is an actuator that is activated by electrical signals transmitted through the peripheral nervous system from the CNS, in order to move or stabilize the skeletal joints. The gross anatomy of a muscle includes its origin, insertion, and its cross-sectional area that characterizes a muscle in terms of its role in the body. At the microanatomical level, skeletal muscles are comprised of cells with the following structural elements: myofibrils, sarcomeres, actins, and myosins. Muscle cells are also called *muscle fibers*, and they are in bundles called fascicles, which then group together to form a muscle. In humans, muscle fibers have a variable length from a few millimeters to several centimeters and a diameter in the range from approximately 10–100 μm. A single nerve fiber, which corresponds to the axon of a motor neuron, transmits its electrical signal from the CNS to a group of muscle fibers constituting the smallest functional unit of the muscle, also known as the *motor unit*. The number of muscle fibers innervated by the same motoneuron is referred to as *innervation ratio*, and this ratio may depend on the motor unit, the muscle, and its role. Muscle fibers can be classified into slow twitch and fast twitch, according to their physiological properties. Slow twitch fibers contract for long periods of time but with little force and are oxidative and fatigue resistant, while fast twitch fibers contract quickly and powerfully but fatigue very rapidly. All fibers of the same motor unit are of the same type; thus, the same classification (slow and fast) can be extended to motor units. The muscle fibers, and motor units, can also be classified in terms of their histological properties in type I and II. The number of type I and type II motor units depends on the muscle and, in the same muscle, on the individual.

10.2.3 Action Potentials

Each excitable cell in the body has a resting potential, which in humans is approximately 70–90 mV, negative inside the cell with respect to the extracellular environment. The resting potential maintains the balance of ions passing through the membrane. This potential can be disturbed because of either physiological events or artificial means, such as using electrical stimulation. The resting potential exists because there is an excess of negative ions inside the cell and an excess of positive ions outside. In neurons, the resting membrane potential is in the range of −40 to −75 mV and around −70 mV in the muscle. For an action potential to occur, the change in the cell membrane potential needs to surpass a threshold value of approximately −55 mV (activation threshold). Chapter 9 describes action potentials in more details. We should note that the electroneurogram (ENG) is the measure of the electric signal from the nerve on either afferent or efferent fibers. The electromyogram (EMG) is the measure of the electric signal from the muscle. The action potential discharged by one motor neuron propagates, in normal conditions, to all the muscle fibers that the neuron innervates. Each muscle fiber is then activated by the release of acetylcholine at the neuromuscular junction. Thus, the amplitude of the signal recorded from a motor unit depends on the number of fibers.

10.2.4 Recording of the Extracellular Potential

A schematic representation of a generic system for measuring extracellular potentials of muscles or nerves is shown in Figure 10.1. The system is composed of the following blocks: the muscle or nerve that generates the signal, the electrode part, the amplifier, the analog-to-digital block, and the computer that handles visualization and data storage.

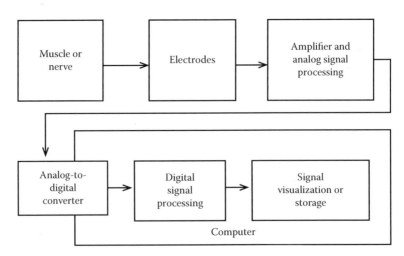

FIGURE 10.1 Schematic representation of a generic system for the measurement of extracellular potentials.

The first block is the physiological media (muscle and nerve) where the ionic current is measured from and converted to electronic current by the electrodes or transducers.

10.2.5 Amplifier Block

The amplifier block can be divided into preamplification and main amplification depending on the system and type of electrodes used. The preamplification is a device that helps in improving the overall quality of the recordings and preparing the signal for further amplification or processing. The preamplifier circuitry may or may not be housed separately from the device for which a signal is being amplified. However, in many applications, the preamplifier is located close to the electrode to avoid signal loss through the wires. Through the help of the preamplifier and the main amplifier, the extracellular potential is not altered in quality, but it is of higher voltage. The main amplifier is usually responsible for the entire analog processing of the signal. This is the stage where the signal is band-pass filtered before the analog-to-digital conversion. The choice of amplifiers depends on the type of recordings and on the electrode configurations. Some types of recording modalities (typical from inside the nerve or muscle) require high input impedance, which in case of surface recordings is a bit lower in order to obtain very high common mode rejection in bipolar configuration.

10.2.6 Analog-to-Digital Convertor Block

The analog-to-digital convertor (ADC) is the transition block between the continuous analog signals recorded from the media and the numerical signals used for processing and storage in computers. The ADC is a device that converts an input analog voltage to a digital number proportional to the magnitude of the voltage. The digital output of the ADC is typically a two's complement binary number that is proportional to the input voltage. An ADC is typically characterized by its resolution, response type, and accuracy and sampling rate.

The computer or processing unit (digital signal processors) is where the signal is handled digitally for further processing and visualization. Visualization of the signals provides the first impression on the quality of the recordings. It allows the experimenter to assess the level of noise, saturation of the amplifier, and so on.

10.3 ELECTRONEUROGRAPHY

The electroneurogram (ENG) is the extracellular potential of an active peripheral nerve recorded at some distance from the nerve or from within the nerve trunk (Sinkjær et al 2006). The recorded potential is composed of the contributions of the superimposed electrical fields of the active sources within the nerve. In general, the extracellular response of a nerve to electrical stimulation is triphasic with amplitude in the lower end of the microvolt scale. This response decays both in amplitude and high-frequency content when the distance from the source increases. Potentials due to changes in the electric field in the extracellular space of nerve fibers can be recorded using different configurations and electrode interfaces.

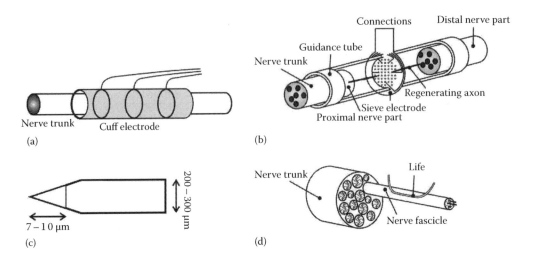

FIGURE 10.2 Different types of electrode for electroneurography: (a) extraneural electrode, circumferential cuff electrode; (b) regenerative electrode, sieve electrode; and (c–d) intraneural electrode, tungsten electrode and position of the intrafascicular electrode. (Adapted from Sinkjær, T. et al., Electroneurography, in: Webster, J. (ed.), *Encyclopedia of Medical Devices and Instrumentation*, John Wiley & Sons, Hoboken, NJ, 2006, pp. 191–232. With permission.)

10.3.1 Neural Interfaces

Advances in electrode design have made it possible to study the peripheral nervous system in detail and often during unconstrained conditions using electrodes placed either outside or inside a nerve bundle. Different types of electrodes that have been proposed can be classified based on their invasiveness (from low to highly invasive) with respect to penetration into the nerve as extraneural, intraneural, and regenerative electrodes (Figure 10.2). The same order is also characterized in terms of low to high selectivity, where selectivity is defined as the ability of recording activity from single nerve cells. Nevertheless, they all record the extracellular potential from one or more peripheral nerve axons simultaneously (also referred to as extracellular electrodes). A fourth type, called the intracellular electrodes, aims at recording the intracellular potential of individual cells where the active recording site is placed inside a cell (Sinkjær *et al* 2006). These electrodes have high risk of killing the cell and are not considered here. The choice of a suitable electrode depends mainly on the application as will be discussed in Sections 10.3.1.1, 10.3.1.2 and 10.3.1.3.

10.3.1.1 Extraneural Electrodes

Extraneural electrodes are placed adjacent to the nerve without penetrating the epineurium. They record the compound activity from a large population of axons and can be subclassified into epineurial, hook, and circumneural electrodes. Epineurial electrodes are placed on the nerve and secured by suturing on the epineurium. They are usually made of platinum or iridium wires fabricated as longitudinal strips holding two or more contacts (Navaro *et al* 2005). Epineurial electrodes have found applications mainly in selective stimulation of particular nerve fascicles, for example, functional electrical stimulation for breathing control by phrenic nerve stimulation and restoration of ankle dorsiflexion in the foot-drop impairment. The hook electrodes are constructed of usually two or more hook-shaped wires typically based on platinum, stainless steel, or tungsten. They are easily handmade at a very low cost, but are not useful for long-term nerve implants and are mainly used in acute animal experiments. Circumneural electrodes are placed around the whole nerve. They are also referred to as cuff electrodes and are composed of an insulating tubular sheath that completely encircles the nerve. They contain two or more electrode contacts exposed at their inner surface and appear with the largest number of variations in designs and configurations. They have gained a great interest in functional neuromuscular stimulation systems both in humans (see Hoffer *et al* 1996 for a review) and animals. Examples of applications are described later in this chapter. Interfascicular electrodes are between extraneural and intraneural, because they penetrate the epineurium without penetration of the perineurium. They are placed between the fascicles to enhance selectivity due to the closer contact with the axons.

10.3.1.2 Intraneural Electrodes

Intraneural electrodes penetrate the endoneurium and the perineurium of the nerve. These electrodes aim at recording from one single nerve fiber or from a small population of nerve fibers, as they are placed in the extracellular space in close proximity with the axons. Intraneural electrodes comprise needle electrodes, intrafascicular electrodes, and silicon-based electrodes.

Needle electrodes are used to measure the so-called microneurography, a special class of ENG, where a needle is inserted directly inside a nerve fascicle. The aim is to study basic neural coding while recording from one fiber or a small population of fibers. There are two types of commercial electrodes: tungsten needle electrode and concentric needle electrode. The technique does not require surgery and is most suitable for short-term recordings. It has proved to be a powerful tool in clinical experiments on conscious humans in several studies.

Longitudinal intrafascicular electrodes (LIFEs) were developed to provide a stable, long-term interface with sufficient selectivity for tracking single unit neural activity. LIFEs are placed within the nerve and in close contact with the nerve fibers allowing for selective recordings and stimulation. Intrafascicular electrodes are implanted longitudinally within individual nerve fascicles by pushing a tungsten guiding needle within the endoneurium parallel to the course of the nerve fibers a few millimeters and then pulling the electrode through the fascicle until the active zone of the electrode is centered in the fascicle. Flexible thin-film penetrating electrodes have the advantage over extraneural cuff electrodes of the possibility of locating a large number of detection sites in a small area (multichannel recordings). The technology of thin-film system has been developed and tested in animal experiments with up to eight channels. The use of LIFE is a suitable trade-off between electrode invasiveness and electrode selectivity (Navarro *et al* 2005). The invasive intrafascicular electrodes provide enhanced selectivity and increased signal-to-noise ratio (SNR) of the recordings compared to extraneural electrodes.

Silicon-based electrodes are designed and manufactured using sophisticated microfabrication techniques. They may be inserted transversely into the nerve as in the case of microneurography or longitudinally as LIFE electrodes. Despite the many electrodes that have been proposed, only the Utah array has been used for peripheral implantation (Sinkjær *et al* 2006). Most of them are mainly used for cerebral cortex implantation.

10.3.1.3 Regenerative Electrodes

Peripheral nerve axons have the capacity to regrow after dissection or traumatic injury. Regenerative electrodes are placed in the path between a proximal and distal nerve stump to interface a high number of nerve fibers by using an array of holes, with electrodes built around them. Due to their appearance, these electrodes are often referred as sieve electrodes. The active sites are placed inside the sidewalls of the holes or in their immediate surroundings to make a physical contact with the axons growing through the holes.

10.4 RECORDING OF NERVE ACTIVITY

10.4.1 Recording Configuration

Various recording configurations are applied in order to maximize the detection of neural signal and minimize the interference pickup. Nerve signals can be recorded in monopolar, bipolar, and tripolar configuration as shown in Figure 10.3. Monopolar is the simplest configuration, where only one electrode is placed in the extracellular environment and the signal is amplified relative to a common reference electrode. Due to the low amplitude of nerve signals, monopolar configuration in peripheral nerves is rarely used as it requires precautions to minimize the interference level. The bipolar configuration on the other hand is the most common configuration for many neural interfaces besides the cuff electrodes. Bipolar configuration requires two recording electrodes and a reference electrode. The two recording electrodes are connected to an amplifier, which amplifies the relative difference of the

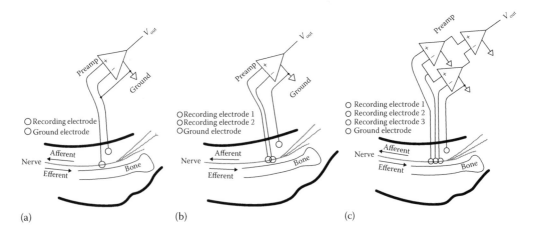

FIGURE 10.3 Different types of nerve recording configurations such as (a) monopolar consisting of one active electrode attached to the nerve and one ground electrode, (b) differential bipolar with two active electrodes, and (c) true tripolar configuration consisting of a double differential amplification circuit. (Adapted from Sinkjær, T. et al., Electroneurography, in: Webster, J. (ed.), *Encyclopedia of Medical Devices and Instrumentation*, John Wiley & Sons, Hoboken, NJ, 2006, pp. 191–232. With permission.)

potentials seen by the two electrodes. Thus, common activities (such as muscle signals) are seen by the differential preamplifier as common mode and are rejected. A good characteristic of differential bipolar amplifier is high common mode rejection ratio (CMRR), which represents the ability of the amplifier to attenuate common interference. Because the neural wave that travels along the nerve fiber is seen on the two electrodes at different time instants, the shape and the amplitude of the recorded action potential will depend on the distance between the two electrodes. Tripolar configuration is the most commonly used configuration for cuff electrodes, because standard cuff electrodes contain three rings surrounding the nerve. In the tripolar case, potentials seen by the two electrodes are averaged to predict the extra cuff interference potential seen at the central electrode, either by directly measuring and averaging the signals, as in the true tripolar configuration, or by shorting the contacts of the outer two electrodes, as in the pseudotripolar configuration. Action potentials recorded with the tripolar recording configuration are triphasic.

10.4.2 Interference

For peripheral nerve recordings, interference can be physiological, that is, signals from other biological sources such as muscle and heart with several orders of magnitude larger than neural signals. Physiological interference may also include activities from distant nerve fibers appearing as a mass background activity. Other sources of interference include the environment (internal or external); artifacts, for example, when the transducer responds to other energy sources; and the electronics noise. Motion artifacts are a well-known problem when the electrodes respond to mechanical movement. Electronic noise has a broad range of frequencies and falls into two classes: *thermal* or *Johnson noise* and *shot noise.* Thermal noise comes mainly from the resistance materials, and shot noise is related to voltage barriers associated with semiconductors. Different approaches can be adopted in order to minimize noise in the measurement system, spanning from modification in the transducer design to proper grounding of the system and analog processing, such as analog filters. Nevertheless, the level of noise is directly related to the adopted recording configuration.

10.4.3 Signal Characteristics

The characteristics of ENG signals depend on the recording configuration and mostly on the type of neural interface used. Neural interfaces can be classified both in terms of invasiveness and selectivity. Invasiveness is related to the penetration and manipulation into the nerve or muscle. Selectivity is defined here as the ability to record action potentials from single nerve cells or fibers. Furthermore, the selectivity of the interface relates to the size of the recording site and the distance from the active nerve cell. Thus, signals recorded from nerves can be characterized by their level of selectivity, amplitude distribution, and frequency content. In terms of selectivity, extraneural electrodes provide signals that are less selective than intraneural electrodes, and signals recorded from intraneural electrodes are less selective than regenerative electrodes. The ENG is a low-amplitude signal and requires amplification between 1,000 and 100,000 to bring the amplitude into the ~1 V range where they can be adequately digitally sampled and stored (Sinkjær *et al* 2006).

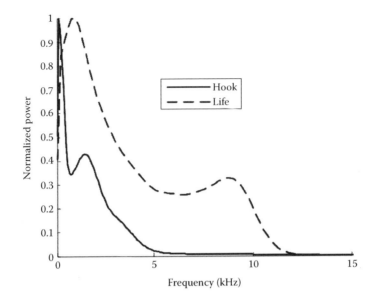

FIGURE 10.4 Examples of frequency contents of recordings from hook (plain line) and LIFE electrodes (dashed line). The signal with hook electrode has been measured in the human vagus nerve, while LIFE signal is obtained from rabbits.

TABLE 10.1   Summary of Signal Characteristics for ENG, Surface, and iEMG

Interfaces	Amplitude (RMS)	Frequency Range (Hz)
Extraneural	~5 μV	100–6,000
Intraneural	~20 μV	100–10,000
Regenerative	~15 μV	100–10,000
iEMG	0–1.5 mV	100–3,000
sEMG	0–1 mV	10–400

It should be noted that the amplitude depends upon how the signal is filtered and how close you are to the source.

The frequency content depends on the interface used. Figure 10.4 depicts examples of signals recorded from cuff and intrafascicular electrodes, and Table 10.1 presents characteristics of the different electrodes.

10.5 PROCESSING AND APPLICATIONS OF NERVE RECORDINGS

10.5.1 Signal Processing

As discussed earlier, nerve signals are of low amplitude and the sources of interference are multiple. Thus, depending on the application of course, digital signal processing is usually necessary in order to improve signal quality. The aim of processing a signal is twofold, signal conditioning and information extraction. Signal conditioning aims at maximizing the SNR. When the noise has frequency components outside the bandwidth of the nerve activities (depending on the electrode used), linear band-pass filtering is sufficient in order to attenuate the effect of noise in the recordings and to remove baseline drifts. However, in the situation where linear band-pass filtering is not sufficient, more advanced techniques are required for improving the SNR. Signal averaging is an example of these techniques. It consists in averaging a known response in order to enhance the energy of the common component. The drawback of this method is that it requires several realizations of the same response to obtain a clean averaged response, which is not feasible in long-term continuous recordings, for example. Another approach that is often used to process signals composed of single fiber action potentials is the wavelet transform. In wavelet-based conditioning, the signal at hand is first transformed into the wavelet domain. Then the wavelet coefficients are subjected to a predefined threshold. The coefficients above threshold are then inverse transformed back to the time domain with theoretically improved SNR. Several versions of the wavelet approach have been proposed in the literature, with the main difference being the applied threshold measure.

Information extraction from nerve recordings is directly associated to the application. For signals recorded with extraneural electrodes, the energy or the amplitude of the signal contains important information. For example, the signal amplitude indicates the intervals when the nerve is active. The CNS modulates the recruitment and discharge frequencies of nerve fibers to achieve information transmission, for example, about movement and force. Thus, the analysis of multiunit signals, such as signals from intraneural and regenerative electrodes, has been focused on decomposing the signals into constituent trains of single fiber action potentials. This process is often referred to as spike sorting, which aims at determining the number of nerve fibers that is active during a time period by associating different shapes and amplitudes of action potentials to different nerve fibers.

10.5.2 Applications Based on Peripheral Nerve Recordings

There are many applications based on nerve activities both at the research and clinical level; therefore, this section will not attempt to provide a complete list of all applications. For many of the previously described nerve interfaces (especially intraneural and regenerative), we can only talk about their intended applications since their performance has mainly been tested in laboratory settings using animal preparations. For example, the sieve electrode has been designed with the aim of providing a chronic interface to monitor peripheral neural activity. Although there is an increased number of reports on long-term recordings, the sieve electrode has, however, never been tested in humans. LIFEs have been designed and developed with the intention to provide stable interface for neuroprosthetic systems. As for sieve electrodes, LIFEs have proven robust for long-term recordings up to 6 months in animal experiments. For example, afferent signals recorded with LIFEs have been used as feedback in closed-loop control of the ankle joint in cats. The use of LIFEs for the control of prosthetic devices has also been demonstrated in humans. Electrical stimulation through the implanted electrodes elicited discrete, graded sensations of touch, joint movement, and position, referring to the missing limb. It was possible to quantify the neuronal firing rate recorded from electrodes placed on motor fibers and use it to control a force actuator in a prosthetic hand and a position actuator of a prosthetic elbow. Until now, the silicon-based electrode array has worked best in acute experiments; it has, for example, been implanted in dorsal root ganglions (through the dura) in acute cat model to evaluate the neural coding of the lower leg position. Despite the good performance of intraneural and regenerative electrodes in recording and stimulation, they have not yet found applications in the clinical practice.

The needle electrode for microneurography is applied in clinical experiments on awake humans for diagnostic purposes, but it is only suitable for short-term recordings, and the procedure may be time consuming (Sinkjær *et al* 2006). In terms of clinical chronic applications, the cuff electrode is to date the most successful and most widely used chronic peripheral interface in both animals and humans. It is the only implanted nerve interface being used in humans with spinal cord injury or stroke to provide functional neuromuscular stimulation systems with natural sensory feedback information. Applications include (1) cutaneous afferent feedback for the correction of foot drop, in which electrical stimulation of the peroneal nerve is used as a method for enhancing foot dorsiflexion in the swing phase of walking, allowing the patient to walk faster and more securely (see Lyons *et al* 2002 for a review); (2) cutaneous afferent feedback for the restoration of hand grasp, where the cuff electrode recordings are used to detect sensation of the skin when objects slip in order to modulate grasping force by modulating the intensity of muscle stimulation; (3) control of bladder function to relieve neurogenic detrusor overactivity; and (4) nerve conduction studies, which will be described later.

10.6 ELECTROMYOGRAPHY

The electromyogram (EMG) is the extracellular potential of an active muscle recorded at some distance from the muscle or from within the muscle. It is a technique for evaluating and recording the electrical activity produced by muscle cells when these cells are electrically or neurologically activated. The electrical activity generated in a muscle can be recorded either on the surface of the skin, on the surface of the muscle, or from inside the muscle. sEMG is usually chosen as a means for studying the behavior of the muscle as a whole, as it measures superimposed interference waveforms from a large number of muscle fibers. Recordings made from inside the muscle are referred to as intramuscular electromyography (intramuscular EMG [iEMG]). Intramuscular recordings allow the study of the physiology and pathology of the motor unit because they are more selective than sEMG.

10.6.1 Intramuscular EMG Electrodes

Recordings of iEMG can be obtained using either needle or wire electrodes. Different needles have been designed for recording extracellular action potential generated by a motor unit. The contribution of each fiber of the motor unit to the signal amplitude depends on its distance from the active electrode. These changes in amplitude with distance vary with the different types of electrodes, especially the recording surface. As for nerve recordings, the selectivity of the electrode to record only from nearby fibers is inversely proportional to the recording surface of the electrode, also called the active site. Figure 10.5 depicts some examples of needle electrodes used for iEMG; however, only the *concentric* needle and *monopolar* needle find application in clinics.

Wire electrodes are also available for intramuscular recordings and are mainly used for research purposes. The advantage of wires with respect to needle electrodes is that they are

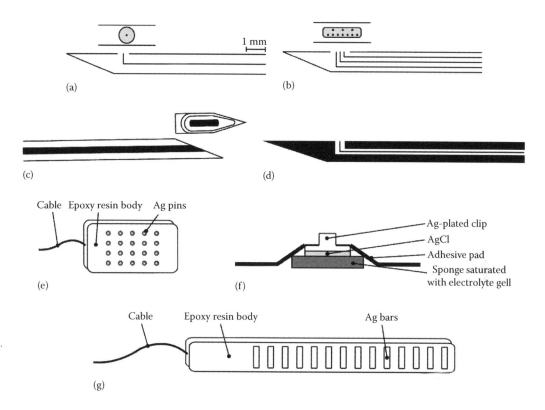

FIGURE 10.5   Different types of intramuscular and surface electrodes: (a) single fiber electrode with one detection surface, (b) multipolar electrode, (c) concentric needle electrode, (d) macro-EMG electrode, (e) silver pin electrode grid, (f) disposable electrode with electrolyte-saturated sponge, and (g) dry silver bar electrode array. (Adapted from Pozzo et al., 2003. With permission.)

flexible and able to remain in place, particularly during movement, and once the electrode is positioned, subjects experience minimal discomfort. Wire electrodes may be made from small diameter, nonoxidizing wires with insulation. They are placed in the cannula of a needle and bent at the tip; the needle is inserted into the muscle and then removed, with the wires left in the muscle. The downside of using wires is that once they are placed, their location cannot be adjusted. Although multichannel needle electrodes for iEMG have also been proposed, they are impractical and uncomfortable for the subject. A multichannel wire-based electrode has recently been proposed for intramuscular recordings. The technology is similar to LIFEs, originally proposed for nerve recordings. Wireless, multichannel implantable sensors are being proposed for targeted iEMG recordings. Some available wires comprise Teflon®-insulated wire (made of gold, silver, platinum, platinum–iridium, stainless steel, and tungsten), Formvar-insulated nichrome wire, and Isonel-insulated platinum wire. Teflon is nontoxic, corrosive resistant, flexible, and suitable for physiological transducers. Teflon also has a high dielectric constant and a low coefficient of friction. Nichrome wire features Formvar insulation. Nichrome wire is 80% nickel/20% chromium. Isonel is a high-temperature polyester enamel that forms a tighter bond to the platinum wire compared to Teflon. Wires are available in different sizes (25–200 μm) and, as in the case of needles, the selectivity of the recording depends on the size of the active site. The most selective recordings are achieved by exposing the wires only at the tip.

10.6.2 Surface EMG Electrodes

The potential generated in the muscle fibers can also be measured by applying conductive elements or electrodes to the skin surface. sEMG is a very common method of measurement of muscle activity, since it is noninvasive with minimal risk to the subject. The characteristics of the recorded signals depend on the electrode type and electrode configuration. Solid silver or gold, sintered silver and silver chloride, carbon, and sponge saturated with electrolyte gel or conductive hydrogel are the commonly used materials in electrode manufacturing. Commercial sEMG electrodes are available mainly as gelled and dry electrodes. Gelled electrodes make use of an electrolytic gel as a chemical interface between the skin and the metallic part of the electrode, allowing for reduction–oxidation chemical reactions to take place in the contact region. They are available as pregelled or aftergelled depending on whether the gel is packed together with the electrode or if the gel has to be applied afterwards. In general, the conductive gel has the aim of facilitating the passage of current from the muscle across the junction between the electrolyte and the electrode. Gelled electrodes provide stable short-term measurements even in the presence of movement. The main drawback of gelled electrodes is that their performance worsens over time because the gel dries; thus, they are not suitable for long-term measurements. Aftergelled electrodes overcome partly this limitation as gel can be filled up with time. The most recent technology offers surface electrodes that are combined in mono- and bidimensional arrays (also referred as *arrays*, *matrixes*, or *grids*) to provide multichannel information with up to hundreds of recording sites on the same grid (for a recent review, see Farina et al. [2010]).

Dry electrodes, as the name indicates, do not make use of conductive gel, but are considerably heavier than gelled electrodes because it is common to have the preamplifier circuitry at the electrode site, due in part to the high electrode–skin impedance associated with dry electrodes. Multichannel dry bar electrodes are also available in arrays but are not suitable for long-time recordings. Figure 10.5 shows some of the surface electrodes available.

10.6.3 Recording of Muscle Activity

Although intramuscular EMG and sEMGs exhibit different signal characteristics, the basic principles for measuring the EMG are the same for both recordings. This section focuses on the technical aspects related to the recordings of the EMG, while signal characteristics will be the topic of the next section.

10.6.4 Recording Configuration

Recording of the EMG can be achieved using several electrode configurations. The most known that are common to both surface and intramuscular is the monopolar and bipolar configurations. In monopolar configuration, there is one active electrode that picks up the

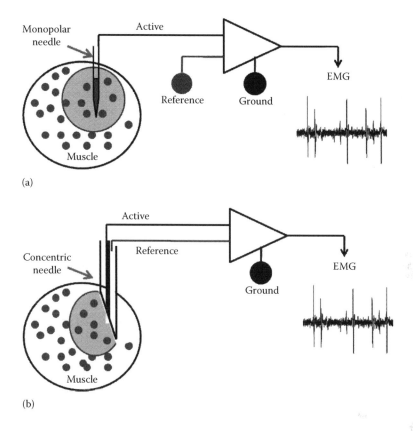

FIGURE 10.6　Schematic representation of (a) monopolar and (b) bipolar recordings for EMG.

electrical signal and a reference electrode or passive electrode. Figure 10.6a is a representation of the recording with a monopolar needle electrode. The concept is the same as for surface electrode as the passive electrode has to be placed away from a muscle activity. The bipolar configuration provides action potentials that are narrower. Another configuration that is often used among neurologists is the belly–tendon configuration for EMG that is somehow in between the monopolar and the bipolar. In the belly–tendon configuration, the active electrode is placed around the motor point, and the reference electrode is placed around the tendon area. Details on EMG signal recordings, including electrode technology, electrode impedance, and spatial filtering, can be found in Pozzo et al. (2003).

10.6.5　Characteristics of Muscle Recordings

Sources of interference in muscle recordings are nearly the same as for nerve recordings, but with different degrees of impacts. We recall that one single nerve fiber innervates 1–1000 muscle fibers depending on the motor unit. Thus, signals recorded from the muscle are usually of greater amplitude than nerve signals. An iEMG record from two pairs of sterilized wire electrodes made of Teflon-coated stainless steel placed in m. extensor carpi radialis of a human volunteer is shown in Figure 10.7 with its corresponding frequency spectrum and a single identified action potential. The muscle unit action potentials have an amplitude in the order of 50–1000 μVpp. The duration of the action potentials is in the range between 3 and 6 ms. The signal power is in the bandwidth 100–3000 Hz. However, the sEMG exhibits different characteristics. This difference is mainly caused by the distance to the source and the volume conductor (tissue between the source and electrode). Depending on the recording modality, the volume conductor may have a negligible or important effect on the acquired EMG signals. The low-pass effect of the volume conductor results in a signal with a frequency content of 10–400 Hz. The distance of the source from the recording system in case of surface methods implies poor spatial selectivity and lower amplitude. Because of the poor spatial selectivity, sEMG recordings are usually influenced by cross talk. Muscle cross talk refers to the signal measured over a muscle but generated by another muscle. Figure 10.7 depicts an example of sEMG recording with its corresponding power spectral density.

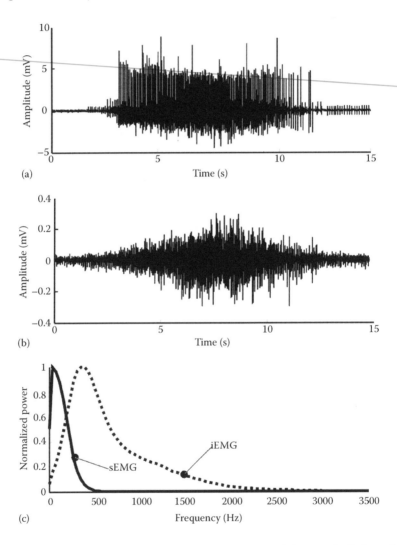

FIGURE 10.7 (a) Raw iEMG and (b) raw sEMG signals measured during a bell-shaped profile and (c) the corresponding distribution in power spectrum for iEMG and sEMG.

10.6.6 Processing and Applications of Muscle Recordings

Preprocessing of iEMG recordings is often limited to linear time-invariant band-pass filters with corner frequencies in the range 0.1–3 kHz (Figure 10.7c), because the SNR in these recordings is relatively high (typically above 10 dB). However, sometimes, equipment interference can decrease the SNR to such extent that advanced techniques such as wavelet or matched filter can be necessary in order to enhance spikes. Analysis of multiunit iEMG has been focused on decomposing the signals into constituent motor unit action potential trains, allowing researchers to study the behavior of individual motor units. Decomposition of iEMG includes a detection stage and a classification stage, which is sometimes followed by the separation of superimposed action potentials. The identification of single motor unit action potential has contributed to the understanding of basic physiological processes, where, for example, motor unit recruitment has been shown to relate to force. Although important advances have been made in iEMG decomposition, full decomposition of iEMG signals remains complex owing to the variability in shape of generated action potentials from the same motor unit and the overlapping in time and frequency of action potentials from different motor units (Merletti and Farina 2009). In addition, full decomposition is time consuming and is limited to low and moderate force levels. Thus, this analysis tool is rarely applied in clinical routine. The actual clinical use of iEMG focuses on the diagnosis of myopathies and diseases of the motor neurons where full decomposition is not necessary. Details on the analysis of iEMG can be found in Merletti and Farina (2009). In other applications, such as quantification of muscle force, signal decomposition is often not needed, and the information on the intensity of muscle activity can be extracted based on the smoothed integrated EMG (SIEMG), the global discharge rate, constraint sample entropy, and Wilson amplitude.

10.6.7 Surface EMG

sEMG provides primarily the global information about the state of the muscle, especially the onset of muscle contraction. Basic processing techniques of the sEMG signals involve linear band-pass filtering for high-frequency and baseline drift removal. More advanced techniques are also applied in some applications. Rectification and integration of the amplitude of these signals are widely used in many applications. sEMG has found applications in the following areas. Gait analysis (GA) or motion analysis is the quantitative laboratory assessment of coordinated muscle functions. Measurement of myoelectric activity provides a clinician with information on movement coordination that cannot be obtained by clinical observation alone, because the muscle activity is invisible. The EMG recordings may help him in forming a more complete picture of patients with gait disorders, in particular those with a central neurological origin. These recordings can assist in choosing the most suitable therapy.

They are, for instance, routinely used in planning corrective surgery for children with cerebral palsy. The overall importance of EMG measurements to the clinical evaluation of neuromuscular disorders is to provide relevant *diagnostic* contributions in terms of nosological classification, localization of focal impairments, detection of pathophysiological mechanisms, and functional assessment. There are several feasible strategies for the final extraction of clinically useful information. See Frigo and Crenna (2009) for a review.

EMG signals are used in many other clinical and biomedical applications. EMG is used as a diagnostic tool for identifying neuromuscular diseases and assessing low-back pain, kinesiology, and disorders of motor control. EMG signals are also used as a control signal for prosthetic devices such as prosthetic hands, arms, and lower limbs. See Pozzo et al. (2003) for specific techniques on amplitude estimation, estimation of muscle activation intervals, and spectral analysis. However, with advances in electrode design, not only temporal patterns can be studied, but it is also possible to investigate motor unit action potential propagation, estimate the motor unit conduction velocity, locate the innervation zones, investigate motor unit recruitment, or decompose the signals to constituent action potentials. Motor unit investigation with sEMG often requires a very low activation level of the muscle or very low force. Classification is also another area where advanced signal processing techniques are applied to the sEMG in order to separate different classes, which can be movements or muscle pathologies.

10.6.8 Systems for Recording Nerve and Muscle Activities

There are many systems that are used to acquire nerve and muscle signals, and many of them are custom based. Here, we list only three systems as examples:

For nerve: AD Instruments: http://www.adinstruments.com

For muscles and nerve conduction: OT Bioelettronica (http://www.otbioelettronica.it) and NeuroMetrix (http://www.neurometrix.com/)

10.7 EVOKED POTENTIALS AND CONDUCTION VELOCITY

10.7.1 Evoked Potentials

An evoked potential is an electrical potential recorded from the neuromuscular system following the presentation of a stimulus. We distinguish between motor-evoked potentials (MEPs) and sensory-evoked potentials. Sensory-evoked potentials include visual-, auditory-, and somatosensory-evoked potentials (SSEPs). This chapter deals only with evoked potentials that are either recorded or elicited from the periphery.

A MEP is recorded from a muscle as a response of direct stimulation of exposed motor cortex or transcranial stimulation (magnetic or electrical) of motor cortex. The most used modality in human is the transcranial magnetic stimulation (TMS). TMS was first described in 1985 as a noninvasive, painless way to stimulate the human brain. TMS works by passing a large, brief current through a wire coil placed on the scalp. The transient current produces a large and changing magnetic field, which induces electric current in the underlying brain. The focus of the activated area is relative to the shape of the coil. MEPs are commonly used in neurological electrophysiology laboratories to diagnose altered tracks in the first and second motoneuron. The stimulation evokes a response from the motor cortex, and the response spreads via the corticospinal tract to the target muscle. The response from the activated muscle can be read using surface electrodes. The results of these evoked potentials are used to assess whether a motor pathway is affected. Several techniques of TMS exist depending on the purpose of the MEPs. MEPs can be used, for example, to assess (1) the motor threshold, which is the lowest TMS intensity capable of eliciting small MEPs; (2) the recruitment curve, known as input–output or stimulus response curve; (3) the central motor conduction time, as an estimate of the conduction time of corticospinal fibers between motor cortex and spinal (or bulbar) motor neurons; and (4) the conduction blocks by means of a collision technique (detailed in the next section). Pathological applications include the diagnosis of myelopathy, amyotrophic lateral sclerosis, and multiple sclerosis. See Chen et al. (2008) for more details on the techniques used for TMS and clinical applications of MEPs.

SSEPs are noninvasive means of assessing somatosensory system functioning and are used to assess the transmission of the afferent volley from the periphery up to the cortex. They are usually obtained by applying short-lasting currents to sensory and motor peripheral nerves and then recorded from the scalp and are mainly used to identify lesions in the somatosensory pathway. Other particular applications include the diagnosis of diseases affecting the white matter like multiple sclerosis and noninvasive studies of spinal cord traumas and peripheral nerve disorders. Recently, SSEPs are being used for monitoring the patient's spinal cord during surgery, giving an early warning of a potential neurological damage in anesthetized patients. Evoked potentials can also be elicited from the peripheral nerves and measured from the innervated muscles (evoked muscle response). Applications of these evoked potentials include the investigation of the neuromuscular jitter, nerve conduction, and muscle conduction velocity.

10.7.2 Nerve Conduction Velocity

The velocity at which nerve fibers conduct action potentials can be measured either directly by placing electrodes on the nerve (invasive techniques) or indirectly by analyzing the resulting muscle response (noninvasive techniques). In invasive techniques, there is the need for two recording sites along the nerve trunk. Typical experimental design includes a stimulation electrode (S1, usually a cuff electrode) and two recording electrodes (E1 and E2) placed at some distance along the nerve fibers. Stimulation at electrode S1 will generate an action potential in different fibers depending on the stimulation intensity, and this potential (referred to as evoked potential) will be seen at electrode E1 and E2 at different time instants ($T1$ and $T2$). Knowing the distance D between E1 and E2 and computing the interval between $T1$ and $T2$, the compound conduction velocity can be calculated as $\Delta D/\Delta T$. It is called compound conduction velocity because it is a measure of the global velocity of all the fibers that are active. Single fiber conduction velocity can also be obtained using intraneural or intrafascicular electrodes; however, the stimulation intensity must be carefully controlled in order to activate very few nerve fibers. Furthermore, single fiber conduction is more important during voluntary activation. During voluntary movement, an action potential from a single fiber will be recorded in ideal cases at different active sites of a LIFE (with several contacts) to obtain the velocity of propagation for a particular fiber. Nevertheless, invasive nerve conduction techniques are not used in clinical settings. Nerve fiber conduction velocities can also be quantified by analyzing the evoked muscle response that resulted from nerve electrical stimulations in order to study the recruitment order of peripheral nerve fibers, for example. In early studies, conduction velocity was computed on the basis of onset time of evoked action potentials; however, the latency is affected by the intensity of stimulation, making it difficult to obtain velocities of slowest-conducting fibers. A possible way of obtaining the velocities of all nerve fibers is to assess the distribution of conduction velocities using the collision techniques.

10.7.3 Collision Techniques

Collision techniques (first introduced by Hopf in 1962) are techniques where paired stimuli are delivered in two different sites, and one can control which nerve fibers are under investigation with the time interval between the pulses. The techniques have not yet found their way into clinical diagnosis due to many possible pitfalls that accompany them. But these techniques seem promising for the assessment of recruitment order, because they can be used to assess the distribution of activated fibers with respect to stimulus intensity. This provides information that is complementary to conventional nerve conduction study, where only the conduction velocity of the fastest-conducting fibers is measured.

Hopf's method makes use of paired supramaximal stimuli, combining a distal stimulus with a delayed proximal stimulus as shown in Figure 10.8. The time delay between the two stimuli is called the interstimulus interval (ISI). When the ISI is short, the orthodromic nerve action potential from the proximal site (S2) is cancelled by collision with the distally evoked antidromic nerve action potential, and only the distally orthodromic impulse gives a muscle response (M1) (Figure10.8a). Incrementing the ISI will move the site of collision, until no more collision takes place. Provided that the nerve is no longer refractory, the proximal orthodromic impulse will evoke a late (or test) muscle response (M2) (Figure 10.8b). This gives the possibility of assessing the activation of different motor units with respect to an increasing size of the compound motor action potential (CMAP). The conduction velocity of the fibers can be determined by dividing the distance between the stimulating electrodes by the values of ISI. Several parameters can be calculated from this distribution. The fastest velocity (V_{max}) corresponds to the ISI at which the test response appears for the first time (ISI_{min}) and the slowest velocity (V_{min}) to the ISI at which the test response reaches its full size (ISI_{max}). The ISI values at which the increment of the test response reaches 5%, 50%, and 95% of the full size are used to compute the $V95$, $V50$, and $V5$, respectively.

The main limitations in Hopf's original method are related to the CMAP distortion and the refractory period. The distortion of the CMAP is thought to be due to the transient slowing of the proximal orthodromic impulse associated with the subnormal period and the transient change in conduction of muscle fibers. The latter is referred to as *velocity recovery effect* by Stålberg (1966), which is mainly caused by the close temporal relationship of the test and conditioning responses. The refractory period of each fiber, which is included in ISI at the stimulation point, has to be corrected when ISI is used to calculate conduction velocity. A third limitation is that the end of the velocity distribution is not easy to determine due to the difficulty of attaining a full size response of the test stimulus that is affected by the velocity recovery effect. To compensate for these limitations, modifications

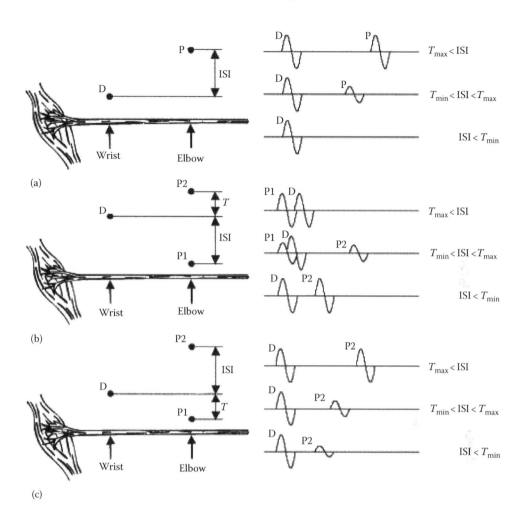

FIGURE 10.8 Schematic illustrations of collision techniques (on the left) and the resulting EMG (on the right). (a—Hopf) In Hopf's collision technique, the two stimuli D and P are used with an ISI between them. Both stimuli are set to supramaximal intensity. The distribution of nerve conduction velocities is estimated by the increase of the test response due to stimulus (P) up to the maximal response. (b—Ingram) In Ingram's collision technique, three stimuli (P1, D, and P2) are used and set to supramaximal intensity. The ISI is the time between stimuli P1 and D. The interval T between D and P2 is constant and set to a value that ensures collision of all the responses from the stimuli D and P2. The distribution of nerve conduction velocities is monitored by the decrease of the test response (P2) until it disappears. (c—Modified Hopf's) In modified Hopf's collision technique, three stimuli (P1, D, and P2) are used. P1 stimulus is submaximal and set to intensities that evoke 20%, 50%, or 80% of the maximal CMAP, while D and P2 are supramaximal. Contrary to Ingram's technique, the time T between P1 and D is constant and set to a value that ensures collision of the responses from the stimuli P1 and D. ISI is therefore the time between D and P2 stimuli. The distribution of conduction velocity (DCV) of the fibers that are not activated by the submaximal stimulus (P1) is estimated by the increase of the test response (P2) up to the maximum response. The DCV of the nerve fibers activated by the stimulus P1 is obtained from this DCV and that of the whole nerve. T_{min} and T_{max} are the fastest and slowest conduction time, respectively, in the nerve between the two stimulating electrodes. (From Hennings, K. et al., *Clin. Neurophysiol.*, 118, 283, 2007. With permission.)

of the technique have been proposed in order to accurately assess the distribution of conduction velocities (Hennings *et al* 2007) and to measure the refractory period distribution.

10.7.4 Muscle Fiber Conduction Velocity

Knowledge of the conduction velocity is physiologically important because it reflects the properties of the membrane (Stålberg 1966). Furthermore, it reflects the modifications of the peripheral properties of the neuromuscular system as a consequence of pathology, fatigue, pain, or exercise. The velocity of propagation of muscle action potentials can be studied at either single fiber, single motor unit, or global muscle level from voluntary activation, direct muscle stimulation, or indirect stimulation through the nerve. Single fiber conduction velocity evaluation is a technique based on the use of a single fiber electromyography electrode. The procedure, which entails recordings from different sites in the muscle, allows the study of the CV of a small sample of axons. Direct muscle stimulation is preferred for this purpose in order to ensure selective fiber activation. The challenge here is to be able to place two recording sites along the same muscle fiber. Conduction velocity measurement using surface recordings is widely used with two or more recording sites. The key element is the determination of the conduction time (delay), which together with the known interelectrode distance is used to calculate the conduction velocity. The necessary feature is to ensure that recording sites are placed along the direction of the muscle fibers from the innervation zones to the tendon regions. There exist different techniques for the assessment of the conduction time or velocity. We distinguish between methods based on a single signal, two signals, and multiple signals as given in Table 10.2. See Farina and Merletti (2004) for further readings.

TABLE 10.2   Summary of the Techniques Used for the Estimation of Muscle Conduction Velocity

Classification of Methods	Specific Method
Methods based on detection of one signal	Characteristics of spectral frequencies
	Estimation of scale factor
	Detection of spectral dips
	Autocorrelation technique
Methods based on detection of two signals along the fiber direction	Delay between reference points in detected waveforms
	Phase difference method
	Distribution function method
	Cross correlation function technique
	Spectral matching
	Generalized spectral matching
Methods based on detection of more than two signals	Generalization of methods for two delayed potentials to multichannel beamforming
	Maximum likelihood without constraint
	Maximum likelihood with constraint
	Maximum likelihood from very short signal windows
	Maximum likelihood from bidimensional recordings

Source: Farina, D. and Merletti, R., *Med. Biol. Eng. Comput.*, 42, 432, 2004.
It should be noted that the estimation of conduction velocity may have different sensitivities depending on several factors such as spatial filter and electrode shape.

10.8 SUMMARY

In this chapter, we have provided the basic anatomy and physiology for nerves and muscles in order to introduce the principles for signal generation. The chapter comprises four main parts: electroneurography, electromyography, evoked potentials, and conduction velocity of nerve and muscle fibers. The electroneurogram (ENG) is the extracellular potential of an active peripheral nerve recorded at some distance from the nerve or from within the nerve trunk. Recording of this extracellular potential can be achieved using different electrode interfaces (extraneural, intraneural, and regenerative) and configurations (monopolar, bipolar, and tripolar). The recorded signal can be conditioned to improve the quality in terms of SNR and processed to retrieve relevant information such as the intervals of activity, recruitment strategy, and knowledge of active cells based on signal decomposition. The retrieved information can then be used for diagnostic or control purposes. The electromyography (EMG) on the other hand is the extracellular potential of an active muscle recorded at some distance from the muscle or from within the muscle. This potential can be recorded using different configurations (monopolar, bipolar, and belly–tendon) from the surface of the skin (sEMG) using surface electrodes or inside the muscle tissue (iEMG) by means of needle or wire electrodes. The processing of EMG signals provides information about the state (passive vs. active) of the muscle and can be used to quantify physiological parameters such as firing pattern and for diagnostic purposes. Evoked potentials are special ENG or EMG signals recorded following the presentation of a known stimulus. They are commonly used in neurological electrophysiological laboratories to diagnose altered tracks in the first and second motoneuron. Knowledge of the conduction velocity of nerve and muscle cells is physiologically important as it reflects the properties of the membrane. Estimation of the CV usually requires signals to be measured from at least two sites along the propagation of the action potential.

ACKNOWLEDGMENTS

The authors thank Marko Jörg Niemeier for his contribution in making some of the illustrations adapted from other sources. We also thank Dr. Ken Yoshida for his scientific contribution.

REFERENCES

Chen R, Cros D, Curra A *et al* 2008 The clinical diagnostic utility of transcranial magnetic stimulation: Report of an IFCN committee *Clin. Neurophysiol.* **119** 504–32

Farina D, Holobar A, Merletti R and Enoka R M 2010 Decoding the neural drive to muscles from the surface electromyogram *Clin. Neurophysiol.* **121** 1616–23

Farina D and Merletti R 2004 Methods for estimating muscle fibre conduction velocity from surface electromyographic signals *Med. Biol. Eng. Comput.* **42** 432–45

Frigo C and Crenna P 2009 Multichannel SEMG in clinical gait analysis: A review and state-of-the-art *Clin. Biomech.* (Bristol: Avon) **24** 236–45

Hennings K, Kamavuako E N and Farina D 2007 The recruitment order of electrically activated motor neurons investigated with a novel collision technique *Clin. Neurophysiol.* **118** 283–91

Hoffer J A, Stein R B, Haugland M K *et al* 1996 Neural signals for command control and feedback in functional neuromuscular stimulation: A review *J. Rehabil. Res. Dev.* **33** 145–57

Lyons G M, Sinkjær T, Burridge J H and Wilcox D J 2002 A review of portable FES-based neural orthoses for the correction of drop foot *IEEE Trans. Neural. Syst. Rehabil. Eng.* **10** 260–79

Merletti R and Farina D 2009 Analysis of intramuscular electromyogram signals *Philos. Trans. A Math. Phys. Eng. Sci.* **367** 357–68

Navarro X, Krueger T B, Lago N, Micera S, Stieglitz T and Dario P 2005. A critical review of interfaces with the peripheral nervous system for the control of neuroprostheses and hybrid bionic systems *J. Peripher. Nerv. Syst.* **10** 229–58

Pozzo M, Farina D and Merletti R 2003 Electromyography, detection, processing, and applications in G. Zouridakis and J. Moore (eds.), *Biomedical Technology and Devices Handbook* (Boca Raton: CRC Press) pp: 4.1–4.60

Sinkjær T, Yoshida K, Jensen W and Schnabel V 2006 Electroneurography in J. Webster (ed.), *Encyclopedia of Medical Devices and Instrumentation* 2nd. Ed. (Hoboken: John Wiley & Sons) pp 191–232

Stålberg E 1966 Propagation velocity in human muscle fibres in situ *Acta Physiol. Scand.* **287** 1–112

CHAPTER 11

Obstetrics and Gynecology

Michael R. Neuman

CONTENTS

11.1 Measurements in Obstetrics 238
   11.1.1   Growth and Development of the Fetus 238
   11.1.2   Observation of the Mother 238
   11.1.3   Fetal Heart Sounds 239
   11.1.4   Fetal Electrocardiogram 240
   11.1.5   Doppler Ultrasound 241
   11.1.6   Fetal Position 243
   11.1.7   Ultrasonic Imaging 244
   11.1.8   Amniotic Fluid Analysis 246
   11.1.9   Fetal Endoscopy 247
   11.1.10 Antenatal Fetal Monitoring 248
   11.1.11 Measurements during Labor and Delivery 249
   11.1.12 Technology of the Fetal Cardiotocogram 251
         11.1.12.1 Direct Fetal Electrocardiography 251
         11.1.12.2 Indirect Fetal Heart Rate Determination by Doppler
                  Ultrasound 252
         11.1.12.3 Direct Uterine Activity by Pressure Catheter ... 253
         11.1.12.4 Indirect Method of Sensing Uterine Activity with a
                  Tocodynamometer 254
         11.1.12.5 Fetal Monitor or Cardiotocograph 255
   11.1.13 Other Biochemical and Biophysical Fetal Measurement Methods ... 257
         11.1.13.1 Fetal Blood Sampling 257
         11.1.13.2 Fetal pH Monitoring 257
         11.1.13.3 Fetal Blood Oxygenation 258
   11.1.14 Monitoring Cervical Dilatation 258
11.2 Measurements in Gynecology 260
   11.2.1   Uterine Contractility 260
   11.2.2   Menstrual Cycling 260
   11.2.3   Patency of Fallopian Tubes 262

11.2.4	Uterine Cervical Compliance	262
11.2.5	Optical Instrumentation in Gynecology	263
11.2.6	Hot Flash Documentation	264
11.3 Summary		265
References		265

Biophysical and biochemical measurements are important in diagnosis, patient monitoring, and research in the medical field of obstetrics and gynecology. Although the reader will find that there are many different types of physiological measurements in this clinical area, the major application of this technology has been during the time period surrounding birth, the perinatal period. Other measurements in both obstetrics and gynecology are primarily involved with diagnosis, and many of them have been limited to the research laboratory or clinical research so far. In this chapter, we will look at applications of physiological measurements in both obstetrics and gynecology.

11.1 MEASUREMENTS IN OBSTETRICS

Obstetrical measurements are concerned with assessment of progress during pregnancy and during labor and delivery. One can start with the diagnosis of pregnancy itself. Biochemical methods are used to detect hormone changes associated with the onset of pregnancy itself. These include measurement of gonadotropins in the maternal blood or urine. Human chorionic gonadotropin (hCG) is the most common factor measured from a blood sample, and this measurement is sensitive enough to give a definitive diagnosis within the first 4 weeks of pregnancy. The measurement is carried out by an analytical biochemical assay usually done in the clinical laboratory. There are also over-the-counter versions of these assays that can be done by the patient herself at home on a blood or urine sample. The details of this assay can be found in the biochemical literature (Cole 2012).

11.1.1 Growth and Development of the Fetus

There are several measurements that can be used to assess the progress of pregnancy. In this section, we shall look at some of these from the perspective of how these measurements are made and the instrumentation involved rather than their significance in the assessment of pregnancy, the latter being more appropriate for a medical textbook.

11.1.2 Observation of the Mother

A simple approach to observing the growth of the fetus is to observe the mother and especially the changes in her abdomen as the fetus and uterus grow. It is usually possible to do this by physical examination of the patient, especially during the second and third trimesters of pregnancy. The growth of the uterus reflects growth of the fetus, and often, the uterine fundus can be felt by a well-trained obstetrician or midwife. Unfortunately, this is not as easily done in obese patients due to the anterior abdominal fat pad. Once the fundus has been located, one can measure the distance from it to the upper margin of the symphysis pubis bone to estimate uterine and fetal growth. This distance can be determined

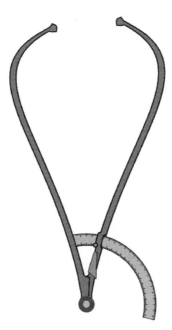

FIGURE 11.1   An obstetrical caliper or pelvimeter that can be used for the measurement of pelvic dimensions as well as measuring uterine fundal height during pregnancy to assess growth of the fetus. (Illustration courtesy of Wikimedia Commons.)

using an ordinary tape measure, usually with a centimeter scale. In the second and third trimesters, this length in centimeters is approximately the same as the duration of pregnancy in weeks. It is important to note, however, that this measurement should not be used to determine the duration of pregnancy. Often, a paper tape is used for this measurement, and it can be marked and discarded following measurement.

For those who desire a more sophisticated method for making this measurement, an obstetrical caliper is available. This instrument as shown in Figure 11.1 consists of two probes with rounded tips that are hinged together at the opposite end from the tips. An arc attached to one of the probes has a scale engraved on it that indicates the displacement of the tips at the point where the second probe intercepts this arc. Although these instruments have been extensively used in the past, the paper tape method seems to be much easier, and because they are disposable, the possibility of cross contamination of pathogens from one patient to the next is greatly minimized.

11.1.3 Fetal Heart Sounds

The fetal heart starts beating at approximately 41 days of gestation, but at this point, it is too small to produce sounds that could be detected by auscultation. Generally, fetal heart sounds cannot be heard transabdominally until about 18 weeks of pregnancy, and then a special type of stethoscope known as a fetoscope needs to be used. Ultrasonic Doppler instruments (Section 11.1.5) can generally pick up the fetal heart at about 10–12 weeks of gestation, and so this type of examination is usually used by obstetricians and nurse-midwives rather than the fetoscope in early pregnancy. An example of a Pinard fetoscope is

FIGURE 11.2 The Pinard horn, an early form of stethoscope for listening to the fetal heartbeat during pregnancy. (Illustration courtesy of Wikimedia Commons.)

illustrated in Figure 11.2. The conical opening of this device is pressed against the maternal abdomen, and the examiner places his or her ear on the earpiece at the opposite end of the device. This device is used more frequently in Europe than in North America with Doppler ultrasound being most frequently used in the New World.

Some investigators have developed phonocardiographic instruments for detecting and recording fetal heart sounds. These devices consist of a sensitive microphone and an amplifier to increase signal levels so that they can be recorded or plotted on a paper chart or computer screen. Although such devices have been used in the research setting, their complexity is such that they are not routinely used in clinical obstetrics.

11.1.4 Fetal Electrocardiogram

The electrocardiogram (ECG) is a time-varying electrical potential found on the surface of the body that is related to various events occurring during each cardiac cycle. In adults, it can be used in the diagnosis of various cardiac conditions by providing information on heart rate, rhythm, and injury currents. It is not as easy to get this information from the fetus during pregnancy because it is not possible to put electrodes on the fetal body surface. Instead electrodes can be placed on the maternal abdomen, which, in turn, is connected to the fetus, but in this case, the electrical signals from the fetus are quite weak, generally in the order of tens of microvolts. This signal is often contaminated by interference from the maternal ECG and other electrical potentials found on the surface of the maternal abdomen. Nevertheless, it is possible to determine fetal R-waves in the signal to establish the fetal heart rate, and in some cases or through the use of signal averaging, one can extract other components of the fetal ECG to determine heart rhythm. It is important to use low-noise, stable biopotential electrodes such as the ones

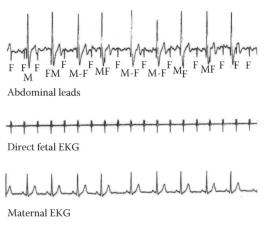

Abdominal leads

Direct fetal EKG

Maternal EKG

FIGURE 11.3   An example of a fetal ECG taken from abdominal leads (top trace) along with a direct recording of the fetal ECG from a fetal scalp electrode (middle trace) and the maternal ECG (bottom trace). (Reprinted from Roux, J.F. *CRC Crit. Rev. Bioeng.*, 2, 119, 1975. With permission.)

made from silver/silver chloride to pick up this signal. Their placement on the maternal abdomen is quite variable, although a good starting point is to have one electrode over the fundus of the uterus with the other electrode over the lower uterine segment. Various investigators have used different types of signal processing routines to extract the fetal ECG from noise and maternal interference, yet it still remains a difficult process to get high-quality fetal ECG recordings in this way (Callaerts *et al* 1986, Ibrahimy *et al* 2003, Kimura *et al* 2006).

An example of a noninvasive echocardiogram taken from abdominal leads is shown in Figure 11.3. It is easy to see the stronger maternal ECG compared to the fetal signal on this lead. Because this patient was studied during active labor, a fetal scalp electrode was also placed so that simultaneous recordings from the fetal presenting part and the abdominal leads could demonstrate the fetal component of the abdominal signal. Often when one is attempting to record an abdominal fetal ECG, a second electrocardiographic channel is recorded simultaneously from electrodes on the maternal chest. These electrodes pick up only the maternal ECG, and either this signal can be used to switch off the abdominal signal during each maternal heartbeat (QRS complex) or the maternal signal can be subtracted from the abdominal signal leaving just the fetal signal. All of these techniques require sophisticated signal processing, which is now possible through microcomputing systems.

11.1.5 Doppler Ultrasound

The most frequently used method of assessing the fetal heart during gestation is by Doppler ultrasound. This measuring technique is based upon the principle that a sonic signal will be reflected when there is an interface where there is a change in the acoustical impedance of the tissue from one side to the other. If this interface is moving with respect to the source and receiver of an ultrasonic signal, the reflected signal will be shifted in frequency according to the Doppler effect that is given by the following equation:

$$f_r = \frac{2f_t u \cos\theta}{c} \tag{11.1}$$

where

f_r is the frequency of the reflected ultrasonic wave
f_t is the frequency of the transmitted ultrasonic wave
u is the velocity of the tissue interface with respect to the ultrasonic source and receiver
c is the velocity of the ultrasound
θ is the angle between the direction of the ultrasonic beam and the velocity of the moving interface

In the case of sensing Doppler frequency shifts due to the fetal heart, there are two types of signals that can be detected through the maternal abdomen. The first is due to the interface between fetal blood cells in major arteries and surrounding tissue. The blood cell walls provide the actual reflecting interface. This interface is moving at blood velocity and is pulsatile such that the Doppler frequency shifts will vary over the fetal cardiac cycle. The second interface that produces a Doppler shift is due to the leaflets of the valves in the fetal heart as they open and close throughout the cardiac cycle. Both types of interface give characteristic signals that can be heard when the frequency differences between the transmitted and reflected waves are determined and played through a loudspeaker. These frequency differences are within the audible range, and most instruments for ultrasonically assessing the fetal heart have the capability of producing this signal so that the clinician and patient can hear the beating of the fetal heart. Some instruments contain electronic circuitry that can detect each heartbeat and determine the fetal heart rate from the signal.

A block diagram of a typical ultrasonic fetal heartbeat instrument is shown in Figure 11.4. It consists of circuitry to excite a transmitting transducer in a probe placed on the maternal abdomen and aimed toward the fetal heart or major arteries. In general for fetal heart instrumentation, the frequency of the ultrasonic signal ranges between 2 and 3 MHz. These frequencies provide a good compromise between depth of penetration and spatial resolution that works well for detecting the fetal heart. A receiving transducer is located in the same probe and picks up the reflected Doppler shifted signals. A circuit then determines the frequency difference between the transmitted and received signals and amplifies this signal so that it can be heard and further processed. A cardiotachometer circuit then determines when each heartbeat has occurred by looking for the maximum frequency difference signal, and the number of these fiduciary marks of the cardiac cycle that occur over a known time interval is counted to determine the fetal heart rate.

In recent years, relatively inexpensive (under USD 50) Doppler ultrasonic instruments for home use have appeared on the market. It is amazing that the same technology that would have cost hundreds of dollars several years ago can now be made so inexpensively.

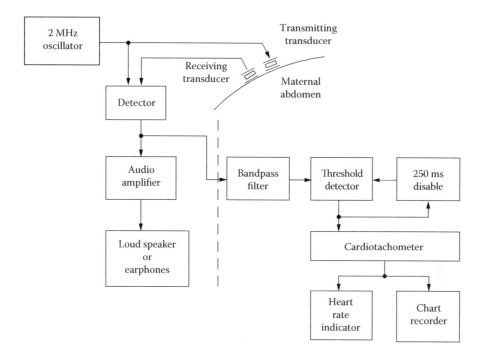

FIGURE 11.4   A block diagram of a Doppler ultrasonic instrument for listening to the fetal heart-beat (left side of the vertical dashed line) and recording the fetal heart rate (the entire diagram) with a fetal monitor.

Such devices are useful for prospective parents to listen to their baby's heartbeat, but they are no substitute for an examination by an experienced medical professional.

Doppler ultrasonic fetal heart measurement systems are measuring blood velocity in the major arteries of the fetus when the ultrasonic beam is appropriately aimed. Since they are, in fact, measuring velocity of the flowing cells, this has led to a new type of measurement from the fetus: fetal velocimetry. In this case, the instrument is more complex than the one used for fetal heart auscultation. Such instruments can create a plot of relative fetal blood velocity as a function of time, but it is relative since the angle between the ultrasonic beam and the blood velocity direction is unknown. Even so this waveform can be used to determine the ratio of systolic to diastolic arterial blood velocity. Such a ratio can be useful in determining cardiovascular system anomalies such as reduced placental blood flow. This feature is often incorporated in ultrasonic imaging devices so the examiner can see the artery being studied during the examination.

11.1.6 Fetal Position

Determining the position of the fetus in utero near term is an important measurement in clinical obstetrics. Before the advent of ultrasonic imaging (see Section 11.1.7), this had to be determined by palpation of the maternal abdomen by an experienced obstetrician. Often, one can feel the location of the fetal head and arms and legs and from this determine the fetal position (or lie) to determine if the fetal head is engaged in the pelvis ready for delivery. These palpation procedures are known as Leopold's maneuvers (Nahum 2002).

This procedure requires a great deal of experience on the part of the obstetrician, and it is not always possible to determine fetal position in this way especially with obese women. A much better method is the use of ultrasonic imaging, which will be described in the next section.

11.1.7 Ultrasonic Imaging

The development of ultrasonic imaging techniques for application in medicine became quite important in the 1960s and early 1970s. As this development was going on, it became clear that there would be significant applications of this imaging and measurement technique in obstetrics, since the mother and fetus met many of the requirements for ultrasonic imaging. First of all, the uterus and its contents are not very deep in the abdomen and pelvis during the latter half of pregnancy, and there is no air or bone between the anterior abdominal surface and the uterine contents that could interfere with imaging, because bone and air are not good conductors of ultrasound. Second, the fetus is suspended in amniotic fluid that is mostly water, an excellent conductor of ultrasonic energy. Thus, ultrasonic imaging of the uterus and its contents became a major application of this new technology. The basic principle behind ultrasonic imaging is that a pulse of ultrasound is transmitted into the body, and some of it gets reflected each time the pulse passes through a tissue interface as previously described for the Doppler ultrasonic instrument. The deeper the interface lies, the longer it takes for the reflected ultrasonic pulse to return to the receiving transducer in the same probe as the transmitting transducer. If this probe is moved around on the abdominal surface or if the ultrasonic direction is modified electronically, one can generate an image based on these reflected pulses that represents a cross-sectional view of the body in the plane defined by the moving ultrasonic beam. This image is known as a B-scan and gives a good cross-sectional representation of the uterus and fetus as illustrated in Figure 11.5. With such an image, it is possible to determine the position of the fetus in

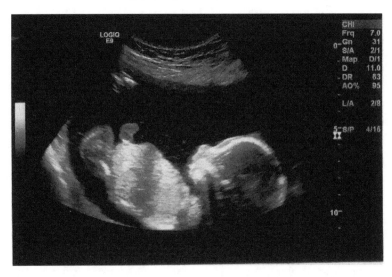

FIGURE 11.5   An example of an ultrasonic B-scan of a 14-week-old fetus in utero. (Illustration courtesy of Wikimedia Commons.)

the uterus and the general anatomy of the fetus and address issues such as if organs and structures are in the correct locations or if congenital anomalies could be present. B-scan images can be used to identify anomalies such as diaphragmatic hernias, missing kidneys, and too much (polyhydramnios) or too little (oligohydramnios) amniotic fluid and even observe the genitalia to determine the gender of the fetus (Callen 2008).

Improvements in ultrasonic imaging technology continued over the years, and new instruments were developed that combined Doppler ultrasound and B-scan imaging so that one could see where the Doppler reflected signals were coming from. The technology was further improved such that Doppler shifted reflections from blood were shown in color with blood moving toward the probe colored red and that moving away from the probe colored blue. With such instrumentation, not only one could identify the blood vessels, but it was possible to do velocimetry of the moving blood as well as identify areas of turbulence. The next major improvement in ultrasonic imaging that occurred in the 1990s was the extension of images to 3D renderings and to even have 4D ultrasonic images where the fourth dimension is time. In addition to making it easier to identify fetal anatomical structures, the 3D imaging has been popular with prospective parents, for they get to see a lifelike image of their baby before birth. An example of such an image is shown in Figure 11.6.

Ultrasonic imaging can be used as a quantitative measure as well as a means of imaging the fetus. A transverse image of the fetal head can be used to measure the displacement between the parietal bones of the skull to give a measure known as the biparietal diameter. Obstetricians use this in determining the age of the fetus, and it is helpful in

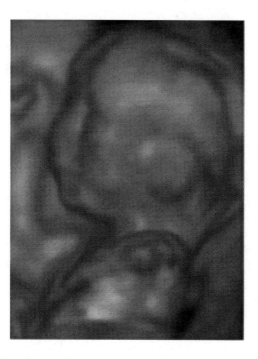

FIGURE 11.6 An example of a 3D ultrasonic image of the face of a 20-week-old fetus in utero. (Illustration courtesy of Wikimedia Commons.)

establishing the estimated date of delivery. Most ultrasonic scanners have displacement measuring software built into the instrument so that one can enter the positions between which the displacement is to be measured, and the ultrasonic machine calculates and displays the displacement between those two points.

As previously mentioned, ultrasonic imaging of the fetus can also be used to examine fetal anatomy and detect any anomalies that are present. A typical ultrasonic examination involves imaging not only the head but also the heart, lungs, intestines, kidneys, bladder, and spine of the fetus. The ultrasonographer looks for normal development of these structures as well as anomalies such as cardiac defects, brain and skull defects, defects of the spinal column, and kidney and bladder problems. In recent years, fetal surgical procedures have been developed that can in some cases correct these problems before birth.

Another physiological measurement that can be useful in identifying fetal problems during gestation is the study of fetal movements. Often, the mother can describe the activity of the fetus in utero from feeling a kick or repositioning of the fetus. Seeing such limb activity on an ultrasonic scan can be valuable in terms of describing normal activity of the fetus as well as identifying times when these activities are absent, since this may be an indication of fetal distress. Fetal breathing movements are also able to be seen on ultrasonic examination. Breathing activity of the fetus, while not essential for respiration, is present especially in the third trimester of gestation and is another variable that can be used to assess fetal well-being. Although the fetus does not demonstrate breathing movements continuously as the newborn would, such movements can be helpful in establishing fetal status. Some investigators have studied the amount of time the fetus spends in breathing movements, and this is related to gestational age (Jansen and Chernick 1991). A reduction in fetal breathing movements can be an indication of compromised fetal status just as a reduction of limb movements is. However, one must be cautious in this assessment because the fetus could be sleeping and therefore not involved in active movements. Some ultrasonographers try to stimulate the fetus with a sound applied on the maternal abdomen to wake the sleeping fetus if the lack of activity is thought to be due to sleep.

11.1.8 Amniotic Fluid Analysis

Measurements in amniotic fluid can provide information on the fetus and pregnancy. A physician using a transabdominal hypodermic needle, the tip of which is placed in the amniotic cavity, can obtain amniotic fluid. It is important in placing the needle not to interfere with the fetus or placenta, and so needle insertion is usually done under ultrasonic imaging guidance. The process of collecting a sample of amniotic fluid in this way is known as amniocentesis. The collection of just a few milliliters of amniotic fluid does not adversely affect the pregnancy.

Measurements on amniotic fluid include identifying excessive amounts of the substance bilirubin in the amniotic fluid and collecting fetal cells for genetic analysis. In the rare case of an intrauterine infection, amnionitis, it is also possible to detect this condition by amniotic fluid analysis. An increase of bilirubin in amniotic fluid can be determined spectrophotometrically. Bilirubin in amniotic fluid gives it a yellow color, but other substances in amniotic fluid can do this too, so observing the color of a sample of amniotic fluid is not a good way to identify relatively high concentrations of bilirubin. Instead this measurement can be made by determining the absorption of light at a wavelength of 450 nm and comparing it to a baseline drawn between the absorptions at 365 and 550 nm (Liley 1961). If the absorption difference between the baseline and the absorption peak at 450 nm is greater than the currently accepted threshold, this is an indication of excessive bilirubin in the amniotic fluid. Such an increase is often related to hemolysis of fetal blood due to an incompatibility with the mother as a result of Rh isoimmunization. Such a condition once detected can be treated by intrauterine blood transfusion.

Genetic analysis of the fetus can be accomplished because amniotic fluid usually contains isolated, free fetal cells that can be collected from the amniotic fluid and the genetic karyotype determined. This measurement is useful clinically in that it can identify some genetic disorders in the fetus such as Down's syndrome allowing parents to make difficult decisions regarding continuing the pregnancy. On a happier note, the genetic analysis can determine the gender of the unborn fetus, but this would not be a reason for performing an amniocentesis, since the same gender information usually can be obtained from ultrasonic imaging.

11.1.9 Fetal Endoscopy

In recent years, the technique of fetal endoscopy, sometimes known as fetoscopy, has been applied to the diagnosis and treatment of fetuses in utero (Watanabe and Flake 2010). This can be done transabdominally or transcervically using a very small, usually no more than 1 mm in diameter, endoscope as illustrated in Figure 11.7. In the former case, the endoscope is introduced into the uterine cavity through a small incision made in the maternal abdomen. The endoscope is introduced under guidance from a real-time ultrasonic image so as not to interfere with or damage the fetus or placenta. This instrument allows the examining physician to observe the fetus and placenta in utero as well as to perform some

FIGURE 11.7 Cartoon of a fetal endoscopic surgical procedure showing the transabdominal endoscope being positioned to perform a surgical procedure to treat twin-to-twin transfusion syndrome. (Illustration courtesy of Wikimedia Commons.)

surgical procedures. These include collecting a fetal blood sample from the umbilical cord, taking a biopsy of fetal tissue, collecting a sample of a chorionic villus from the placenta, coagulating aberrant circulation, or treating congenital abnormalities such as a diaphragmatic hernia or urinary tract obstruction in the fetus. These intrauterine treatments are often referred to as fetal endoscopic surgery. Just as with amniocentesis, by obtaining samples of fetal blood or fetal cells, it is possible to do genetic analysis to determine if the fetus has severe genetic disorders. Many of these procedures involving fetal endoscopy are still experimental and only performed at a few medical centers. Although it is a minimally invasive way to get access to the fetus, the procedure has significant risks associated with it and should only be used when the benefits outweigh the risks.

11.1.10 Antenatal Fetal Monitoring

The antenatal period is the time of pregnancy before birth. During this time, there are measurements described in the previous sections to help assess fetal well-being during pregnancy. A specific problem during the antenatal period is the detection and prevention of premature labor and delivery. This has been a major concern in the field of obstetrics for many years, and physicians and biomedical engineers have tried to find ways to detect the early onset of premature labor so that attempts can be made to stop it and allow the pregnancy to continue to term.

There are additional problems that can compromise the pregnancy during the antenatal period. Placental insufficiency or a low-lying placenta can be seen where the fetus, especially near term, is unable to exchange enough oxygen and nutrients with the mother to sustain normal growth and development. There can also be other cardiovascular related problems resulting in circulatory insufficiency that can compromise the fetus and need to be detected.

Several measurements have been explored to detect these kinds of problems during the antenatal period, but these have not been successful in identifying fetal problems that can be corrected. Probably the most frequently applied method is periodic fetal cardiotocogram (CTG) recordings from patients at increased risk during the antenatal period. The fetal CTG will be described in more detail in the following section, but for the present time, it can be considered as a continuous recording over time of the instantaneous fetal heart rate and uterine activity. Such recordings are usually made on a computer screen or paper chart. They can yield information regarding how the fetal heart rate responds to fetal movements, uterine activity, and even external stimuli such as loud sounds. Such recordings are generally made in a clinical facility, although there have been commercial efforts to allow patients to make these recordings in their homes and transmit them to a central facility over the Internet or a telephone link. For the most part, these approaches have not been successful in terms of their contribution to arresting premature labor and preventing premature delivery.

The prospective mother herself can do a much simpler measurement. This involves monitoring fetal movement activity where the mother counts the times she feels a fetal kick or other movement during the same period of time on successive days. Although this is not definitive, a reduction in fetal activity could be an indicator of problems, and the prospective parents should seek advice from their provider. Of course, the reduction in activity may just be the result of quiet period when the fetus is sleeping, and the fetus may resume normal activity during a time of wakefulness. Thus, it is better to make several measurements throughout the day before being concerned about reduced fetal activity. It is also possible to get this information using instrumentation. Continuous ultrasonic imaging can provide indications of fetal movement, and some investigators have been able to detect fetal movements using a tocodynamometer on the anterior abdomen (Timor-Tritsch 1979). The tocodynamometer is a sensor for detecting uterine activity through the maternal abdomen, but it also can be used to detect fetal intrauterine movements in some cases. This sensor will be described in more detail in the following section.

There are other methods that can be used to assess the fetus during the antenatal period. The fetal cardiotocograph is also used clinically in what is known as a fetal nonstress or a stress test. In the case of the nonstress test, the mother is asked to lie or sit with an ultrasonic Doppler sensor and a tocodynamometer attached to her anterior abdomen. The ultrasonic Doppler instrument picks up the fetal cardiac activity and can present this as a heart rate plotted as a function of time. The tocodynamometer picks up spontaneous uterine contractions or fetal movement activity, and its instrumentation can provide that signal as a function of time. Both signals are shown simultaneously on a computer screen or paper chart record. A fetus in good condition will respond to fetal movements or a spontaneous uterine contraction by accelerating its heart rate for a short period of time after the event occurs. This is similar to the adult response of increased heart rate when there is some kind of external stimulus such as a startle from a loud noise or increased activities such as during exercise. If the fetal heart rate accelerates during this stimulus, this is considered a normal response, and it is referred to as a responsive nonstress test.

In the case of the stress test, uterine activity is induced by giving oxytocin, a drug that stimulates uterine contractions. Again, the normal fetus would respond to the contraction by an increased heart rate during and immediately following a contraction with the heart rate than returning to baseline. If, on the other hand, the fetal heart rate does not accelerate or perhaps it even decelerates during and immediately after the uterine contraction, this is indicative of a nonreassuring result. Therefore, this might stimulate further investigation by the health-care provider.

11.1.11 Measurements during Labor and Delivery

The most frequently used measurement in clinical obstetrics is intrapartum fetal monitoring. The monitor, known as a fetal cardiotocograph, displays and records the instantaneous fetal heart rate and uterine activity as a function of time. The recording is known as the fetal CTG and is routinely used in hospitals around the world even though its ability to detect fetal distress is limited. There are two types of fetal monitoring: indirect and direct. In the indirect case, sensors are placed on the maternal abdomen, and there is no direct contact with the fetus or uterus; thus, it is noninvasive. The direct method involves placing sensors on the fetus and within the uterine cavity.

Although there had been previous work on sensing the fetal ECG and uterine activity, fetal monitoring got its start in the early 1960s through the pioneering work of Edward

Hon (Hon and Hess 1957, Hon 1960a,b). Dr. Hon who was both an obstetrician/gynecologist and an electrical engineer developed an electrode for sensing the fetal ECG during labor by contact with the fetal presenting part. He measured uterine activity by means of a pressure catheter and external pressure sensor. The catheter was filled with a normal saline solution, and its distal tip was introduced into the uterine cavity in contact with the amniotic fluid. Dr. Hon identified three different characteristic patterns of the CTG that he and others thought were indicative of fetal status. These patterns involved the response of the fetus to the stress uterine contraction during active labor. These are illustrated from his early work in Figure 11.8. The first pattern involved a deceleration of the instantaneous fetal heart rate in response to the uterine contraction with the deceleration pattern appearing similar to the uterine contraction pattern but inverted. This pattern was referred to as an early deceleration, and the fetal heart rate returned to baseline at the end of the contraction. This pattern along with patterns that showed no change or acceleration of the fetal heart rate during the contraction was thought to indicate a normal response of the fetus.

The second pattern, known as a late deceleration, consisted of a fetal heart rate deceleration during or near the end of the contraction with a slow recovery following it. This

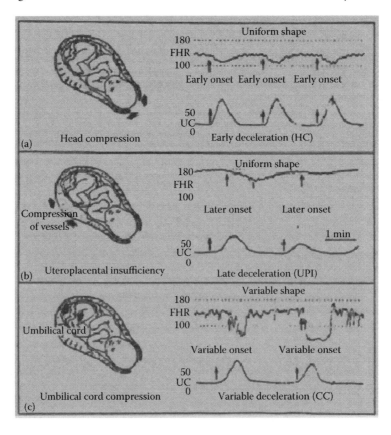

FIGURE 11.8 Fetal heart rate patterns of head, vessels, and cord compression as originally defined by Edward H. Hon. (Reprinted with permission from Hon, E.H., *An Atlas of Fetal Heart Rate Patterns*, Harty Press, New Haven, CT, 1968. Copyright 1968 by Harty Press.)

was described by Dr. Hon as an *ominous pattern* as the current thought at that time was that it was associated with serious fetal distress. The third pattern was known as a variable deceleration where the instantaneous fetal heart rate would rapidly decelerate and then accelerate back to baseline at different times during and following the uterine contraction pattern. This characteristic curve was thought to be associated with possible fetal distress especially as it could be a result of temporary obstruction of umbilical cord blood flow due to the cord being caught around the fetal neck.

The association of these patterns with fetal distress was the result of many anecdotal or small clinical studies done by Dr. Hon and others in the field (including the author of this book), and this was the accepted norm for fetal CTG patterns for many years. Fetal monitoring was considered essential in high-risk pregnancies and thought to be useful to catch the early onset of fetal distress in apparently normal pregnancies. Thus, an industry to produce fetal cardiotocographs developed and these instruments were found in just about every hospital with an active OB/GYN service. Although these monitors were thought to improve obstetrical outcomes, the only thing that was evident during their use was an increase in the cesarean section rates in that when the monitor showed the late and sometimes variable deceleration patterns, obstetricians were anxious about fetal distress and opted to do a cesarean section rather than risk fetal mortality or morbidity.

In the 1990s, several investigators embarked on larger-scale clinical trials of fetal CTG monitoring (Brown *et al* 1982, Spencer 1993). An evaluation of 13 of these reports was done for medical legal purposes and is known as the Cochrane report (Alfirevic *et al* 2013). Its conclusion was that patients who had CTG monitoring during labor had no significant difference in fetal outcome from those who did not use the monitor and the fetal heart and uterine activity was manually assessed periodically by clinical staff. The only statistically significant difference noted in the Cochrane report was the increase in cesarean section rate when CTG monitors were used. These studies resulted in the late and variable deceleration fetal heart rate patterns being reclassified as nonreassuring patterns rather than ominous. Today, fetal CTG monitors are still in use, and in many hospitals, they are routinely used, but they are more for reassurance than for detecting fetal distress.

11.1.12 Technology of the Fetal Cardiotocogram

11.1.12.1 Direct Fetal Electrocardiography

The direct method of fetal electrocardiography involves placing a biopotential electrode on the fetal presenting part that is usually the fetal head during active labor. This electrode is placed through the cervix and currently is a small, stainless steel double helical spiral as illustrated in Figure 11.9. The spiral is such that it can only penetrate the fetal skin, but by doing so, it can measure biopotentials below the fetal stratum corneum. The reference electrode on the opposite side of the fetal scalp electrode from the helical spiral makes contact with the vaginal fluid pool. Thus, the fetal scalp electrode is in essence measuring the potential across the stratum corneum of the fetal scalp. Since the stratum corneum has a relatively high resistivity compared to other fetal tissues, this electrode picks up the voltage drop across this outer skin layer resulting from the fetal ECG and

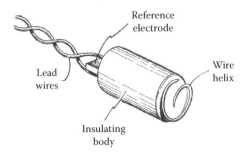

FIGURE 11.9   Fetal scalp electrode for direct fetal ECG recording during labor.

other potentials. This gives a much weaker ECG signal than would be seen from normal chest or limb electrodes on the newborn; thus, greater amplification is needed than would be used with a conventional electrocardiograph or cardiac monitor. The amplified signal generally shows the R-wave quite well, but often, the other components of the ECG are buried in the noise.

The instantaneous or beat-to-beat heart rate is determined by a cardiotachometer circuit that identifies the peak of the R-wave and measures the time interval between successive R-wave peaks: the R–R interval. The fetal heart rate in beats per minute is then determined by the following formula:

$$HR = \frac{60,000}{RR} \tag{11.2}$$

where
   HR is the heart rate in beats per minute
   RR is the R–R interval in milliseconds and 60,000 refers to 60,000 ms in a minute

This instantaneous heart rate is plotted as a function of time as one channel on a two-channel computer screen and/or a paper chart recorder. The second channel is used to display uterine activity that will be described in the following. Most fetal monitors also have the capability to show the actual fetal ECG on a computer screen, and this can help to indicate if the fetal scalp electrode is correctly positioned and in some cases to show the configuration of the fetal ECG.

11.1.12.2 Indirect Fetal Heart Rate Determination by Doppler Ultrasound

The indirect method for determining fetal heart rate involves the use of Doppler ultrasound as previously described. The ultrasonic transducer is placed on the maternal abdomen such that it illuminates the fetal heart or major artery and a high-quality reflected Doppler signal is obtained. This signal varies in frequency over the fetal cardiac cycle, and so the signal is converted to a voltage proportional to the frequency using a frequency-to-voltage converter circuit. This voltage varies in amplitude over the cardiac cycle with a peak determining each heartbeat as is the case with the fetal ECG. The difference between the two signals is the ultrasound peak representing a heartbeat as broader than a sharp R-wave of the ECG, so the exact time of the peak in the Doppler ultrasonic signal is not as precise as it is with the fetal ECG. Nevertheless, this peak also can be passed into the instantaneous cardiotachometer circuit where the peak-to-peak time interval is determined and used to calculate the instantaneous heart rate.

The advantage of the ultrasonic signal is that it is noninvasive and it can be obtained any time during labor and certainly before the occurrence of active labor and rupture of the fetal membranes that has to occur before direct fetal monitoring can be started. For the indirect method of fetal heart rate monitoring, it is necessary that the area of the fetal heart and great vessels is under constant ultrasonic illumination and a strong reflected Doppler shifted signal can be picked up by the transducer. This is generally not a problem unless the mother is significantly obese in which case only the direct method can be used.

Ever since the ultrasonic indirect fetal heart rate monitoring method was instituted, clinicians have been concerned about the effect of the constant ultrasonic illumination on the fetus. To this date, although studies have been carried out, there is no indication that the ultrasonic signal has any adverse effect on the fetus (Sheiner 2012). An additional limitation of the indirect method of determining fetal heart rate is that because it involves Doppler ultrasound, it is sensitive to maternal and fetal movements as well as the activity of the fetal heart. This often leads to artifactual recordings of heart rate data, but it can also provide indication of fetal movement activity. If the fetus does move during labor, it may be necessary to reposition the ultrasonic sensor on the maternal abdomen to obtain optimal fetal heart rate signals. Thus, since there are advantages and limitations to this method of measuring the fetal heart rate, clinicians and research investigators will need to consider these when determining the best method of sensing the fetal heart.

11.1.12.3 Direct Uterine Activity by Pressure Catheter

Uterine activity can be determined by measuring the pressure in the amniotic fluid once the mother is in active labor and fetal membranes have been ruptured. During uterine contractions, the pressure in the amniotic fluid increases, and the increase in pressure is related to the strength and duration of the contraction. During the early days of fetal monitoring, an external pressure sensor such as a Statham-type wire strain gage sensor was coupled to the external end of the catheter, and a means of keeping the catheter filled with sterile physiologically normal saline solution was used. The catheter had to be periodically flushed so that no gelatinous or solid material would block it and hence affect the pressure readings; and it was always important to keep the proximally located pressure sensor at the same elevation as the distal tip of the catheter so that there would be no hydrostatic pressure errors.

Today, inexpensive single-use catheters with a miniature semiconductor pressure sensor at their tip can be used to obtain the intrauterine pressure. This method is preferred because this approach eliminates the need to compensate for hydrostatic pressure and the need to flush the catheter to keep it clear of debris is greatly diminished.

An amplifier within the fetal monitor increases the strength of the signal from the pressure sensor and presents it as the second channel on the computer screen and/or the paper chart record. Thus, we have a display of the full fetal CTG.

11.1.12.4 Indirect Method of Sensing Uterine Activity with a Tocodynamometer
Although the use of an intrauterine catheter or pressure-sensing probe provides a good-quality measure of uterine activity, it has limitations that make a noninvasive technique more attractive. The primary limitation is that this technique can only be used when the mother is in active labor and fetal membranes have been ruptured. The introduction of a foreign object into the uterine cavity always presents the risk of intrauterine infection, and this can put both mother and fetus at risk. A noninvasive method of measuring uterine contractions would be much more desirable, for it would eliminate these problems.

The tocodynamometer is a device that has been developed to meet this need. Its basic structure is illustrated in Figure 11.10. A plunger or probe presses against the anterior abdominal skin just as an examining clinician would do with her or his hand. When there is no contraction of the uterine muscle, the abdominal wall is more compliant and the probe can be displaced by depressing the abdominal wall by a small amount. When a uterine contraction occurs, however, the uterine muscle becomes stiff, and hence the abdominal wall becomes less compliant. This causes the probe in the tocodynamometer to be pushed back into the sensor decreasing its displacement from the noncontraction state. A displacement sensor such as a strain gage or a linear variable differential transformer (LVDT) in the tocodynamometer can measure the displacement of the probe and produce an electrical signal proportional to its displacement. This electrical signal is amplified by the fetal monitor and displayed and/or recorded on the second channel of the instrument.

Although a tocodynamometer can sense and display uterine contractions, this sensor is not as good as the intrauterine pressure devices in terms of quantifying contractions. In some cases where simultaneous intrauterine pressure and tocodynamometer recordings were made, it has been demonstrated that what appears to be a uterine contraction shows up on the uterine pressure recording but occasionally not on the tocodynamometer's trace. The tocodynamometer must be firmly attached to the maternal abdomen to record uterine contractions. Often, this means that a tight belt needs to be used to keep the tocodynamometer firmly in place. Women frequently complain that this belt is uncomfortable and adds discomfort to what can also be considered as an uncomfortable situation.

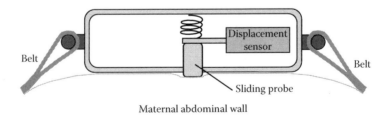

FIGURE 11.10   Cross-sectional view of a tocodynamometer for sensing uterine contractions during pregnancy and labor. The sliding probe is pushed into the sensor's package during a contraction, and the displacement sensor detects this motion and converts it to an electrical signal representing the contraction.

11.1.12.5 Fetal Monitor or Cardiotocograph

An example of a complete fetal monitor is illustrated in Figure 11.11. The block diagram illustrates the major functions that make up the monitor, and an example of a commercially available monitor is also shown in this illustration. The two channels can be used for either direct or indirect fetal monitoring, and some machines actually have the capability to monitor twins indirectly by using two ultrasonic sensors. One can see from Figure 11.11 that it is possible to display the monitored information on a computer screen on the

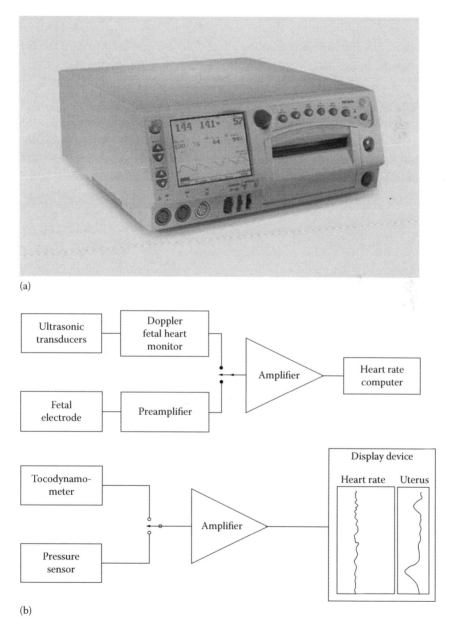

(a)

(b)

FIGURE 11.11   Example of a fetal monitor (a) with a general functional block diagram of its architecture (b).

monitor and on a paper chart that can be attached to the patient's medical record. In many labor and delivery units in hospitals, fetal monitors are also connected to a central station where clinicians can watch several patients in labor at the same time. The central station also has the capability of permanently storing the monitored information for future reference. In the United States, regulations require that fetal monitoring data be kept for five years beyond the age of majority of the infant. This means $18 + 5 = 23$ years in many states. Needless to say, computer storage methods are more useful in this application than filing paper charts (Roux *et al* 1975).

When using fetal monitoring technology, one should also be aware of its limitations as well as its advantages. We have already discussed the Cochrane report and the limitations of the clinical utility of the CTG recording. There are additional clinical issues that need to be considered. When one looks at fetal monitoring, one sees only a very simple aspect of a rather complex system. Looking at the fetal heart rate response to the stress of the uterine contraction can give one example of the physiological status of the fetus, but it is only one variable to assess a rather complex system. In the next section, we will look at other physiological variables that have been studied for expanding the assessment of the fetal response to this stressor. Another limitation of the device is that it ties the mother to a machine while she and her fetus are being monitored. Thus, she has limited mobility that may also add to the discomforts of labor.

This problem has been addressed by many investigators and manufacturers of fetal monitoring equipment over the years by establishing wireless connections between patient's sensors and the signal processing and recording equipment. An example of such a system was a wireless device developed by the author and colleagues in the early 1970s that was very small and could be attached to the fetus for direct fetal monitoring and worn entirely within the mother (Neuman 1970). The sensors and transmitter unit for such a system are illustrated in Figure 11.12 and demonstrated the possibility of a completely internal system. Although

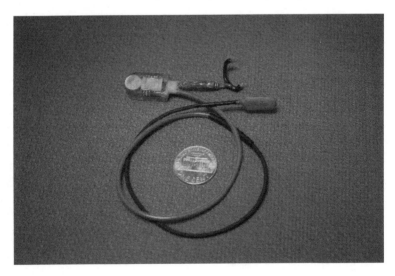

FIGURE 11.12   Miniature wireless telemetry device for cardiotocographic monitoring during labor. The entire transmitting device can be placed within the mother's body during monitoring.

this showed the possibility of retaining the entire sensing system in the mother's reproductive tract, more practical approaches were also developed (Neuman and O'Connor 1980). Monitoring equipment manufacturers developed devices where either direct or indirect monitoring sensors could be used, but the mother wore a transmitting unit as one might wear a mobile telephone, and the sensors were connected to it by wires. Although there are still some doubts about the use of electronic fetal monitoring in obstetrical practice, it is clear that this technology is here to stay until better approaches to obstetrical monitoring are developed.

11.1.13 Other Biochemical and Biophysical Fetal Measurement Methods

Since the early days of fetal monitoring, biomedical engineers and clinical investigators have been looking for additional variables that could be used to assess fetal condition, especially in the period of time surrounding labor and delivery. These adjuncts to fetal heart rate monitoring have been mostly biochemical in nature, but some have used physical techniques to assess the chemistry. For the most part, the techniques described in this section have been experimental although in a few cases commercial products have appeared. Many of these, however, were subsequently removed from the market.

11.1.13.1 Fetal Blood Sampling

Although access to fetal blood has only been available during active labor, as mentioned previously in Section 11.1.9, one can also get fetal cord blood samples under fetal endoscopic examination. In this section, we will limit our consideration to determining fetal acid–base status.

Fetal microblood analysis was first developed and reported by Eric Saling in 1964 (Saling 1964). It is based on the fact that fetal acidemia can be an indicator of prolonged fetal distress. In fetal microblood sampling, a small incision is made in the skin of the fetal presenting part, and a 20–40 μL blood sample is collected in a capillary tube. The pH of this sample can be measured in a microblood analyzer often used in critical care units for blood gas analysis or by using a special fetal blood pH analyzer. In either case, the blood pH is determined using a small glass electrode that has been designed especially for this application. In some cases, additional blood chemistries can be determined from this small sample, but the approach has been primarily based on measuring fetal blood pH. The fetus is considered to be acidotic if the pH value is less than 7.20.

11.1.13.2 Fetal pH Monitoring

There are limitations to fetal blood sampling even if one accepts that acidotic fetal capillary blood is an indicator of fetal distress. Fetal blood sampling is a rather complicated procedure and uncomfortable to mother and fetus, so it only can be done just a few times during active labor. Thus, it is just a few samples of the fetal pH during labor. It would be better if a continuous measure of fetal blood pH could be done during labor, since variations could occur between samples. In 1974, Stamm reported a miniature glass pH electrode that was incorporated into a larger than usual fetal scalp electrode for direct fetal heart rate monitoring (Stamm *et al* 1976). The enlarged fetal scalp electrode had a hole along its axis such that the miniature glass pH electrode could fit in it. Once the fetal scalp electrode was in place, the obstetrician could visualize the electrode and make a small incision in the fetal skin through the hole in the scalp electrode. The pH sensor could then be placed in the incised tissue through the electrode hole, and measurements could be made continuously. This was done by measuring the potential difference between the pH electrode and a silver/silver chloride reference electrode in contact with the vaginal fluid through the back of the pH sensor. This device was later commercialized, but it was ultimately withdrawn from the market.

11.1.13.3 Fetal Blood Oxygenation

Another important physiological variable that can be measured in active labor is oxygen tension or saturation in the fetal blood. Although in some cases it is possible to make this measurement in fetal microblood samples (this is capillary blood), being able to measure arterial blood would be more desirable. It would also be more desirable to be able to make the measurements continuously and noninvasively.

A technique for measuring the partial pressure of oxygen in fetal arterial blood during active labor was developed in the Huch laboratory in 1979 (Huch *et al* 1980, Schneider *et al* 1979). This work involved using a transcutaneous partial pressure of oxygen sensor that was developed for use on neonates and modifying it for obstetrical applications. One of the issues that had to be overcome was to maintain good attachment to the fetal presenting part, usually the head. Some fetuses could have a lot of hair that would further complicate the attachment. To minimize this problem, a small razor was developed that could shave about a square centimeter of hair from the fetal head to allow attachment of the sensor. Although this was a useful and safe technique, parents were not very keen about their babies having a small bald spot. The method for attachment of the sensor involved the use of a cyanoacrylate tissue adhesive that was available in European countries at the time. This technique of oxygen monitoring of the fetus and labor turned out to only be used for research purposes and is no longer applied.

A second method for fetal oxygen determination is the use of pulse oximetry. A small, thin reflectance pulse oximetry probe can be slid between the fetal head and the lower uterine segment during active labor once fetal membranes have been ruptured. This device works on the principle of reflectance pulse oximetry, and it determines arterial blood hemoglobin oxygen saturation by measuring reflected light from the tissue at two wavelengths. Light-emitting diodes produce illumination at an infrared wavelength near the isosbestic point where the optical properties of oxygenated and reduced hemoglobin are similar and a red wavelength where there is a substantial difference in the optical properties between these two hemoglobins. Light at both of these reflected wavelengths will have a pulsatile component due to arterial blood entering the illuminated capillaries, and the ratio of these pulsatile components is related to the hemoglobin oxygen saturation of the arterial blood. More details on this technique can be found in Chapter 16 (Colditz *et al* 1999, Dildy *et al* 1996).

11.1.14 Monitoring Cervical Dilatation

The progress of labor is generally determined by periodically measuring the cervical dilation and fetal descent into the pelvis. These measurements are made by the clinical staff by feeling the opening of the cervix and the position of the fetal head with respect to the pelvic outlet using two fingers. Needless to say, such an examination can only give a few discrete samples at times throughout the progression of labor, and they can be unpleasant as far as the mother is concerned. A means of continuously measuring the cervical dilation during labor along with the CTG would allow clinical staff to view how well labor is progressing and evaluate the effectiveness of uterine contractions in terms of progress of labor and their effect on the fetal heart rate. Several investigators have worked on devices to allow this to be done. One of the earliest devices developed was a caliper cervimeter that had a scale external to the body with the caliper tips attached to the cervix at the three and nine o'clock positions (Friedman and VonMicsky 1963). Although being a bit awkward, this allowed one to watch changes in cervical diameter during a uterine contraction. The device was later improved by replacing the visual scale on the caliper with an electronic displacement sensor so that the cervical dilatation could be recorded along with the CTG. Even with this improvement, the measurement was somewhat awkward since the mother had to have this device both within and outside the vagina.

A different approach was the use of an ultrasonic displacement instrument to monitor cervical dilatation (Zador *et al* 1976). This instrument consisted of small 1 mm × 1 mm × 7 mm ultrasonic crystals that were attached to small clips. These clips, in turn, were attached to the cervix at the three and nine o'clock positions. Each ultrasonic transducer was connected to an electronic circuit that would periodically generate a pulse of ultrasound at one side of the cervix that would propagate through the cervix and fetus to the other side of the cervix where it would be detected by the second transducer. Since ultrasound propagates through biologic tissue at a velocity of 1.5 mm/μs, the electronic circuit would measure the time delay between transmission and reception of the ultrasonic pulse, and this time would be proportional to the cervical dilatation. The electronic circuit then converted the time to an electrical signal that could be recorded as the third channel of a CTG. An example of such a record is shown in Figure 11.13.

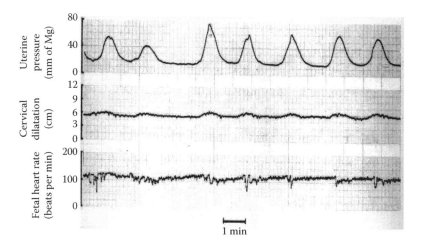

FIGURE 11.13 Monitoring cervical dilatation (center trace) during active labor by measuring ultrasonic transit time between sensors at three o'clock and nine o'clock on the uterine cervix.

In this illustration, one sees that the cervix dilated during a uterine contraction but returned to close to its baseline following the contraction.

11.2 MEASUREMENTS IN GYNECOLOGY

Many types of physiological measurements are made in gynecology, yet most of these are done for research purposes rather than routine clinical care. There are a few exceptions, and one of the most basic is the clinical approach to determining approximate size and position of the nonpregnant uterus. This is done by a manual examination whereby the examiner can feel the position of the cervix in the vaginal fornix and also feel the fundus of the uterus transabdominally. If the examiner is interested in determining the size of the uterine cavity, a small device known as a uterine sound is used. It is a thin probe with a blunt tip that can be passed through the cervix until the tip contacts the fundus of the uterine cavity. The probe has a centimeter scale engraved on it so that the examiner can determine length of the probe from the fundus to the external os of the cervix. In this way, the clinician gets a rough idea of the extent of the uterine cavity.

11.2.1 Uterine Contractility

Uterine contractions occur regularly in nonpregnant women as well as during pregnancy up to its end when they are forceful and are an important part of the labor and delivery process. In some cases, women experience cramping associated with the nonpregnant uterine contractions. This often occurs in women during their time of menstruation. Uterine contractions can be measured using a similar technique to the direct measurement of uterine activity during active labor. A catheter or probe can be introduced into the uterine cavity through the cervix and used to measure the intrauterine pressure. Since the cervix is open by a small amount in this case, one is no longer measuring pressure in a closed chamber as would be the case during pregnancy when the fetal head acts as a *stopper* as the cervix dilates. To avoid this problem in the nonpregnant state, one can put a small distensible balloon at the distal tip of the catheter and fill the catheter and balloon with a sterile saline solution after the catheter tip and balloon are introduced into the uterus. Now there is a closed system, namely, the balloon, and pressure can be measured either by a pressure sensor located at the proximal end of the catheter or a catheter with a pressure sensor at its distal tip. In either case, the pressure within the balloon will increase with uterine contractions, and the amplitude of this variation can be a measure of the strength of the contractions as they relate to symptoms of cramping.

11.2.2 Menstrual Cycling

It is important to know if a woman is experiencing normal menstrual cycles when one is assessing reproductive endocrinology. Of course, the simplest way to evaluate menstrual cycles is to ask whether she is having regular menstrual periods or not. Although this is a way of determining normal menstrual cycles, when irregularities are present, more quantitative measurements are desirable. These methods include biochemical and biophysical measurements. The biochemical measurements involve the periodic determination of female reproductive hormone levels such as estrogen, progesterone, follicle-stimulating hormone, and luteinizing hormone, to name a few. Excellent clinical laboratory assays are available for these measurements in blood, but there is no practical technique to measure these quantities continuously over the month-long menstrual cycle. For this reason, several biophysical techniques have developed that measured quantities that are related to these circulating hormone levels.

In normal cycling women, the reproductive hormone progesterone is produced by the ovaries following ovulation to protect pregnancy should fertilization occurs. This hormone also causes the average body core temperature to increase by as much as 0.5 °C, and this is a change that can often be measured. The simplest measurement is with a clinical thermometer that can estimate core temperature. This is often done by patients first thing in the morning after waking, and they keep a daily chart where these temperatures are plotted against the menstrual cycle day from the end of menstruation of the previous period to its next onset. Electronic devices have been developed to measure and record these temperatures (Mordel *et al* 1992). The book's author and colleagues developed a wireless system with a temperature sensor and radio-frequency transmitter that was worn intravaginally to do this (Neuman *et al* 1984).

Temperatures can be taken anywhere on the body, but core temperatures are the most reliable since they are only minimally affected by environmental conditions. It is also important to take the temperature at the same time of the day each day to avoid effects of diurnal variations. It is also important to take the temperature during times of relatively little activity since exercise can also affect core body temperature.

Although basal body temperature measurements from the female breast have been reported, perfusion measurements from the breast have been found to be more reliable. A qualitative measure of perfusion can be made by thermal methods. A heated probe is placed in good thermal contact with the breast, and it should be insulated from its environment. It is heated to a temperature slightly higher than the temperature of the breast, and some of this heat is withdrawn by convection of the capillary blood flowing in the tissue under the sensor. The larger the sensor, the deeper into the breast this convective effect will occur. Either one can electronically maintain the sensor probe at a fixed temperature and measure the changes in power needed to do this or one can hold the power constant and measure the variations in temperature probe as the blood perfusion cools it. It is important to point out that these temperature-related techniques have only been used for research purposes and should not be considered to be routine methods to biophysically follow the menstrual cycle in women.

In some cases, researchers are interested in measuring the quantity of menstrual fluid that is discharged by a woman. A simple way to do this, although very inconvenient for the patient, is to weigh a tampon or pad prior to its use and after it has been removed. The weight change in this case is an indicator of the amount of fluid absorbed by the pad or tampon. Another technique has been to measure the change in electrical conductivity between electrodes distributed in the pad or tampon. Once again, these techniques have been limited to research activities, and both are very inconvenient for the patient and researcher. The weighing technique is only a sampling method to study menstrual flow, while the electrical impedance technique can be measured continuously to establish flow patterns.

11.2.3 Patency of Fallopian Tubes

The evaluation of infertility problems in women often involves many different measurements in an effort to identify the cause. Determining abnormalities in the menstrual period, as discussed in the previous section, is often one of the first measurements to be done. A second measurement that is often carried out in such an evaluation is to determine the patency of the Fallopian tubes. If they are blocked, the sperm cannot reach the ovum in the Fallopian tube to fertilize it, and the fertilized ovum cannot continue its path to the uterus. The Rubin test was developed as a simple means to demonstrate Fallopian tube patency (Hill 1983). It consists of a probe that can be placed against the external os of the cervix to maintain a seal, and carbon dioxide is slowly blown into the uterine cavity while its pressure is being measured and recorded. If the tubes are blocked, the pressure will rise quickly indicating the blockage, while if the tubes are open, the carbon dioxide will escape from the fimbriated ends of the tubes, and there will not be a great pressure rise. Although this test has been used for many years and is a relatively simple office procedure, today, imaging techniques make it possible to inject contrast medium into the uterine cavity and tubes so that their lumens can be visualized on a CT scan.

One of the mysteries that remain in gynecology today is understanding the biomechanics of how sperm and ova are transported in the Fallopian tube. This is a significant problem in that making sure that they meet at the right time and right place in the Fallopian tube is important in enhancing chances for conception. It may also be possible to alter the biomechanics of tubal transport as a means of contraception. Studies have been done by several investigators over the years to better understand this process; but these studies have had to be done in animal models. Blandau et al. (1975) and colleagues developed a technique for observing ovum transport in rabbits. They placed a fluid-filled, transparent chamber over the Fallopian tube and ovary and stained the ovum and its cumulus mass with a blue dye. It was then possible to see the ovum and cumulus through the thin wall of the Fallopian tube with a low-power microscope. This allowed one to document movements of the ovum within the Fallopian tube by cine or video recording through a dissecting microscope.

Other investigators have attempted to understand ovum and sperm transport in the Fallopian tube by measuring the frequency of the beating cilia, the flow of oviductal fluid, and oviductal contractions or measuring electrical impedance changes along the oviduct. In the case of oviductal impedance changes, it was possible to place an array of fine-wire electrodes along the oviduct and uterine horn of rabbits and observe the propagation of contractions along these structures (Schnatz and Neuman 1981).

11.2.4 Uterine Cervical Compliance

Cervical insufficiency during pregnancy is a condition where the uterine cervix becomes compliant and can spontaneously open during the second trimester of pregnancy. This can result in a spontaneous abortion of the fetus if its increasing weight can cause the cervix to open. Although there are corrective measures that can be taken to prevent this abortion, the diagnosis of this condition during pregnancy may be difficult to make until the cervix is already open. For this reason, investigators have looked at ways to identify those women at higher risk for cervical insufficiency before they become pregnant (Kiwi *et al* 1988).

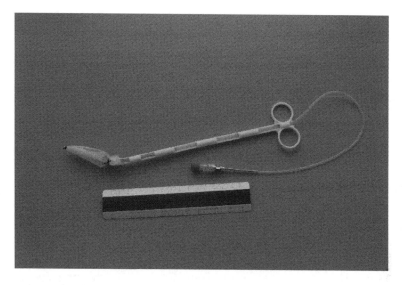

FIGURE 11.14   Example of a balloon for measuring cervical compliance.

An example of a simple biomechanical test to do this is illustrated in Figure 11.14. A cylindrical balloon that is either 3 or 4 cm long and 1 cm in diameter when there is no wall tension is placed at the tip of a probe that can be placed in the cervical canal when the balloon is completely collapsed. The ends of the balloon are constrained so that the balloon is only able to expand in diameter when fluid is infused into it. Once the balloon is placed in the cervical canal, it is slowly inflated with a sterile saline solution by an infusion pump, and a pressure sensor coupled to the infusing fluid measures the balloon pressure. This generates a pressure–volume curve. The slope of this curve is the reciprocal of the cervical compliance; thus, a low slope indicates a compliant cervix and a patient who might be at higher risk for cervical insufficiency.

11.2.5 Optical Instrumentation in Gynecology

There are some optical measurement techniques that are used in gynecology primarily for the identification of cervical cancer or precancerous lesions. A well-established method in clinical gynecology is to do a Pap smear of cervical mucus that contains exfoliated cells and to identify any histologically abnormal cells using standard cytopathological techniques. If unusual cells are found, often the next step in the diagnosis, cervical colposcopy, involves observing the cervix through a long-focal-length, moderate-power microscope during a pelvic examination (Pretorius and Belinson 2012). Specific changes in the architecture of the external os of the cervix as seen through the microscope in the natural state or when the cervix is stained with an iodine stain help to identify sites of pathology. One can also look at the pattern of vascular capillaries in the cervical epithelium to help identify these sites. The examiner can then determine if a small biopsy of any of these areas is necessary to make a histologic diagnosis.

An optical technique that is still under development involves looking at the fluorescence of cervical tissue when excited at specific wavelengths (Follen *et al* 2005). This approach involves the use of a probe that is coupled to a light source at a known wavelength and a detector consisting of a spectrophotometer. Wavelengths in the visible spectrum have been found to be associated with cervical squamous intraepithelial lesions. Investigators working in this area report that clinical studies show improved performance over the Pap smear, but the sensitivity and specificity is not as good as colposcopy. Perhaps future improvements in this measurement will make it possible to reliably use this instrumentation as a screening method for cervical cancer as the Pap smear is currently being used.

11.2.6 Hot Flash Documentation

Hot flashes are an annoying sensation of warmth and sweating experienced by many menopausal women. Although current treatment methods are not ideal, they can reduce the frequency of occurrence. Yet, to be able to obtain a better understanding of this phenomenon as well as to develop better treatments, it is essential to have good records of the occurrence and duration of hot flashes. Presently used techniques involve the patient herself keeping a diary of the occurrence and perceived severity of hot flashes. This method of measurement has been found to be not very reliable, and there can be considerable variability between individuals in terms of the threshold for detecting and establishing the intensity of a hot flash.

To date, various physiological measurement techniques have been used to detect and record hot flashes for research purposes, and the most prominent of these have been the measurement of skin temperature and skin sweating as an indication of the occurrence of a hot flash. Skin temperature measurement devices are straightforward applications of temperature measurement technology with a small thermistor attached to the skin as the temperature sensor. Sweating is determined by an electrical impedance measurement technique that does not have some of the environmental temperature constraints seen with the thermistor instrument. Bahr et al. (2014) has reported a miniature self-contained hot flash sweating detector based on the impedance technique. Their device and an example of the data it collects are illustrated in Figure 11.15. These investigators have demonstrated that their device has reasonable sensitivity and specificity when compared to patient self-reports.

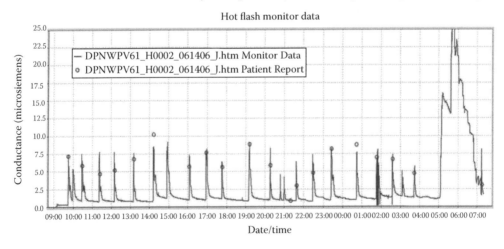

FIGURE 11.15   A sample of raw data collected by the hot flash monitor. The hot flash events about every hour can be clearly seen with the circles showing where the subject marked the data when experiencing hot flash symptoms. The skin conductance increased from about 1 μS to about 5 μS. The large event near the end of the trace shows the effects of exercise.

11.3 SUMMARY

There are many different types of physiological measurements that are made in the field of obstetrics and gynecology. This chapter has illustrated a few of these concentrating on clinical measurements. The most frequently made measurements are associated with obstetrics and monitoring the process of labor and delivery. Although the efficacy of such measurements is still controversial, they are frequently used to reassure clinicians managing the obstetrical patient. Other measurements in both obstetrics and gynecology have been important as research instrumentation, but a few have made it to routine clinical practice as fetal monitoring has. Nevertheless, physiological measurements in obstetrics and gynecology represent important tools for studying, understanding, and ultimately treating women's health issues.

REFERENCES

Alfirevic Z et al 2013 Continuous cardiotocography (CTG) as a form of electronic fetal monitoring (EFM) for fetal assessment during labour *Cochrane Database Syst. Rev.* **5** CD006066

Bahr D E et al 2014 Miniature ambulatory skin conductance monitor and algorithm for investigating hot flash events *Physiol. Meas.* **35** 95–110

Blandau R J et al 1975 Methods for studying oviductal physiology *Gynecol. Invest.* **6** 123–45

Brown V A et al 1982 The value of antenatal cardiotocography in the management of high-risk pregnancy: a randomized controlled trial *Br. J. Obstet. Gynaecol.* **89** 716–22

Callaerts D et al 1986 An adaptive on-line method for the extraction of the complete fetal electrocardiogram from cutaneous multilead recordings *J. Perinat. Med.* **14** 421–33

Callen P W (ed.) 2008 *Ultrasonography in Obstetrics and Gynecology* (Philadelphia: Saunders)

Colditz P B et al 1999 Fetal pulse oximetry. Instrumentation and recent clinical experience. *Clin. Perinatol.* **26** 869–80

Cole L A 2012 The hCG assay or pregnancy test *Clin. Chem. Lab. Med.* **50** 617–30

Dildy G A et al 1996 Intrapartum fetal pulse oximetry: past, present, and future *Am. J. Obstet. Gynecol.* **175** 1–9

Follen M et al 2005 Optical technologies for cervical neoplasia: update of an NCI program project grant *Clin. Adv. Hematol. Oncol.* **3** 41–53

Friedman E A and VonMicsky L I 1963 Electronic cervimeter: a research instrument for the study of cervical dilatation in labor *Am. J. Obstet. Gynecol.* **87** 789–92

Hill E C 1983 The Rubin test *JAMA* **250** 2366–8

Hon E H 1960a Apparatus for continuous monitoring of the fetal heart rate *Yale J. Biol. Med.* **32** 397–9

Hon E H 1960b The instrumentation of fetal heart rate and fetal electrocardiography. I. A fetal heart monitor *Conn. Med.* **24** 289–93

Hon E H 1968 *An Atlas of Fetal Heart Rate Patterns* (New Haven: Harty Press)

Hon E H and Hess O W 1957 Instrumentation of fetal electrocardiography *Science* **125** 553–4

Huch A et al 1980 Experience with transcutaneous PO_2 ($tcPO_2$) monitoring of mother, fetus and newborn *J. Perinat. Med.* **8** 51–72

Ibrahimy M I et al 2003 Real-time signal processing for fetal heart rate monitoring *IEEE Trans. Biomed. Eng.* **50** 258–62

Jansen A H and Chernick V 1991 Fetal breathing and development of control of breathing *J. Appl. Physiol.* **70** 1431–46

Kimura Y et al 2006 Measurement method for the fetal electrocardiogram *Minim. Invasive Ther. Allied Technol.* **15** 214–7

Kiwi R et al 1988 Determination of the elastic properties of the cervix *Obstet. Gynec.* **71** 568–74

Liley A W 1991 Liquor amnii analysis in the management of the pregnancy complicated by rhesus sensitization *Am. J. Obstet. Gynecol.* **82** 1359–70

Mordel N *et al* 1992 The value of an electronic microcomputerized basal body temperature measurement device (Bioself) in in vitro fertilization cycles *Gynecol. Endocrinol.* **6** 283–6

Nahum G G 2002 Predicting fetal weight. Are Leopold's maneuvers still worth teaching to medical students and house staff? *J. Reprod. Med.* **47** 271–8

Neuman M R *et al* 1970 A wireless radiotelemetry system for monitoring fetal heart rate and intrauterine pressure during labor and delivery *Gynecol. Invest.* **1** 92–104

Neuman, M.R *et al* 1984 Telemetry of basal body temperatures in women and respiration in neonates *Biotelemetry VIII* Dubrovnik, Yugoslavia, 137–140

Neuman M R and O'Connor 1980 A two channel radiotelemetry system for clinical fetal monitoring *Biotelemet. Patient Monit.* **7** 104–21

Pretorius R G and Belinson J L 2012 Colposcopy *Minerva Ginecol.* **64** 173–80

Roux J F *et al* 1975 Monitoring intrapartum phenomena *CRC Crit. Rev. Bioeng.* **2** 119–58

Saling E 1964 Microblood studies on the fetus. Clinical application and first results *Z. Geburtshilfe Gynakol.* **162** 56–75

Schnatz P T and Neuman M R 1981 An electrical impedance technique for recording oviductal and uterine activity *Med. Biol. Eng. Comput.* **19** 64–5

Schneider H *et al* 1979 Correlation between fetal scalp $tcPO_2$ and microblood samples *Birth Defects Orig. Artic. Ser.* **15** 235–43

Sheiner E 2012 A symposium on obstetrical ultrasound: is all this safe for the fetus? *Clin. Obstet. Gynecol.* **55** 188–98

Spencer J A 1993 Clinical overview of cardiotocography *Br. J. Obstet. Gynaecol.* **100**(Suppl 9) 4–7

Stamm O *et al* 1976 Development of a special electrode for continuous subcutaneous pH measurement in the infant scalp *Am. J. Obstet. Gynecol.* **124** 193–5

Timor-Tritsch I E 1979 Human fetal respiratory movements: a technique for noninvasive monitoring with the use of a tocodynamometer *Biol. Neonate* **36** 18–24

Watanabe M and Flake A W 2010 Fetal surgery: progress and perspectives *Adv. Pediatr.* **57** 353–72

Zador I *et al* 1976 Continuous monitoring of cervical dilatation during labor by ultrasonic transit time measurement *Med. Biol. Eng.* **14** 299–305

CHAPTER 12

Ophthalmology

Ayyakkannu Manivannan and Vikki McBain

CONTENTS

12.1 Introduction   268
12.2 Parameters of Human Visual Apparatus (Eye)   268
12.3 Electrophysiology   268
    12.3.1 Electrooculogram   269
    12.3.2 ERG   269
    12.3.3 VEP   271
        12.3.3.1 VEP Measurement Technique   271
12.4 Equipment for Electrodiagnostic Testing in the Eye   272
    12.4.1 Stimulators   272
    12.4.2 Recorders   272
    12.4.3 Patient   273
    12.4.4 Electrodes   273
12.5 ERG System   274
    12.5.1 Amplifiers   274
    12.5.2 Analog-to-Digital Converters   275
12.6 Ganzfeld Stimulator   275
12.7 Clinical Applications   275
12.8 Equipment   276
    12.8.1 Multifocal and Standard Electroretinography Systems   276
12.9 Tonometry   276
    12.9.1 Introduction   276
    12.9.2 Physiology   276
    12.9.3 Clinical Need to Measure IOP   276
    12.9.4 Principles of Measurement of Pressure in the Eye   276
        12.9.4.1 Schiøtz Tonometer   277
        12.9.4.2 Procedure to Use a Schiøtz Tonometer   279
        12.9.4.3 Goldmann Tonometer   279
        12.9.4.4 Procedure to Use of a Goldmann Tonometer   281
        12.9.4.5 Other Types of Goldmann Tonometer   282
        12.9.4.6 Noncontact or Puff Tonometer   282
References   285

12.1 INTRODUCTION

Visual sensation occurs when physical phenomena (photons) impinge upon the receptor organ, the eye, and give rise to physiological changes within various layers of the retina. The information resulting from these changes (mainly electrical) is processed and passed on to the brain for further processing. Final stimulation in the brain provides the visual sensation.

Even with many new techniques helping us in the understanding of physiological functions of many organs in the human body, knowledge of the processes in the eye is incomplete. Our current understanding and hypothesis of human visual function are mainly based on extrapolation of data from studies conducted on animals. The main limitation is that invasive techniques are either not practical or not permitted to extract the information.

Currently, visual function is assessed using relatively simple tests. The results of these tests highly depend on the cooperation of the patient and some other parameters that will be discussed later in this section.

12.2 PARAMETERS OF HUMAN VISUAL APPARATUS (EYE)

Unlike many other animals, humans have evolved to a point where vision has many more functions than the purposes of hunting for food and defense. The prime function of the eye is detection that contributes to the development of the whole body in early years. Performance in later part of life highly depends on various parameters of the visual detection.

The main parameter of the visual system is resolution. The visual acuity depends on the *spatial resolution,* and the color vision performance is affected by the *spectral resolution. Adaptability* is another interesting parameter. Human eyes can remarkably adapt to illumination levels that vary by a factor of 10^9! The speed of adaptation is slow especially at low light levels. Rapid or large changes in the illumination levels lead to the sensation of *dazzle.*

Spatial distribution of receptors in the eye is another interesting parameter. There are two different kinds of receptors in the eye: cones and rods. Cones provide the color vision (photopic), whereas rods provide night vision (scotopic). Humans have evolved to have a fovea, the center part of the eye where concentration of cones is the highest. The density of cones dramatically reduces toward the periphery, whereas the rod density reaches maximum.

The eye has an amazing *absolute sensitivity.* In very low light levels, a dark-adapted eye is able to detect an unbelievable level of about 1 pW/cm^2 of incident energy. It is believed that a rod needs only one photon to excite it!

Several factors such as age, state of adaptation, wavelength of the light input, and position of stimulation in the eye affect the *response time* of the eye. It can respond to small changes in the light level very rapidly but integration of light input from millions of rods and cones requires time resulting in us not able to see changes occurring within few milliseconds.

12.3 ELECTROPHYSIOLOGY

Electrophysiology in the eye measures the electrical activity of various cells in the retina. These cells include the photoreceptors (rods and cones), inner retinal cells (bipolar and pigment epithelium cells), and the ganglion cells. In a dark status, the photoreceptors are unstimulated. In this condition, there is a constant flow of current from the inner to the outer segment of the retina. Absorption of light by the outer segment of the photoreceptor causes a slow hyperpolarization and reduces the current flow. This change in current flow is proportional to the intensity of the stimulation. In electrical terms, we could say that the photoreceptors are active (*on*) in dark condition and light turns them *off*. The hyperpolarization and depolarization of photoreceptors during light stimulation cause the generation of an action potential. This action potential is propagated to the nerve axons through ganglion cells. These action or spike potentials have constant amplitude, but their frequency depends on the intensity of stimulation. This whole process could be considered as analog-to-digital conversion and frequency modulation to transfer the information to the brain.

Common clinical electrophysiology measurements are the electrooculogram (EOG), electroretinogram (ERG), multifocal ERG (MfERG), pattern ERG (PERG), and visual evoked potential (VEP).

12.3.1 Electrooculogram

There is a standing electrical potential difference between the front and back of the eye. This is about 6 mV, cornea being positive and the fundus negative. The major component of the eye that contributes to the EOG is the pigment epithelium. The EOG varies with the intensity of illumination. As the EOG measures the difference between the front and back of the eye, any eye movement induces a change in potential. By simply measuring the voltage difference between two electrodes placed on skin on either side of the eye, any disease relating to the integrity of the interface between the layers of retinal pigment epithelium (RPE) and photoreceptors can be studied. The EOG is also useful to diagnose pathologic and physiologic nystagmus relating to eye movement. The EOG signal ranges from 10 μV to 6 mV. The frequency ranges from dc to 100 Hz. A normal EOG signal is shown in Figure 12.1.

12.3.2 ERG

Stimulation of the retina by a flash of light produces an action potential of about 0.5 mV. The ERG can be detected between an electrode placed on the cornea and another electrode placed on the skin nearby, although it is possible to record the ERG from skin electrodes.

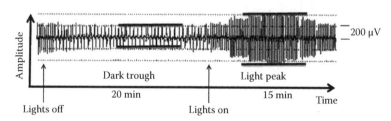

FIGURE 12.1   A normal EOG signal recorded over a period of 35 min. The signal is derived from two electrodes placed at medial and lateral canthi. The patient makes a fixed 30° lateral eye movement by alternately looking at two fixation lights. The eye movements are made every 1–2 s for about 10 s/min. The ratio of the light peak and dark trough (Arden index) gives an indication of health of interaction between RPE and photoreceptors. (Waveform courtesy of Dr. Vikki McBain, Electrophysiology Department, Eye OPD, ARI, Aberdeen, U.K.)

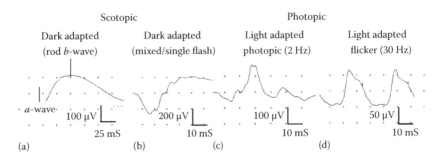

FIGURE 12.2 Normal ERG waveforms under various testing conditions. (a) Rod-specific response to a low luminance level stimulus (0.01 cd/sm²) from a dark-adapted eye. There is insufficient photoactivation resulting in flat a-wave. (b) Mixed rod–cone response to a high luminance level stimulus (3 cd/sm²) from a dark-adapted eye. Since the eye is dark adapted, rod response contributes to about 75% of the signal; (c) and (d) are responses mainly from cones as the rods are saturated with a background luminance of 3 cd/sm². (Waveform courtesy of Dr. Vikki McBain, Electrophysiology Department, Eye OPD, ARI, Aberdeen, U.K.)

The ERG is generally used to study the photopic and scotopic responses of the retina and is often used to investigate unexplained defective vision that is not diagnosed by visual examination. Different cell types within the retina contribute to the ERG. The ERG highly depends on the stimulus conditions such as flash or pattern stimulus, background light, and the colors of the stimulus.

Different types of ERG tests can be performed. Clinically, the most commonly performed tests are PERG and MfERG.

The ERG signal contains many parts (see Figure 12.2) and their signal level highly depends on the testing conditions. A detailed description of signal levels and their frequencies can be found in the electrophysiology book by Hechenlively and Arden (2006). The amplitude of the ERG signal ranges from 0.5 µV to 1 mV. The oscillatory potential has a lower range of 0.1–100 µV, whereas the PERG has the lowest range of 0.1–10 µV. Due to its very small signal range, the PERG requires the highest amplification. Very high amplification may result in an increase in noise in the output signal. Averaging of multiple recording of PERG signals can help to increase the signal-to-noise ratio (SNR). The signal frequency range of the a and b waves in PERG is 0.2–100 Hz. The c wave is a slow varying signal ranging from dc to 2 Hz. Oscillatory potentials vary from 100 to 300 Hz allowing the use of a band-pass filter to separate this signal from other ERG signals.

If a flash ERG is performed on a dark-adapted eye, the response is primarily from the rod system. Flash ERGs performed on a light-adapted eye will reflect the activity of the cone system. For sufficiently bright flashes, the ERG will contain an a-wave (initial negative deflection) followed by a b-wave (positive deflection). The leading edge of the a-wave is produced by the photoreceptors, while the remainder of the wave is produced by a mixture of cells including photoreceptors, bipolar, amacrine, and Muller cells or Muller glia. The PERG, evoked by an alternating checkerboard stimulus, primarily reflects activities of retinal ganglion cells.

12.3.3 VEP

The VEP is a recorded potential from the nervous system following a visual stimulus. The VEP is of low voltage (5–50 µV) and is usually extracted from the EEG and EMG (see elsewhere in the book) by signal processing and averaging. The VEP is useful to assess the optic nerve head and also as an objective means of estimating visual acuity and even for measuring refractive errors.

The VEP depends on the functional integrity of the visual pathway that includes the retina, optic nerve, and occipital cortex. Since there are variations in the method of measuring the VEP, standards have been developed and are followed by many eye clinics around the world (Odom *et al* 2010). This standard covers only the basic clinical VEP recording. Variations such as motion VEP and multifocal VEPs are not covered in this standard (see Table 1 in the standard; Odom *et al* 2010). In this chapter, we will describe the technique as specified in this standard.

Recorded VEPs from humans can range from 0.1 to about 20 µV. The signal frequency is up to approximately 300 Hz. The use of ac coupling after the first-stage amplification can avoid output saturation due to dc signal shift.

With pattern-reversal VEP, the patient sees a checkerboard stimuli with large 1° (60 min of arc) and small 0.25° (15 min of arc) checks from a display monitor. All checks should be isoluminant, square, and of equal size. A fixation point is used for the patient to fixate.

12.3.3.1 VEP Measurement Technique

The skin is prepared by cleaning the area of contact and either a silver–silver chloride or gold disc electrode is placed on the skin using a suitable paste or gel. Impedance of the electrode should be less than 5 kΩ (as measured between 10 and 100 Hz). A high-contrast black-and-white stimulus is presented to the patient. The viewing distance is from 50 to 150 cm. This distance is adjusted based on the field of view required, the physical size of the screen, and the check size. A normal VEP waveform is shown in Figure 12.3. The main

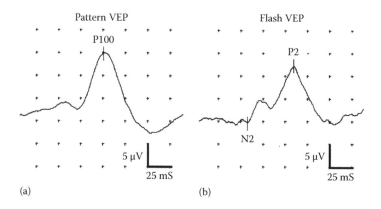

FIGURE 12.3 Normal VEP waveform for two different testing conditions: (a) PVEP showing a prominent peak at 100 ms, known as P100; (b) with a flash stimulation, the amplitude of P2 and its latency from N2 are used as a measure of response. (Waveform courtesy of Dr. Vikki McBain, Electrophysiology Department, Eye OPD, ARI, Aberdeen, U.K.)

factor affecting the VEP waveform is age. Other factors such as contrast and size luminance levels of the projected pattern also affect the waveform.

12.4 EQUIPMENT FOR ELECTRODIAGNOSTIC TESTING IN THE EYE

Equipment for electrodiagnostic testing in the eye consists of two major components: stimulator and recorder.

Equipment parameters for PERG, ERG, EOG, and VEP measurements in clinical electrodiagnostic service centers are similar. Some commercial equipment is capable of performing all three tests. Some research laboratories in university and teaching hospitals built their own system to cater for their research needs. See Figure 12.4.

12.4.1 Stimulators

Stimulators provide a controlled light source capable of producing spots and patterns at various intensities, color, size, and frequency. It should be possible to alter the background intensity and color. They are usually controlled in synchronization with the recorders. Stimulators are commonly known as ganzfeld bowl or pattern stimulators. The required stimulation is usually projected on a white surface or produced by an LCD screen. It is also essential that these stimulators are calibrated at frequent intervals. Details of stimulus calibration are provided in the International Society for Clinical Electrophysiology of Vision (ISCEV) guidelines (Marmor *et al* 1989).

12.4.2 Recorders

In general terms, the recording equipment should be able to record and amplify electrical signals generated from various structures in the visual system. These signals range from 0.01 to 1000 µV, with frequency range from 0.1 to 300 Hz. The instrument should be able to provide patient isolation and work in a noisy environment.

As signal levels are very small, a high–input impedance and high-gain amplifier are used. The input impedance should be higher than 10 MΩ. The common-mode rejection ratio

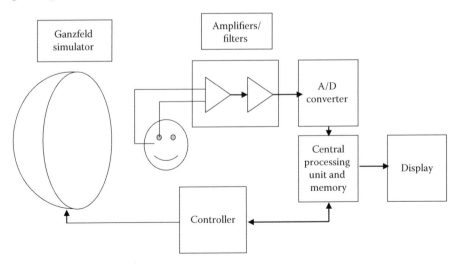

FIGURE 12.4   Equipment for electrodiagnostic testing in the eye.

(CMRR) should exceed 120 dB. The electrode and the first amplifier should be electrically isolated and must meet local electrical safety standards. After amplification, most commercially available ERG recording systems employ an analog-to-digital converter (ADC) and record the signal as a digital signal. The sampling rate of the ADC should be more than 500 Hz with a resolution of 12 bits, giving 4096 sampling levels. Filters are usually employed to reject signals below 1 Hz and above 100 Hz. Since VEPs are very small and are prone to noise, most systems allow the user to measure the VEP by taking the average of multiple sweeps. The number of sweeps depends on the age of the patient and their ability to fixate for a long period of time. It is recommended to take two measurements to verify the reproducibility.

12.4.3 Patient

Testing conditions are different for adult and children and babies of age 3 months. Patients must be positioned comfortably at the correct distance from the stimulus. The stimulus field size is adjusted to suit the testing conditions. The patient must be able to maintain the testing position for the duration of examination as some tests may take a long time. Patient cooperation is critical for good reproducible results. A typical measurement electrophysiology setup is shown in Figure 12.5.

12.4.4 Electrodes

Conductive fiber, gold foil, and Dawson, Trick, Litzkow (DTL) electrodes (named after Dawson et al. [1979]) are commonly used in the United Kingdom, whereas corneal

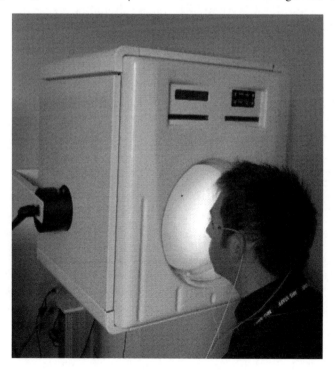

FIGURE 12.5   A volunteer viewing stimulation in a ganzfeld stimulator.

contact electrodes are preferred in the United States. DTL electrodes are ready to use and single use (disposable) and pose minimal discomfort to the patient. In some applications, conjunctival loop electrodes and corneal wicks are used. Contact lens electrodes are made from medical grade silver/nylon and provide the highest amplitude and very good SNR due to their ultralow impedance.

The corneal surface should be protected with a nonirritating and nonallergenic ionic conductive gel/paste that is relatively nonviscous as viscous solutions can attenuate signal amplitude. A good skin condition is also necessary. Gentle abrasion is sometimes carried out to clean the contact area. The input impedance of a baby's skin is high but by cleaning the skin, the impedance can be brought down to less than 5 kΩ. Users should be aware that signal amplitude depends on the type of electrode and that the amplitude reduces as point of ocular contact moves away from the corneal apex. Topical anesthesia is required for the contact lens electrode but is not required for foil or DTL electrodes. It is important that all electrophysiologists are able to choose the appropriate electrodes for each test and ensure good contact with skin or cornea so that the measured signal has the highest SNR and the resulting waveforms are comparable to the standard signal.

12.5 ERG SYSTEM

12.5.1 Amplifiers

Most electrophysiology systems use a differential amplifier to detect and amplify signals from electrodes. Gains of 1,000 to 10,000 are common. They have a high input impedance, typically about 10 MΩ. CMRR at 50 or 60 Hz should be above 100 dB. Due to their high gain, even very small dc voltage difference at the electrode will result in a distorted or clipped amplified signal . To avoid this situation, the amplifier is ac coupled. Some commercial systems such as the UTAS + SunBurst ERG system from LKC Technologies do allow dc to avoid ac lockup. The gain of each stage of amplifiers should be linear in the input range that they amplify. After the first-stage amplification, an electrophysiology system will further offer filtering of signal at a specific frequency range (low pass, band pass, and high pass). The filtering could be performed either as analog filtering or by digital signal processing (DSP). If using the DSP technique, the signal is filtered after it is converted to a digital form by ADCs. Bessel, Butterworth, and Chebyshev filters have been reported. Bessel filters due to their better time domain characteristics are used in many commercial systems. In addition, some systems use a band rejection or a notch filter to reject supply frequency interference (50 or 60 Hz). Notch filters are not recommended by some physiologists as they will mask electrode problems. Correct filtering of unwanted frequencies improves the SNR of the signal recorded. It is possible to improve the SNR of the recorded signal by some image processing techniques such as averaging, three-point smoothing, and Fourier transforms. Patient isolation is achieved by driving the initial stage amplifier with batteries. The output of the first-stage amplifier is connected to the next stage amplifier through opto isolators. Amplifiers must also have very low-noise characteristics. The use of amplifiers with ultralow noise levels of 1 μV RMS is reported commercially (Espion E3 Electroretinography Console, Diagnosys Aug 2014) (http://www.diagnosysllc.com/products/product1.php).

12.5.2 Analog-to-Digital Converters

In a modern electrophysiology system, the recorded signal is stored in a digital form. This is done using an ADC. Minimum acceptable resolution of an ADC is 12 bits, giving a range of 4096 levels of amplitude. If we assume an input range of ±0.5 V, a 12-bit conversion will result in a bit resolution of 240 µV. Now, commercial systems offer a 24-bit resolution resulting in 16 million levels. With 24 bits, a resolution of 0.06 µV is feasible. Signal sampling frequency depends on the signal range. Sampling frequency of commercial systems is adjustable between 300 Hz and 5 kHz, which is sufficient for all electrophysiology signals.

12.6 GANZFELD STIMULATOR

The purpose of a ganzfeld stimulator is to provide a high-intensity flash of light in a uniform background of illumination. It is an integrating hemisphere, typically with a diameter of 50 cm, the inner surface of which is coated with white matte paint. The patient sits in front of the stimulator in a darkened room and looks into the stimulator during the test (Figure 12.5). An incandescent light bulb is used for background illumination, and a xenon flash lamp is used to create the light stimulus. The timing, color, and intensity of the stimulus are controlled electronically. Various color filters are used to create the required background. Modern systems such as ColorDome from Diagnosys LLC now are able to provide lightweight ganzfeld stimulators with a LED light source as both background and stimulus. They have the capability of self-calibration for both color and intensity. Most commercial systems offer ISCEV standard tests for clinical examination as well as the capability to design their own tests for research purposes.

12.7 CLINICAL APPLICATIONS

Electrophysiology is used to diagnose a variety of acquired and inherited eye conditions. PERG is usually used to diagnose macular disorders of the retina such as macular edema and serous detachment at the macula. In reduced macular function, the amplitude of the P50 component of the PERG waveform is reduced. Many eye clinics prefer to perform both PERG and ERG to test the entire retina. Combined results are used to study the functionality of the central retina (PERG) and the peripheral retina (ERG). Pattern VEP (PVEP) is a sensitive indicator of optic nerve function. Reduced amplitude of VEP is noticed in case of optic nerve head damages such as optic atrophy and ischemic optic neuropathy. VEP is used in the evaluation and management of pituitary tumors (Egemen *et al* 1991). In cases such as noncooperative patients, patients with media opacities, and babies, PVEP is not feasible. A simpler test of flash VEP (FVEP) is performed to extract valuable information on the extent of remaining optic nerve functions. The main use of the EOG in clinical practice is in the diagnosis of Best disease (Jarc-Vidmar *et al* 2001). The EOG from patients with Best disease shows a severely reduced Arden index. Retinitis pigmentosa is a disorder that affects rod photoreceptor function and can be diagnosed with a reduced Arden index in the EOG (Vingolo *et al* 2006).

12.8 EQUIPMENT

12.8.1 Multifocal and Standard Electroretinography Systems

Currently, there are few multifocal and standard electroretinography systems available on the market. These systems provide a wide range of research and clinical functions. They can perform current standard clinical visual function tests while visualizing the fundus area that is being stimulated.

The E^2 Electrophysiology System from Diagnosys is a modular suite of hardware, software, and stimulators capable of performing all ISCEV standard tests (Odom *et al* 2010). The system is capable of measuring response of the whole retina or from photoreceptor and inner nuclear layer. A separate functional assessment of the photopic and scotopic systems is possible.

With the system, ColorBurst and ColorDome stimulators can be bought to carry out standardized and user-definable protocols for full-field ERG testing. Standard protocols for each stimulator include EOG, maximal combined response, on/off response, 30 Hz flicker, oscillatory potentials, FVEP, single-flash cone ERG, and single-flash rod response.

12.9 TONOMETRY

12.9.1 Introduction

The eye is under pressure due to liquid in the two chambers—anterior chamber and posterior chamber. The *aqueous humor* is a thick jelly that fills the anterior chamber (space between the lens and the cornea). Normal intraocular pressure (IOP) ranges from 12 to 18 mmHg.

12.9.2 Physiology

Aqueous humor is produced by the ciliary body at a rate of approximately 3 mL per day. This fluid normally drains out of the eye through microscopic pores (trabecular meshwork) in the angle between the cornea and iris. Unlike the aqueous humor that is continuously replenished, the vitreous humor in the vitreous chamber is stagnant.

12.9.3 Clinical Need to Measure IOP

Glaucoma is a disease where progressive damage occurs to the optic nerve head causing loss of field of vision. An increase in IOP is one of the most important factors related to this damage. IOP has been measured for over 120 years. IOP could be lowered by various medical and surgical methods. Accurate and rapid IOP measurement can prevent permanent visual loss.

12.9.4 Principles of Measurement of Pressure in the Eye

Tonometry is a method to measure the IOP. The Imbert–Fick law states that for an ideal thin-walled sphere, the pressure inside the sphere equals the force from a plane surface necessary to flatten the sphere's surface divided by the area that is flattened (Figure 12.6):

$$\text{Pressure}\,(P) = \frac{\text{Force}\,(F)}{\text{Area}\,(A)}$$

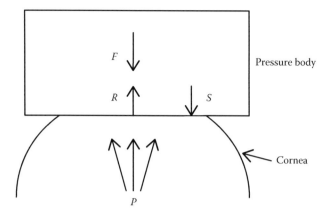

FIGURE 12.6 Principle of measurement of pressure in the eye. Force is applied by a pressure body on the cornea of the eye. From the area of the cornea flattened, the pressure of the eye (*P*) could be measured.

Other factors to consider are surface tension (*S*) and corneal rigidity (*R*). The surface tension is due to the tear film and aids the force applied, whereas the rigidity of the cornea opposes the force.

Two basic methods are used in tonometry. With the *impression principle*, a plunger 3 mm in diameter deforms the cornea and the indentation is measured (Schiøtz 1920). With the *applanation principle*, a plane surface is pressed against the cornea. Force required to flatten a defined area gives the pressure (Goldmann and Schmidt 1957).

12.9.4.1 Schiøtz Tonometer

The design of a tonometer was first described by its inventor Schiøtz (1920). The design of current Schiøtz tonometer has not changed for the past 90 years. The tonometer consists of a cylinder, the bottom of which is shaped with a concave structure so that it can sit on the cornea. A plunger slides gently through the cylinder. The top end of the plunger is connected to an arm of a lever that moves another arm acting as a pointer over a scale. This way, the vertical movement of the plunger body is translated to side movement on the scale. The footplate has a radius of curvature of about 14 mm. This radius of curvature allows only a small portion of the plunger to have contact with a cornea that typically has a radius of curvature of 8 mm. The contact point of the footplate acts as a reference point from which the movement of the plunger is measured. Most instruments come with a test block so that the calibration can be checked. The plunger weight is 5.5 g, which is sufficient to cause indentation for a normal range of 10–21 mmHg. See Figure 12.7. Additional weights are provided to give a total plunger weight of 7.5, 10, and 15 g so that measurements ranging up to 120 mmHg could be measured. All instruments come with a calibration chart to convert the displacement to pressure.

A plunger is used as the pressure body and is gently lowered on the cornea. The pressure of the eye is measured from the scale (see Figure 12.8).

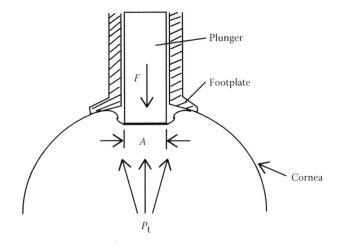

FIGURE 12.7 Principle of impression tonometer. (Adapted from Perkins, E.S. and Hill, D.W., *Scientific Foundations of Ophthalmology*, Butterworth-Heinemann Elsevier, Waltham, MA, 1977.)

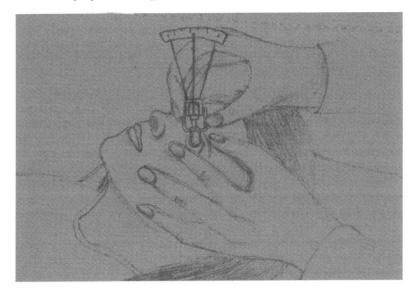

FIGURE 12.8 Use of a Schiøtz tonometer on a human patient. (Hand drawn by Sneha Manivannan.)

Initial Schiøtz tonometers correlated the meter reading to the pressure as measured. In 1954, the American Academy of Ophthalmology and Otolaryngology investigated the relationship between extent of indentation and pressure measured. They reported two pressure terms: P_t, the measured pressure, and P_0, the actual pressure in the eye. P_t will be higher than P_0 due to additional stress caused by the tonometer and indentation by the plunger:

$$P_t = \frac{F}{a + bR}$$

where
    R is the scale reading
    a and b are the constants

Friedenwald (1957) used a corneal rigidity factor K and derived an equation to measure P_0:

$$\log P_0 = \log \frac{F}{a+bR} - KV_t$$

where
  P_t is the pressure with the tonometer
  P_0 is the actual IOP
  R is the tonometer reading
  K is the rigidity coefficient
  V_t is the indentation volume
  a and b are the constants

To avoid measurement of a, b, and K, Friedenwald suggested a conversion table of pressure/scale division based on average values of statistically calculated a, b, and K.

12.9.4.2 Procedure to Use a Schiøtz Tonometer

Before taking a pressure on a patient, the plunger must be checked for free movement. The footplate and the tip of the plunger must be cleaned with isopropyl alcohol. The zero position is then checked by placing the tonometer on the test block that comes with the tonometer.

The patient is then explained about the procedure and asked to lie down or sit comfortably with face looking up. The position is very important as the plunger force highly depends on gravity. One drop of proxymetacaine 0.5% eye drop is instilled on the eye to create a local anesthesia.

The patient is asked to keep his or her eye open and the tonometer is gently lowered on the cornea (see Figure 12.8). The tonometer should be in a vertical position, and the footplate should evenly sit on the cornea. The plunger should be centered to the cornea. The reading is noted and the IOP is obtained from the conversion chart.

A Schiøtz tonometer may not be accurate when compared to the Goldmann tonometer, but has the greatest advantage of being portable and simple to use and is commonly used by ophthalmologists visiting patients at their home and to measure pressure in remote areas where it is not possible to use a Goldmann tonometer.

Even though it is still possible to buy a Schiøtz tonometer (http://www.optivision2020.com/tonometer-schiotz.html), its use has slowly disappeared in the developed world. It is reported of some use in few developing world countries.

Figure 12.9 shows a Schiøtz tonometer. It is supplied with additional weights and a calibration test block.

12.9.4.3 Goldmann Tonometer

The first Goldmann tonometer was reported in 1954. The modern Goldmann tonometer as reported in 1957 (Goldmann and Schmidt 1957) that is widely used in eye clinics throughout the world gives a reliable measurement of IOP within 0.5 mmHg.

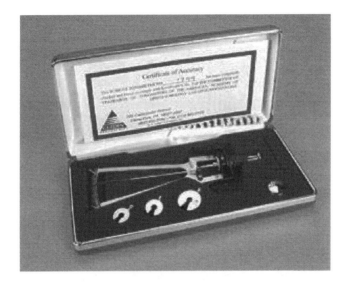

FIGURE 12.9 A Schiøtz tonometer. It is supplied with additional weights and a calibration test block.

The Goldmann tonometer considers two physical properties of the cornea and tear film in the accurate measurement of IOP. Figure 12.10 shows that the rigidity of the cornea opposes the force applied, whereas the surface tension due to tear film aids the force:

$$P_t = \frac{F}{A} - P_M + P_N$$

where
    P_M is due to rigidity
    P_N is due to surface tension

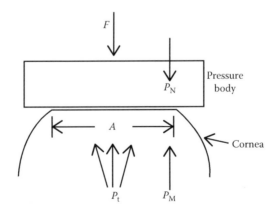

FIGURE 12.10 Principle of applanation tonometer. (Adapted from Perkins, E.S. and Hill, D.W., *Scientific Foundations of Ophthalmology*, Butterworth-Heinemann Elsevier, Waltham, MA, 1977.)

FIGURE 12.11   AT900 Goldmann tonometer. (Courtesy of Haag-Streit AG, Koeniz, Switzerland.)

Goldmann proved that when the diameter of the applanation area is 3.06 mm, the actual IOP, P_0, is approximately equal to 0.98 times P_t. Figure 12.11 shows the Goldmann tonometer.

The instrument consists of the applanation head that is mounted on a spring-loaded lever connected to a slit-lamp fundus microscope. The applanation head is a measuring prism of organic glass with a diameter of 7.0 mm. Since the prism is in contact with the cornea during the measurement, its edge is rounded to avoid any damage to the cornea. The tension on the spring is adjustable, and the adjustment knob is calibrated in terms of IOP, in mm of mercury. The measuring prism splits the contact point image with the cornea into two semicircles. The operator sees the cornea through the applanation head (Figure 12.12a). The force applied is adjusted until the semicircles just overlap. At this point, the diameter of contact is 3.06 mm and the IOP is read off the scale (Figure 12.11).

12.9.4.4 Procedure to Use of a Goldmann Tonometer

The patient's head is positioned on the chin rest of the slit lamp. One drop of proxymetacaine 0.5% eye drop is instilled on the eye to create a local anesthetic effect. Once the anesthetic is active, the conjunctiva is gently touched with a strip of fluorescein-impregnated paper. The fluorescein with the help of blue light allows the ophthalmologist to see the cornea that is otherwise transparent. The tonometer head is then brought very close to the cornea. The slit lamp is positioned at 45° to the eye so that the slit illumination illuminates the area between the cornea and the tip of the applanation head. A blue filter is then inserted in the beam path of the white light of the slit lamp. A green filter in the detection pathway of

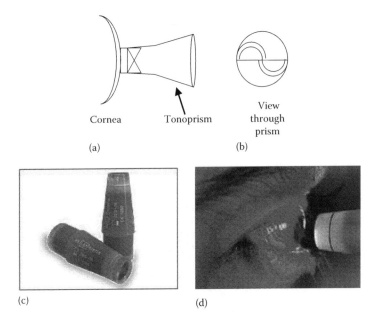

Cornea   Tonoprism

View
through
prism

(a)   (b)

(c)   (d)

FIGURE 12.12   Use of a Goldmann tonometer. (a) The tonoprism is in contact with the cornea. (b) When a correct area of 3.06 mm is flattened, the inner lines of the circles touch each other. (c) A commercially available tonoprism (Courtesy of Haag-Streit AG, Koeniz, Switzerland.) shown in use (d).

the eye of the ophthalmologist allows him or her to see the fluorescent cornea when the tip of the applanation head contacts the cornea. The prism splits the circle into two semicircles. The dial on the tonometer is rotated until the inner edges of the semicircles exactly align (Figure 12.12b). At this point, the diameter of the contact with the cornea is 3.06 mm. The tonometer reading is then read and is taken as the IOP.

12.9.4.5 Other Types of Goldmann Tonometer
The previously described Goldmann tonometer is commonly used in eye clinics. Using the same principle, Perkins developed a handheld tonometer (Figure 12.13) (Perkins and Hill 1977). Instead of using a spiral spring of a Goldmann tonometer, a tonopen uses a semiconductor pressure sensor to measure the force required to flatten the cornea when the diameter of contact is 3.06 mm (Figure 12.14).

12.9.4.6 Noncontact or Puff Tonometer
The puff tonometer reported first in 1972 (Grolman 1972) measures the IOP by measuring the time taken for a standard jet of air to flatten a known area of the cornea (Figure 12.15). An air pump blows a jet of air at the cornea of the patient. An infrared light source and a photodiode positioned at 45° to the eye are used. The light source projects a collimated beam of light on the cornea. Since the cornea is curved, the light is scattered from the cornea and very little light reaches the detector. As the puff of air continues to blow on the cornea, the cornea gets flattened and with the tear film's help acts as a reflecting mirror. The detector at this stage gets maximum light input. Since the air pressure is linearly increased over time,

FIGURE 12.13   Perkins handheld tonometer. (Courtesy of Haag-Streit AG, Koeniz, Switzerland.)

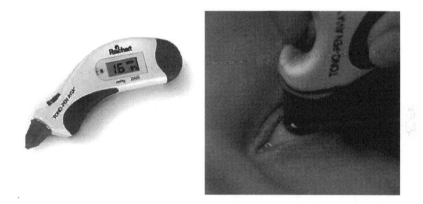

FIGURE 12.14   Tono-Pen AVIA® Applanation Tonometer. (Courtesy of Reichert Technologies, Depew, NY.)

the time taken to get the maximum output from the detector is directly proportional to the pressure. The width of the air stream is designed in such a way that the cornea flattened area has a diameter of 3.06 mm.

Figure 12.16 shows a typical relationship between the air pump pressure, time, and detector signal. In this example, the cornea is flattened in about 24 ms. The air pressure at this point is 25 mmHg, which will be the same as the IOP.

Noncontact tonometers (Figure 12.17) also known as Puff tonometers are regularly used as a screening tool to measure IOP in the general population. Measuring IOP is now performed as part of an eye test when one visits an optometrist. Since there is no need for anesthesia, the test can be performed quickly with minimum discomfort for the patient. It is important to remember that any irregularities of the corneal surface (e.g., scars after LASIK surgery for vision correction) will prevent accurate measurement.

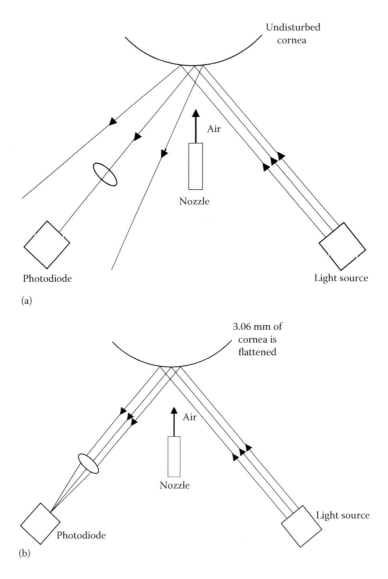

FIGURE 12.15 Noncontact or puff tonometer. (a) Condition before the measurement is made. Curved cornea diverges the beam of light and so the photodiode receives little light. (b) When an area of cornea is flattened, the tear film on cornea acts as a mirror reflecting the beam of light in the path of the photodiode. A lens focuses this beam of light to the photodiode thus maximizing the output. (Adapted from Perkins, E.S. and Hill, D.W., *Scientific Foundations of Ophthalmology*, Butterworth-Heinemann Elsevier, Waltham, MA, 1977.)

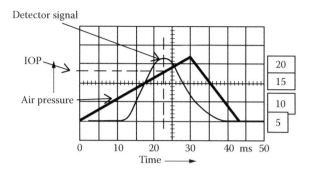

FIGURE 12.16 Air pressure and photodiode signal of a puff tonometer. Detector signal reaches maximum when the cornea is flattened by 3.06 mm (see Figure 12.12b). (Adapted from Perkins, E.S. and Hill, D.W., *Scientific Foundations of Ophthalmology*, Butterworth-Heinemann Elsevier, Waltham, MA, 1977.)

FIGURE 12.17 A noncontact tonometer **NT-530** (http://www.nidek-intl.com/products/examination/nt-530.html).

REFERENCES

American Academy of Ophthalmology and Otolaryngology 1954 Standardisation of tonometers Decennial Report by the committee on standardization of tonometers

Dawson W W, Trick G L and Litzkow C A 1979 Improved electrode for electroretinography *Invest. Ophthalmol. Vis. Sci.* **18** 988–91

Diagnosys 2014 http://diagnosysllc.com/products/classic-system/

Egemen N, Gokalp H Z, Kulcuoglu A, Naderi S and Zorlutuna A 1991 Visual evoked potentials in evaluation and management of pituitary adenomas *Turkish Neurosurg.* **2** 64–77

Friedenwald J S 1957 Tonometer calibration. An attempt to remove discrepancies found in the 1954 calibration scale for Schiøtz tonometers *Am. Acad. Ophthal. Otolaryng.* **61** 108

Goldmann H and Schmidt T H 1957 Uber applanations-tonometrie *Ophthamologica* **134** 221–42

Grolman B 1972 A new tonometer system *Am. J. Optom.* **49** 646

Hechenlively J R and Arden G B 2006 *Principles and Practice of Clinical Electrophysiology of Vision* (Cambridge: The MIT Press)

Jarc-Vidmar M, Popović P, Hawlina M, and Brecelj J 2001 Pattern ERG and psychophysical functions in Best's disease *Doc. Ophthalmol.* **103** 47–61

Odom J V, Bach M, Brigell M, Holder G E, McCulloch D L, Tormene A P and Vaegan 2010 ISCEV standard for clinical visual evoked potentials (2009 update) *Doc. Ophthalmol.* **120** 111–9

Optivision 2014 http://www.optivision2020.com/tonometer-schiotz.html

Perkins E S and Hill D W 1977 Chapter 34. New types of tonometers by J Draeger & K Jessen. In *Scientific Foundations of Ophthalmology* London: Butterworth-Heinemann Elsevier)

Schiøtz H 1920 Tonometry *Brit. J. Ophthalmol.* **4** 201–49

Marmor M F, Arden G B, Nilsson S-E, and Zrenner E 1989 Standard for clinical electroretinography *Doc. Ophthalmol.* **73** 303–11

Vingolo E M, Livani M L, Domanico D, Mendonça R H, and Rispoli E 2006 Optical coherence tomography and electro-oculogram abnormalities in x-linked retinitis pigmentosa *Doc. Ophthalmol.* **113** 5–10

CHAPTER 13

Orthopedics

Miriam Hwang, Ubong Udoekwere, and Gerald Harris

CONTENTS

13.1 Introduction 288
13.2 Clinical Assessment 289
    13.2.1 Physical Examination 289
        13.2.1.1 Inspection and Palpation 290
        13.2.1.2 Motion and Strength 290
        13.2.1.3 Test Maneuvers 292
    13.2.2 Imaging Studies 292
        13.2.2.1 Plain Radiograph (X-Ray Images) 293
13.3 X-Ray Machines (Radiography) 296
    13.3.1 Operation and Components 297
13.4 CT 299
    13.4.1 CT Scanners 299
    13.4.2 Operation and Components 299
13.5 Radionuclide Bone Scan 300
    13.5.1 Radionuclide Bone Scan Machines (Nuclear Medicine Scanner) 301
    13.5.2 Operation and Components 301
13.6 MRI 303
    13.6.1 MRI Machines 304
    13.6.2 Operation and Components 306
13.7 Real-Time Musculoskeletal Ultrasound 307
    13.7.1 Real-Time Musculoskeletal Ultrasound Machine 308
        13.7.1.1 Operation and Components 308
13.8 Functional Studies 309
    13.8.1 Nerve Conduction Study and Electromyography 309
    13.8.2 Electrodiagnostic Instrumentation 311
        13.8.2.1 EMG Instrumentation 311
        13.8.2.2 NCS Instrumentation 311

13.9 Gait/Motion Analysis   313
    13.9.1 Clinical Gait Instrumentation   313
    13.9.2 Video System   314
    13.9.3 Force Platform   315
References   315

13.1 INTRODUCTION

The orthopedic assessment is based on the functional anatomy and physiology of the musculoskeletal system. The bony skeleton, joints and their connective tissue (i.e., articular cartilage, ligaments, capsules, synovial membrane), muscles and their tendons and fasciae, and bursae comprise the musculoskeletal system.

Bones function to provide structural support for the body, allow movement by providing levers for muscles, maintain mineral homeostasis, and serve as a hematopoietic organ. The long bones of the extremities are composed of the tubular shaft (diaphysis), the rounded epiphysis at both ends, and the metaphysis between the diaphysis and epiphysis, just below the growth plate. Bone tissue is classified as cortical and cancellous bone, and the relative proportion of these two types of bony tissue in any bone depends on the specific bone and its skeletal location. Cortical bone is solid and dense, usually surrounds the marrow space, and constitutes the diaphysis and surface of long bones, as well as the surface of vertebrae and other short or irregular bones. Cancellous or trabecular bone is less compact than cortical bone, consists of a meshwork of trabecular plates and rods interspersed within the marrow space of the metaphysis and epiphysis, and is highly vascular. Cells responsible for bone remodeling, the osteoblasts and osteoclasts, are situated both in the endosteum lining the bone marrow space and the periosteum covering the outer cortical bone surface, and these sites of metabolic activity can be indirectly assessed through blood chemistry and radionuclide imaging studies. Bone pathologies frequently seen in orthopedics include traumatic or pathologic fractures and dislocation, infection, avascular necrosis, tumors, osteoporosis, metabolic disorders, and congenital bone disease (Clarke 2008, Gardner 1963).

The joints of the skeleton are classified into three types: (1) fibrous joint, where the articulating surfaces are connected by fibrous tissue and minimal movement is allowed; (2) cartilaginous joint, in which the articulating surfaces are united by a fibrocartilage such as the intervertebral disc in the spinal column; and (3) synovial joint, in which ligaments and a synovial membrane-lined capsule connect the articulating bones and a joint cavity exists between the articulating surfaces allowing for a wide range of movement (Gardner 1963). Synovial joints allow movements in different planes of motion, depending on the orientation of the adjacent bone surfaces. These movements are categorized as gliding, angular (flexion/extension or abduction/adduction), or rotatory (pronation/supination or internal/external rotation) and are brought about by the contraction of muscles traversing the joints, with the overlying fibrous retinacula acting as fulcrums to the long tendons crossing the wrist and ankle joints. The ligaments and capsules within and surrounding the joint function to protect the joint within its functional range of motion. When these fibrous structures of the joint are damaged due to inflammation or stretched beyond the physiologic range of motion, active and passive motion of the joint can induce pain and instability. Joint pathologies range from malalignment to arthritides (inflammatory, infective, degenerative, traumatic) and from ankylosis to ligament laxity/rupture to dislocation.

The contractile property of the skeletal muscle is the main source of body movement. The orientation of a muscle's attachment to bones and the number of joints it crosses determine the direction and type of joint movement, while the muscle fiber orientation within the muscle belly determines the force of muscle contraction as well as direction of pull. To produce a specific determined movement, muscles contract in synergy with agonist muscles, while antagonist muscles remain silent or contract eccentrically for a coordinated joint movement. The muscle fascia is a connective tissue that surrounds and separates individual muscles as well as compartmentalizes muscles into functional groups with their respective supplying vessels and nerves. The polar ends of the fasciae surrounding the individual muscles combine together to form dense cords or bands called tendons or flat and broad aponeuroses that insert onto the distal bone of the joint. Long tendons are surrounded by a synovial sheath to protect against friction with the bony surfaces as they pass through narrow grooves such as the tendon of the biceps' long head or beneath fibrous retinacula in the wrist and ankle. Pathologies pertaining to muscle and tendon include denervation and atrophy; contusion, tears, or ruptures; inflammation; and compartment syndrome.

A bursa is a synovium-lined sac situated in sites of friction in joints, between muscle and bone, or subcutaneous areas subject to continuous pressure. Common sites where bursae are present are around the knee, the shoulder deep to the acromion, and around bony prominences such as the olecranon, ischium, and greater trochanter. The major pathology pertaining to the bursa is inflammation, bursitis, in which the pressure effects of the acute inflammation or adhesive changes in chronic inflammation determine its signs and symptoms.

Each component of the musculoskeletal system can be assessed individually or in association with adjacent structures through physical, radiologic, and instrumental examinations. The subsequent sections of the chapter will review important principles in clinical assessment through physical examination, interpretation of imaging studies (plain radiograph, computed tomography [CT], magnetic resonance imaging [MRI], radionuclide bone scan, real-time ultrasound [US]), and functional evaluation (nerve conduction study [NCS], electromyography [EMG], gait/motion analysis). The final section of this chapter will introduce several validated instruments currently in use in the functional orthopedic assessment, including the dynamometer, compression pressure monitor, balance system, electric goniometer, and motion analysis system.

13.2 CLINICAL ASSESSMENT

13.2.1 Physical Examination

A unique feature in orthopedic practice is the fact that the pathology is often visible and/or palpable on examination, and numerous manipulative tests have been developed and validated for the diagnosis of various orthopedic conditions. On the other hand, a variety of metabolic, circulatory, and neurologic disorders manifest as orthopedic symptoms and deformities. It is beyond the scope of this chapter to detail the various examination procedures and thus will focus on the principles of the orthopedic physical examination. The importance of a thorough medical history, patient's complaint, and present illness is emphasized as these are critical in focusing the physical examination and establishing a working differential diagnosis. Furthermore, the information gathered from the history and physical examination guides the clinician in selecting the appropriate laboratory test, imaging study, or functional assessment to confirm the diagnosis.

The orthopedic physical examination is comprised of inspection and palpation, motion and strength evaluation, and test maneuvers.

13.2.1.1 Inspection and Palpation

Inspection begins as the patient enters the examination room. Gait and posture should be noted for any asymmetries, atypical gait patterns, or abnormal posturing of the spine and extremities, as well as for facial expressions of pain or discomfort. Observation of any skin discoloration or bruising, joint or soft tissue swelling, deformities, atypical callous formation or skin abrasion, and asymmetry in limb contour or skin crease pattern should be noted. Palpation of distal arterial pulses and surface skin temperature in association with capillary filling time provide information on the vascular function of the area in question, especially in cases of circulatory compromise such as a vascular necrosis, compartment syndrome, or orthopedic changes secondary to peripheral neuropathies or systemic inflammatory diseases. The underlying causes of either edema or joint effusion can be differentiated by its consistency (pitting vs. nonpitting, tense vs. fluctuant), distribution (generalized vs. focal), and whether it is accompanied by tenderness, erythema, heat, or bruising. Tenderness that is elicited by mild pressure on a joint, bony surface, tendon, or muscle, as well as the type of pain that is produced, such as dull aching pain or electric shock-like sensation, should also be noted.

13.2.1.2 Motion and Strength

Joint range of motion is tested bilaterally to assess symmetry in both active and passive motions. The asymptomatic side is tested first to establish a baseline norm within the individual, and active motion is tested prior to testing passive range to detect limited range or pain. Range of motion of the extremity is measured with a goniometer for all applicable planes of motion for the specific joint, including flexion/extension, abduction/adduction, pronation/supination, medial/lateral rotation, and inversion/eversion. The standard goniometer consists of a protractor with two extended arms. Joint range is measured with the fulcrum of the goniometer aligned to the joint axis, the stationary arm of the goniometer stabilized along the longitudinal axis of the proximal segment of the joint, and the other arm aligned with the distal mobile segment. The joint angle at each endpoint is noted, and the difference between the endpoints represents the range of motion. Spinal ranges of motion include flexion/extension, rotation, and lateral flexion and are measured with inclinometers. An inclinometer consists of a 360° disc protractor set on a level base, in which the fluid level within the disc indicates the range of movement in degrees. For spinal range of motion, two inclinometers are placed longitudinally parallel to the long axis of the spinal column at specified points for the particular spinal segment being evaluated, and the change in degrees of movement is read between the endpoint of movement. For example, to measure lumbar flexion, one inclinometer is placed on the T12 spinal process, the other on the S1 spinal process, and both inclinometers are set to 0° in the resting standing position. Following active flexion, the measurements are read in the flexed position; the upper reading represents total flexion, the lower reading represents sacral flexion, and the difference between the upper and lower inclinometer measurements is calculated as the lumbar flexion range of motion. For spinal rotation measurement, a single inclinometer is placed perpendicular to the spinal column in the resting position and the change in degrees to endpoint of rotation is noted (Magee 1997).

In addition to range of motion, active movement should also be assessed for its quality and rhythm, the ability to sustain a posture for the evaluation of potential muscle weakness or joint pathology, as well as compensatory movements. During assessment of passive motion, change in muscle tone or resistance throughout the range and joint end feel at end range is assessed. Increase in muscle tone during passive movement may be the result of involuntary muscle spasms or voluntary muscle guarding in apprehension of possible pain. Joint end feel is a specific sensation that is transmitted to the examiner's hand when applying gentle overpressure at the end of a range. There are three types of normal end feel: (1) *bone to bone*, a painless unyielding sensation such as that felt at the end of elbow extension; (2) *soft tissue approximation*, a painless yielding compression that stops further movement as occurs in end of elbow flexion; and (3) *tissue stretch*, a firm elastic resistance that is felt toward the end range when the joint capsule or ligaments are the primary restraints to movement, such as in knee extension or shoulder lateral rotation. Pathologic patterns are determined when these end-feel types occur prior to reaching the end range or where it is not expected to occur. For example, a bone-to-bone end feel at the end range of elbow flexion would be abnormal, as would a tissue stretch end feel in the midrange of shoulder rotation. A *springy block* end feel is an abnormal end feel similar to the tissue stretch pattern but occurs in joints containing menisci and indicates internal derangement, such as when the knee is locked in the presence of a meniscal tear. An *empty* end feel describes the condition when significant pain is elicited by passive movement in the absence of any sense of resistance and can indicate acute bursitis, abscess, or tumors (Cyriax and Cyriax 2003).

The muscle strength grading system (Table 13.1) developed by the Medical Research Council is widely used to assess muscle function. The grading system is based on muscle performance in relation to range of motion and the magnitude of manual resistance

TABLE 13.1   Muscle Strength Grading System

Grade	Movement
5	Muscle contracts normally throughout a full range of motion against maximum resistance.
4	Muscle strength is reduced but is able to contract throughout range of motion against some resistance.
3	Muscle contracts throughout range of motion against gravity, but not against resistance.
2	Muscle contracts throughout a full range of motion with gravity eliminated, but not against gravity.
1	Minimal muscle contraction or twitch with no discernible joint movement.
0	No visible or palpable muscle contraction.

applied by the examiner. Scores are ranked from no contraction to a contraction that can be performed against gravity to a contraction that can overcome maximal resistance by the examiner (John 1984).

In addition to muscle strength, response to resisted movement can guide in the diagnosis of disorders involving the contractile structures, that is, the muscles, tendons, and their bone attachments. The procedure is tested with the joint and extremity positioned to avoid joint motion or contraction of muscles that are not being tested while the subject exerts maximal contraction against resistance. A *painless strong* contraction implies normal contractile structure, whereas a *painful strong* contraction indicates a minor lesion in a portion of the muscle, tendon, or its bony attachment, a *painless weak* contraction can be attributed to complete rupture of the contractile tissue or a complete nerve injury, and a *painful weak* contraction indicates a severe lesion around the joint such as a fracture, as the weakness is most likely the result of reflex inhibition secondary to pain.

13.2.1.3 Test Maneuvers

The morphology of the articular surfaces and the type, orientation, and location of its stabilizing structures determine joint stability as well as form the basis of the maneuvers testing its integrity. The orientation of ligaments, tendons, vessels, and nerves in relation to skeletal structures should be considered in performing any test maneuver involving manipulation of a joint or provocation of symptoms. It is particularly important to maintain the joint in a position that will not put undue stress on the pathologic joint or adjacent intact structures during the test maneuver. Joints that are prone to instability are those with a wide functional range of motion and those that bear body weight, such as the shoulder and knee. A number of tests are performed to assess the structural integrity of such joints by applying pressure in a particular plane of movement and determining the amount of displacement from the physiologic range of movement. The knee in particular has numerous ligamentous and fibrocartilaginous structures in and around the joint, which all contribute to its movement and stability but are also susceptible to trauma due to their superficial location directly under the skin. The shoulder joint, although encapsulated by muscles of the rotator cuff, is a relatively free-floating structure as the humeral head articulates with the shallow glenoid of the scapula, which itself is a very mobile bone anchored only to the clavicle with the acromioclavicular and coracoclavicular ligaments in addition to muscular attachments to the chest wall, and thus is prone to dislocation as well as impingement syndromes. Commonly used test maneuvers in the orthopedic setting for the evaluation of joint stability are presented in Table 13.2; this list however is by no means complete and the reader is referred to texts of musculoskeletal physical examination for a more comprehensive list of test maneuvers.

13.2.2 Imaging Studies

The development of radiological technology and other imaging studies has been invaluable in visualizing and confirming orthopedic lesions in relation to neighboring anatomic structures as well as in determining the physiologic nature of the disease process. This section will focus on the clinical applications and interpretation of each imaging modality, while the technical aspects will be referred to Chapter 18.

TABLE 13.2   Test Maneuvers for Joint Stability

Joint	Test	Procedure	Instability
Shoulder	Apprehension/ crank	Arm abducted to 90° and slowly rotated laterally; positive if patient resists further movement or has facial expression of apprehension.	Anterior instability
	Load and shift	Examiner stabilizes shoulder with one hand over scapula and clavicle, while the other hand grasps the thumb over the posterior humeral head and fingers over the anterior humeral head. The humerus is gently pushed into the glenoid to *load*, pushing the humeral head forward or backward to observe the amount of translation; positive if head translates more than 25% of diameter of the humeral head.	Anterior instability (with forward push) Posterior instability (with backward push)
Knee	Lachman	Patient supine, knee flexed to 20°–30°, and with one hand of the examiner stabilizing the distal femur, the other hand grasps the proximal tibia and pulls it up (anteriorly); positive with soft end feel and anterior translation.	Anterior cruciate ligament
	Drawer	Patient supine, hip flexed 45°, knee flexed 90°, and foot stabilized with the examiner's body. The examiner's hands are placed around the proximal tibia and drawn forward or pushed backward on the femur; positive if translation exceeds 6 mm.	Anterior cruciate ligament (forward translation); posterior cruciate ligament (backward translation)
	Valgus stress	Abduction of the tibia relative to the femur in slight knee flexion (20°–30°).	Medial collateral ligament; posterior cruciate ligament; posterior oblique ligament
	Varus stress	Adduction of the tibia relative to the femur in slight knee flexion (20°–30°).	Lateral collateral ligament; posterolateral capsule; iliotibial band
	McMurray	Patient supine with knees fully flexed. Examiner medially or laterally rotates the tibia; positive if pain is elicited or a snap or click is audible.	Lateral meniscus (with medial tibial rotation); medial meniscus (with lateral tibial rotation)

13.2.2.1 Plain Radiograph (X-Ray Images)

The plain radiograph image is produced as x-rays pass through body tissues onto a flat panel detector comprised of a cesium iodide scintillator and amorphous silicon- or selenium-based readout matrix (Greenspan 2010, Reiff 1997). The inverse relationship between tissue density and amount of x-ray penetration through tissues results in images of increased opacity with tissues of higher density. Six basic densities are identified in the orthopedic assessment, which are, in ascending order, air, fat, water, soft tissue, bone, and metal.

X-ray images are used for the diagnosis and monitoring of fractures, dislocations, joint and spinal alignment, degenerative joint disease, tumors, and infection of bones. Anteroposterior (AP) and lateral views are taken to examine long bones, and an oblique view is added when examining joints and the spinal column. Reading of bone radiographs should focus on the size and shape of bones, including the cortical thickness, trabecular pattern, and any changes in bone density, as well as side-to-side symmetry. Details such as disruption of cortical margins or periosteal changes, morphology, and location of a lesion within the bone, in addition to changes in soft tissue densities surrounding the bone, give information on the pathologic process and assist in the differential diagnosis, ranging from stress fractures to infections to tumors. Significant radiographic bone abnormalities include *radiolucencies* such as cysts; *osteolytic lesions* that raise suspicion of osteomyelitis, rickets, rheumatoid arthritis, or generalized osteopenia, among others; *spur formation* that indicates enthesopathy or degenerative changes; *periosteal reactions* such as the *sunburst appearance* indicating cortical destruction with elevation of periosteum, commonly seen in osteogenic sarcoma; and *sclerotic lesions* that are attributed to numerous etiologies including exostosis, renal osteodystrophy, or metastatic tumor. The pattern or distribution of bony abnormalities should also be noted as it provides information for diagnosing conditions such as malignancies, metabolic bone disease, and physical abuse. In joint areas, changes in bone surface such as irregularities or spur formations and increases or decreases in joint space area should be noted and interpreted with pertinent clinical findings. Observation of radiopacities in nonosseous regions or within joint spaces indicates heterotopic ossification and should be followed with bone scans or laboratory tests including alkaline phosphatase levels to assess metabolic activity.

Whole spine AP images in the standing or sitting position are used for assessment of scoliosis or kyphosis with the angles of curvatures measured with the Cobb method (Figure 13.1). Angle measurement is done by first identifying the apex of the curve and the uppermost and lowermost vertebrae that comprise the curve. A line perpendicular to the upper margin of the upper end vertebra and a line perpendicular to the lower margin of the lower end vertebra are drawn and extended so the two lines intersect. The angle of intersection between the lines is the Cobb angle. Angles less than 10°–15° require no treatment but

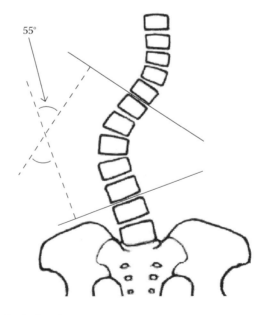

FIGURE 13.1   Cobb method of scoliosis measurement.

with regular follow-up until skeletal maturation; curves 20°–40° may require braces to prevent progression of curves until skeletal maturation; curves greater than 40°–45° may require surgical correction.

Lower extremity alignment is assessed with standing radiographs including the pelvis with proper positioning such that the patellae are facing forward. Examining the position of a joint or limb in relation to the mechanical axis of the lower extremity (i.e., a vertical line passing through the center of the femur head and center of the ankle joint) allows the determination of alignment as well as the presence of rotational deformities (Figure 13.2). Deviation of the knee joint center from the mechanical joint axis of the lower extremity can be assessed with the tibiofemoral angle (TFA) for genu valgum/knock-knees and the metaphyseal–diaphyseal angles (MDAs) of the proximal tibia and distal femur for genu varum and Blount's disease (Figure 13.2). Specific angles and lines for radiologic assessment of hip joint alignment include the femur neck–shaft angle (NSA) and Hilgenreiner–epiphyseal angle for coxa valga/vara, as well as the Perkin's, Hilgenreiner's, and Shenton's lines, and the acetabular angle for the assessment of developmental hip dysplasia (Figure 13.3) (Morrisy and Weinstein 2006).

The plain radiograph is the most routinely used imaging modality from which significant information can be obtained. However, there are limitations in the sensitivity of detecting minor bone lesion, especially in the early stages of injury, and visualization of

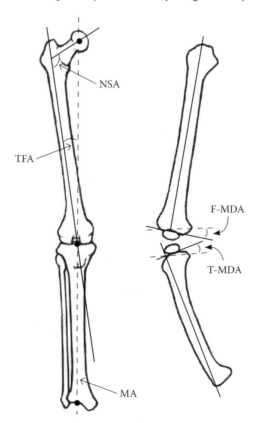

FIGURE 13.2   Assessment of lower extremity alignment.

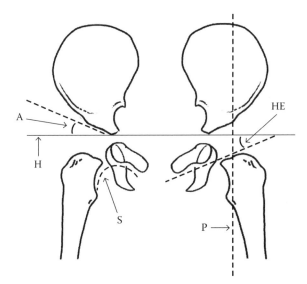

FIGURE 13.3 Pelvic landmarks used in the assessment of developmental dysplasia of hip (DDH). H, Hilgenreiner's line; P, Perkin's line; S, Shenton's line; A, acetabular angle; HE, Hilgenreiner–epiphyseal angle. The femur head is normally in the lower medial quadrant of the area formed by the crossing of H and P. Displacement of the femur head beyond the lower medial quadrant, discontinuity of Shenton's line, or an acetabular angle greater than 25° are indicative of DDH. The HE angle is an indicator of coxa vara when greater than 40° in the developing child.

musculoskeletal tissue other than bone or assessing metabolic or hemodynamic processes within a bony lesion is very limited as well. The next sections outline other imaging modalities that complement the plain radiograph in the assessment of orthopedic conditions.

The normal range for NSA in the adult is 120°–135°: coxa vara if NSA < 120° and coxa valga if NSA > 135°. In the developing child, NSA is approximately 160° at birth, gradually decreases to ~140° at age 5, and decreases to adult values by adolescence. Coxa vara is also assessed with the Hilgenreiner–epiphyseal angle as shown in Figure 13.2. TFA is the angle formed by the longitudinal axes of the femur and tibia. The TFA changes during childhood development from genu varum in infancy to neutral around 18 months, then to 12°–15° of valgum 3–4 years, and reaches the normal adult angle of 5°–7° in slight genu valgum by age 7 (Salenius and Vannaka, 1975). Femoral MDA (F-MDA) is formed by a line perpendicular to the long axis of the femur and a line parallel to the distal femoral physis; tibial MDA (T-MDA) is formed by a line through the beaks of the tibial metaphysis and a line perpendicular to the lateral border of the tibial diaphysis (Sabharwal 2009). MDA < 10° is considered physiologic bowing, whereas MDA > 16° is indicative of Blount's disease. An F-MDA/T-MDA ratio < 1 also suggests a greater contribution of the tibial segment to the bowing, indicating Blount's disease; F-MDA/T-MDA ratio > 1 indicates physiologic bowing.

13.3 X-RAY MACHINES (RADIOGRAPHY)

An x-ray machine is made up of a high vacuum tube (*x-ray tube*) with two electrodes within the tube on either ends: a filament cathode electrode on one end and an anode electrode on the other (Figure 13.4). In addition, x-ray machines also contain the necessary supporting

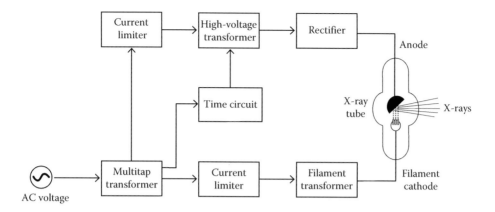

FIGURE 13.4   Block diagram of an x-ray machine.

circuitry to regulate efficiency and safety of x-ray production. The supporting circuitry allows for control of x-ray by regulating current, voltage, and exposure times that ultimately determine the amount of x-ray emission, amount of x-ray energy delivered to the patient, and duration of delivery, respectively.

13.3.1   Operation and Components

1. *Cathode and filament circuit*: The cathode is typically a tungsten filament that is heated via current (filament current) to emit electrons. The filament current is generated by the *filament transformer* (Figure 13.4), which is powered by another transformer called the multitap autotransformer. The multitap ac line autotransformer is the main power supply for the x-ray machine. This combined circuitry (i.e., cathode filament and filament transformer) makes up the *filament circuit*.

2. *Anode and high-voltage circuit*: The generated electrons are accelerated across a high-voltage difference from cathode to the anode. The anode is typically a tungsten target on the opposite end of the tube. X-ray photons are consequently created from the energy transfer processes as the high-speed electrons collide with the anode target. The *high-voltage circuit* produces the high dc voltage applied across the tube such that the anode is relatively positive to the cathode, so that electrons are accelerated from cathode to anode. The high-voltage circuit is also powered by the multitap ac line autotransformer (Chatterjee and Miller 2010).

3. *Timer circuit*: Finally, the last component of the x-ray machine is the *timer circuit*, and it functions to control the duration of x-ray exposure to the patient. This circuit controls the switch that activates or deactivates the high-voltage transformer, thus regulating the duration of the voltage applied across the tube. The timing circuit is also powered by the multitap ac line autotransformer (Chatterjee and Miller 2010).

4. *X-ray detector*: The generated x-rays are directed and focused at the patient. The x-rays that penetrate the patient (called transmitted x-rays) are then detected and measured by a detector to produce x-ray images. Typically, the detector measures

the attenuation of the transmitted x-ray by transducing or measuring the amount of energy of the x-ray photons as they strike the detector. The x-ray photons that make up the x-ray beam experience different degrees of attenuation as they pass through varying densities of tissue in the body. X-ray photons pass through relatively denser structures (e.g., bones or foreign dense objects lodged in tissue) and are more attenuated or absorbed by the denser structure.

In the past, a photosensitive film placed on the opposite side of the patient was used as a medium to detect transmitted x-rays, ultimately producing an image that reflected the different degrees of attenuation of the x-ray photons. The images of internal body structures on the film appear white relative to less dense tissues (e.g., muscles) that attenuate photons to a lesser degree, which appear black on the film.

In contemporary radiography, solid-state detectors that convert x-ray photon energy to electrical signals, which are then digitized for interpretation and reconstruction by the data acquisition computer (DAC) system, are typically employed. Solid-state detectors make use of an array of scintillating crystals connected to an array of photodiodes. Scintillating crystals are inorganic crystals (e.g., cadmium tungstate or sodium iodide) that fluoresce when struck by ionizing radiation. Each scintillating crystal in the array fluoresces when struck by an x-ray photon, thus converting x-ray photon energy to visible light. The generated light energy is further converted to an electrical signal via the photodiode connected to that crystal. Consequently, the amount of the electrical energy produced is determined by the degree of fluorescence (intensity of the light) of the crystal, which is in turn proportional to the energy of the transmitted x-ray photon (Bronzino 1995). See block diagram of the x-ray detector operation in Figure 13.5.

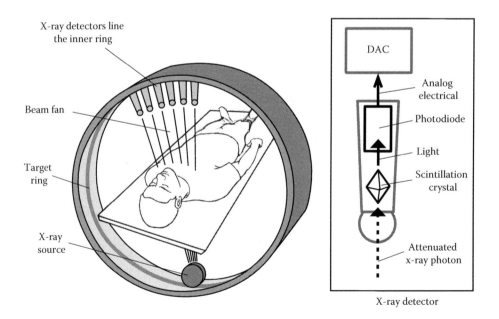

FIGURE 13.5 CT scanner and block diagram of the x-ray detector component embedded in CT scanner.

13.4 CT

CT images are obtained by penetration of multiple collimated x-ray beams through a cross-sectional slice of tissue at different angles during different time intervals. The most significant benefit of the CT image is the spatial resolution and anatomic detail in providing a 2D representation of the bone and allows for accurate localization of a lesion as well as determination of its extent. Conventional CT images present bone cortex as opaque high-density signals, muscles as homogenous intermediate-density signals, and fluid and gas as progressively darker low-density signals. The bone window setting of CT allows for more detailed images delineating cortical and cancellous bone architecture, allowing increased sensitivity for visualizing lesions within the bone such as bone cysts, tumors, infections, and fractures. The high-resolution cross-sectional images afforded by CT are particularly useful for visualization of lesions involving the small articulating bones of the wrist and foot, the tibial plateau, intervertebral and facet joints of the spinal column, and the major joints of the extremities, in which the anatomical orientation of bones may impede accurate visualization on plain radiographs. Slice thickness and intervals of the CT images can be modified to clarify or evaluate in detail suspected skeletal lesions. In contrast to the high-resolution images afforded for bone, soft tissue structures such as tendons and ligaments are less discernible on CT; thus, imaging of such is better accomplished with MRI or US studies.

13.4.1 CT Scanners

Basic CT scanners consist of a large gantry that houses the major electronic components and circuitry needed to operate the scanner. The gantry assembly is a large ring-shaped structure with a table, on which the patient lays (Figure 13.5). The table is capable of sliding into the ring's center. CT scanners function takes multiple cross-sectional x-ray images at different angles and different focal planes of a tissue and then renders a tomographic composition of the tissue. CT scanners typically consist of three main operational components: an x-ray source, an x-ray detector, and a DAC system (Figure 13.5).

13.4.2 Operation and Components

1. *X-ray source*: The x-ray source of most CT scanners consists of similar components to those in an x-ray machine, that is, x-ray tube and supporting circuitry to regulate x-ray production. The x-ray source is housed in the internal rim of the ring.

2. *X-ray detectors*: During operation, the internal rim rotates within the ring allowing the x-ray source to rotate around the patient, thus focusing x-rays at different angles around the patient. Contemporary CT scanners have an array of detectors that line the entire internal rim so that the rotating x-ray source is constantly being detected as the source rotates. The x-ray source rotates about a fixed axis around the patient at different angles and different focal planes to obtain a series of cross-sectional x-ray images. Furthermore, the characteristic rotational motion used in the CT scanner makes the images of the organ/tissue being scanned much clearer while blurring the surrounding area. This improves resolution and tissue differentiation when compared

to plain radiographic x-ray images (Bronzino 1995). Conventional CT scanners typically employ a solid-state scintillating crystal x-ray detectors identical to what is used in contemporary x-ray machines.

3. *DAC system*: The electrical signal generated from the detector is sent to the DAC where it is first amplified and then digitized via an analog-to-digital processor. The digital information is then processed by the DAC that utilizes a tomographic reconstruction software algorithm to generate 3D rendering from the multiple cross-sectional x-ray images collected.

13.5 RADIONUCLIDE BONE SCAN

Bone scans are performed by intravenous injection of a radioisotope tracer, technetium-99m (Tc-99m) diphosphonates, and images are obtained with a gamma camera. The Tc-99m-labeled diphosphonates are taken up rapidly by bones depending mainly upon the rate of new bone formation and blood flow. Approximately 2–4 h after injection, 50% of the isotope is accumulated in the skeletal system, while the tracers in the soft tissues have cleared out, and the resulting image is presented as a general outline of grossly uniform uptake patterns in the normal skeleton. Focal areas of normally increased uptake are areas of the sternomanubrial angle and medial clavicle, acromion and coracoid process of the scapula, and the sacral alae. In developing children, increased isotope uptake is seen symmetrically in the growth center of the long bones, as well as the marrow-containing facial bones.

Bone scans usually present areas of skeletal lesions as increased or *hot* uptake areas relative to the rest of the skeleton. These areas of hot uptake are very sensitive in detecting abnormalities, especially in the early or acute stage of a pathologic process, making it a very useful screening modality. For instance, bone scans can detect minor fractures within 24 h of occurrence, when plain radiographs are negative or equivocal. Transaxial and coronal plane images, in addition to anterior, posterior, and lateral images, can be obtained for view, and three different phases of scan (dynamic, blood pool, and bone) are also available to observe the anatomic distribution and dynamics of the pathologic process. Its usefulness has been proven in the diagnosis of fractures, inflammation, infections, metabolic bone diseases, tumors, and metastases. However, bone scans present only areas of abnormalities and cannot differentiate among the different disease processes, so interpretation of findings is based on the pattern of uptake as well as the associated clinical findings and temporal changes in subsequent bone scan images, if available.

Evaluation of bone scans should always begin by assessing for symmetry, and any asymmetry in uptake is noted. The size and distribution of hot uptake areas assist in the differential diagnosis. Metastatic lesions usually present as multifocal randomly distributed areas of hot uptake that are of variable shape, size, and intensity. In cases of spinal involvement, observing the uptake to extend from the vertebral body into the pedicle will determine it to be a metastatic lesion as opposed to a nonmetastatic uptake pattern, which is confined to the posterior elements. Multifocal rib fractures can be distinguished from metastatic lesions by their uniform size, shape, intensity, and usually linear pattern of increased uptake. The increased bone turnover rate in metabolic bone diseases results in areas of increased uptake depending on the specific disease process. In a general metabolic condition affecting the skeletal system, the uptake pattern is uniform in intensity and extends into the distal appendicular skeleton with disproportionally intense calvarial uptake; in cases of focal disorder of bone metabolism, the increased uptake will correspond to the focal area and shape of the affected bone. Triphasic bone scan is particularly useful in the diagnosis of osteomyelitis by the pattern of focal hyperperfusion, focal hyperemia, and focal increased bone uptake on the three sequential phases, respectively. Increased uptake in normally nonosseous areas can lead to the diagnosis of heterotopic ossification, the activity of which can be monitored as well with follow-up scans at timely intervals. On the other hand, patterns of decreased or *cold* uptake can also be observed in cases of acute bone ischemia, osteoporosis, or certain tumors and cystic lesions (Love *at al* 2003).

As bone scan demonstrates a high sensitivity in detecting abnormal lesions and their anatomic distribution, but low specificity in discriminating among distinct disease processes, it is considered most useful therefore in guiding the clinician in deciding the subsequent diagnostic work-up, including laboratory tests or further imaging studies.

13.5.1 Radionuclide Bone Scan Machines (Nuclear Medicine Scanner)

Nuclear medicine scanners function by detecting and then imaging radiation from small amounts of radioactive materials (e.g., radionuclide) that have been introduced into the body intravenously, by inhalation, or by ingestion. Depending on what structures are to be imaged, the radionuclide introduced into the body is first chemically attached to a biologically active molecule called a tracer. The tracer is what actually targets the particular tissue to be investigated. In radionuclide bone scans, Tc-99m radionuclide is chemically attached to a diphosphonate tracer molecule because the diphosphonate molecule is specifically absorbed by bone tissue (Bronzino 1995).

Basic nuclear medicine scanners are composed of a detector–photomultiplier module positioned over a patient on a table, an amplifier–analyzer module, and a DAC system.

13.5.2 Operation and Components

1. *Detector*: The detector detects radiation emitted from the patient due to the radionuclides that have been introduced into the body. The nuclear machine detector principle is identical to that of the CT scanner. The detector is made up of an array of scintillating crystals. Each scintillating crystal fluoresces when struck by an ionizing radioactive particle from the radionuclide. The light photons generated by the crystal array are directed into an array of coupled photomultiplier tubes (PMTs). Each crystal in the array is connected to a PMT (Figure 13.6). As in the CT scanner, the light intensity (degree of fluoresce) produced by a crystal is proportional to the energy of the incident radioactive particles (Carr and Brown 1981).

2. *PMT*: The PMT is a vacuum tube with a special cathode material (photocathode) on one end, an anode on the other end, and a series of electrodes called dynodes that are daisy-chained in the tube between the anode and cathode (Figure 13.6)

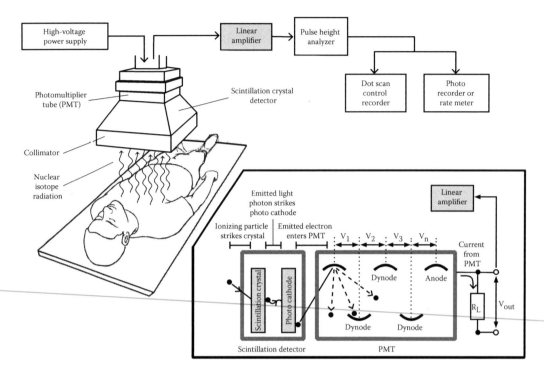

FIGURE 13.6   Block diagram and operation of bone scanner. (Insert): Block diagram and operation of scintillation crystal and PMT.

(Bronzino 1995). A high voltage is applied across the tube such that the anode is relatively positive to the photocathode. Each dynode in the chain is held at a higher voltage relative to the next such that the dynode closest to the anode is more positive than that closest to the photocathode. The photocathode is a cathode material that is capable of emitting electrons when struck with light photons. This phenomenon is called *photoelectric effect*. The dynode electrodes are coated with a special material that permits the emission of lower-energy electrons when struck by higher-energy electrons (Bronzino 1995). In simple terms, the main function of the photomultiplier is to intensify or multiply the photoelectric effect produced by the light signal.

As light photons exit the detector, they first strike the photocathode consequently producing electrons in the PMT. Given the potential differences in the tube, the generated electrons are accelerated toward the anode. Along the way, the electrons strike the dynodes placed in their path, producing lower-energy electrons, thus multiplying the initial number of electrons. Newly emitted low-energy electrons then gain energy as they are accelerated from one dynode to the next, also multiplying with each dynode collision until they reach the anode. When the electrons finally reach the anode, the avalanche of electrons due to the dynode collisions causes an accumulation charge at the anode resulting in a sharp current output that flows to an external resistor to produce a voltage output (Bronzino 1995).

3. *Linear amplifier, position, and pulse height analyzer*: The voltage output from the photomultiplier is amplified by a linear amplifier circuit (Figure 13.6). The output voltages generated by each PMT tube in the array are first amplified by a linear amplifier and then sent to a position and pulse height analyzer circuit (Bronzino 1995). This circuit extracts spatial information about where the fluorescence was produced within the scintillating crystal array in the detector, as well as the light intensity of each crystal's fluorescence in the array. The information about the light intensity of each crystal should be proportional to the output of the PMT it is coupled with. The information from the position and pulse height analyzers contains a spatio intensity map of the radionuclide location in the tissue.

4. *DAC system*: The electrical signal output from the analyzer circuits is sent to the DAC where it is digitized and displayed on a monitor to reveal an image showing a 2D spatio intensity distribution of the radionuclide location in the body. For radionuclide bone scans, an image of the bone is seen.

13.6 MRI

MRI uses exposure to magnetic fields and their effect on the hydrogen atom to produce high-resolution images of the musculoskeletal system. The excellent tissue contrast properties of MRI make it the modality of choice in visualizing pathologies of muscle, ligament, tendon, and cartilage, in addition to bone. MR images have the additional advantage of being able to present the anatomy in multiple planes (sagittal, transverse, and coronal), thus providing an optimal plane of view for specific joints and structures and 3D information of the shape and size of a lesion. Its detailed representation of the regional anatomy also allows for establishing a diagnosis in the acute stage of injury when hemorrhage and edema render a joint or limb unstable for physical manipulation.

MR images are categorized into three basic types: T1-, T2-, and proton-density-weighted images (T1WIs, T2WIs, and PDWIs), depending on the strength of the magnetic field, the imaging technique, and tissue characteristics such as the density of mobile protons. Basic principles in viewing the three weighted images are required for accurate interpretation. T1WIs present water as dark low signal intensities, while fat tissue is presented in bright high signal intensity; such properties make the T1WI useful for the detection of fat, and the signal intensity of fat can be set as a reference point for other lipid-containing tissues. T1WIs are most valuable in observing anatomic detail of musculoskeletal structures and, in particular, for the evaluation of meniscal and bone marrow abnormalities. T2WIs produce images in which fat is presented as low signal intensity, whereas water is presented in a bright high signal intensity, such as for cerebrospinal fluid in the spinal canal, and is thus valuable in the detection of edema, inflammation, infections, tumors, or cystic lesions. T2WIs are also useful in the evaluation of tendons, ligaments, and cartilage pathologies, as these tissues contrast well with surrounding bony and muscular structures. Bone cortex is typically of very low signal on both T1WI and T2WI, while the fatty marrow substance appears as low signal on T2WI, and muscles with their water content display relatively higher signal intensity compared to the fibrous ligaments, tendons, and cartilage. An intact intervertebral disc, composed of an outer fibrous annulus with an inner nucleus pulposus, is visualized on a sagittal plane T2WI as a high-signal center encapsulated by a low–signal intensity disc. Thus, a basic knowledge of the structural composition of musculoskeletal tissues is important in MRI evaluation. PDWIs display a contrast intensity, which is between the T1WI and T2WI, and are useful for visualizing anatomic detail and areas that can be obscured by the bright signals of T2WI. Contrast enhancement with gadolinium in T1WI with fat suppression is used in MRI to depict areas of increased vascularity as in acute inflammation or tumor, while direct percutaneous injection of the contrast agent into a joint (i.e., MR arthrogram) allows visualization of articular structures and aids in the diagnosis of joint pathologies (Farhoodi *et al* 2010).

Due to the three different weighted images and the three different planes of visualization available with MRI, a systematic approach to reviewing the images should be implemented for accurate interpretation:

1. Identify the MRI as either T1WI or T2WI: This is accomplished by identifying areas that normally contain fluid; based on its signal intensity, it is possible to determine the image as T2 weighted if the fluid is presented as a bright signal and T1 weighted when presented as a dark signal.

2. View the T2WI for areas of increased signal intensity: The images should be reviewed in sequential order in each of the planes of view, and note any areas of atypical signal.

3. Compare areas of abnormal signal intensity found on T2WI to the corresponding area on T1WI and note any differences in regional anatomy or signal intensities.

4. Compare with remaining PDWIs or contrast-enhanced images for additional information such as increased vascularity of the lesion, rim enhancement of a mass, or negative filling.

5. Combine the MRI observations with pertinent clinical findings to develop a differential diagnosis.

13.6.1 MRI Machines

Unlike CT scanners or plain radiograph x-rays, MRI machines do not use ionizing radiation. An MRI machine generates images of internal body structures by utilizing the natural magnetic characteristic of hydrogen atoms, which are abundant in body tissue since the body is composed mainly of water and lipid molecules. Since hydrogen protons have a spin and carry a charge, they exhibit a magnetic behavior and hence act like tiny biological magnets. As a result, the hydrogen protons have randomly oriented directions of magnetization, also known as magnetic moments. When placed in a much stronger external magnetic field, the magnetic moments of the hydrogen protons become aligned in the direction of the external field and continue to precess about the direction or axis of the external magnetic field at a constant frequency. This precessional frequency is called Larmor frequency and is determined by the strength of the external magnetic field.

While still exposed to the external magnetic field, subsequent exposure of the hydrogen protons to a radio-frequency (RF) pulse of identical Larmor frequency results in absorption of electromagnetic radiation energy by hydrogen protons, consequently causing the protons to increase their energy state and alter their magnetic moment alignment. The RF pulses also cause the protons to precess in phase with each other. If a conductor is placed in the vicinity, this change in net magnetic moment caused by the energy absorption induces a current in the conductor (Faraday's law of induction). When the RF exposure is turned off, the induced current rapidly decreases from the maximal value that was induced during RF exposure. Furthermore, energy is released as hydrogen protons return to their initial lower-energy states and precession phases, realigning their magnetic moments with the external magnetic field again. In addition to the induced current, the energy released by the transitional processes has a characteristic resonance frequency that reflects the magnetic characteristics of the hydrogen nuclei (Chatterjee and Miller 2010).

The ability of hydrogen nuclei, or any magnetic nuclei, to absorb electromagnetic radiation and release energy in the presence of a magnetic field and gated RF pulse is called nuclear MR (NMR). NMR is the major principle of operation of an MRI machine. An MRI machine uses a large circular magnet (i.e., external magnetic field source) in conjunction with an RF generator (i.e., electromagnetic radiation source) to induce NMR phenomenon from hydrogen nuclei within the tissue. The MRI machine then records the energy emitted from the hydrogen nuclei in the tissue due to NMR and uses the information to construct an image to visualize the tissue.

Basic MRIs are designed as hollow cylinders that completely surround and house a single patient lying on a bed in a fashion similar to a CT scanner (Figure 13.7). The major

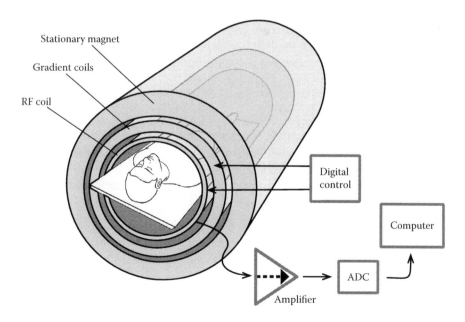

FIGURE 13.7   MRI machine with block diagram of operation of its main components.

components of an MRI include a field magnet, gradient coils, RF coils, and a DAC system (Figure 13.7). Most of the MRI's electronic components mentioned are housed in the body of the cylinder.

13.6.2 Operation and Components

1. *Field magnets*: The field magnet is usually circular in nature and is housed in the internal portion of the cylinder. The field magnet creates a strong but stable external magnetic field around the patient. The magnetic field acts on hydrogen nuclei in tissue, aligning the magnetic moments of the hydrogen nuclei in the direction of the external magnetic field.

2. *Gradient coils/magnets*: The gradient magnets are relatively smaller, less powerful magnets arranged just below the stronger main field magnet. MRIs typically use three gradient magnets, one each for the *x*, *y*, and *z* axis orientation. Simply turning on these gradient magnets will superimpose additional magnetic fields around the patient, in addition to that of the main magnetic field. But by controlling the rapid activation and termination of the gradient magnets, a net magnetic field with strength gradients can be created and precisely controlled across the length of the patient. Using the three axis-oriented gradient magnets, the magnetic field can be varied linearly in each plane across the specific region being imaged. This allows for spatial localization and encoding of the hydrogen protons in the targeted tissue during imaging.

3. *RF coils*: The RF coils are used to emit RF pulses within the cylinder of the MRI machine. The RF coils are also used to detect induced currents and energy transitions via electromagnetic energy processes described earlier. In other words, the RF coils act both as an RF transmitter and receiver. The RF coil emits radio waves directed at the patient while the external magnetic field is on, consequently causing the hydrogen nuclei within the tissue to absorb energy of the incident radiation and change magnetic alignment. Once the radio waves are turned off, the energized hydrogen nuclei slowly return back to their previous magnetic alignment within the magnetic field and reemit the radiation absorbed from the RF pulses (i.e., NMR effect). The receiver feature of the RF coils detects the electromagnetic energy inductions after the RF exposure is turned off.

4. *DAC system*: The signals detected in the RF coil are characteristic of the spatial distribution of hydrogen nuclei within a tissue. The hydrogen nuclei spatial distribution information is embedded in the frequency component of the RF signal. Furthermore, contained in the detected signal is information about the induced current and rate of decay of the current, the time taken for the hydrogen protons to return to the previous precession phase, and the time taken to return to the previous magnetic alignment. Once the signal is detected/received in the RF coil, it is sent to the DAC to render the MR image. The DAC first amplifies the signal and then digitizes it via an analog-to-digital processor. The DAC finally applies a reconstruction algorithm and Fourier transform to the digital information to generate an MR image.

13.7 REAL-TIME MUSCULOSKELETAL ULTRASOUND

US involves the transmission of high-frequency sound waves into tissues through a transducer, and the time for the echo to return is calculated to determine the depth of a structure from which the image is formed. Linear array transducer probes provide optimal images throughout the imaging field, and different transducer frequencies are used for tissues at different depths. Superficial structures such as muscle, tendons, ligaments, and small or upper extremity joints are well visualized with higher-frequency transducers (7.5–20 MHz), while lower frequencies (3.5–5 MHz) are used for deeper structures such as the hip joint. Each structure can be discriminated by its unique echogenic texture, which is determined by the degree of reflection of the US waves on the tissue surface. Echogenicity is described as hyperechoic (white signal), isoechoic, hypoechoic, and anechoic (black); bone surface appears hyperechoic and fluids appear hypo- or anechoic. The US of each structure varies depending on its tissue properties, and the image varies with transducer orientation in regard to the longitudinal axis of the structure. The clinician performing the US exam should be familiar with the normal echogenicity of musculoskeletal structures (Table 13.3) as well as their functional anatomy in interpreting significant findings (van Holsbeek and Intracaso, 2001; Hashefi, 2011).

The advantage of real-time US in diagnosing musculoskeletal conditions is in the fact that the limb or joint can be visualized in various positions including active and passive movements that trigger or aggravate symptoms, thus allowing for a functional assessment in addition to determining the anatomic derangement. The most common clinical application of musculoskeletal US is in the evaluation of joint functions and pathologies of tendon and muscle. The ligaments, joint capsule, the muscles and tendons that cross the joint, and the articulating bones, cartilaginous structures, and any bony attachments are viewed in various positions and during movement. Tears of ruptures in ligaments are presented as hypoechoic irregularities within the ligament or near their bony attachments, occasionally accompanied by hypo- or anechoic pockets of inflammatory fluid or

TABLE 13.3 Normal Echogenic Properties of Musculoskeletal Structures

Structure	Normal Echogenicity
Bone surface	Bright and hyperechoic with posterior shadowing
Hyaline cartilage	Anechoic
Fibrocartilage	Hyperechoic
Muscle	Hypoechoic parallel bundles surrounded by echogenic fibroadipose septa
Tendon	Fine hyperechoic fibrillar pattern running in parallel in longitudinal view and round to oval hyperechoic stippled pattern on transverse view; subject to anisotropy, the scattering of beams that are not perpendicular to the tendon surface
Ligament	Hyperechoic fibrillar pattern, sometimes running in different directions
Synovium	Medium echogenicity, usually seen around long tendons of the extremities surrounding an anechoic synovial space or within the joint capsule
Bursa	Hypo- or anechoic, situated between bone and muscles in joints and nonvisible in normal condition
Nerve	Similar to tendon, but less echogenic, less fibrillar, and no anisotropy
Connective tissue	Midechoic and irregular, similar to subcutaneous fat that is slightly less echogenic

hemorrhage. Abnormal friction of connective tissue structures against bone can be visualized during active or passive motion, while bony abnormalities indicating degenerative change can be observed in irregular cortical surfaces or discontinuities. Bursae, usually situated between muscles and bones to decrease joint friction, are nonvisible in normal conditions, but can be visualized as larger than normal well-delineated hypoechoic spaces in the setting of bursitis. The synovium that lines the bursal sac also lines the proximal and distal portions of long tendons of the extremities and when inflamed leads to an increase in the area of the hypoechoic rim surrounding the tendons. Inflammation of the tendon itself can manifest as loss of the typical fibrillar pattern with lower echogenicity. Tendon tears or ruptures can be evident by disruption of the normal continuity and loss or movement. Similarly, muscle trauma and ruptures can also be evaluated statically for structural integrity and dynamically by observing loss of movement and measurement of muscle retraction. Individual muscles can be identified by their anatomic location and orientation with respect to adjacent muscles, and the hyperechoic fascial sheath can be visualized as it forms the boundaries between muscles. Intramuscular hematomas present as a hypoechoic area with a regular border within the muscle substance in the acute stage and with time increase in echogenicity and decrease in size with irregular borders. Finally, peripheral nerves can be visualized along their course throughout the extremities, and US is particularly useful in observing nerves as they pass superficially between skin and bone or beneath ligaments, which increase the risk of compression or entrapment neuropathies. The US findings of peripheral nerve lesions can be correlated to electrodiagnostic (Edx) findings to localize the lesion in order to guide subsequent therapeutic interventions.

13.7.1 Real-Time Musculoskeletal Ultrasound Machine

13.7.1.1 Operation and Components

Real-time medical US utilizes sonar principles similar to how submarines send out high-frequency sound waves (US) and detect resultant echo reflections from different surfaces. Medical ultrasonic devices consist of a transducer that houses a piezoelectric material. The material's piezoelectric property enables the probe to serve both as an ultrasonic generator and an ultrasonic receiver. As an ultrasonic generator, the piezoelectric material converts electrical energy to mechanical vibrations that produce ultrasonic sound waves, and as a receiver, it converts mechanical vibrations caused by reflected ultrasonic echo vibrations back to electrical pulses.

During operation, the transducer is placed against the skin on the selected region to be imaged and sends ultrasonic pulses into the region. As the ultrasonic sound waves penetrate into the body, they strike different internal structures and are reflected as echoes back to the transducer (Figure 13.8). The reflected echoes return to the transducer at different latencies, directions, and intensities. These patterns of echo signatures inherently contain information about the distances, locations, and material consistencies of the tissues they reflected from. The signature patterns are measured, digitized, and translated by a computer as different tissues displayed on a monitor.

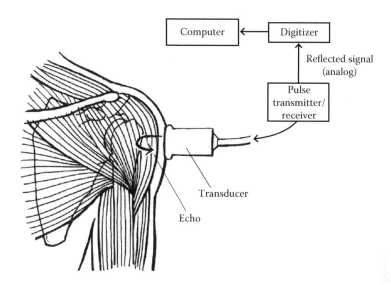

FIGURE 13.8   Simple block diagram of medical musculoskeletal US components and operation.

13.8 FUNCTIONAL STUDIES

13.8.1 Nerve Conduction Study and Electromyography

Edx studies are employed to evaluate peripheral nerve injuries accompanying traumatic skeletal injury and radiculopathies commonly encountered in orthopedics. The findings obtained from NCS and EMG can determine the location, severity, and chronicity of insult, as well as the viability and potential for regeneration of the pathologic nerve. The clinician conducting these studies must have a thorough knowledge of the functional anatomy and physiology of muscles and the peripheral nervous system. This section will focus on the clinical applications of NCS and EMG, and the technology behind the Edx systems is presented in detail in Chapter 10.

An NCS consists of applying a percutaneous electric current over a peripheral nerve and recording the evoked action potential distally from the tissue that a particular nerve innervates. In motor NCS, the compound muscle action potential (CMAP) is recorded from muscles and represents the summation of the motor unit responses; the sensory component of a nerve is assessed by obtaining a sensory nerve action potential (SNAP) that is recorded from a superficial nerve or a sensory distribution area of the stimulated nerve. Nerve conduction velocity (NCV) is calculated by dividing the distance between nerve stimulation and recording sites by the latency for sensory nerves; for motor NCV, the action potential transmission time across the neuromuscular junction needs to be eliminated; thus, a second stimulation is given proximal to the initial distal stimulation, and the distance between the proximal and distal stimulation sites is divided by the latency difference between the CMAPs obtained by the two stimulations for NCV calculation. The amplitude of the action potentials reflects the number of axons in the stimulated nerve, while the conduction velocity reflects the integrity of the myelin sheath covering of the axons. Stimulation of a nerve on multiple sites along its course and observing changes in action potential amplitude and conduction velocity are useful in localizing the lesion, which is commonly performed in detecting peroneal nerve compression at the fibular head and ulnar nerve compression around the medial epicondyle. EMG is performed by recording electrical potentials from skeletal muscles and complements NCS to form a comprehensive evaluation of peripheral nerve functions. Usually, a needle electrode is inserted into a number of muscles to record membrane potentials at rest and motor unit action potentials (MUAPs) during volitional contraction. The size, shape, duration, and recruitment patterns of MUAPs allow differentiation between neurogenic and myogenic etiologies of weakness, as well as determine chronicity of injury and nerve regeneration. The distribution of muscles with abnormal findings on EMG enables the clinician to differentiate the cause of motor weakness, from isolated or multiple nerve injuries to radiculopathy to generalized neuropathy to myopathy. It is also helpful in localizing the lesion within a nerve with a long anatomic course such as the radial nerve or to specific sections of a major nerve plexus or isolated spinal root segments. For instance, if wrist extension weakness resulted following a distal humerus comminuted fracture, and EMG revealed normal findings in the triceps, anconeus, brachioradialis, and extensor carpi radialis longus muscle but abnormal findings in the extensor carpi radialis brevis and all other radial innervated muscles distal to it, the clinician could conclude the nerve lesion to be at the radial nerve following branching to the extensor carpi radialis longus. Electrodiagnosis is also invaluable in differentiating between pathologies with similar clinical presentations. When presented with finger abduction weakness and sensory change in the medial aspect of the upper extremity, combined NCS and EMG findings allow differentiation between ulnar neuropathy, lower trunk or medial cord lesion of the brachial plexus, and C8 radiculopathy.

The timing of performing Edx studies is important in determining the severity and type (axonal vs. demyelinative) of nerve injury as well as prognosticating its outcome (Dumitru *et al* 1999; Quan and Bird 1999). When an axon is transected, a degeneration process (Wallerian degeneration) occurs in the distal fragment of the axon over several days, during which time the distal nervous tissue remains electrically active. Thus, if the nerve is stimulated distal to the site of injury within the first several days of injury, a nearly normal action potential will be elicited, while nerve stimulation proximal to the injury cannot evoke a response as the axonal connection is disrupted. However, this finding is very similar to nerve conduction findings of conduction block, in which action potential propagation is interrupted due to focal demyelination without disruption of the axon itself. Hence, axonal injury and demyelination cannot be differentiated by NCS within several days of nerve injury; motor nerves remain excitable up to 7 days, while sensory nerves retain excitability for 10 days postinjury. After about 10 days, during which Wallerian degeneration has occurred, nerve stimulation distal to the injury site will result in an action potential of low amplitude or absent response. The same principle holds true for EMG, but the time for pathologic findings to be observable is about 14–20 days postinjury depending on the length of the involved nerve axon. Thus, the optimal time for initial NCS is between 10 and 14 days postinjury, to determine presence of axonopathy, and at least 18–20 days for EMG findings to reveal muscle manifestations of neuropathy. Follow-up examinations can be scheduled every 3–6 months to detect electrophysiological signs of nerve regeneration or reinnervation of distal muscles.

13.8.2 Electrodiagnostic Instrumentation

13.8.2.1 EMG Instrumentation

During gait, the *on/off* activity patterns of skeletal muscles across joints are detected and acquired using EMG instrumentation devices. The EMG recording devices primarily consist of electrode wires, differential amplifiers, and a DAC system.

The purpose of the metallic electrode wire is to detect small voltage potentials (millivolts) that are generated during muscular contractions. To acquire the voltage signals across a muscle, two surface electrodes are placed on the surface of the skin some distance apart where the muscle is located (Figure 13.9). The electrical activity generated by the muscular contraction is then picked up as a potential difference between the two electrode wires. Surface electrodes are usually preferred for superficial muscles. For deeper muscles, fine wire electrodes called intramuscular electrodes are inserted directly into the muscles.

A differential amplifier is then used to amplify the strength of the voltage signal patterns picked up by the electrode pair (Figure 13.9). The distal terminals of the electrode pair from a muscle are connected to a differential amplifier. Each differential amplifier functions by amplifying the potential differences between each electrode pair. In addition, differential amplifiers have the added advantage of rejecting any common noise in the voltage signal being amplified.

The amplified analog voltage signal from the amplifier is sent to the DAC where it is digitized via an analog-to-digital processor. The digital information is then processed by the DAC software, which then renders the amplitude variations in electrical activity due to muscular contractions as digital temporal traces on a computer monitor.

13.8.2.2 NCS Instrumentation

NCS instrumentation generally consists of recording electrodes, stimulating electrodes, differential amplifiers, and a DAC system.

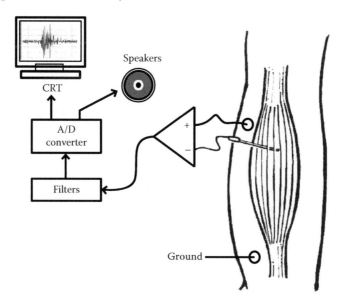

FIGURE 13.9   Block diagram of EMG instrumentation.

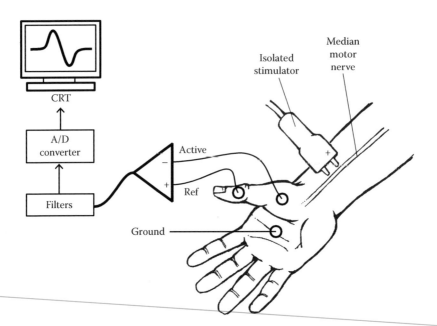

FIGURE 13.10   Block diagram of nerve conduction instrumentation.

Stimulating electrodes are a pair of electrodes, an anode electrode and a cathode electrode, used to stimulate the nerve being studied (Figure 13.10). The cathode electrode applies negative electrical impulses, while the anode applies positive impulses. The nerve bundle is thus stimulated by the application of voltage between the two electrode terminals. This electrical stimulation produces an action potential in the nerve bundle, which then travels along the sensory and motor nerves in the nerve bundle. The effect of the stimulation is then recorded as a CMAP as a result of motor nerve conducting the stimulation to the muscle motor end plate, or as a SNAP at the distal end of the sensory nerve, as a result of the sensory nerve conduction.

Recording electrodes are a pair of metallic electrodes that picks up the sensory nerve action potential (SNAP) or compound muscle action potential (CMAP). For recording a SNAP, the recording electrodes are placed on the skin surface innervated by the sensory nerve a set distance from the stimulating electrodes; for CMAP, the recording electrodes are placed on the surface of the skin directly over the motor end plate of a muscle that is directly innervated by the nerve being studied or stimulated Figure 13.10. When the nerve bundle being studied is electrically stimulated, an action potential is generated and travels along the nerve to the site of the recording electrode pair. Consequently, as the action potential travels between the two electrodes of the recording electrode, it is picked up as a voltage potential difference between the two electrodes.

A differential amplifier is then used to amplify the strength of the voltage signal's patterns picked up by the recording electrodes. The amplified analog voltage signal from the amplifier is sent to the DAC where it is digitized and processed, eventually displaying SNAP or CMAP electrical activity as digital temporal traces on a computer monitor.

13.9 GAIT/MOTION ANALYSIS

Human motion analysis including gait analysis can be carried out on many levels. The unaided eye of a clinician, for instance, is used to evaluate a patient's gait or walking ability in a simple hallway. To be effective, such an analysis requires extensive training and experience. Clinicians with an appropriate background use a structured approach to focus on specific joints and body segments as they analyze the motion patterns. A problem with this method is that the observer is not able to observe simultaneous motions or a multilevel motion that occurs simultaneously in the different anatomical planes. While the majority of motion in gait occurs in the sagittal plane, coronal and transverse plane motion can significantly affect the sagittal pattern. These problems are compounded when attempting an analysis of patients with gait pathologies and patients at the extreme of the age spectrum. Antalgia and fatigue are other limiting factors for many of these patients, preventing extended periods of walking and aggravating existing gait deviations.

Typical gait analysis classically refers to the orientation of the pelvis and distal extremities. The pelvis is usually described with reference to the global or laboratory reference system, while the distal segments are referenced to the next proximal segment, for example, knee motion as tibia with respect to the femur.

Gait analysis is employed widely as a treatment planning tool in pediatric cerebral palsy (Cook *et al* 2003, Etnyre *et al* 1993, Gage *et al* 1984, Gage and Novacheck 2001). Joint motion and loading patterns are usually combined with measures from a comprehensive physical exam (ranges of motion, strength, spasticity, selective motor control), as well as EMG and temporal–spatial measures (walking speed, stride length, cadence).

A number of treatment interventions rely on gait analysis to determine the course of treatment and subsequent follow-ups. As an example, the rectus transfer that involves transferring the distal rectus femoris to the posterior portion of the knee is indicated with limitations in swing phase knee flexion and activity of the rectus during the swing phase of gait (Carney *et al* 2006, Gage *et al* 1987, Perry 1987). Nonoperative interventions can also be directed through gait analysis results, such as botulinum toxin-A (Botox) injection to spastic muscles. Gait analysis can be used to target specific muscles and disregard spastic muscles with a lesser effect.

13.9.1 Clinical Gait Instrumentation

The quantitative approach to gait assessment requires tracking the displacement of body segments as a patient walks naturally along a walkway. Combining these data with limb muscle activity (EMG) and ground reaction force data (force plates) associated with gait generates a wealth of information about the gait biomechanics. These data include a tri-axial description of the joint motion (kinematics) and internal reaction forces (kinetics). The analysis of these data provides the clinician with detailed information about spatial and temporal joint patterns of motion, joint reaction forces, and muscle firing patterns. An analysis of these data facilitates the selection of appropriate therapeutic and/or surgical interventions and follow-up care (Harris and Smith 1996).

13.9.2 Video System

Today, the conventional method for quantifying clinical gait involves the use of a *video system* to record and monitor external markers placed on specific bony landmarks on the body segments being observed. The video system typically consists of two or more cameras that capture a video at very fast frame rates (i.e., high-speed cameras) and a CPU computer. The markers used can be either retroreflective markers (passive markers) or LED markers (active markers). The light reflective or fluorescent nature of the markers creates a significant contrast between each marker and its nonreflective surroundings. This contrast makes it possible for the high-speed cameras to distinctly capture and identify the location of each marker in each video frame as the body segments move through space during gait. The cameras are positioned systematically around the patient's workspace such that at least two cameras will have a view of each marker from different perspectives (Figure 13.11). Each video frame recorded by each camera contains a 2D image of the markers (Harris and Smith 1996). The video system's CPU then uses mathematical coordinate transformation algorithms to triangulate the relative 3D coordinate of a particular marker based on the 2D images of the marker from multiple cameras. Accordingly, CPU frame-by-frame analysis allows the video system to keep track of the 3D position of the markers relative to a fixed world reference frame, thus tracking the motion of the segments during gait. Furthermore, each body segment is typically defined by a minimum of three markers that allow the video system to use a biomechanical stereometric technique to represent body segments as solid 3D objects moving in space.

Relative movements between body segments can easily be decoded by the CPU to generate joint angular trajectories. For instance, the relative movement between the markers that define the thigh segment and markers that define the shank segment is interpreted by the CPU as angular changes in the knee joint (Harris and Smith 1996). Similarly, by

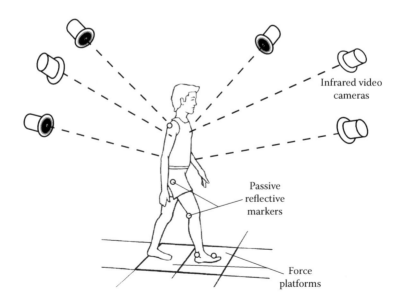

FIGURE 13.11   Sample gait lab instrumentations and layout.

tracking and analyzing the movement of the foot segment markers, the CPU can easily obtain temporal–spatial parameters such as stride duration, stance duration, foot trajectory, and stride length.

13.9.3 Force Platform

Force platforms are complex transducers that measure the ground reaction forces exerted between the foot and the ground as the patient steps on it during a natural gait. The force platforms are typically embedded mounted level with the floor along the walkway on which the patient walks (Figure 13.11). The interaction between the patient and the force plate(s) also provides information about the location of the center of pressure (COP) of the foot during gait, as well as the vertical and shear force vector components of the ground reaction force during gait. These data ultimately reveal how the force or load of the patient is distributed across the foot's plantar surface during the ground contact phase of gait (Harris and Smith 1996). Due to the multisegmented nature of the body, these force vectors generated by foot contact also produce torques (moments) across joints between limb segments. Therefore, analyzing the loading effect on the patient with the force plates allows the clinician to identify potentially harmful torque and load effects across the joints (Harris and Smith 1996).

Furthermore, to maintain posture and balance during gait, the joint torques produced by these load interactions are counteracted by equivalent torque forces generated by skeletal muscles across the joints. Therefore, using a force platform to determine the torque across a joint also quantifies the muscular force generated by the muscles spanning that joint.

Force platforms typically utilize a piezoelectric or strain gage sensor to convert any load exertion to electrical output. Each force platform is designed as a rectangular plate with sensors placed under the plate, one at each corner of the plate, such that any load applied on the top of the plate is transmitted to all four sensors. Piezoelectric force sensors utilize a piezoelectric material that generates an electrical signal when mechanical stress is applied by an external force. Strain gage sensors utilize strain gages. Applying force to a strain gage produces a strain on a resistive material within a Wheatstone bridge circuit, thus changing its resistance and the subsequent electrical output of the circuit.

REFERENCES

Bronzino J D 1995 *The Biomedical Engineering Handbook* (Boca Raton: CRC Press; IEEE Press)

Carney B T, Oeffinger D and Gove N K 2006 Sagittal knee kinematics after rectus femoris transfer without hamstring lengthening *J. Pediatr. Orthop.* **26** 265–7

Carr J J and Brown J M 1981 *Introduction to Biomedical Equipment Technology* (New York: Wiley)

Chatterjee S and Miller A 2010 *Biomedical Instrumentation Systems* (Clifton Park: Delmar Cengage Learning)

Clarke B 2008 Normal bone anatomy and physiology *Clin. J. Am. Soc. Nephrol.* **3** S131–9

Cook R E, Schneider I, Hazlewood M E, Hillman S J and Robb J E 2003 Gait analysis alters decision-making in cerebral palsy *J. Pediatr. Orthop.* **23** 292–5

Cyriax J H and Cyriax P J 2003 *Cyriax's Illustrated Manual of Orthopedic Medicine*, 2nd edition (Edinburgh: Butterworth Heinemann)

Dumitru D, Amato A A and Zwartz M 2001 *Electrodiagnostic Medicine*, 2nd edition (Philadelphia: Hanley Belfus)

Etnyre B, Chambers C S, Scarborough N H and Cain T E 1993 Preoperative and postoperative assessment of surgical intervention for equinus gait in children with cerebral palsy *J. Pediatr. Orthop.* **13** 24–31

Farjoodi P, Addisu M, Carrino J A and Khanna A J 2010 Magnetic resonance imaging of the musculoskeletal system *J. Bone Joint Surg. Am.* **92** S2 105–16

Gage J R, Fabian D, Hicks R and Tashman S 1984 Pre- and postoperative gait analysis in patients with spastic diplegia: a preliminary report *J. Pediatr. Orthop.* **4** 715–25

Gage J R and Novacheck T F 2001 An update on the treatment of gait problems in cerebral palsy *J. Pediatr. Orthop. Part B* **10** 265–74

Gage J R, Perry J, Hicks R R, Koop S and Werntz J R 1987 Rectus femoris transfer to improve knee function of children with cerebral palsy *Dev. Med. Child Neurol.* **29** 159–66

Gardner E D 1963 Physiology of joints *J. Bone Joint Surg. Am.* **45** 1061–6

Greenspan A 2010 *Orthopedic Imaging: A Practical Approach*, 5th edition (Philadelphia: Lippincott Williams & Wilkins)

Harris GF and Smith PA 1996 Human motion analysis: current applications and future directions (New York: IEEE Press).

Hashefi M 2011 Ultrasound in the diagnosis of noninflammatory musculoskeletal conditions *Semin. Ultrasound CT MRI* **32** 74–90

John J 1984 Grading of muscle power: comparison of MRC and analogue scales by physiotherapists. Medical Research Council *Int. J. Rehabil. Res.* **7** 73–181

Love C L, Din A S, Tomas M B, Kalapparambth T P and Palestro C J 2003 Radionuclide bone imaging: an illustrative review *Radiographics* **23** 341–58

Magee D J 1997 *Orthopedic Physical Assessment*, 3rd edition (Philadelphia: W.B. Saunders)

Morrisy R T and Weinstein S L 2006 *Pediatric Orthopedics*, 6th edition (Philadelphia: Lippincott Williams & Wilkins)

Perry J 1987 Distal rectus femoris transfer *Dev. Med. Child Neurol.* **29** 153–8

Quan D and Bird S J 1999 Nerve conduction studies and electromyography in the evaluation of peripheral nerve injuries *Univ. Pennsylvania Orthop. J.* **12** 45–51

Reiff K J 1997 Flat panel detectors—closing the (digital) gap in chest and skeletal radiology *Eur. J. Radiol.* **31** 125–31

Sabharwal S 2009 Blount disease *J. Bone Joint Surg. Am.* **91** 1758–76

Salenius P and Vannka E 1975 The development of the tibiofemoral angle in children *J. Bone Joint Surg. Am.* **57** 259–61

van Holsbeeck M T and Introcaso J H 2001 *Musculoskeletal Ultrasound*, 2nd edition (St. Louis: Mosby)

CHAPTER 14

Physiologic Measures in Otology, Neurotology, and Audiology

Il Joon Moon, E. Skoe, and Jay T. Rubinstein

CONTENTS

14.1	Routine Methods	318
14.2	ABR	319
	14.2.1 Technique	320
	14.2.2 Clinical Application	321
14.3	Stacked ABR	321
	14.3.1 Technique	321
	14.3.2 Clinical Application	322
14.4	Chirp ABR	323
14.5	Automated ABR	323
	14.5.1 Technique	323
	14.5.2 Clinical Application	324
14.6	Electrically Evoked ABR	324
	14.6.1 Clinical Applications	325
14.7	ABR to Complex Sounds (cABR)	325
	14.7.1 Technique	326
	14.7.2 Clinical Application	326
14.8	Auditory Steady-State Response	326
	14.8.1 Technique	326
	14.8.2 Clinical Application	327
14.9	Electrocochleography	328
	14.9.1 Technique	328
	14.9.2 Clinical Application	329
14.10	Intraoperative Monitoring	330
14.11	OAE	331
14.12	Spontaneous OAEs	332

14.13 Transient-Evoked OAE   332
14.14 Distortion Product OAEs   333
     14.14.1 Clinical Application   333
14.15 Cochlear Implant Telemetry   333
     14.15.1 Cochlear Implant Electrode Impedance Measurement   333
14.16 Electrically Evoked Compound Action Potential   334
     14.16.1 Technique   335
     14.16.2 Clinical Application   335
14.17 Vestibular-Evoked Myogenic Potential   336
14.18 Cervical VEMP   336
     14.18.1 Technique   336
14.19 Ocular VEMP   337
     14.19.1 Technique   338
     14.19.2 Clinical Applications   339
14.20 Electroneurography   339
     14.20.1 Technique   339
     14.20.2 Clinical Application   339
14.21 Nonroutine Methods   340
     14.21.1 Subcortical/Cortical Auditory ERPs   340
     14.21.2 Exogenous ERPs   342
     14.21.3 P1 (P100) and P1–N1–P2 Complex   342
         14.21.3.1 Clinical Application   342
14.22 Endogenous (Cognitive) ERPs   343
     14.22.1 Acoustic Change Complex   343
         14.22.1.1 Clinical Application   343
     14.22.2 Cognitive Potential (P300)   344
         14.22.2.1 Technique   345
     14.22.3 Mismatch Negativity   345
         14.22.3.1 Technique   346
         14.22.3.2 Clinical Application   347
     14.22.4 N400   347
         14.22.4.1 Technique   347
         14.22.4.2 Clinical Application   347
14.23 Conclusion   347
References   348

14.1 ROUTINE METHODS

Auditory evoked responses have been used as an important tool in audiological and neurotological diagnosis. Technical developments and improved procedures in this field have increased reliability. This chapter ranges from short-latency response, such as electrocochleography (ECoG), to the long-latency responses, such as P300 and N400, and divides them into routine and nonroutine use methods. Although generators, time epochs, stimulus, and response dependencies are different among these responses, there are common principles.

They use similar acoustic stimuli and amplification as well as signal averaging techniques to detect meaningful signals from noisy background electroencephalographic activity.

14.2 ABR

One of the most routinely used methods in audiology and neurotology is the auditory brain stem response (ABR). The ABR is an averaged, scalp recording of neural activities, which is composed of a series of voltage deflections within the first 15 ms after the onset of a transient auditory stimulus. It reflects a highly synchronous activation of the auditory pathway from the cochlea to the brainstem (Deupree and Jewett 1988). The ABR has a well-established utility in otology and neurotology since its introduction to clinical medicine in the 1970s. The primary measures of the ABR are the latency and amplitude of its peaks, a series of six waves. The generators of the ABR peaks span the cochlear (eighth) nerve, pons, and midbrain: wave I, ipsilateral distal eighth nerve; wave II, ipsilateral proximal eighth nerve; wave III, ipsilateral cochlear nucleus/superior olivary complex; wave IV, bilateral multiple brainstem origins; and wave V, contralateral distal lateral lemniscus/inferior colliculus (Stone *et al* 2009) (Figure 14.1). The most

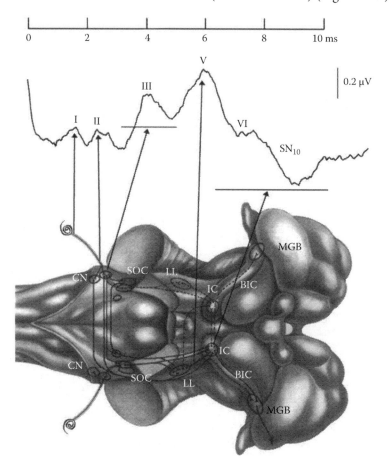

FIGURE 14.1   Schematic summary of the anatomical location of the neural generators of the ABR. (From Bahmer, A. et al., *J. Neurosci. Methods*, 173, 306, 2008.)

prominent positive peak, and the one with the greatest clinical utility, is wave V with its characteristic following negativity, VI (Hughes and Fino 1985).

The latency of a component peak in the ABR is widely used in clinical applications because the peaks are robust and reliable. The amplitude is typically measured from the peak of the component to its succeeding trough, and it is determined by not only the amount of neural activity generated but also the degree of neural synchronization.

14.2.1 Technique

The ABR can be generated with a variety of stimuli. However, conventional ABR recordings use transient stimuli, such as click sound or tone bursts. A variety of commercial systems (e.g., Navigator Pro [Bio-logic, IL], SmartEP [Intelligent Hearing Systems, FL]) have been used to generate click sounds or tone bursts. Click sound is produced by exciting a transducer with a brief-duration electrical pulse (around 100 μs) (Katz 2009). For tone bursts, the American National Standards Institute (ANSI) recommends that a tone used for audiometric purpose must be presented for a duration of not less than 200 ms and have a rise time and fall time ranging between 25 and 100 ms. ABRs to tone bursts may yield better predictions of peripheral sensitivity than the click-evoked responses (Stapells *et al* 1995). At moderate intensity levels, the abrupt onset of the click activates a large portion, but not the entirety, of the cochlea. The typical stimulus paradigm is a moderately high-intensity transient stimulus, at rates of roughly 8–24 s^{-1}. Generally, two replicable trials are needed from each ear. To obtain ABRs, at least three and sometimes four scalp (surface) electrodes including a ground electrode are needed. Needle electrodes, which are placed under the skin, can also be used, but this is more invasive than surface electrodes. Various types of surface electrodes (disposable or reusable) are commercially available. The surface electrodes are usually from 5 to 10 mm in diameter. A ground electrode is often placed on the forehead, at least 3 cm apart from other electrodes.

Auditory evoked potential amplitude is typically much smaller than that of the background noise. When obtaining ABRs, this noise is minimized by the following methods: (1) ensuring low electrode impedance (EI) (i.e., less than 5 kΩ); (2) reducing noise from muscle activities by making the subject comfortable or using sedation; (3) differential amplification, such as using three electrode leads per recording channel; (4) filtering the output of the bioamplifier; (5) signal averaging; and (6) artifact rejection (Katz 2002). For differential amplification, three electrodes are usually used that are referred to as the noninverting (positive), the inverting (negative), and the common (ground) leads. The voltages recorded in the noninverting and inverting leads are relative to the common lead. The difference (voltage) from the inverting common channel is subtracted from the difference (voltage) of the noninverting common channel, and only the remainder is amplified. By differential amplification, common mode noise, which appears similarly at both the noninverting and inverting electrodes, can be reduced. Filtering eliminates noise that is outside the frequency range of the desired response. For click-evoked ABR, the response is in the 100–3000 Hz frequency range. So, selectively eliminating electrical activity below 100 Hz and above 3000 Hz will reduce the background noise. About 2000–4000 repetitions are averaged within a 10–15 ms recording window in ABR measurement. In addition, to minimize artifact, single data trials are rejected when they contain out-of-bounds potential values at single electrodes. For example, if the electroencephalography (EEG) noise reaches a certain amplitude, for example, 20–40 µV, then the entire sweep is ignored and is not added to the accumulating average. Low artifact rejection level causes a high rejection rate and requires additional test time.

14.2.2 Clinical Application

ABR has been used for detecting retrocochlear pathology because the peak latencies can be affected by nearly all cochlear and retrocochlear processes. Retrocochlear lesions affecting the auditory pathway often show latency delay in ABR measures. For example, retrocochlear schwannoma exerts pressure on the cochlear nerve, influencing the transmission time of neural activity. This change leads to poorer neural synchronization and delays in neural activation (Stephens and Thornton 1976, Zappia *et al* 1997, Shih *et al* 2009). However, more recently, newer methods, such as stacked ABR, have been proposed due to the lack of sensitivity and specificity of standard ABR (see below Sections 14.3 and 14.4).

The need for objective measurement of hearing has grown dramatically with the advent of newborn hearing screening programs. The click- or transient-evoked ABR is the most widely used electrophysiologic procedure for estimating auditory sensitivity in infants and children. Beyond newborn hearing screening programs, there is a general need for objective hearing measurement. Patients with severe cognitive and/or motor deficits and persons who falsify their hearing loss cannot be tested with a standard pure-tone audiogram. ABR has several advantages in predicting hearing threshold, including that it does not require active cooperation, task comprehension, or appropriate behavioral responses. However, it should be clear that the ABR does not measure hearing: ABRs reflect sound-initiated neural activity, while hearing is the conscious perception of sound. However, ABR detection threshold is close to psychophysical threshold, and *threshold* ABRs have been shown to be reliable predictors of behavioral threshold within certain constraints (Brantberg *et al* 1999).

14.3 STACKED ABR

To successfully detect small acoustic tumors, measuring neural activity from the entire cochlea is mandatory. However, conventional ABR is limited by being a measurement of neural activity mainly from the high-frequency cochlear regions. Small tumors that do not affect these high-frequency auditory nerve fibers cannot be detected using conventional ABR. Hence, the value of conventional ABR for acoustic tumor screening has been questioned. To overcome this limitation, stacked ABR was developed (Don *et al* 1997, 2005). By compensating for the time delay in the cochlea, increased temporal synchrony can be obtained, resulting in a larger compound neural response. Time delay compensation can be achieved in two different ways: either by input compensation using a chirp stimulus (Bell *et al* 2002) or by output compensation via the stacked ABR (Elberling and Don 2010).

14.3.1 Technique

The stacked ABR, using a derived-band technique, measures the activity of essentially all the eighth nerve fibers, not just a subset. The derived-band ABR technique consists of

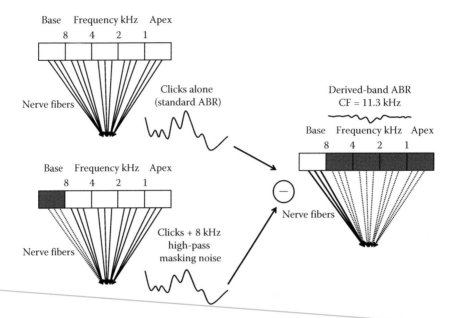

FIGURE 14.2   Illustration of the derived-band response: the response to the clicks + 8 kHz high-pass masking noise is subtracted from the response to clicks alone to form the derived-band ABR with a CF = 11.3 kHz. (From Don, M. et al., *Audiol. Neurootol.*, 10, 274, 2005.)

recording ABRs first to broadband clicks presented alone. Then, click stimuli are delivered in the presence of high-pass masking pink noise with varying cutoff frequencies. The cutoff frequency of the high-pass noise is lowered from one run to the next in octave steps from 8 kHz to 500 Hz. This process masks progressively lower-frequency areas of the cochlea. Subtracting the response for one run from the previous one forms a derived-band response. By repeating this response subtraction process for different center frequencies, a series of derived-band ABRs can be obtained (Figure 14.2).

Figure 14.3 displays how the stacked ABR is derived from the derived-band ABRs. The stacked ABR is obtained by time-shifting the derived-band waveforms to different center frequencies so the peak latencies of wave V in each derived band coincide and adding then these shifted derived-band ABR waveforms. Theoretically, by temporally aligning the peak activity from each segment of the cochlea, the synchronized total activity of the cochlea can be measured.

14.3.2 Clinical Application

The stacked ABR has high sensitivity for small acoustic tumors as compared with standard ABR (Don *et al* 2005). Although magnetic resonance imaging with gadolinium enhancement is the most sensitive diagnostic technique for detecting acoustic tumors, it is still a costly screening tool. Therefore, the stacked ABR is a sensitive, cost-effective, and noninvasive tool for screening small acoustic tumors, which may not be detected by standard ABR measures. Enhanced magnetic resonance though is still the standard of care for younger patients where detection of smaller tumors is mandatory.

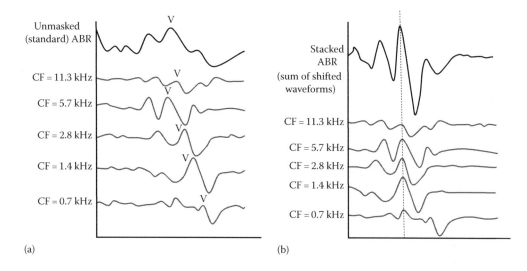

FIGURE 14.3 The stacking method. (a) The standard ABR to clicks (top trace) and the five derived-band ABRs. (b) Construction of the stacked ABR (top trace) from the derived-band ABRs by (1) temporally aligning the derived-band ABR waveforms to align the peak latencies of wave V in each band and (2) adding these aligned derived-band ABR waveforms. (From Don, M. et al., *Audiol. Neurootol.*, 10, 274, 2005.)

14.4 CHIRP ABR

The chirp stimulus is designed to compensate for the cochlear traveling wave delay and increase the temporal synchrony between the neural elements, in order to give rise to larger compound neural responses than conventional ABR (Elberling and Don 2010). In response to a brief stimulus, such as a click, the cochlear traveling wave takes some time to reach from the base to the apex of the cochlea (cochlear traveling wave delay). Therefore, with a click stimulus, the different neural units along the cochlea cannot be stimulated at the same time. The chirp stimuli can increase temporal synchrony more than brief stimuli, in which higher-frequency components are delayed relative to lower-frequency components. Chirps are designed using models of the cochlear–neural delay, the time delay between an acoustic stimulus at the tympanic membrane (TM), and the neural activity in the auditory nerve (Fobel and Dau 2004, Elberling *et al* 2007, Elberling and Don 2008).

14.5 AUTOMATED ABR

Conventional ABR has long been known to be the most sensitive method of assessing the auditory function of newborns. However, it is relatively costly and more time consuming for initial screening (Barsky-Firkser and Sun 1997). An automated ABR (AABR) method has been developed for newborn hearing screening to overcome these limitations.

14.5.1 Technique

Natus Medical Incorporated (San Carlos, CA) was the first company to offer a commercially available AABR hearing screener, the ALGO®. A central microprocessor, an EEG system, a stimulus generating system, ambient noise and myogenic activity detection systems, and a template-matching detection algorithm are integrated into a handheld device. Three disposable electrodes are placed on the patient's forehead (noninverting), nape of the neck (inverting), and shoulder (ground) to obtain the ABR. The device delivers 37 clicks/s sequentially to both ears at an intensity level of 35 dB nHL, applying a band-pass frequency range between 700 and 5000 Hz. Responses are measured in an automated way from each ear separately. ALGO's software algorithm compares the measured responses with the grand average obtained in 35 normally hearing children (Erenberg 1999). By statistical comparison, the likelihood ratio (LR) is calculated resulting in a *pass* or *refer* response, which is easily interpreted by an untrained professional. In every 500 sweeps of clicks, the ALGO® updates the LR, and testing automatically terminates when an LR of greater than or equal to 160 is reached, indicating a *pass*; if after 15,000 sweeps the LR is under 160, this indicates a *refer*. A *pass* is indicative of normal hearing with at least 99% certainty. A *refer* indicates that the collected data could not be discriminated from no response.

14.5.2 Clinical Application

AABR has been widely used as a primary screening tool for newborn hearing loss. Since the goal of newborn hearing screening is to identify infants with significant hearing impairment, a low false-positive rate is critical for developing reliable screening programs. AABR gives a lower false-positive rate than other objective methods (Benito-Orejas *et al* 2008). In addition, AABR has several advantages over other screening methods including (1) no requirement of a trained audiologist; (2) high specificity, meaning that no infant who passes the AABR screen has significant hearing loss at the time of the screen; (3) artifact rejection system for ambient noise and myogenic activity; (4) robustness to middle ear fluid or debris in the ear canal; (5) short testing time; (6) noninvasiveness; and (7) portability (Erenberg 1999). However, AABR also has significant limitations in that it is not sensitive to low-frequency or very high-frequency hearing loss and that it must be used in conjunction with other objective methods to identify auditory neuropathy. However, concerns remain regarding accurate diagnosis with AABR (Guastini *et al* 2010).

14.6 ELECTRICALLY EVOKED ABR

The electrically evoked ABR (EABR) is an objective measurement for cochlear implant (CI) users. In most cases, EABR can be recorded from CI users without the need to electrically access intracochlear electrodes. The EABR reflects the response of the central auditory system to electrical stimulation. Electrical pulses, which are delivered via an implanted intracochlear electrode array, stimulate spiral ganglion neurons. A multielectrode intracochlear array (with direct electrical access to the intracochlear electrodes) used in current CI devices allows us to record the response from electrodes of the array not used for stimulation. Just as in the ABR, this early auditory response reflects the activity of the auditory nerve, cochlear nucleus, lemniscus lateralis, and the inferior colliculus and is characterized by five prominent peaks. EABR wave I is quite difficult to obtain since it has a latency of approximately 0.35 ms, due to the absence of traveling wave delay, and is therefore embedded in stimulus artifact. Waves III and V can be easily recorded and used in clinical applications.

14.6.1 Clinical Applications

A review of recent publications indicates a declining interest in the EABR as a diagnostic measure, because electrically evoked compound action potential (ECAP) is easier and faster to record in the clinical setting. Most clinical uses for the EABR can now be accomplished more easily using the ECAP. However, there are still applications for which EABRs are appropriate such as when the integrity of the central auditory pathways through the brainstem is in question (Brown 2003, Miller *et al* 2008). In addition, recently, intracochlear EABR has been used to predict long-term outcomes in patients with a narrow internal auditory canal and likely hypoplasia of the auditory nerve (Song *et al* 2010). Preoperative EABR may also be useful in determining CI candidacy in children with inner ear malformations (Kim *et al* 2008).

14.7 ABR TO COMPLEX SOUNDS (cABR)

Conventional ABR generally uses simple stimuli such as clicks and sinusoidal tones to assess the integrity of the auditory pathway. However, clicks and tones are poor proxies for behaviorally relevant sounds encountered outside the laboratory or clinic (e.g., speech, music, environmental sounds). In addition, the response to a complex sound is not always predictable from the response to click and sinusoidal tones (Johnson *et al* 2008). This has led to the development of the cABR—a broad class of ABRs elicited by complex, behaviorally relevant sounds such as speech. The complexity of the human soundscape is represented by precise temporal and spectral neural codes within the auditory brainstem. In addition to the transient-evoked responses considered earlier, the brainstem also produces frequency-following responses (FFRs), sustained, phase-locked responses to periodic stimuli (Galbraith *et al* 2004, Russo *et al* 2004, Banai *et al* 2005, Sinha and Basavaraj 2010). Figure 14.4 illustrates these two distinct features (Skoe and Kraus 2010).

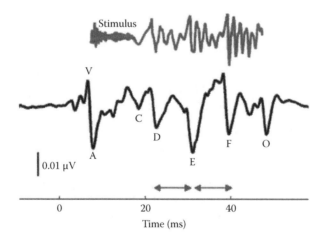

FIGURE 14.4 Transient and sustained features in response to the speech syllable /dɑ/. Time domain representation of a 40 ms stimulus /dɑ/ (gray) and response (black). This stimulus evokes seven characteristic response peaks (transient V, A, C, O; sustained D, E, F, and O). (From Skoe, E. and Kraus, N., *Ear Hear.*, 31, 302, 2010.)

14.7.1 Technique

cABRs can be recorded using the same data acquisition procedures as click ABRs. The two most widely used stimuli are the consonant–vowel (CV) syllable /dɑ/ (Kraus and Nicol 2005, Russo *et al* 2005) and mandarin syllables with lexical tones (Xu *et al* 2006, Krishnan and Gandour 2009), although musical and environmental stimuli have also been employed (reviewed in Skoe and Kraus 2010). The syllable /dɑ/ is an acoustically complex sound that begins with a high-frequency stop burst followed by a spectrally dynamic transition between the /d/ and /ɑ/ and terminates in a stable, yet harmonically rich vowel. The transient onset response to the stop burst is similar to the click ABR, and the sustained response to the vowel is equivalent to a tone-evoked FFR. The stimulus is typically presented monoaurally with alternating polarity at 80 dB sound pressure level (SPL) with a repetition rate between 4 and 10 s^{-1}. For cABRs, a vertical one-channel montage is the most common configuration. This configuration requires three electrodes corresponding to the active (noninverting), reference (inverting), and ground electrodes. In addition, 100–3000 Hz band-pass filtering is usually used, and 1000–6000 sweeps are used to collect cABRs.

14.7.2 Clinical Application

cABR is viewed as a sensitive biological marker of maturation and auditory training (Russo *et al* 2005, Johnson *et al* 2008). As with other variants of ABRs, cABRs are highly replicable across test sessions and can be reliably measured under passive listening conditions (Russo *et al* 2005, Song *et al* 2011, Hornickel *et al* 2012a). Moreover, in normal-hearing individuals, there is a relationship between cABRs and higher-level language process such as reading and speech perception in noise (Banai *et al* 2009, Parbery-Clark *et al* 2009), suggesting that cABRs could be used to help identify individuals at risk for language and communication disorders and target individuals who may benefit most from auditory training (Hayes *et al* 2003, Hornickel *et al* 2012b, Song *et al* 2012).

14.8 AUDITORY STEADY-STATE RESPONSE

A steady-state response is a brain potential that has a constant phase and amplitude at each of its frequency components (Picton *et al* 2003). Auditory steady-state responses (ASSRs) use continuous stimuli to trigger periodic neuronal responses. Sound stimuli for generating ASSR are usually amplitude-modulated (AM) pure tones with modulation frequencies (MFs) between 70 and 110 Hz. Neural generators are triggered according to the depth and frequency of the AM.

14.8.1 Technique

Continuous stimuli often include trains of tone bursts, with each tone burst producing its own responses. For example, a 125 ms train of modulated tone bursts with a two-cycle rise–fall time and a one-cycle plateau, at 2000 Hz carrier frequency (CF), with an interstimulus interval (ISI) between each burst at 25 ms (MF = 40 Hz) can produce a series of response peaks every 25 ms. As ASSR is periodic, frequency-domain methods can be used to automatically derive the response spectrum. A peak in the spectrum occurs at the MF; however, the presence of this peak depends on the integrity of the auditory pathway at the CF.

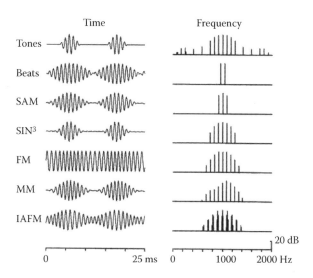

FIGURE 14.5   Various types of stimuli usually used to evoke steady-state responses. (From Picton, T.W. et al., *Int. J. Audiol.*, 42, 17, 2003.) The stimulus waveforms are plotted in the time domain on the left, and the spectra of the stimuli are displayed on the right. SAM, sinusoidally AM tone; SIN^3, modulation envelope based on the third power of the usual sinusoidal envelope; FM, frequency modulation; MM, mixed modulation (100% amplitude modulation and 25% frequency modulation); IAFM, independent amplitude and frequency modulation.

Auditory sensitivity varies as a function of CF, with the 2 kHz CF stimuli usually producing more reliable steady-state response and giving lower thresholds. Various kinds of stimuli have been used to evoke ASSR (Figure 14.5), including modulated white noise.

The neural origin of the ASSR depends on the AM frequency: AM frequencies of 20 Hz or less trigger generators in the primary auditory cortex and association areas. Response characteristics to 20–50 Hz modulation rates are similar to those found in the auditory midbrain, thalamus, and primary auditory cortex. MFs greater than 50 Hz are similar to evoked potentials from the midbrain (Kuwada *et al* 1986, Picton *et al* 1987). In clinical applications, modulation rates between 70 and 110 Hz are used to assess hearing thresholds; because at these higher modulation rates, the ASSR is not affected by sleep (Lins and Picton 1995).

14.8.2 Clinical Application

ASSR results can be interpreted objectively using regression formulae. That means ASSR allows an automated analysis by statistical evaluation of summating periodic neural responses generated by continuous stimuli. The primary clinical application of ASSR has been estimating hearing thresholds in infants and children. Infant hearing loss should be screened and treated as early as possible to prevent delays in the development of speech and language. Conventionally, for newborn hearing screening, either the click-evoked ABR or otoacoustic emissions (OAEs) are used (Norton *et al* 2000). However, these tests can be time consuming. Moreover, by these tests, obtaining frequency-specific auditory thresholds is limited. ASSR can overcome these drawbacks. In addition to being time efficient, ASSR can provide an estimate of hearing threshold across the audiometric range, which is essential for adjusting hearing aid amplification across frequencies. In addition, it has been reported that the number of ASSR components detected at each stimulus level is significantly related with the word recognition score (Dimitrijevic et al 2004). A possible explanation for this result is that like speech, the multiple modulated tones used for ASSR contain information that varies rapidly in intensity and frequency. ASSR in response to suprathreshold modulated noise is also correlated with temporal gap and modulation detection (Purcell et al 2004). One of the most important disadvantages of ASSR for clinical application is the within/across subject variability in responses.

14.9 ELECTROCOCHLEOGRAPHY

ECoG is the recording of the synchronous electrical activity produced by the cochlea and auditory nerve. For ECoG, recording electrodes are usually placed in close proximity to the cochlea or auditory nerve. Generally, closer placement of the recording electrode to the cochlea and auditory nerve will result in improved signal-to-noise ratio. However, the closer the electrode is placed, the more invasive it is. Direct nerve recording or intracochlear recording has been used in animal recordings. Transtympanic or extratympanic electrodes have been used successfully to record the ECoG in clinical settings.

Three electrical potentials, namely, the endocochlear potential (EP), cochlear microphonic (CM) potential, and summating potential (SP), make up the ECoG. The EP is a positive DC voltage maintained in the scala media relative to the surrounding tissue, whose source is the stria vascularis (Hudspeth 1989). Unlike the other three potentials, the EP is not stimulus induced and can be recorded without acoustic stimulation.

Basilar membrane vibration results in bending or shearing of the stereocilia on the top of each hair cell, which produces electrical changes in the surrounding tissue. The AC component of this potential is termed the CM. The CM is a summation of the receptor potentials of many individual hair cells, and it mimics the stimulus waveform. The SP is the DC component of the response. It has been widely used in a clinical setting, since the SP can be abnormal in patients diagnosed with Meniere's disease (MD). The compound action potential (AP) is a summation of the synchronous APs of the ~30,000 neurons that form the auditory nerve. The AP is characterized by a relatively large negative peak (N1) followed by a positive peak (P2).

14.9.1 Technique

There are a variety of methods for recording ECoG, which vary according to the electrode placement, stimulus and recording parameters, and data analysis techniques. In general, placing the recording electrode as close as possible to the source of the potential can improve signal-to-noise ratios. However, there is a trade-off that positioning the electrode close to the cochlea is usually invasive. Direct nerve recordings or intracochlear recordings are possible in animals but require invasive surgery in humans and are therefore typically used only as adjuncts when such surgery is needed for other purposes. Transtympanic electrodes have been used to obtain the ECoG in clinical settings. The electrodes are passed through the TM and positioned on either the round window or the promontory.

The responses are typically large and stable, but the relative invasiveness limits the routine use of transtympanic recording. The least invasive method for recording ECoG is using extratympanic electrodes. This is widely used in clinics, although responses recorded from the ear canal are small and require a lot of averaging. For extratympanic recording of ECoG, many electrodes have been developed, such as the wick electrode, the TIPtrode, and Eartrode (Katz 2009). TIPtrode, for example, consists of a foam tip for an insert earphone that has been covered with gold foil, which makes contact with the ear canal. TIPtrode offers both sound delivery and recording functions in one unit. Eartrode has small *wings* of flexible plastic, and a ball electrode is attached to the tip of one wing. As plastic wing expands in the ear canal, it presses the ball electrode gently into the wall.

For extratympanic recordings, the active electrode (e.g., TIPtrode or Eartrode) is typically positioned in the ear canal, on the TM, or on the promontory, while the reference electrode is positioned on the contralateral earlobe, mastoid, or vertex. The ground electrode can be placed on the forehead or other far-field surface location. Reference and ground electrodes are surface electrodes. Click or tone burst stimuli are typically used for ECoG recording. For click stimuli, relatively short time windows of about 4–5 ms are sufficient to record ECoG. For longer tone burst stimuli, the CM and SP potentials occur across the duration of the tone, so a longer time window is needed. Band-pass filtering is typically used to record the ECoG. A band pass of 100–3000 Hz is used to record the N1 peak of the ECoG in response to click stimuli. To record both SP and AP portions of the ECoG, a wider band pass is necessary because the SP is a DC (i.e., low frequency) potential.

14.9.2 Clinical Application

MD is a chronic illness defined as the idiopathic syndrome of endolymphatic hydrops, attacks of vertigo, and fluctuating hearing loss. ECoG has been considered as an adjuvant parameter in the diagnosis of MD. In endolymphatic hydrops, the basilar membrane is deflected toward the scala tympani, generating a larger-than-normal dc component (Gibson *et al* 1977). This large response results in an increased SP and consequently elevated SP/AP ratios in patients with MD that is associated with greater hearing loss (Ge and Shea 2002). However, ECoG may not play a decisive role in diagnosing MD owing to its lack of sensitivity (Nguyen *et al* 2010). Despite this, ECoG may occasionally be useful in patients with atypical MD (Kimura *et al* 2003) as well as a predictor of poor hearing outcome in Meniere's patients (Moon *et al* 2012).

In clinical practice, ECoG is also a tool for monitoring the status of the peripheral auditory system in the operating room. ECoG is not affected by anesthesia, and both transtympanic and tympanic ECoG can be recorded during general anesthesia. The large amplitude of ECoG can allow rapid feedback about the status of the auditory system during surgery, because it gives a better signal-to-noise ratio than other methods for recording wave I of the ABR (Simon 2011). In addition, direct monitoring of the auditory nerve by positioning an electrode, such as a ball-tipped or cotton wick–type electrode, on the auditory nerve is also possible. The wick electrode consists of a silver electrode wire that is insulated along its length and has a soft mesh (or cotton) that is sutured on its active tip. That mesh is then impregnated with an electrolyte before being placed. During this procedure, the auditory nerve is usually exposed, and an electrode can be placed directly on the auditory nerve. This ECoG method can provide the surgeon with almost real-time feedback. However, this direct monitoring is only possible when the auditory nerve is exposed and the electrode can be kept in place during the surgery.

14.10 INTRAOPERATIVE MONITORING

Neurotological and neurosurgical procedures involving the skull base and internal structures carry a significant risk of damage to the cranial nerves (CNs), including the vestibulocochlear nerve (CN VIII). Thus, if a dysfunction can be identified early, permanent damage to these structures may be avoided by modifying the surgical management promptly. ABR, ECoG, and compound nerve action potential (CNAP) have been used for intraoperative monitoring of CN VIII.

The ABR is resistant to sedative medications and general anesthesia and is not sensitive to the level of consciousness (Stone *et al* 2009). This characteristic allows ABR to be used as an intraoperative monitoring tool during brainstem surgery or acoustic tumor surgery. For ABR, the most useful information in CN VIII neuromonitoring is delivered by waves I and V.

Sudden loss of or changes in ABR waveforms result from damage to the labyrinth or CN VIII, injury to the labyrinthine artery, and/or cerebellar retraction that causes neurophysiologic dysfunction of CN VIII (Figure 14.6).

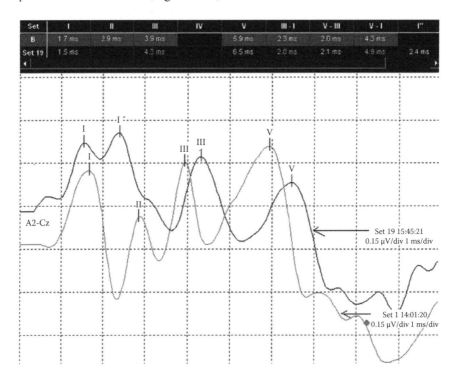

Set	I	II	III	IV	V	III-I	V-III	V-I	I"
B	1.7 ms	2.9 ms	3.9 ms		5.9 ms	2.3 ms	2.0 ms	4.3 ms	
Set 19	1.5 ms		4.3 ms		6.5 ms	2.8 ms	2.1 ms	4.9 ms	2.4 ms

FIGURE 14.6 Cerebellar retraction in cerebellopontine angle surgery stretching CN VIII. Besides an increase in wave V absolute latency, this stretching causes the appearance of a secondary wave I between waves I and II, a sign of impaired conduction. (From Simon, M.V., *J. Clin. Neurophysiol.*, 28, 566, 2011.)

Intraoperative ABR can also be used as a predictor of postoperative hearing outcome. For example, whereas an increase in wave V latency is not a good predictor for postoperative hearing deficit, the persistence of wave V at the end of surgery is highly predictive of preserved hearing (Gouveris and Mann 2009).

ECoG is another method used for CN VIII monitoring but requires direct recording of the cochlear action potentials (CAPs). An advantage of ECoG over ABR is that it is a higher amplitude potential, because it is a more direct reflection of the function of the cochlea and CN VIII (Simon 2011). Thus, it represents the most peripheral potentials, as a near-field CAP that is the equivalent of wave I of the ABR. The better signal-to-noise ratio of ECoG offers faster responses than ABRs. However, if the auditory nerve is directly damaged, ECoG can be preserved, while ABRs are lost. ECoG has been widely used as a predictor of postoperative hearing outcome. A permanent decrease of 75% or more in the CAP amplitudes is related to postoperative hearing loss, while just a temporary decrease is likely associated with a more favorable outcome (Gouveris and Mann 2009).

Cochlear CNAPs can be recorded directly from the CN VIII through an electrode placed on the nerve during surgery. As with the CAP recorded by ECoG, CNAP also has higher signal-to-noise ratio than ABR, which is advantageous in noisy environments like the operating room. In addition, it requires minimal averaging, giving faster feedback to the surgeon than ABR (Yamakami *et al* 2009). Abrupt loss of CNAPs is most likely associated with a vascular injury, while a gradual loss is usually associated with a mechanical injury. Loss or decreased amplitude in CNAP can also present when the CN VIII nerve is compressed, because desynchronization of the individual nerve APs occurs (Simon 2011).

14.11 OAE

OAEs, first described by Kemp in 1978, are low-level sounds that originate within the cochlea and propagate through the middle ear into the ear canal. OAEs are evoked by a variety of types of stimuli, and they can be measured using a sensitive microphone placed in the ear canal (Kemp 1978), when both the cochlea and middle ear systems are functioning normally or near normally. Typically, the sensitive, small microphone is located in a small probe that is coupled to the ear with a foam or rubber tip. The probe has one or two speakers for presentation of sound stimuli. OAEs are a preneural phenomenon, which are unaffected by stimulus rate or polarity, and a sensitive indicator of peripheral auditory function.

Outer hair cells (OHCs) contribute to active cochlear amplification, a process that enhances the vibration of the basilar membrane at the peak of the traveling wave (Davis 1983). A protein called prestin is thought to be the molecular motor responsible for active OHC motility (Santos-Sacchi *et al* 2001). In addition to OHC movement, hair cell stereocilia bundles also play a significant role in providing amplification (Ricci 2003), and both OHC and stereocilia are involved in the production of OAEs (Liberman *et al* 2004).

OAEs are classified into two types, spontaneous and evoked. For obtaining evoked OAEs, transients (transient-evoked OAEs [TEOAEs]) or tone pairs (distortion product OAEs [DPOAEs]) are widely used. Both types of OAEs are abnormal in the presence of mild hearing loss (Harris and Probst 1991).

14.12 SPONTANEOUS OAEs

Spontaneous OAEs (SOAEs) are measured in the absence of an external stimulation and appear as pure-tone-like signals in the ear canal. SOAEs do not appear in ears with hearing loss greater than 30 dB HL. However, they are not a useful clinical test because they can be measured in only 50%–60% of normal-hearing ears (Burns *et al* 1992, Penner and Zhang 1997).

14.13 TRANSIENT-EVOKED OAE

Transient stimuli such as clicks can evoke emissions from a large portion of the cochlea. These low-level emissions are sampled and time averaged and reflect the frequency composition of the cochlear output. Figure 14.7 represents TEOAE results for the right (normal) and left (abnormal) ears of a 6-week-old infant.

Two averaged TEOAE traces (labeled *A* and *B* in Figure 14.7) are generated for each ear. The two waveforms are superimposed in the left side of each panel. The cross-sectional correlation between two waveforms was quantified as *reproducibility*, an important component for the analysis of TEOAE. Generally, a reproducibility of 50% or greater indicates a normal TEOAE response. For the right ear in Figure 14.7, waveforms A and B are robust

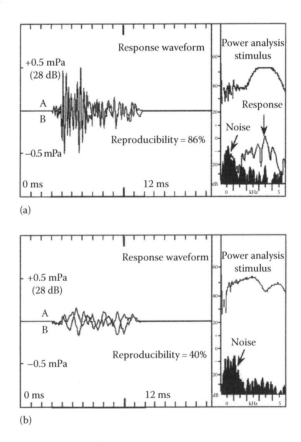

FIGURE 14.7 TEOAEs recorded in a 6-week-old infant. (a) Right ear and (b) left ear. (From Johnson, K.C., *Otolaryngol. Clin. North Am.*, 35, 711, 2002.)

and highly reproducible. In addition, the power spectrum of both the stimulus and the response is displayed in the right side of the panel. The open histogram plots the frequency spectrum common to both waveforms, considered to be the true OAE response, and the filled histogram provides an estimate of the noise remaining in the averaged waveform. For the right ear, robust emissions are well observed above the noise floor. In contrast, the TEOAE waveforms for the left ear are not reproducible, and the response does not exceed the floor.

14.14 DISTORTION PRODUCT OAEs

When the cochlea is stimulated with two different tones, the OAE includes the two stimulus tones along with multiple tones that are not present in the stimulus. These additional tones, called distortion products (DPs), result from the nonlinear processes of the cochlea. The most robust of DPs occurs at the frequency equal to $2f_1 - f_2$, where f_1 indicates the lower-frequency stimulus and f_2 indicates the higher-frequency stimulus. Using spectral analysis, the SPL at $2f_1 - f_2$ can be derived from the averaged DPOAE for each tonal pair. The noise floor is determined by calculating the SPL in a narrow frequency band above and below each DP frequency. Generally, DP and noise floor amplitudes are plotted as a function of f_2, with DPOAE amplitudes ranging from 15 to 25 dB SPL throughout the f_2 frequency range in normal-hearing subjects.

14.14.1 Clinical Application

A common use of OAE recording is to evaluate the function of OHCs. OAEs, especially TOAEs, have been widely used to screen for hearing problems in newborns or very young children because they can be recorded quickly and reliably in those populations. However, the use of DPOAEs for hearing screening is comparatively limited because of the large variability in emission levels (Prieve and Stevens 2000, Krueger and Ferguson 2002, Johnson *et al* 2005).

14.15 COCHLEAR IMPLANT TELEMETRY

Today, all implants are capable of sending the internal information, such as the status of the implanted electronics, back to the speech processor. This process is generally referred to as telemetry. As the needs for assessing the electrical fields induced in the cochlea as well as neural responsiveness are increasing, telemetry has become important. Recently, CI manufacturers have released telemetry systems that can automatically measure ECAP thresholds intraoperatively and postoperatively (van Dijk *et al* 2007, Gartner *et al* 2010).

14.15.1 Cochlear Implant Electrode Impedance Measurement

Optimal intracochlear electrode placement is a prerequisite for obtaining a maximal outcome after cochlear implantation. If an incorrect electrode placement or malfunctioning electrodes are detected during CI surgery, immediate action can be taken to correct them. Therefore, to evaluate device functionality and electrode placement, various electrophysiologic methods have been used, such as EI, ECAP, and spread of excitation (Cosetti *et al* 2012).

Using EI, the electrode array function as well as the resistive characteristics of the fluid or tissue surrounding each electrode can be evaluated. EI provides the status of individual electrode integrity, such as short or open circuits. However, abnormal intraoperative impedances can be transient, caused by air bubbles around inserted electrodes (Cosetti *et al* 2012). The details vary with the electrode array and the CI manufacturer, but generally, intracochlear impedance should be in the 2–12 kΩ range. In the future, EI measurements may play a role in atraumatic electrode insertion and preservation of residual hearing. By continuously monitoring impedance during surgical electrode insertion, insertional trauma can theoretically be prevented (Campbell *et al* 2010).

14.16 ELECTRICALLY EVOKED COMPOUND ACTION POTENTIAL

In a deaf cochlea, due to the absence of hair cells, CMs and SP cannot be generated. The only potential generated in response to electrical stimulation is the AP of the auditory nerve. Most of the current CI systems stimulate the nerve using the implanted electrode array and also measure the ECAP using the same electrode array. This technique is referred to as neural response imaging (NRI; Advanced Bionics Corporation, Los Angeles, CA), neural response telemetry (NRT; Cochlear Corporation, Sydney, New South Wales, Australia), or auditory nerve response telemetry (MED-EL, Innsbruck, Austria).

Typically, a biphasic electrical pulse is delivered to one intracochlear electrode, and the resulting neural response is measured on an adjacent electrode (Mens 2007). Using various stimulus amplitudes, a response growth function is obtained from which the ECAP threshold can be estimated. Figure 14.8 shows examples of nerve trunk and intracochlear ECAPs.

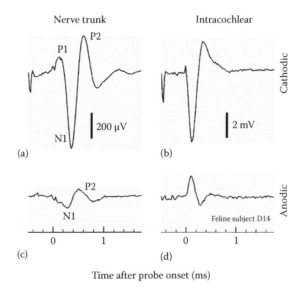

FIGURE 14.8 ECAPs obtained from an acutely deafened cat stimulated with a monopolar intracochlear electrode. ECAPs displayed in panels (a) and (c) were recorded using a ball electrode positioned on the nerve trunk, while those displayed in panels (b) and (d) were recorded using an intracochlear electrodes. (From Miller, C.A. et al., *Hear. Res.*, 242, 184, 2008.)

The ECAP is an indicator of the responsiveness of the auditory nerve. The practical advantage of ECAP recordings over the EABR is its relatively large amplitude. In general, ECAP recording is rapid, not degraded by motion artifact and unaffected by type or depth of anesthesia (Cosetti *et al* 2012). However, there is still a considerable variability of ECAP measurement.

14.16.1 Technique

The ECAP typically includes both a negative peak N1 and a positive peak P1. N1 occurs within a fraction of a millisecond after stimulation. The ECAP amplitude is much larger as compared with the EABR, because the recording electrodes are inside the cochlea. In addition, noise created by muscle activity is also greatly reduced. Therefore, for measuring the ECAP, the number of averages can be as few as 50, in contrast to EABRs, which require at least hundreds of sweeps. The most important challenge to obtain a reliable response is the electrical stimulus. The large electrical stimulus precedes the N1 response, so it is necessary to separate the genuine neural response from the preceding stimulus. To reduce recording artifacts, various techniques have been used, such as alternating stimulus polarity and forward masking (Brown *et al* 1990, Mens 2007).

In alternating stimulus polarity, only the electrical stimulus changes polarity. By alternating stimulus polarity in consecutive stimuli, the electrical stimulus cancels out in the averaging process. In the forwarding masking technique, a high-amplitude *masker* pulse is followed by a *probe* pulse after a small delay (e.g., 0.5 ms) that is shorter than the refractory period of the auditory neurons. If the amplitude of the masker is high enough to bring all neurons within the excitation area into a refractory period, the recording of the *probe* pulse contains no neural response. Both a probe-alone response and masker-alone response are recorded. By time-shifting and adding the recordings, the electrical stimulus can be removed to reveal the probe-alone neural response (Battmer *et al* 2004). An optimized version of this forward masking technique is now widely used.

14.16.2 Clinical Application

Measuring ECAP provides a number of clinical benefits. ECAP can be used during CI operation to verify implant and auditory nerve integrity. Postoperatively, it can be used to monitor recipient progress; however, how intraoperative ECAP thresholds relate to postoperative speech performance is still unclear (Cosetti *et al* 2010). Spread of excitation can also be a useful intraoperative assessment tool of proper CI electrode placement (Grolman *et al* 2009).

Intraoperative ECAP measurement can provide useful information for guiding initial fitting of the speech processor, especially in very young children (Gordon *et al* 2004). However, ECAP and EI generally do not affect intraoperative decision making whether to use a backup device or not (Cosetti *et al* 2010, 2012).

During fitting procedure of CI, a clinician must subjectively determine the individual's hearing dynamic range on each electrode: from low to high current levels. This mapping is difficult and time consuming, especially with young children who cannot cooperate. Hence, objective fitting methods are necessary. For this purpose, ECAP thresholds are measured at a number of locations on the intracochlear electrode array and are used during fitting procedure. Although a direct correlation between ECAP thresholds and psychophysical measures is too low to allow direct prediction of map levels (Brown *et al* 2000), ECAP threshold profiles across the array correlate well with psychophysical profiles (Brown *et al* 2000, Franck 2002). Moreover, stimuli delivered at the ECAP threshold current levels almost always elicit sound percepts, usually in the upper half of the dynamic range (Cafarelli *et al* 2005).

14.17 VESTIBULAR-EVOKED MYOGENIC POTENTIAL

Evoked potentials are widely used to evaluate vestibular function as well as the functional status of the auditory system. Vestibular-evoked myogenic potentials (VEMPs) are biphasic, short-latency, inhibitory electrical changes. Cervical VEMP (cVEMP), which is measured at the sternocleidomastoid muscles (SCMs), results from sound stimulation of the saccular portion of the vestibular system.

To stimulate vestibular receptors and afferents, air-conducted sound (ACS), bone-conducted vibration (BCV), and galvanic pulses have been used (Curthoys 2010). ACS and BCV activate receptors and afferents from both the utricular and saccular maculae (Curthoys 2010). One small group of primary vestibular afferents—otolith irregular afferents—is extremely sensitive to vibration stimuli and is selectively activated by 500 Hz BCV (Narins and Lewis 1984, Fernandez *et al* 1990). In addition, many of these afferents are also activated by ACS (Murofushi *et al* 1995). In contrast, semicircular canal neurons are insensitive to these 500 Hz stimuli. Therefore, the 500 Hz stimulus frequency can ensure that otolithic neurons are selectively activated. P13 of a cVEMP to ACS reflects saccular function, because the SCM response being measured depends predominantly on saccular activation. Similarly, the n10 of the ocular VEMP (oVEMP) to 500 Hz BCV reflects utricular function, because the eye-movement response is mainly determined by utricular activation. To deliver galvanic stimulation, large surface electrodes (approximately 1000 mm²) are used (MacDougall *et al* 2005). A custom-designed, battery-driven, isolated current stimulator or commercial electrical stimulator (Anima Co. Ltd., Tokyo, Japan) has been used to deliver the desired current to patients (Monobe and Murofushi 2004).

14.18 CERVICAL VEMP

The cVEMP is characterized by a positive and negative peak occurring at latencies of approximately 13 and 23 ms, termed P13 and n23, respectively (Figure 14.9).

The pathway activated by cVEMP testing includes the saccule, inferior vestibular nerve, vestibular nucleus, and medial and lateral vestibulospinal tract to the ipsilateral SCM (Rosengren *et al* 2010). Therefore, cVEMP testing has been used for unilateral assessment of the saccule and the inferior vestibular nerve.

14.18.1 Technique

Generally, cVEMPs are recorded using surface electrodes over the tensed SCM during ACS stimulation. Because cVEMPs are inhibitory to the SCM, electrical activity must be present

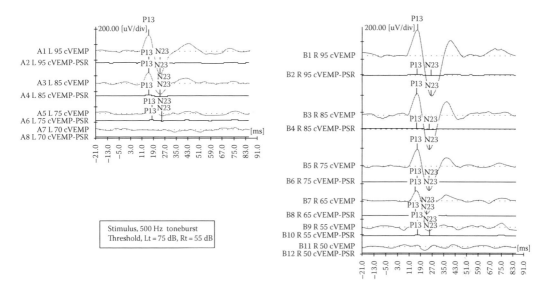

FIGURE 14.9 VEMPs recorded from the right and the left SCMs in response to sound stimulation of the right ear. (From Brantberg, K., *Semin. Neurol.*, 29, 541, 2009.) Negative is plotted up in this figure.

within the muscle to obtain a response. Two methods are generally used for cVEMP: the head elevation nonrotation method and the head rotation method (Mudduwa *et al* 2010). In the first method, the patient lies in a supine position and elevates his or her head by 30°. The second method requires the patient to rotate his or her head toward one shoulder. This method is better tolerated but less reliable and produces smaller potentials (Wang and Young 2006). VEMP can be evoked by clicks presented at 90–100 dB normal-hearing level (140–145 dB SPL). If patients have a conductive hearing loss, then bone-conducted tones and skull taps can be used to elicit the response. However, the VEMP response from skull tap results from the activation of utricular receptors (Rosengren *et al* 2009). Galvanic stimuli can also elicit VEMPs by delivering short-duration (1–2 ms), pulsed current (3–4 mA) to the mastoid (Watson and Colebatch 1998).

14.19 OCULAR VEMP

oVEMPs are potentials recorded from electrodes beneath the eyes in response to sounds, vibration, skull taps, or galvanic stimulation. It is generally accepted that oVEMP testing measures the myogenic activity of the inferior oblique muscle (Chihara *et al* 2009). The initial short-latency negative potential (n10) of the oVEMP is considered to be of vestibular origin and is an indicator of utricular function. Figure 14.10 shows oVEMP results for a healthy subject and patient with unilateral vestibular dysfunction.

As patients with superior vestibular neuritis (SVN) show small or absent n10 responses, the n10 of BCV indicates mainly utricular function (Iwasaki *et al* 2009). These same patients show normal cVEMPs to ACS, indicating that saccular macula and inferior vestibular nerve were intact and functioning normally. The oVEMP is a crossed reflex, so in patients with unilateral vestibular dysfunction, oVEMPs are abnormal on the contralateral

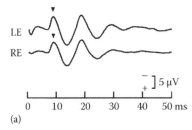 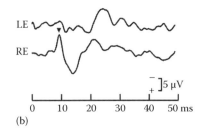

(a)   (b)

FIGURE 14.10 Examples of average oVEMPs. BCV stimulation in a healthy subject and a patient with known unilateral vestibular dysfunction. (a) Healthy subject and (b) patient with right uVD. In the unilateral vestibular dysfunction patient, the contralateral n10 response (arrowheads) is absent, whereas the n10 beneath the ipsilesional eye is of normal amplitude. This asymmetry in n10 is the indicator of utricular dysfunction. (From Curthoys, I.S. et al., *Clin. Neurophysiol.*, 121, 132, 2009.)

side. Although the oVEMP is much smaller than cVEMP, it is easily recordable. In contrast to cVEMP, which is an inhibitory response, oVEMPs are excitatory responses.

14.19.1 Technique

The subject is in a supine position on a bed with the chin close to the chest. Commercially available, surface electrodes (e.g., DE-2.1, DeISys, Landenberg, PA) are applied to record the potentials from beneath both eyes. To record the myogenic potentials, the active (+) recording electrodes are located on the infraorbital ridge 1 cm below the lower eyelid, and the reference (–) electrodes are placed about 2 cm below the first electrodes. The ground electrode is on the chin or sternum. The subject is instructed to look up at a distant target exactly at midline. The elevated position of the eye, in the orbit, is crucial since the size of n10 decreases substantially as the gaze is lowered (Rosengren *et al* 2005).

To generate an effective BCV, a Bruel and Kjaer 4810 minishaker is usually used. In addition, gentle taps with a tendon hammer are also a very effective way to elicit n10. In contrast, to generate an oVEMP response, a standard clinical bone oscillator is ineffective because the magnitude of the linear acceleration is too small to activate otolith afferents.

The stimulation of 500 Hz BCV at Fz, the junction of the midline at the hairline, can assess the symmetry of otolith function, because 500 Hz vibration leads to linear acceleration stimulation of both ears simultaneously and about equally (Iwasaki *et al* 2007). Therefore, in a clinical setting, the asymmetry ratio (AR) is an excellent indicator of unilateral otolithic loss. The AR is defined according to the amplitude of n10 response to Fz BCV stimulation:

$$\text{Asymmetry ratio (AR)} = \frac{(\text{Larger n10} - \text{Smaller n10})}{(\text{Larger n10} + \text{Smaller n10})} \times 100$$

The average AR for healthy subjects is approximately 11.73% ± 8.26%, and none of the 67 asymptomatic healthy subjects showed ARs >40% (Iwasaki *et al* 2008).

14.19.2 Clinical Applications

Because VEMP testing is very well tolerated by patients and relatively easy to perform, VEMP has been widely used in the clinic and research laboratory (Rosengren *et al* 2010). VEMP testing permits evaluation of saccule (cVEMP) and utricle (oVEMP) functions in the vestibular organ and thus provides clinically relevant information in cases of peripheral vestibular disorders. cVEMP has been used in MD to identify saccular dysfunction (Hong *et al* 2008). In addition, in patients with superior semicircular canal dehiscence syndrome, decreased thresholds and increased amplitudes in VEMP are usually observed. Given the relative ease of VEMP testing and high level of reproducibility of results, VEMP has widely been used in the diagnosis of this syndrome (Streubel *et al* 2001, Brantberg and Verrecchia 2009). In patients with vestibular neuritis, cVEMP testing can be used to assess the involvement of the inferior vestibular nerve (Ochi *et al* 2003).

14.20 ELECTRONEUROGRAPHY

Facial nerve palsy is a devastating condition with huge emotional impact and subsequent physical limitations. In a clinical setting, evaluating facial nerve viability accurately is extremely important to manage facial nerve disorders. Although history and physical examination give useful information in such patients, physiologic testing is also needed. Electroneurography (ENoG) is a relatively objective tool to assess the facial nerve integrity, which electrically stimulates the facial nerve near the stylomastoid foramen and measures the response of the facial muscles.

14.20.1 Technique

The stimulus electrodes are located near the stylomastoid foramen, where the facial nerve trunk passes by, with the cathode electrode posterior and anode electrode anterior. The recording electrodes are placed along the nasolabial groove to detect the motoric response of the facial expression muscles (Figure 14.11).

The amplitude of the response is usually over 2 μV. Typically, the normal side is tested first and the ENoG value is used as a reference. The distal facial nerve responses to maximal electrical stimulation of the two sides (ipsilesional and contralesional sides) are compared, and the value is quantified as percentages (Figure 14.12). Therefore, the same stimulation and recording technique should be used in both sides. Generally, a 30% or greater asymmetry (or change over time) is considered significant.

It is important to deliver the lowest possible stimulus that produces a maximal response. Usually, the stimulus intensity is between 20 and 40 mA (Beck and Hall 2001). The largest possible facial nerve response should be obtained from each side, and it does not mean that the amplitude level from one side to the other should be symmetric.

If the stimulating electrode is placed anteriorly ahead of the ear, then the masseter muscle can be stimulated, inadvertently leading to contamination and misinterpretation of the response.

14.20.2 Clinical Application

ENoG has been widely used as a prognostic indicator in patients with Bell's palsy. Severe reduction of ENoG amplitude (less than 10%–18% of the normal side) is highly correlated

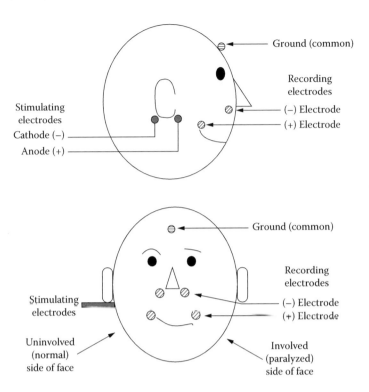

FIGURE 14.11 Conventional sites for stimulating and recording electrodes used in ENoG measurement. Commercially available surface electrodes are used as stimulating and recording electrodes. Stimulating electrodes are placed near the stylomastoid foramen, and recording electrodes are ipsilaterally placed along the nasolabial groove. (From Beck, D.L. and Hall, J.W., *Hear. J.*, 54(3), 36, 2001.)

with incomplete recovery in patients with Bell's palsy (May *et al* 1983, Fisch 1984, Chow *et al* 2002), because the proportion of neural injury in the facial nerve can determine the final outcome. However, it is of note that Wallerian degeneration in the distal segment takes 72 h to emerge after an acute facial nerve injury. Therefore, it is recommended that ENoG should be delayed until 72 h after the onset of facial palsy. Besides predicting the prognosis of Bell's palsy, ENoG has been used in selecting patients for decompression surgery for traumatic facial nerve paralysis (Gantz *et al* 1999, Quaranta *et al* 2001, Yetiser 2012).

14.21 NONROUTINE METHODS

14.21.1 Subcortical/Cortical Auditory ERPs

Auditory evoked potentials generated at the cortex can be a useful tool for studying higher cognitive processing of sound and for understanding of the underlying neuronal network. Middle-latency auditory potentials occur approximately 15–70 ms after the stimulus. The most commonly used middle-latency auditory potentials are positive peaks Pa and Pb (P1), which originate in subcortical structures, such as the hippocampus (Woods *et al* 1987). Long-latency auditory potentials, elicited more than 50 ms after the stimulus, reflect the

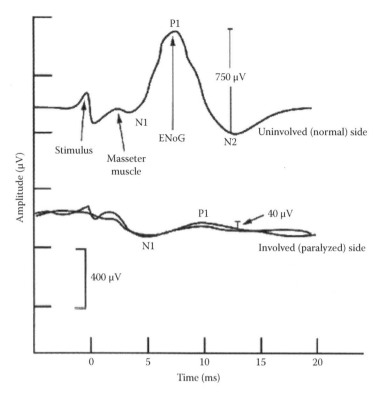

$$\text{Percent degeneration} = 100 - \frac{\text{Amplitude on involved side of face (μV)}}{\text{Amplitude on uninvolved side of face (μV)}} \times 100$$

FIGURE 14.12 An example of a significant degeneration in facial nerve response. As compared with normal ENoG response (top), abnormal ENoG response in the paralyzed side (bottom) shows delayed and very small (almost absent) peaks. Facial nerve degeneration is calculated by comparing ENoG amplitude for the paralyzed side to the normal side, as illustrated in the equations at the bottom of this figure. Significant degeneration requiring surgical intervention is >90%. (From Beck, D.L. and Hall, J.W., *Hear. J.*, 54(3), 36, 2001.)

activity of the cortical structures in the processing of sound. Cortical evoked potentials reflect not only the reception and processing of sensory information but in some cases also the higher-level processing, including selective attention, memory updating, and semantic comprehension.

There are two types of cortical evoked potentials—obligatory (exogenous) and cognitive (endogenous) (Cone-Wesson and Wunderlich 2003). As with subcortical evoked potentials (e.g., ABRs), the response characteristics of obligatory cortical event-related potentials (ERPs) depend primarily on stimulus parameters, such as sound intensity or presentation rate and the integrity of the generators. In contrast, cognitive ERPs vary with the listener's attention to and performance on cognitive tasks assigned. The cognitive ERPs include the mismatch negativity (MMN), the P3a, the P300 (P3 or P3b), the N400, error-related anterior negativity, reorienting negativity, and P600 (Fellman and Huotilainen 2006). For these tests, an *oddball* stimulus paradigm is usually used in which a standard stimulus is presented with an 80%–90% probability and the oddball (deviant) stimulus is presented with a 10%–20% probability. The deviant stimulus may differ from the standard ones in a simple feature (e.g., difference in frequency or intensity) or complex feature. ERPs are obtained separately for the standard and target stimuli. Sound delivery occurs via headphones, insert earphone, or soundfield, which can also be used to elicit ERPs (Jirsa 1992, Tremblay *et al* 2004). For some of these cognitive ERPs to emerge, active participation of the subject in a task is required. Therefore, the subject is trained to detect the target stimulus during testing. Some ERPs, such as MMN and P3a, can be elicited without any specific participation, and thus they can be recorded in young children or infants (Wunderlich *et al* 2006).

These ERPs are measured with EEG or magnetoencephalography (MEG). ERPs are typically extracted from the scalp-recorded EEG by signal averaging, when the auditory stimulation of ~70 dB is delivered in a quiet room. To record multichannel cortical ERPs, the international 10–20 EEG electrode positioning system is usually used. Electrode caps are now available in pediatric and adult different sizes, making it possible to record hundreds of channels across the scale. The digitized EEG recording is amplified and processed, usually online band pass between 0.05 and 70 Hz, with a sampling rate of 500–1000 Hz ERP components are defined by their positive or negative polarity, latency from the onset of stimulus, and scalp distribution. The sequence and latencies of ERP components reflect the temporal dynamics of neural activity, whereas the amplitudes represent the extent of neural resource allocation to specific cognitive processes. Abnormalities of these components can provide useful diagnostic information.

14.21.2 Exogenous ERPs

Exogenous ERPs are composed of (P) or negative (N) waves at certain latencies. They are named the P50, the P1 (or P100), the N100 (or N1), the P2, the N2, and the sustained potential SP (Näätänen 1992).

14.21.3 P1 (P100) and P1–N1–P2 Complex

The P1 is a robust positivity occurring at around 100–300 ms in children. The neural generator of the P1 is the thalamocortical projections to the auditory cortex and may represent the first recurrent activity in the auditory cortex (Sharma *et al* 2009). The P1–N1–P2 complex is typically evoked by brief stimuli and recorded passively (Figure 14.13). To clicks, P1 has a latency of approximately 50 ms after sound onset in normal-hearing adults. N1 follows the positive peak P1 and appears approximately 100 ms after sound onset. Therefore, N1 is sometimes referred to as the N100. In addition, speech-evoked P1–N1–P2 complex has been described for decades, and it has been used to probe different aspects of auditory processing in both adults and children (Cunningham *et al* 2000, Martin *et al* 2008).

14.21.3.1 Clinical Application

Latency of the P1 reflects the sum of synaptic delays throughout the peripheral and central auditory pathways (Eggermont *et al* 1997). The P1 peak latency varies as a function

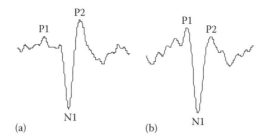

FIGURE 14.13 The P1–N1–P2 complex evoked by a click (a) and a synthesized speech stimulus (/ba/) in an adult (b). A positive peak (P1) is followed by a negative component (N1) and another positive peak (P2). (From Martin, B.A. et al., *Ear Hear.*, 29, 285, 2008.)

of age, and a normal range latency has been established for different ages (Sharma *et al* 2002). P1 responses have also been measured in deaf children who received CI. Of note, children who received CI stimulation early in their childhood (<3.5 years) showed normal P1 latencies, whereas those who received CI late in their childhood (>7 years) had abnormally long P1 latencies (Sharma and Dorman 2006). For the majority of late-implanted children, P1 latencies did not reach normal limits even after several years of experience with the CI. This finding suggests a sensitive period for central auditory development of about 3.5 years.

P1–N1–P2 complex has been used to estimate hearing sensitivity, because it provides a reasonable estimate of behavioral thresholds (within 5–10 dB) (Lightfoot and Kennedy 2006). However, because it is affected by attention, sleep, and sedation, ABR is more commonly used for this purpose.

14.22 ENDOGENOUS (COGNITIVE) ERPS

14.22.1 Acoustic Change Complex

Acoustic change complex (ACC) is a cortical auditory evoked potential (P1–N1–P2) elicited by a change within an ongoing sound stimulus (Figure 14.14) (Näätänen 1992, Martin *et al* 2008). When elicited, the ACC indicates that the brain, at a cortical level, has detected change(s) within a speech sound and the patient has the neural capacity, given intact higher neural centers, to discriminate the sounds (Martin *et al* 2008).

14.22.1.1 Clinical Application
There are several advantages of ACC in a clinical setting. ACC shows a good correlation with behavioral measures of intensity and frequency discrimination (Martin and Boothroyd 2000, Sharma *et al* 2002). In addition, test–retest reliability of ACC is excellent in adults (Tremblay *et al* 2003). Furthermore, nowadays, electrically evoked ACC can be recorded in patients with CIs (Brown *et al* 2008).

However, ACC is a clinical tool in development. The ACC shows a reasonable agreement with behavioral psychophysical discrimination threshold (Martin and Boothroyd 2000). Therefore, the ACC gives the greatest promise in infants and young children, particularly those with hearing loss, for whom appropriate behavioral tests are not

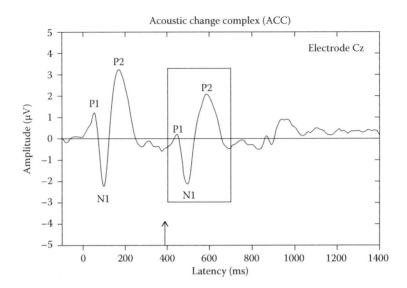

FIGURE 14.14 The response to the 800 ms duration diphthong stimulus /ui/ is shown. A P1–N1–P2 complex is elicited by the onset of stimulation, and a second P1–N1–P2 complex is elicited by the acoustic change from /u/ to /i/. (From Martin, B.A. et al., *Ear Hear.*, 29, 285, 2008.)

available (Martin *et al* 2010). In addition, ACC provides important insights into the brain's ability to discriminate acoustic changes, such as frequency, amplitude, and periodicity cues.

Recently, continuous alternating stimulus presentation has been proposed as a way to elicit ACC more efficiently, a requisite for the clinic (Martin *et al* 2010). In this stimulus paradigm, the silent period between sounds is minimized, allowing for the number of acoustic changes presented in a given time period to be increased.

14.22.2 Cognitive Potential (P300)

The P300 (also known as P3 or P3b) is a large, broad, positive component in the ERP that usually peaks around 300 ms after the onset of a task-relevant stimulus. The P300 is commonly elicited in the oddball paradigm, with participants being asked to categorize the stimuli, either by counting or by pressing a button when a deviant is heard. The P300 is diminished or abnormal in listeners with attention, memory, or cognitive disorders.

After an initial sensory processing, an attention-driven process ascertains whether the current stimulus is either the same as the previous stimulus or not. If no change in the stimulus attribute is detected, the neural model of the stimulus environment is unchanged, and only sensory-evoked potentials (N100, P200, N200) are recorded after signal averaging. If a new stimulus is detected and the subject allocates attentional resources to the target, a P300 potential is generated in addition to the sensory-evoked potentials. Figure 14.15 schematically presents the P300 context-updating model (Polich 2007).

The neural generators for P300 include the primary auditory cortices, association areas of the centroparietal cortex, the frontal lobe, and the hippocampus, which are areas of the brain involving memory (Picton 1992).

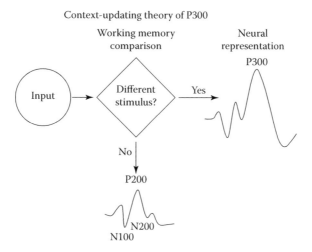

FIGURE 14.15   Schematic illustration of the P300 context-updating model. (From Polich, J., Clin Neurophysiol., 118, 2128, 2007.)

14.22.2.1 Technique

The oddball paradigm is frequently used for eliciting P300. The amplitude of P300 depends on the probability of the attended stimulus; the lower the probability, the larger the amplitude of P300 (Duncan-Johnson and Donchin 1977). The more complex the stimulus processing required by the task, the longer the latency of P300 latency. For example, a difficult task, such as discriminating 2 tones close in frequency, elicits a longer latency than tones widely spaced in frequency. The presence, latency, and amplitude of the P300 vary with the attention level as well as the cognitive demands of the listening task. As P300 is an index of a particular cognitive process, the demands of the task should be adjusted to the ability level of the participants, whether children or elderly adults. When active discrimination task cannot be applied, a passive oddball or single-stimulus task can be used (Mertens and Polich 1997).

14.22.3 Mismatch Negativity

The MMN is a brain response to violations of a rule and reflects the brain's ability to detect small changes in the acoustic environment (Näätänen and Alho 1995, Näätänen and Winkler 1999). A common procedure in eliciting the MMN involves the presentation of a series of identical stimuli ($p = 0.8$) with occasional oddball stimuli ($p = 0.2$). The oddball paradigm is thought to activate a subliminal memory trace for the features of the standard stimulus, from which future auditory input is predicted (Näätänen and Winkler 1999). When a deviant sound is presented, it is automatically compared to the memory trace. If there is a difference between memory trace and the incoming stimulus, MMN is elicited. The MMN reflects the brain's ability to automatically monitor and preconsciously detect changes in ongoing auditory stimulations, and it is associated with preattentive cognitive function and a *primitive intelligence* involved in the early stages of perception of stimulus differences in the brain (Duncan *et al* 2009).

14.22.3.1 Technique

The MMN is elicited in the presence of any small change in the repetitive aspect of auditory stimulation. This discriminable change can be of different types: frequency, duration, intensity, perceived sound-source location, silent gap, or phoneme replacement. In a clinical setting, the probability of the deviant sound ranges between 0.05 and 0.20. As MMN is a small component, at least 150 deviant stimuli of each type are presented, and evoked potentials are averaged in order to get an adequate signal-to-noise level. The standard and deviant stimuli are usually presented at relatively short ISIs, such as 500 ms to 1 s. The MMN can even be elicited in the absence of the participant's attention (passive listening), which gives a great advantage in the assessment of very young or impaired participants. When a participant ignores the auditory stimuli, the MMN recording, in fact, can be optimal. Therefore, a visual distraction task, such as watching a movie (with no sound levels) or reading a book, is commonly used to eliminate other cognitive components elicited during active attention. Therefore, it is usually considered an obligatory ERP. Like P300, its neural generators include the primary auditory cortices, secondary association areas, frontal lobe, and hippocampus (Näätänen 2000).

The MMN is elicited by sudden changes in stimulation. It is observed as an increase in negativity in the latency range of 100–250 ms (the latency range of N1–P2) after change onset, which is strongest in temporal and frontal areas of the scalp map (Garrido et al 2009). This increased negativity is observed when subtracting the response waveform for standard stimuli from the response waveform for deviant stimuli. Amplitude, latency, and duration criteria are components of the derived MMN, and these values vary with the stimulus parameters. Larger stimulus acoustic differences lead to larger and earlier MMNs.

Typically, it is sufficient to use 5–10 active electrodes to record MMN, which always include Fz, Cz, C3, C4, and mastoid locations. The preferred reference is the nose. A recording band pass of 0.1–30 Hz is recommended, along with a sampling rate of no less than 200 Hz.

For MMN to become a useful tool for clinical practice, MMN recording and analysis methodology have recently been simplified and standardized (Näätänen et al 2004, Pakarinen et al 2007). The *optimal* MMN paradigm uses five deviant tones that differ from the standard tones on one of five possible attributes: duration, frequency, spatial location, intensity, or by having a sound gap in the middle of the tone. Each of the deviants ($p = 0.10$ each) is presented in the same series with standard tones ($p = 0.50$). Using the *optimal* MMN paradigm, a recording time can be shortened to approximately 15 min.

Although MMN looks similar to ACC, they are different neural responses. ACC is elicited by an acoustic change within a sound stimulus, whereas an acoustic difference between different stimuli or stimulus patterns elicits MMN. Moreover, the stimulus for eliciting ACC cannot be used to elicit MMN because the acoustic change within the stimulus occurs with a high probability. The advantage of the ACC paradigm over MMN is that responses to every stimulus can be used in analysis. This is important not only for the signal-to-noise ratio of the averaged waveform but for minimizing recording time.

14.22.3.2 Clinical Application

At present, the clinical applications of the MMN are limited due to large variability. The MMN has been recommended as an electrophysiologic index of perceptual and discriminative abilities (Näätänen 2003).

14.22.4 N400

The N400 is a negative wave in the average ERP approximately 400 ms after stimulus onset that is elicited to meaningful stimuli such as sentences. For example, semantically incongruent words within a sentence elicit a negative peak (N400) from 200 to 600 ms, while predictable sentence-final words elicit a broad positive waveform in this time range. However, smaller N400s are also obtained by the second words of semantically related as compared with semantically unrelated pairs. Thus, N400 amplitude is sensitive to lexical characteristics in addition to contextual factors. Generally, N400 amplitude is related to the difficulty of retrieving stored conceptual knowledge associated with a word, which is dependent on both the stored representation itself and the retrieval cues provided by the preceding context (Duncan *et al* 2009). The neural generator of N400 is the left temporal lobe with a smaller contribution from the right temporal lobe (VanPetten and Luka 2006).

14.22.4.1 Technique

Maximal N400 differences are observed between highly predictable and wholly incongruent sentence-terminal words, and in this case, as few as 20 trials per condition are possible (Kutas and Hillyard 1980). Generally, 40–120 trials per condition are used. Stimuli should not be repeated for a given participant, because N400 amplitude is markedly reduced by stimulus repetition (Olichney *et al* 2000). However, a task involving an overt response is not necessary to elicit robust N400 responses (Connolly *et al* 1990). To analyze N400, age matching clinical and control group is mandatory, because the amplitude of N400 semantic context effects declines continuously from childhood onward (Kutas and Iragui 1998). Visual sentences are conventionally presented one word at a time, with a sufficient intertrial interval of 3–5 s. To record N400 properly, four to five midline locations (Fpz, Fz, Cz, Pz, Oz) and two lateral pairs over the anterior and posterior temporal lobes (T3, T4, T5, T6) are required. Generally, a 32-channel recording montage is adequate to get more information about the scalp distribution of N400.

14.22.4.2 Clinical Application

The presence of N400 is an indicator of semantic comprehension and possible semantic awareness. Therefore, N400 has been used to examine normal language development and changes over time. In addition, it has been employed in patients with a variety of developmental, neurological, and psychiatric disorders (Johnson *et al* 2008).

14.23 CONCLUSION

This chapter reviewed a variety of concepts and clinical applications regarding physiologic measures in otology, neurotology, and audiology fields. Some of these various measures have become routine in otology and audiology clinics around the world, and some are predominately used for research purposes. By integrating clinical and research practices, these physiologic measures can be used optimally and maximally.

REFERENCES

Bahmer A, Peter O *et al* 2008 Recording of electrically evoked auditory brainstem responses (E-ABR) with an integrated stimulus generator in Matlab *J. Neurosci. Methods.* **173** 306–14

Banai K, Hornickel J *et al* 2009 Reading and subcortical auditory function *Cereb. Cortex* **19** 2699–707

Banai K, Nicol T *et al* 2005 Brainstem timing: Implications for cortical processing and literacy *J. Neurosci.* **25** 9850–7

Barsky-Firkser L and Sun S 1997 Universal newborn hearing screenings: a three-year experience *Pediatrics* **99** E4

Battmer R D, Dillier N *et al* 2004 Evaluation of the neural response telemetry (NRT) capabilities of the nucleus research platform 8: initial results from the NRT trial *Int. J. Audiol.* **43**(Suppl 1) S10–5

Beck D L and Hall J W 2001 Electroneuronography (ENoG): neurophysiologic evaluation of the facial nerve *Hear. J.* **54**(3) 36–44

Bell S L, Allen R *et al* 2002 An investigation of the use of band-limited chirp stimuli to obtain the auditory brainstem response *Int. J. Audiol.* **41** 271–8

Benito-Orejas J I, Ramirez B *et al* 2008 Comparison of two-step transient evoked otoacoustic emissions (TEOAE) and automated auditory brainstem response (AABR) for universal newborn hearing screening programs *Int. J. Pediatr. Otorhinolaryngol.* **72** 1193–201

Brantberg K 2009 Vestibular evoked myogenic potentials (VEMPs): usefulness in clinical neurotology *Semin. Neurol.* **29** 541–7

Brantberg K, Fransson P A *et al* 1999 Measures of the binaural interaction component in human auditory brainstem response using objective detection criteria *Scand. Audiol.* **28** 15–26

Brantberg K and Verrecchia L 2009 Testing vestibular-evoked myogenic potentials with 90-dB clicks is effective in the diagnosis of superior canal dehiscence syndrome *Audiol. Neurootol.* **14** 54–8

Brown C J 2003 Clinical uses of electrically evoked auditory nerve and brainstem responses *Curr. Opin. Otolaryngol. Head Neck Surg.* **11** 383–7

Brown C J, Abbas P J *et al* 1990 Electrically evoked whole-nerve action potentials: data from human cochlear implant users *J. Acoust. Soc. Am.* **88** 1385–91

Brown C J, Etler C *et al* 2008 The electrically evoked auditory change complex: preliminary results from nucleus cochlear implant users *Ear Hear.* **29** 704–17

Brown C J, Hughes M L *et al* 2000 The relationship between EAP and EABR thresholds and levels used to program the nucleus 24 speech processor: data from adults *Ear Hear.* **21** 151–63

Burns E M, Arehart K H *et al* 1992 Prevalence of spontaneous otoacoustic emissions in neonates *J. Acoust. Soc. Am.* **91** 1571–5

Cafarelli Dees D, Dillier N *et al* 2005 Normative findings of electrically evoked compound action potential measurements using the neural response telemetry of the Nucleus CI24M cochlear implant system *Audiol. Neurootol.* **10** 105–16

Campbell A P, Suberman T A *et al* 2010 Correlation of early auditory potentials and intracochlear electrode insertion properties: an animal model featuring near real-time monitoring *Otol. Neurotol.* **31** 1391–8

Chihara Y, Iwasaki S *et al* 2009 Ocular vestibular-evoked myogenic potentials (oVEMPs) require extraocular muscles but not facial or cochlear nerve activity *Clin. Neurophysiol.* **120** 581–7

Chow L C, Tam R C *et al* 2002 Use of electroneurography as a prognostic indicator of Bell's palsy in Chinese patients *Otol. Neurotol.* **23** 598–601

Cone-Wesson B and Wunderlich J 2003 Auditory evoked potentials from the cortex: audiology applications *Curr. Opin. Otolaryngol. Head Neck Surg.* **11** 372–7

Connolly J F, Stewart S H *et al* 1990 The effects of processing requirements on neurophysiological responses to spoken sentences *Brain Lang.* **39** 302–18

Cosetti M K, Shapiro W H *et al* 2010 Intraoperative neural response telemetry as a predictor of performance *Otol. Neurotol.* **31** 1095–9

Cosetti M K, Troob S H *et al* 2012 An evidence-based algorithm for intraoperative monitoring during cochlear implantation *Otol. Neurotol.* **33** 169–76

Cunningham J, Nicol T *et al* 2000 Speech-evoked neurophysiologic responses in children with learning problems: development and behavioral correlates of perception *Ear Hear.* **21** 554–68

Curthoys I S 2010 A critical review of the neurophysiological evidence underlying clinical vestibular testing using sound, vibration and galvanic stimuli *Clin. Neurophysiol.* **121** 132–44

Curthoys I S, Manzari L *et al* 2009 A review of the scientific basis and practical application of a new test of utricular function—Ocular vestibular-evoked myogenic potentials to bone-conducted vibration *Acta Otorhinolaryngol. Ital.* **29** 179–86

Davis H 1983 An active process in cochlear mechanics *Hear. Res.* **9** 79–90

Deupree D L and Jewett D L 1988 Far-field potentials due to action potentials traversing curved nerves, reaching cut nerve ends, and crossing boundaries between cylindrical volumes *Electroencephalogr. Clin. Neurophysiol.* **70** 355–62

Dimitrijevic A, John M S *et al* 2004 Auditory steady-state responses and word recognition scores in normal-hearing and hearing-impaired adults *Ear Hear.* **25** 68–84

Don M, Kwong B *et al* 2005 The stacked ABR: a sensitive and specific screening tool for detecting small acoustic tumors *Audiol. Neurootol.* **10** 274–90

Don M, Masuda A *et al* 1997 Successful detection of small acoustic tumors using the stacked derived-band auditory brain stem response amplitude *Am. J. Otol.* **18** 608–21; discussion 682–5

Duncan C C, Barry R J *et al* 2009 Event-related potentials in clinical research: guidelines for eliciting, recording, and quantifying mismatch negativity, P300, and N400 *Clin. Neurophysiol.* **120** 1883–908

Duncan-Johnson C C and Donchin E 1977 On quantifying surprise: the variation of event-related potentials with subjective probability *Psychophysiology* **14** 456–67

Eggermont J J, Ponton C W *et al* 1997 Maturational delays in cortical evoked potentials in cochlear implant users *Acta Otolaryngol.* **117** 161–3

Elberling C and Don M 2008 Auditory brainstem responses to a chirp stimulus designed from derived-band latencies in normal-hearing subjects *J. Acoust. Soc. Am.* **124** 3022–37

Elberling C and Don M 2010 A direct approach for the design of chirp stimuli used for the recording of auditory brainstem responses *J. Acoust. Soc. Am.* **128** 2955–64

Elberling C, Don M *et al* 2007 Auditory steady-state responses to chirp stimuli based on cochlear traveling wave delay *J. Acoust. Soc. Am.* **122** 2772–85

Erenberg S 1999 Automated auditory brainstem response testing for universal newborn hearing screening *Otolaryngol. Clin. North Am.* **32** 999–1007

Fellman V and Huotilainen M 2006 Cortical auditory event-related potentials in newborn infants *Semin. Fetal Neonatal Med.* **11** 452–8

Fernandez C, Goldberg J M *et al* 1990 The vestibular nerve of the chinchilla. III. Peripheral innervation patterns in the utricular macula *J. Neurophysiol.* **63** 767–80

Fisch U 1984 Prognostic value of electrical tests in acute facial paralysis *Am. J. Otol.* **5** 494–8

Fobel O and Dau T 2004 Searching for the optimal stimulus eliciting auditory brainstem responses in humans *J. Acoust. Soc. Am.* **116** 2213–22

Franck K H 2002 A model of a nucleus 24 cochlear implant fitting protocol based on the electrically evoked whole nerve action potential *Ear Hear.* **23**(1 Suppl) 67S–71S

Galbraith G C, Amaya E M *et al* 2004 Brain stem evoked response to forward and reversed speech in humans *Neuroreport* **15** 2057–60

Gantz B J, Rubinstein J T *et al* 1999 Surgical management of Bell's palsy *Laryngoscope* **109** 1177–88

Garrido M I, Kilner J M et al 2009 The mismatch negativity: a review of underlying mechanisms Clin. Neurophysiol. **120** 453–63

Gartner L, Lenarz T et al 2010 Clinical use of a system for the automated recording and analysis of electrically evoked compound action potentials (ECAPs) in cochlear implant patients Acta Otolaryngol. **130** 724–32

Ge X and Shea J J Jr 2002 Transtympanic electrocochleography: a 10-year experience Otol. Neurotol. **23** 799–805

Gibson W P, Moffat D A et al 1977 Clinical electrocochleography in the diagnosis and management of Meniere's disorder Audiology **16** 389–401

Gordon K, Papsin B C et al 2004 Programming cochlear implant stimulation levels in infants and children with a combination of objective measures Int. J. Audiol. **43**(Suppl 1) S28–32

Gouveris H and Mann W 2009 Association between surgical steps and intraoperative auditory brainstem response and electrocochleography waveforms during hearing preservation vestibular schwannoma surgery Eur. Arch. Otorhinolaryngol. **266** 225–9

Grolman W, Maat A et al 2009 Spread of excitation measurements for the detection of electrode array foldovers: A prospective study comparing 3-dimensional rotational x-ray and intraoperative spread of excitation measurements Otol. Neurotol. **30** 27–33

Guastini I, Mora R et al 2010 Evaluation of an automated auditory brainstem response in a multistage infant hearing screening Eur. Arch. Otorhinolaryngol. **267** 1199–205

Harris F P and Probst R 1991 Reporting click-evoked and distortion-product otoacoustic emission results with respect to the pure-tone audiogram Ear Hear. **12** 399–405

Hayes E A, Warrier C M et al 2003 Neural plasticity following auditory training in children with learning problems Clin. Neurophysiol. **114** 673–84

Hong S M, Yeo S G et al 2008 The results of vestibular evoked myogenic potentials, with consideration of age-related changes, in vestibular neuritis, benign paroxysmal positional vertigo, and Meniere's disease Acta Otolaryngol. **128** 861–5

Hornickel J, Knowles E et al 2012a Test-retest consistency of speech-evoked auditory brainstem responses in typically-developing children Hear. Res. **284** 52–8

Hornickel J, Zecker S et al 2012b Assistive listening devices drive neuroplasticity in children with dyslexia PNAS **109** 16731–6

Hudspeth A J 1989 How the ear's works work Nature **341** 397–404

Hughes J R and Fino J J 1985 A review of generators of the brainstem auditory evoked potential: contribution of an experimental study J. Clin. Neurophysiol. **2** 355–81

Iwasaki S, Chihara Y et al 2009 The role of the superior vestibular nerve in generating ocular vestibular-evoked myogenic potentials to bone conducted vibration at Fz Clin. Neurophysiol. **120** 588–93

Iwasaki S, McGarvie L A et al 2007 Head taps evoke a crossed vestibulo-ocular reflex Neurology **68** 1227–9

Iwasaki S, Smulders Y E et al 2008 Ocular vestibular evoked myogenic potentials to bone conducted vibration of the midline forehead at Fz in healthy subjects Clin. Neurophysiol. **119** 2135–47

Jirsa R E 1992 The clinical utility of the P3 AERP in children with auditory processing disorders J. Speech Hear. Res. **35** 903–12

Johnson J L, White K R et al 2005 A multisite study to examine the efficacy of the otoacoustic emission/ automated auditory brainstem response newborn hearing screening protocol: introduction and overview of the study Am. J. Audiol. **14** S178–85

Johnson K C 2002 Audiologic assessment of children with suspected hearing loss Otolaryngol. Clin. North Am. **35** 711–32

Johnson K L, Nicol T et al 2008 Developmental plasticity in the human auditory brainstem J. Neurosci. **28** 4000–7

Katz J 2002 Handbook of Clinical Audiology (London: Lippincott Williams & Wilkins)

Katz J 2009 *Handbook of Clinical Audiology* (Philadelphia: Wolters Kluwer)

Kemp D T 1978 Stimulated acoustic emissions from within the human auditory system *J. Acoust. Soc. Am.* **64** 1386–91

Kim A H, Kileny P R *et al* 2008 Role of electrically evoked auditory brainstem response in cochlear implantation of children with inner ear malformations *Otol. Neurotol.* **29** 626–34

Kimura H, Aso S *et al* 2003 Prediction of progression from atypical to definite Meniere's disease using electrocochleography and glycerol and furosemide tests *Acta Otolaryngol.* **123** 388–95

Kraus N and Nicol T 2005 Brainstem origins for cortical "what" and "where" pathways in the auditory system *Trends Neurosci.* **28** 176–81

Krishnan A and Gandour J T 2009 The role of the auditory brainstem in processing linguistically-relevant pitch patterns *Brain Lang.* **110** 135–48

Krueger W W and Ferguson L 2002 A comparison of screening methods in school-aged children *Otolaryngol. Head Neck Surg.* **127** 516–9

Kutas M and Hillyard S A 1980 Reading senseless sentences: brain potentials reflect semantic incongruity *Science* **207** 203–5

Kutas M and Iragui V 1998 The N400 in a semantic categorization task across 6 decades *Electroencephalogr. Clin. Neurophysiol.* **108** 456–71

Kuwada S, Batra R *et al* 1986 Scalp potentials of normal and hearing-impaired subjects in response to sinusoidally amplitude-modulated tones *Hear. Res.* **21** 179–92

Liberman M C, Zuo J *et al* 2004 Otoacoustic emissions without somatic motility: can stereocilia mechanics drive the mammalian cochlea? *J. Acoust. Soc. Am.* **116** 1649–55

Lightfoot G and Kennedy V 2006 Cortical electric response audiometry hearing threshold estimation: accuracy, speed, and the effects of stimulus presentation features *Ear Hear.* **27** 443–56

Lins O G and Picton T W 1995 Auditory steady-state responses to multiple simultaneous stimuli *Electroencephalogr. Clin. Neurophysiol.* **96** 420–32

MacDougall H G, Brizuela A E *et al* 2005 Patient and normal three-dimensional eye-movement responses to maintained (DC) surface galvanic vestibular stimulation *Otol. Neurotol.* **26** 500–11

Martin B A and Boothroyd A 2000 Cortical, auditory, evoked potentials in response to changes of spectrum and amplitude *J. Acoust. Soc. Am.* **107** 2155–61

Martin B A, Boothroyd A *et al* 2010 Stimulus presentation strategies for eliciting the acoustic change complex: increasing efficiency *Ear Hear.* **31** 356–66

Martin B A, Tremblay K L *et al* 2008 Speech evoked potentials: from the laboratory to the clinic *Ear Hear.* **29** 285–313

May M, Blumenthal F *et al* 1983 Acute Bell's palsy: prognostic value of evoked electromyography, maximal stimulation, and other electrical tests *Am. J. Otol.* **5** 1–7

Mens L H 2007 Advances in cochlear implant telemetry: evoked neural responses, electrical field imaging, and technical integrity *Trends Amplif.* **11** 143–59

Mertens R and Polich J 1997 P300 from a single-stimulus paradigm: Passive versus active tasks and stimulus modality *Electroencephalogr. Clin. Neurophysiol.* **104** 488–97

Miller C A, Brown C J *et al* 2008 The clinical application of potentials evoked from the peripheral auditory system *Hear. Res.* **242** 184–97

Monobe H and Murofushi T 2004 Vestibular testing by electrical stimulation in patients with unilateral vestibular deafferentation: galvanic evoked myogenic responses testing versus galvanic body sway testing *Clin. Neurophysiol.* **115** 807–11

Moon I J, Park G Y *et al* 2012 Predictive value of electrocochleography for determining hearing outcomes in Meniere's disease *Otol. Neurotol.* **33** 204–10

Mudduwa R, Kara N *et al* 2010 Vestibular evoked myogenic potentials: review *J. Laryngol. Otol.* **124** 1043–50

Murofushi T, Curthoys I S *et al* 1995 Responses of guinea pig primary vestibular neurons to clicks *Exp. Brain Res.* **103** 174–8

Näätänen R 1992 *Attention and Brain Function* (Hillsdale: L Erlbaum Associates)

Näätänen R 2000 Mismatch negativity (MMN): perspectives for application *Int. J. Psychophysiol.* **37** 3–10

Näätänen R 2003 Mismatch negativity: clinical research and possible applications *Int. J. Psychophysiol.* **48** 179–88

Näätänen R and Alho K 1995 Mismatch negativity—a unique measure of sensory processing in audition *Int. J. Neurosci.* **80** 317–37

Näätänen R, Pakarinen S *et al* 2004 The mismatch negativity (MMN): towards the optimal paradigm *Clin. Neurophysiol.* **115** 140–4

Näätänen R and Winkler I 1999 The concept of auditory stimulus representation in cognitive neuroscience *Psychol. Bull.* **125** 826–59

Narins P M and Lewis E R 1984 The vertebrate ear as an exquisite seismic sensor *J. Acoust. Soc. Am.* **76** 1384–7

Nguyen L T, Harris J P *et al* 2010 Clinical utility of electrocochleography in the diagnosis and management of Meniere's disease: AOS and ANS membership survey data *Otol. Neurotol.* **31** 455–9

Norton S J, Gorga M P *et al* 2000 Identification of neonatal hearing impairment: summary and recommendations *Ear Hear.* **21** 529–35

Ochi K, Ohashi T *et al* 2003 Vestibular-evoked myogenic potential in patients with unilateral vestibular neuritis: abnormal VEMP and its recovery *J. Laryngol. Otol.* **117** 104–8

Olichney J M, Van Petten C *et al* 2000 Word repetition in amnesia. Electrophysiological measures of impaired and spared memory *Brain* **123** 1948–63

Pakarinen S, Takegata R *et al* 2007 Measurement of extensive auditory discrimination profiles using the mismatch negativity (MMN) of the auditory event-related potential (ERP) *Clin. Neurophysiol.* **118** 177–85

Parbery-Clark A, Skoe E *et al* 2009 Musical experience limits the degradative effects of background noise on the neural processing of sound *J. Neurosci.* **29** 14100–7

Penner M J and Zhang T 1997 Prevalence of spontaneous otoacoustic emissions in adults revisited *Hear. Res.* **103** 28–34

Picton T W 1992 The P300 wave of the human event-related potential *J. Clin. Neurophysiol.* **9** 456–79

Picton T W, John M S *et al* 2003 Human auditory steady-state responses *Int. J. Audiol.* **42** 177–219

Picton T W, Skinner C R *et al* 1987 Potentials evoked by the sinusoidal modulation of the amplitude or frequency of a tone *J. Acoust. Soc. Am.* **82** 165–78

Polich J 2007 Updating P300: An integrative theory of P3a and P3b *Clin. Neurophysiol.* **118** 2128–48

Prieve B A and Stevens F 2000 The New York State universal newborn hearing screening demonstration project: introduction and overview *Ear Hear.* **21** 85–91

Purcell D W, John S M *et al* 2004 Human temporal auditory acuity as assessed by envelope following responses *J. Acoust. Soc. Am.* **116** 3581–93

Quaranta A, Campobasso G *et al* 2001 Facial nerve paralysis in temporal bone fractures: outcomes after late decompression surgery *Acta Otolaryngol.* **121** 652–5

Ricci A 2003 Active hair bundle movements and the cochlear amplifier *J. Am. Acad. Audiol.* **14** 325–38

Rosengren S M, McAngus Todd N P *et al* 2005 Vestibular-evoked extraocular potentials produced by stimulation with bone-conducted sound *Clin. Neurophysiol.* **116** 1938–48

Rosengren S M, Todd N P *et al* 2009 Vestibular evoked myogenic potentials evoked by brief interaural head acceleration: properties and possible origin *J. Appl. Physiol.* **107** 841–52

Rosengren S M, Welgampola M S *et al* 2010 Vestibular evoked myogenic potentials: past, present and future *Clin. Neurophysiol.* **121** 636–51

Russo N, Nicol T *et al* 2004 Brainstem responses to speech syllables *Clin. Neurophysiol.* **115** 2021–30

Russo N M, Nicol T G *et al* 2005 Auditory training improves neural timing in the human brainstem *Behav. Brain Res.* **156** 95–103

Santos-Sacchi J, Shen W *et al* 2001 Effects of membrane potential and tension on prestin, the outer hair cell lateral membrane motor protein *J. Physiol.* **531** 661–6

Sharma A and Dorman M F 2006 Central auditory development in children with cochlear implants: clinical implications *Adv. Otorhinolaryngol.* **64** 66–88

Sharma A, Dorman M F *et al* 2002 A sensitive period for the development of the central auditory system in children with cochlear implants: implications for age of implantation *Ear Hear.* **23** 532–9

Sharma A, Nash A A *et al* 2009 Cortical development, plasticity and re-organization in children with cochlear implants *J. Commun. Disord.* **42** 272–9

Shih C *et al* 2009 Ipsilateral and contralateral acoustic brainstem response abnormalities in patients with vestibular schwannoma *Otolaryngol. Head Neck Surg.* **141** 695–700

Simon M V 2011 Neurophysiologic intraoperative monitoring of the vestibulocochlear nerve *J. Clin. Neurophysiol.* **28** 566–81

Sinha S K and Basavaraj V 2010 Speech evoked auditory brainstem responses: a new tool to study brainstem encoding of speech sounds *Indian J. Otolaryngol. Head Neck Surg.* **62** 395–9

Skoe E and Kraus N 2010 Auditory brain stem response to complex sounds: a tutorial *Ear Hear.* **31** 302–24

Song J H, Nicol T *et al* 2011 Test-retest reliability of the speech-evoked auditory brainstem response *Clin. Neurophysiol.* **122** 346–55

Song J H, Skoe E *et al* 2012 Training to improve hearing speech in noise: biological mechanisms *Cereb. Cortex* **22** 1180–90

Song M H, Bae M R *et al* 2010 Value of intracochlear electrically evoked auditory brainstem response after cochlear implantation in patients with narrow internal auditory canal *Laryngoscope* **120** 1625–31

Stapells D R, Gravel J S *et al* 1995 Thresholds for auditory brain stem responses to tones in notched noise from infants and young children with normal hearing or sensorineural hearing loss *Ear Hear.* **16** 361–71

Stephens S D and Thornton A R 1976 Subjective and electrophysiologic tests in brain-stem lesions *Arch Otolaryngol.* **102** 608–13

Stone J L, Calderon-Arnulphi M *et al* 2009 Brainstem auditory evoked potentials—a review and modified studies in healthy subjects *J. Clin. Neurophysiol.* **26** 167–75

Streubel S O, Cremer P D *et al* 2001 Vestibular-evoked myogenic potentials in the diagnosis of superior canal dehiscence syndrome *Acta Otolaryngol. Suppl.* **545** 41–9

Tremblay K L, Billings C *et al* 2004 Speech evoked cortical potentials: Effects of age and stimulus presentation rate *J. Am. Acad. Audiol.* **15** 226–37 quiz 264

Tremblay K L, Friesen L *et al* 2003 Test-retest reliability of cortical evoked potentials using naturally produced speech sounds *Ear Hear.* **24** 225–32

van Dijk B, Botros A M *et al* 2007 Clinical results of AutoNRT, a completely automatic ECAP recording system for cochlear implants *Ear Hear.* **28** 558–70

Van Petten C and Luka B J 2006 Neural localization of semantic context effects in electromagnetic and hemodynamic studies *Brain Lang.* **97** 279–93

Wang C T and Young Y H 2006 Comparison of the head elevation versus rotation methods in eliciting vestibular evoked myogenic potentials *Ear Hear.* **27** 376–81

Watson S R and Colebatch J G 1998 Vestibulocollic reflexes evoked by short-duration galvanic stimulation in man *J. Physiol.* **513** 587–97

Woods D L, Clayworth C C *et al* 1987 Generators of middle- and long-latency auditory evoked potentials: Implications from studies of patients with bitemporal lesions *Electroencephalogr. Clin. Neurophysiol.* **68** 132–48

Wunderlich J L, Cone-Wesson B K *et al* 2006 Maturation of the cortical auditory evoked potential in infants and young children *Hear. Res.* **212** 185–202

Xu Y, Krishnan A *et al* 2006 Specificity of experience-dependent pitch representation in the brainstem *Neuroreport* **17** 1601–5

Yamakami I, Yoshinori H *et al* 2009 Hearing preservation and intraoperative auditory brainstem response and cochlear nerve compound action potential monitoring in the removal of small acoustic neurinoma via the retrosigmoid approach *J. Neurol. Neurosurg. Psychiatry* **80** 218–27

Yetiser S 2012 Total facial nerve decompression for severe traumatic facial nerve paralysis: a review of 10 cases *Int. J. Otolaryngol.* **2012** 607359

Zappia J J, O'Connor C A *et al* 1997 Rethinking the use of auditory brainstem response in acoustic neuroma screening *Laryngoscope* **107** 1388–92

CHAPTER 15

Pathology

Chemical Tests

Arlyne B. Simon, Brendan M. Leung, Tommaso Bersano-Begey,
Larry J. Bischof, and Shuichi Takayama

CONTENTS

15.1	Introduction to Clinical Pathology	356
15.2	Chemical Pathology Laboratory	357
15.3	Immunology Laboratory	358
15.4	Microbiology Laboratory	358
15.5	Molecular Pathology Laboratory	358
15.6	Hematology Laboratory	358
15.7	Transfusion Medicine	359
15.8	Coagulation Laboratory	359
15.9	Tissue Typing Laboratory	359
15.10	Point-of-Care Testing	359
15.11	Design Considerations for Automated Systems in Clinical Pathology	360
15.12	Sample Management Instrumentation	360
15.13	Reagent Handling Instrumentation	361
15.14	Quality Control	362
15.15	Clinical Cytometry	362
	15.15.1 Fluidics Systems	363
	15.15.2 Optical System	365
	15.15.3 Electronics and Detection System	366
	15.15.4 Data Acquisition System	367
	15.15.5 Electrostatic and Mechanical Cell Sorting	368
	15.15.6 Slide-Based Cytometry	369
15.16	Immunoassays	369
	15.16.1 Colorimetric Assays	370
	15.16.2 Lateral Flow Immunoassays	372
	15.16.3 Fluorescence Intensity Reading–Based Assays	374
	15.16.4 Fluorescence Instrumentation for Singleplexed Immunoassays	376

15.16.5   Fluorescence Instrumentation for Multiplexed Immunoassays   377

15.16.6   Luminescent Assays   379

15.16.7   Bead-Proximity Assay Instrumentation   379

15.16.8   Electrochemiluminescence Instrumentation   379

15.16.9   Chemiluminescent Instrumentation for Singleplexed and Multiplexed Immunoassays   379

15.16.10   Refractive Index–Based Label-Free Detection Methods   380

15.17   Nucleic Acid Assays   380

15.17.1   Enzyme-Mediated Target Amplification Technologies   381

15.17.2   Real-Time PCR Amplification Technology   382

15.17.3   Digital PCR–Based Amplification Technologies   383

15.17.4   Transcription-Based Amplification Technologies   384

15.17.5   Strand Displacement Amplification Technology   384

15.17.6   Rolling Circle Amplification Technology   384

15.17.7   Signal Amplification Technologies   385

15.17.8   Fluorescence In Situ Hybridization   385

15.17.9   Peptide Nucleic Acid–Based FISH   386

15.17.10   Chromogenic In Situ Hybridization   386

15.17.11   Nucleic Acid Microarray Platforms   387

15.17.12   Planar-Based DNA Microarray Technologies   387

15.17.13   Bead-Based DNA Microarrays   388

15.17.14   Next-Generation DNA Sequencing Technologies   389

15.18   Conclusions and Future Challenges   389

References   390

15.1 INTRODUCTION TO CLINICAL PATHOLOGY

Clinical pathology is the medical discipline that utilizes laboratory-based testing to provide data to aid clinicians in the care of patients. For almost every type of health maintenance exam as well as disease state, a clinical pathology laboratory can provide diagnostic or prognostic information. This field is also referred to as laboratory medicine since the testing is routinely performed in laboratories. The laboratories are directed either by clinical pathologists, typically with additional fellowship training, or by an individual with a PhD who has completed a clinical fellowship in a specific area of laboratory medicine.

Since clinical pathology impacts on essentially all disease states, the number and diversity of tests performed is tremendous. Due to this variety of tests and the need for expertise in specific medical areas, clinical pathology is typically divided into multiple specialty laboratories. While the number and division of the laboratories may vary depending on the scope and size of the institution, a general division and focus of the laboratories is shown in Table 15.1. More specific information regarding disease diagnosis and the types of instruments for each laboratory will be provided in the following sections.

The basis for instrumentation in clinical pathology relates both to the need to provide highly accurate and reproducible results for patient care and to the need to perform these tests in a high-throughput fashion. There is a constant push to obtain higher quality

TABLE 15.1   Overview of Clinical Laboratories and Relevant Laboratory Assays

Laboratory	Laboratory Assays Relevant to
Chemical pathology and immunology	Body chemistry, organ function, antibody detection
Microbiology	Infectious disease (bacteriology, virology, mycology, and parasitology)
Hematology and coagulation	Red and white blood cell and platelet disorders and blood clotting
Transfusion medicine	Blood product transfusion
Tissue typing	Stem cell and solid organ transplantation
Molecular pathology	Molecular mutations in inherited disease and oncology
Cytogenetics	Chromosomal alterations in inherited disease and oncology

and quantity of information at ever-increasing speeds and lower costs. Due to the diversity of assays in clinical pathology, describing the large variety of instruments to encompass all of these assays would not be possible in a review of this size. By design, this chapter is less comprehensive in terms of description of all the types of commercially available machines. There are other texts that fill this need well already (Kost 1996). Instead, instrument categories and assay mechanisms that are used for multiple assays, potentially even in different laboratories, will be discussed in separate sections to help describe major instrumentation types. This chapter provides a general overview of how testing is organized and distributed across instruments. An important trend is the increased use of molecular diagnostics that identify disease states more precisely and quickly. The chapter is thus more focused on these small molecule–based, protein-based, and nucleic acid–based tests and cytometry rather than culture tests, histology, and other tests that are also valuable but are time consuming or qualitative in nature. We also describe the practical needs and requirements in addition to the sensing and quantification mechanisms used in the instruments. Thus, we include a section that explains key design considerations for automated instruments. The specific methods for detecting protein, nucleic acids, and cell-based assays are summarized primarily in table format. We have chosen this combination of practical and technological perspectives to help fill a gap in understanding between instrument developers and clinical users so that the transition of new instruments from the bench to bedside can be accelerated.

15.2 CHEMICAL PATHOLOGY LABORATORY

The chemical pathology laboratory is perhaps the most diverse of the clinical laboratories as the testing performed in this laboratory provides physiological information about nearly all of the organ systems and most pathophysiological conditions. Some analytes are detected directly, such as the measurement of electrolytes by ion-selective electrodes. Other analytes, such as many proteins, are detected indirectly by an immunological assay or by the product of an enzymatic reaction that is detected by spectrophotometry. There is also often more than one method that can be used to detect a specific analyte. Due to the large volume of these tests, instruments performing the core of the chemical assays are automated. There are also automated delivery systems that can centrifuge patient samples and distribute the tubes to the proper instrument through a conveyor-belt-type line.

15.3 IMMUNOLOGY LABORATORY

The immunology laboratory may function within a chemical pathology laboratory or be a separate laboratory. Most of the assays within the immunology laboratory are designed to detect and, in some cases, quantify either total antibodies or antibodies that recognize a specific antigen. Detection of antibodies against specific antigens is useful in the assessment of an infectious disease, such as Epstein–Barr virus in infectious mononucleosis, as well as autoimmune diseases where the antibody recognizes a self-antigen from the patient. Identification of these autoantibodies has been traditionally performed by fluorescent microscopic examination of cells and tissues that contain target antigens for the autoantibodies. However, new instruments that utilize fluorescent beads and automated detection are now being incorporated into these and other related testing procedures.

15.4 MICROBIOLOGY LABORATORY

The microbiology laboratory identifies infectious organisms and provides information regarding drug susceptibility to guide appropriate therapy. Attempts to identify pathogenic organisms include microscopy, culture, serology, and nucleic acid testing. Traditionally, diagnostic testing within the microbiology laboratory has been labor intensive due to the need to appropriately process the various specimens that enter the laboratory as well as the hands-on effort required to set up and evaluate cultures. Several advances in diagnostics have reduced the manual labor in this type of laboratory. These include instruments that continuously monitor blood culture bottles for evidence of microbial growth and instruments capable of performing automated biochemical analysis and antibiotic susceptibility testing.

Molecular diagnostics and its associated instrumentation are rapidly redefining microbiology laboratories. For example, nucleic acid–based testing is used both qualitatively to detect nucleic acids for a specific organism, such as *Herpes simplex virus*, and quantitatively to detect copy numbers of a specific nucleic acid, such as for monitoring HIV therapy. DNA sequencing can be used to aid in the identification of microorganisms.

15.5 MOLECULAR PATHOLOGY LABORATORY

Molecular pathology or molecular diagnostics is the use of information contained within nucleic acids, both DNA and RNA, to provide diagnostic, prognostic, and therapeutic results. This area is rapidly growing and impacts all areas of pathology, including inherited disease, oncology (both solid tumor and hematologic malignancies), infectious disease, and other areas where genetic information may impact treatment or risk assessment. Examples of this latter category include pharmacogenetic testing to predict drug metabolism and genotyping of red blood cell antigen genes to aid blood transfusions.

15.6 HEMATOLOGY LABORATORY

The hematology laboratory functions to provide cell counts within blood and to identify abnormalities within red blood cells, white blood cells, and platelets. This laboratory may also provide diagnostic information regarding other body fluids, such as urine, cerebrospinal fluid, and bronchoalveolar lavage fluid. While manual microscopic examination is performed for select cases, the majority of cell counting and identification is performed by automated high-throughput instruments. These instruments combine powerful imaging techniques with advance pattern recognition algorithms to identify rare target cells within the sample population.

Another significant function of the hematology laboratory is to perform immunophenotyping by flow cytometry to aid in the diagnosis of leukemias and lymphomas as well as provide cell counts for specific populations. An example of this latter testing includes determining the number of CD4+ cells in an HIV/AIDS patient.

15.7 TRANSFUSION MEDICINE

There are two distinct services that a transfusion medicine laboratory may perform. The first is the safe administration of blood products, including red blood cells, platelets, and plasma, to patients. This function requires accurate identification of a patient's blood type and the presence of antibodies. Manual methods may still be required to identify some specific antibodies against red blood cells. However, automated instruments are routinely used to identify the ABO and Rh antigens on a patient's red blood cells as well as to perform an initial screen for antibodies against red blood cells. The second major function of a transfusion medicine lab includes therapeutic apheresis, red blood cell exchange, and collection of stem cells for transplantation.

15.8 COAGULATION LABORATORY

The role of the coagulation laboratory is to provide information regarding the ability of an individual to properly regulate the homeostasis of blood clotting. This laboratory monitors clotting factors within the plasma as well as the function of platelets. Clotting factor assays are performed on instruments that detect in vitro clot formation, while platelet assays utilize instruments that can detect ATP release from the platelet as well as platelet aggregation.

15.9 TISSUE TYPING LABORATORY

A tissue typing laboratory provides vital information that is required for successful solid organ and stem cell transplantation. This laboratory identifies human leukocyte antigens of a potential transplant recipient as well as antibodies that individual may possess against other human leukocyte antigens. This information is utilized to help select the best available donor for the recipient to maximize the likelihood of successful transplantation. Due to the increasing number of transplants performed, instrumentation is utilized to aid in the identification of a patient's human leukocyte antigen complex (HLA) as well as potential antibodies. Analysis of a patient's DNA provides more accurate information regarding a patient's HLA genes than serological assays. High-throughput fluorescent-based instruments are being utilized to identify potential antibodies a patient may have.

15.10 POINT-OF-CARE TESTING

The goal of point-of-care testing is to provide a real-time, rapid result that a caregiver can utilize to make a clinical decision. An example of routine point-of-care testing is glucose monitoring for individuals in the hospital or clinic as well as at home. Attempts are also being made to provide point-of-care molecular testing, with such an example being identification of a patient with methicillin-resistant *Staphylococcus aureus* (MRSA) upon admission to the hospital to provide appropriate isolation requirements.

15.11 DESIGN CONSIDERATIONS FOR AUTOMATED SYSTEMS IN CLINICAL PATHOLOGY

Automation of clinical pathology labs is relatively recent. The benefits of implementing automated systems in clinical pathology laboratories started to be well demonstrated and well documented in the mid-1990s (Kost 1996), and some key advantages include

- Increased sample throughput

- Enhanced assay precision by removing operator variability

- Faster turnaround times

- Improved efficiency through continuous operation

- Better quality control by automatic logging of data and collection and interlaboratory tracking of specimens

- A user-friendly computerized interface

- Improved safety by reducing technologist exposure to infectious specimen and hazardous chemicals

Since then, the availability and implementation of various task-targeted to total automation systems has increased dramatically. When designing biochemical analysis systems for clinical use, it is helpful for the tool developers to keep these types of automation requirements in mind. There is increased pressure for all laboratorians to perform reliable clinical tests using smaller volumes of patient specimen, to accurately detect low concentrations of target antigens or biomarkers in the specimen using highly sensitive assays, and to demonstrate faster turnaround times so that patients spend less time in the emergency room or can potentially be discharged from the hospital more rapidly. At the same time, the specific situation for each health-care facility may be different. Large hospitals, for example, with large numbers of samples may benefit from total automation, whereas smaller clinics may find task-targeted automation more practical. In this section, we will briefly discuss some important instrument design considerations to facilitate pathology workflow from specimen collection to data acquisition, including specimen bar coding to provide well-documented tracking of samples during intra- and interlaboratory transport.

15.12 SAMPLE MANAGEMENT INSTRUMENTATION

Sample management encompasses the sample collection and testing process in hospital laboratories. Highly efficient sample management systems are required to reduce the turnaround time between test order and results. Under these pressures and constraints, the sample collection process can easily become error prone: mislabeled specimen, miscommunicated test orders, and lost or misrouted samples are examples of adverse events. A commonly used solution to minimize these potentially clinically significant errors is it to affix barcode identification tags that contain a list of tests that were meant to be performed onto the collected specimen samples. Such barcode medical administration (BCMA) systems ensure that the five rights of patients are preserved (right patient, right medication, right dose, right time, and right route).

Automated systems that maximize the number of specimens processed per instrument run or maximize the number of tests that can be performed on a given specimen are advantageous for both stat (urgent) and routine tests. Stat (derived from the Latin word *statim*, which means *immediately*) laboratory tests are required urgently to manage medical emergencies, and these tests are given the highest priority during test processing, analyzing, and reporting. Examples of stat tests include assays for hemolytic transfusion and for myocardial infarction. Stat test–compatible instruments should be designed with minimum delay and minimum queuing to ensure fast time to first result. Besides stat instrumentation, automated systems that operate in batch modes are needed for high-volume routine testing in clinical labs.

15.13 REAGENT HANDLING INSTRUMENTATION

To improve assay precision and ensure real-time results, the number of manual steps or time taken to perform necessary assay steps should be minimized. Automation can optimize reaction kinetics by reducing sample volume, maintaining the ideal temperature, or agitating reagent–sample mixtures. Instruments like immunoassay analyzers must also implement between-sample washes or other methods to prevent contamination between successive samples, particularly in heterogeneous assays where extensive washing is required to separate bound from unbound antigens. While automation of these complex processes in large total automation systems has become widespread, platforms that cater to the needs of smaller hospitals with lower throughput that may only need task-targeted automation are still lacking. Recent examples that aim to fill this need include the centrifugal microfluidic platform that performs sample and reagent handling for in vitro diagnostic assays. In this miniaturization technology, microfluidic channels are embedded into plastic discs (standard 12 in. compact disc size). Fluid samples are loaded through ports located near the center of the discs. When discs are rotated, the centrifugal force generated will cause the samples to flow radially away from the center. The sample flow rate can be controlled by disc rotation speed, minimizing need for external pumps and other complicated controls. Centrifugal microfluidics can automate various operations commonly used in clinical pathology labs including cell lyses, sample separation, washes, reagent dosing, incubations, and signal detection. Currently, several types of clinical assays have been successfully implemented on the centrifugal microfluidic platform, including immunoassays and pathogen detection by polymerase chain reaction (PCR) (Gorkin *et al* 2010). The main advantages of centrifugal microfluidics include system integration and automation for targeted tasks in a small overall footprint at low cost.

15.14 QUALITY CONTROL

Instruments that provide stable, long-term assay calibration curves are essential in the clinical lab. Factors that affect assay calibration include reagent stability, reagent evaporation, and cross contamination. In order to improve the stability of reagents in the clinical analyzer, reagents can be stored as liquids, suspensions, or lyophilized powders (Wild 2005). Reagent evaporation can be avoided by storing assay reagents in self-sealing packages that automatically reseal when tests are not performed on the instrument. Additionally, use of disposable tips or low-adsorption Teflon-coated probes can help prevent sample-to-sample cross contamination.

The constraints of the sample collection process alone require an interdisciplinary team of medical physicists and chemical, electrical, and mechanical engineers to design and develop the instrumentation. Furthermore, to develop new clinical instruments, a fundamental understanding of assay principles and detection methods of some common cell, protein, and nucleic acid clinical measurements is essential. The goal of the following three main sections is to provide the reader with a concise description of key instrumentation needed to accurately perform small molecule and cell-based clinical diagnostic and prognostic tests.

15.15 CLINICAL CYTOMETRY

Common in clinical immunology and hematology labs, the flow cytometer is used for DNA content analysis in solid tumors as well as rapid immunophenotyping on cells in whole blood, bone marrow, and tissue samples. More specific examples of flow cytometric analyses include (1) monitoring CD4 T-cell levels in the diagnosis and prognosis of HIV-infected patients, (2) quantifying fetal hemoglobin for detecting fetomaternal hemorrhage, and (3) identifying specific cell surface proteins for the diagnosis and classification of leukemias.

Flow cytometry uses the principles of light scattering and fluorescence to simultaneously quantify size, internal complexity, and fluorescence markers of individual cells or particles in a heterogeneous cell population. The key advantage of flow cytometers is their ability to measure fluorescence per cell or particle, whereas spectrophotometers measure the absorbance or transmission of the bulk specimen. This unique technological feature enables the identification and separation of rare cells from patient specimen while simultaneously excluding dead cells and debris from data analyses. There are two types of flow cytometers, those that only detect light scattering and fluorescence from single cells and those that have the added capability of sorting cells and physically placing the cells into discrete locations for subsequent analysis.

A fluorescence-based detection method, flow cytometry, is comprised of four interrelated systems, namely, fluidics, optics, electronics, and data acquisition systems. These four systems are generally incorporated into all cytometers irrespective of instrument manufacturer and whether the instrument possesses sorting capability. Briefly, the fluidics system transports patient specimen via hydrodynamic focusing to the laser beam for interrogation. The laser-based optical system causes light scattering and excites antibody-conjugated fluorochromes, which bind specifically to target antigens on cell subpopulations. The electronics system collects and digitizes emitted fluorescence, scattered and reflected light.

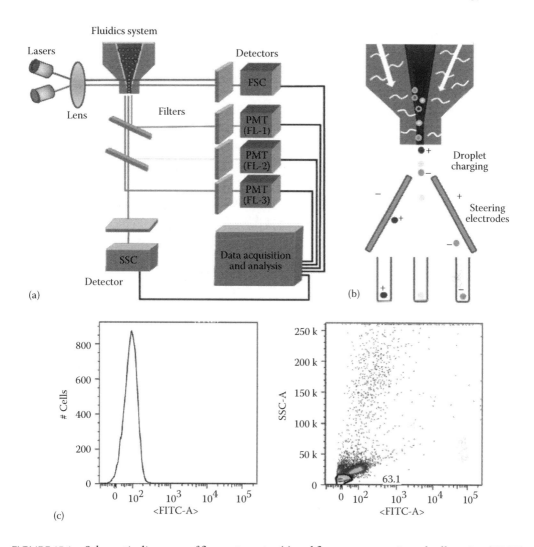

FIGURE 15.1 Schematic diagrams of flow cytometry (a) and fluorescence-activated cell sorting (FACS) (b). In a conventional flow cytometer, cells flowing in single file stream are interrogated by focused lasers. The scattered lights, namely forward scatter (FSC) and side scatter (SSC), as well as emitted lights from fluorescent cells in the stream are separated and measured with individual detectors, and the intensities and frequencies were plotted (c). Based on these signals, cells can be physically sorted and separated by applying charge on the droplet. Downstream from the optical detection module, charged cells were steered in parallel plate electrodes and guided into the appropriate sample tube (b).

Lastly, the resulting data are analyzed by the data acquisition system. The general design of a flow cytometer is illustrated in Figure 15.1.

15.15.1 Fluidics Systems

The main purpose of the fluidics system is to separate and align cells in a fluid stream so that they can be interrogated individually downstream. In conventional flow cytometers, cell suspension aliquots are injected into the closed-flow chamber and are forced into the core of a fast-flowing liquid stream called sheath fluid. The width of the core fluid stream that contains the cells can be controlled by changing the differential pressure between sheath and core flow. This hydrodynamically focused core stream enable cells to travel in single file at thousands of cells per second through a narrow orifice (typically 70 μm). To ensure this orderly flow of cells, the dimensionless Reynolds number, which provides the ratio of inertial forces (resistant to motion) to viscous forces in the fluid flow, should not be too large, typically less than 100 (Robinson 2004). The Reynolds number is defined as $Re = \rho v D/\mu$ where Re is the Reynolds number, ρ is the fluid density (kg/m^3), v is the mean velocity of the sample stream (m/s), D is the diameter of the core sample stream (m), and μ is the dynamic viscosity of the fluid (kg/(m·s) or Pa·s).

Despite its widespread use, hydrodynamic focusing has several drawbacks. First, the small orifice needed to maintain laminar flow means that current methods have difficulty handling particle larger than 70 μm in diameter. This limits the potential for studying cell–cell interactions or cell agglomerates. Also the small volumetric flow rate of the sample core limits the throughput. Typically, a single-sheath flow-based fluidics system can handle 50,000 cells/s. However, this is not sufficient for high-throughput applications, for example, isolation and investigation of rare circulating tumor cells, where the frequency of target cells in blood can be much lower than one in a billion cells. In addition, the control of sheath fluid and sample injection requires complex mechanisms and bulky equipment, thus limiting the potential of flow cytometry in terms of size and cost reduction as well as parallel analysis.

Several methods have been explored in microfluidic flow systems as a means to focus particles in a flow stream without the need for sheath fluid, that is, sheathless focusing. These methods can be categorized as either field-based or flow-assisted methods. In field-based methods, external forces exerted by electrical, acoustic, or optical fields are used to focus particles into specific regions within the flow stream. Dielectrophoretic (DEP) focusing uses an electrode pair or an array of electrodes to generate an electric field and drive particles toward the center of the flow stream. The efficiency of DEP focusing depends on the field strength, size, and charge of the particle (Srivastava *et al* 2010). In contrast, acoustic focusing uses pressure gradients generated by acoustic standing waves within the flow stream to either the node or antinode position (Goddard *et al* 2006a,b). The acoustic wave generates by a piezoelectric crystal attached to the external surface of flow capillary can focus particles (diameter ~40 μm) within the bulk medium steam. In soft and acoustically nonreflective polymer channels such as polydimethylsiloxane (PDMS), pairs of transducers can be arranged to generate standing surface acoustic waves that interfere constructively to focus particles into 10 μm stream within the PDMS channels (Shi *et al* 2008). Some commercial cytometers have incorporated these technologies as ways to reduce the need for sheath flow, thus reducing the bulk and cost associated with flow control, pumping, and fluid storage needed for current flow cytometer platforms.

In contrast to field-based focusing, flow-assisted sheathless focusing methods capitalize upon the steric hindrance between particles and static microstructures to separate particles based on size. One example of flow-assisted sheathless focusing is the deterministic lateral displacement (DLD) array, which consists of a staggered postarray to steer particles toward the center of the channel. The steric hindrance effect of DLD may also be achieved by using splitting and merging microchannels, where large particles stay within the main channel and smaller particles readily move into branching channels. Another form of deflection strategy uses anisotropic V-shaped grooves, which generate pressure fields within the stream, and particles are deflected into an equilibrium position in a process called hydrophoresis. These obstacles cause rotational streamlines to form within the fluid while simultaneously inducing steric hindrance to the particles in the stream. Large particles are more likely to be knocked off the streamlines due to hindrance than smaller particles, and in general, particles with a diameter comparable to the obstacle gap will be steered to the center of the channel and remain in this position due to particle–wall interaction. Using this technique, Choi et al. (2008) were able to focus Jurkat cell within a 1 mm channel with a standard deviation of 8.7 μm.

Even in the absence of anisotropic features on channel walls, particles can still be focused using inertial lift forces, including shear-gradient force and wall-induced lift force. These forces are normally negligible at low particle Reynolds number, defined as $Re_p = \rho D^2 U / \mu D^h$, where D is the particle diameter, U is the maximum fluid velocity, μ is the dynamic fluid viscosity, and ρ is the fluid density. The hydraulic diameter, D^h, is defined as $D^h = 2wh/(w + h)$, where w and h are the width and height of the channel, respectively. In cases where particles such as blood cells flow within microchannel, Re_p can exceed the order of 1, where inertial lift begins to dictate particle movement. Inertial lift results from the balance between wall effect lift forces and shear-gradient forces acting on the particle. Using curved channels, Di Carlo (2009) demonstrated that inertial forces can focus particles (10 μm polystyrene beads) within a standard deviation of 80 nm. He has also developed tuning parameters based on the Reynolds number and channel dimensions so that one can estimate the length of channel and fluid velocity required to achieve desired degrees of alignment for a given particle size. An aspect of inertial lift focusing is that it is particle size dependent, and this has led to its application as a cell sorting mechanism.

Regardless of which method is used, stable positioning of the focused sample stream is critical for accurate downstream measurements. With the exception of acoustic focusing, there are few implementations of sheathless particle focusing methods to reliably align particles in 3-D (i.e., lateral as well as vertical focusing with respect to flow) regardless of particle size. The consistencies and accuracies of some of these sheathless methods also depend on flow rate and particle size, thus limiting their throughput and range of application. To date, sheathless flow systems still have lower throughput compared to conventional sheath flow systems, but this gap is closing with better focusing technologies. These challenges will have to be resolved before sheathless flow focusing can be widely adopted commercially.

15.15.2 Optical System

Prior to sample injection into the flow chamber, cells are incubated with fluorescently tagged antibodies that bind specifically to antigens on target cells. As cells pass through the flow chamber, the cells deflect incident light and fluorescent probes are excited by single or multiple lasers. Individual cells deflect incident light in two major ways: in the forward direction (forward scatter light) and at a right angle to the laser beam's axis (right-angle scatter light). The forward scatter light is proportional to the surface area or size of each cell, whereas the right-angle scatter light is proportional to the granularity and internal complexity of the cell. The choice of the laser is primarily dependent upon its excitation wavelength because the excitation wavelength should be within the absorption spectra, ideally close to the maxima, of the fluorescent probes used. In this regard, the argon ion laser is the most commonly used excitation source because the coherent 488 nm light that it emits can excite multiple fluorochromes. Two of these fluorochromes are fluorescein isothiocyanate (FITC) and phycoerythrin (PE), which emit light at 530 and 570 nm, respectively. If the emission maxima of distinct fluorochromes are sufficiently far apart as is the case with FITC and PE, a single laser can excite both fluorescent probes simultaneously, and resulting signals are directed through a network of mirrors and optical filters to different detectors.

To maximize the amount of information collected from a small volume patient specimen as well as to better characterize the immunophenotype of each cell, many clinical flow cytometers perform multicolor analysis, evaluating greater than four fluorescent signals per cell in addition to cell size and granularity. Such multicolor analysis is enabled by use of optical filters. Long pass filters transmit light above a specified wavelength. Conversely, short pass filters transmit light below a cutoff wavelength. Bandpass filters, however, transmit light just within a well-defined narrow range of wavelengths (bandwidth). Minimizing or eliminating filter cross talk is an important design consideration. Even with the use of optical filters, problems arise during multicolor analysis due to emission spectral overlap between different fluorochromes. For example, FITC-labeled cells will have strong fluorescence signals in the green fluorescent channel but will also have weak signals in the orange-red channel. Fluorescence compensation control calculations help minimize the detection of false-positives.

Recent advances in probe design and detection methods have helped mitigate the problem of channel cross talk and made it possible to increase the number of multiplexed assays for the simultaneous detection of multiple antigens. Semiconductor nanoparticles, also known as quantum dots, have a much narrower excitation and emission spectrum compared to conventional organic fluorophores, thus reducing optical cross talk. Another approach employs Raman scattering as a detection modality to further improve spectral resolution. Using specially designed probes and sensitive charge-coupled device (CCD) imaging systems, surface-enhanced Raman spectroscopy (SERS) can be incorporated into conventional flow cytometers. The advantage here is that one excitation source can yield many distinct signals, which can all be collected with one detector. However, to the author's knowledge, this technology is still in proof-of-concept phase and yet to be commercially implemented.

15.15.3 Electronics and Detection System

Emitted forward scatter, right-angle scatter light, and fluorescent signals are routed to highly sensitive silicon photodiode or photomultiplier tube (PMT) detectors. Of these detector types, the more sensitive PMTs are generally used to detect fluorescence signals as well as the weaker right-angle scatter signals, which only account for 10% of the emitted light per cell. Measurements from each detector are referred to as parameters, for example, forward scatter, right-angle scatter, or fluorescence. Data obtained from each parameter are called events and represent the number of cells that exhibit the target properties or markers. Each voltage pulse from an event has three key parameters: pulse height, pulse width, and pulse area or integral. These voltage parameters determine cell size, granularity, and fluorescence intensity.

With the view of maximizing the information that can be enumerated from each sample preparation, microscopy components can be incorporated into the signal acquisition path to capture images of individual cells as they pass through the detection point. Known as imaging flow cytometry, fluorescence signal as well as phase contrast images are captured using a combination of high-sensitivity CCD sensors, optical filtration systems, and digital imaging algorithms. Current commercial units of these so-called imaging flow cytometers are able to image 300 cells/s, at 40–60× magnification, while capturing 6 images per cell, including dark-field (side-scatter), bright-field (transmitted light), and four fluorescence images corresponding to FL1 to FL4 spectral band in a conventional flow cytometer. When coupled with an automated image recognition algorithm (up to 40 features per image), these images can be used to quantitatively express cell parameters that are not easily captured with conventional flow cytometers, including cell shape, distribution, and colocalizations of probes. These images can also be used to verify the consistency of immunostaining protocols. The imaging capability can extend the diagnostic power of flow cytometer in areas such as detection of circulating tumor cells and high-throughput fluorescence in situ hybridization (FISH) staining.

15.15.4 Data Acquisition System

Digitized fluorescent and light signals are often stored within the flow cytometry data acquisition system in standard flow cytometry formats, such as the Society of Analytical Cytology FCS 4.0 standard (Lee *et al* 2008). Briefly, data sets within each data file are organized with three required *header*, *text*, and *data* sections and one optional *data analysis* segment. The *header* contains text byte offsets necessary for locating the other segments within the data file. The *text* contains keyword pairs that describe the experiment, instrument used, specimen, data set, and any additional information the clinician wishes to include. The *data* segment contains the actual multicolor flow cytometry data sets. Lastly, the optional *analysis* segment can contain texts that describe clinician analyses of the generated data.

A key advantage of flow cytometry is gating—the ability to eliminate dead cells and debris from cell analyses. Typically, cells are gated by their size, which is estimated by forward scatter light. Often, dead cells have lower forward scatter and greater right-angle scatter than living cells. To analyze cytometric data, multiple parameter histograms are generated. As illustrated in Figure 15.1, a commonly used format is the dual-parameter histogram used to compare two parameters such as light scatter and fluorescence or two different wavelengths of fluorescence, where one parameter is plotted on the horizontal axis and the second parameter is plotted on the vertical axis.

15.15.5 Electrostatic and Mechanical Cell Sorting

Although the previous sections described the design principles needed for nonsorting flow cytometers, most of these principles can be also applied to cell sorting flow cytometers. Differences between clinical cytometry techniques are tabulated in Table 15.2. In this section, we will describe the distinct design considerations for cell sorting flow cytometers. Currently, electrostatic and mechanical sortings are the two main methods for cell sorting via flow cytometry. During electrostatic sorting, the cells are ejected through a nozzle with

TABLE 15.2   Comparison of Clinical Cytometry Techniques

Assay Mechanism		Flow Cytometry		Image Cytometry
Signal detection mode	Light scattering (FSC and SSC)	Fluorescence (immunofluorescence)	Raman scattering[a]	Fluorescence
Cell labeling method	N/A	Antibody conjugated with fluorescent dye of semiconductor quantum dots	Metal nanoparticle probes	Fluorescent or chromatic
Detectable parameters	Cell shape, cell size, granularity	Cell viability, cycle analysis, immunotyping, tumor cell identification	Immunotyping, cell surface marker expression	DNA damage, apoptosis, cell cycle, immunotyping, protein expression
Sensitivity	0.5 μm	750 MESF[b]	200 probes	<1000 MESF
Resolution	1.0 μm FSC, 0.5 μm SSC	3% of peak area	N/A	0.25 μm
Throughput	Very high	Very high	Very high	High
Sample manipulation	No	Immunostaining	Immunostaining	Chromogenic or fluorescent staining
Multiparametric capability	2	10–15 (with three lasers)	>200	5
Advantages	No labeling required	Compatible with well-established dyes and fluorophores	Enhanced spectral resolution compared to optical detection Simple optical setup (one light source and one detector)	Require less sample Can generate morphometric data Subcellular spatial resolution Compatible with chromatophores
Disadvantages	Limited readout Cannot distinguish cells with similar size and shape	Does not retain spatial information	Requires high spectral resolution for data analysis	Limited multiplex capacity compared to flow cytometry
References	[BD Biosciences; Watson et al 2008]	[BD Biosciences; Watson et al 2008]	[Goddard et al 2006; Lerner et al 2010; Tuchin 2011]	[Henriksen et al 2011]

[a] Not FDA approved.

[b] MESF, molecule of equivalent soluble fluorescein.

vibration of the core liquid stream that results in droplet breakup with the encapsulation of individual cells within. An electrical charge is applied to droplets that contain target cells, and these charged droplets are deflected from the core stream into collection tubes. For mechanical sorting, target cells are deflected via an acoustic pulse or via insertion of a motor-driven syringe into the sample stream. Since physical manipulation is required for cell isolation in mechanical sorters, the sorting speed is slower than electrostatic sorters. However, mechanical sorters are enclosed and thus are less prone to sample contamination than electrostatic sorters, where droplet formation occurs in air.

The maximum rate of cell sorting depends upon the frequency of formation of stable and uniform droplets. The frequency of droplet formation is directly proportional to fluid velocity but inversely proportional to orifice diameter as described by $f = v/4.5d$, where f is the droplet formation frequency, v is the stream velocity, and d is the orifice diameter (Robinson 2004). Since the orifice diameter is usually about 70 μm in clinical cytometers, fluid velocity can only be altered by changing the sheath fluid pressure. For example, operating with a sheath pressure of 60 psi can enable the formation of up to 100,000 droplets/s (Macey 2007) However, if the sheath fluid pressure is too high, cell viability is compromised. Thus, the speed of cell sorting is limited by the sheath pressure.

15.15.6 Slide-Based Cytometry

An emerging field of cytometry is slide-based cytometry (SBC). The basic principle behind SBC is similar to microscopy techniques commonly used for cell analysis. The idea is to create both a uniform illumination source and a sensitive detection system to reliably quantify signals from tissue (or cell) samples mounted on microscope slides. Like microscopy, the signals can be either chromatic or fluorescence. On the other hand, optics employed in SBC generally have a much greater depth of field to sufficiently capture the dynamics between different parts of the slides. The data acquisition and analysis software should also provide sufficient resolution to distinguish cell boundaries and subcellular compartments.

An obvious advantage of SBC over conventional flow cytometry is the addition of spatial information in combination with cytometric data. This capability may allow pathologists to make more accurate diagnoses, especially in areas such as cancer detection and cell cycle analysis within tissue sample, where the target cells may be too rare to be detected by flow cytometry. The use of robotics and other automation technologies has also improved the throughput SBC to a level that approaches flow cytometry. So far, the multiplex capability of SBC still lags behind flow cytometry, but the combination of alternate detection modality as the afore mentioned may close this gap in the near future.

15.16 IMMUNOASSAYS

Clinical immunoassay platforms have multiple critical diagnostic functions, including the detection of analytes such as pathogens, allergen-specific immunoglobin antibodies, haptens, autologous antigens, or disease-associated biomarkers in patient specimen. In clinical immunoassays, the target analyte forms a stable complex with highly specific antibodies, facilitating diagnoses of infectious, immune, and other diseases and physical conditions. Typically, only one analyte can be measured during an assay (singleplex assays); however, recently, assay developers have manufactured and validated multiplex assays for the simultaneous detection of multiple analytes in a given patient sample.

Immunoassays can be classified as heterogeneous or homogeneous (Figure 15.2). Most are heterogeneous and require wash steps to separate bound from unbound analytes and labels. In planar-based heterogeneous assays, capture ligands are immobilized onto a solid surface (glass, plastic), and target antigens or antibodies within patient specimen bind to these ligands. In other bead-based assay formats, the capture ligands are immobilized onto microbeads. Automated clinical analyzers that perform all wash steps consistently can help reduce assay variability. In contrast to heterogeneous assays, homogeneous assays employ simple *mix-and-read* techniques and do not require separation and wash steps. The lack of wash steps reduces handling errors and speeds up assay times. Not washing, however, can cause high background signals in homogeneous assays due to the presence of interfering species in the bodily fluid being analyzed.

As biomarker concentrations in body fluids can be very low, there is a need for many immunoassays to be highly sensitive. Assay sensitivity, however, can be limited by several factors such as the intrinsic affinity and specificity of antibodies to the biomarker being measured, washing procedures, background noise, and signal detection sensitivity. Even the best antibodies have affinity limits of ~1×10^{-12} M^{-1} (Grossman 1998), and developers of new immunoassays should be aware of this potential constraint. Detection methods, in general order of low to high sensitivity, include colorimetric, fluorescent, and luminescent readouts. Another important consideration is the dynamic range of the assay. The high quantum efficiency and signal amplification provided by fluorescence and chemiluminescence techniques help achieve lower detection limits and larger dynamic ranges (ranges over which there is a linear relationship between the antigen concentration and detected signal) than the more conventional colorimetric assays. More recently, label-free detection platforms like surface plasmon resonance (SPR) have also been applied to immunoassays. Although used extensively in research, currently, no SPR-based diagnostic assays have received FDA approval. In this section, we will discuss four detection methods for heterogeneous- and homogeneous-based clinical assays: colorimetric, fluorescent, luminescent, and SPR. Radioimmunoassays are not covered due to decreasing use and replacement by luminescent methods.

15.16.1 Colorimetric Assays

In colorimetric assays, the detection antibody is labeled with an enzyme conjugate—usually peroxidase or alkaline phosphatase. Addition of the appropriate chromogenic substrate generates a colored reaction product that absorbs light in the visible range (400–700 nm). The intensity or optical density of the colored product is directly proportional to the concentration of antigen being measured. Clinical colorimetric assays are quantified using single-wavelength colorimeters or multiwavelength spectrophotometers. These instruments are typically designed to quantify signal in the center of each well of a microplate or tube. However, in the absence of shaking, the colored reaction product can remain unevenly distributed and concentrated near the enzymes, which are surface immobilized.

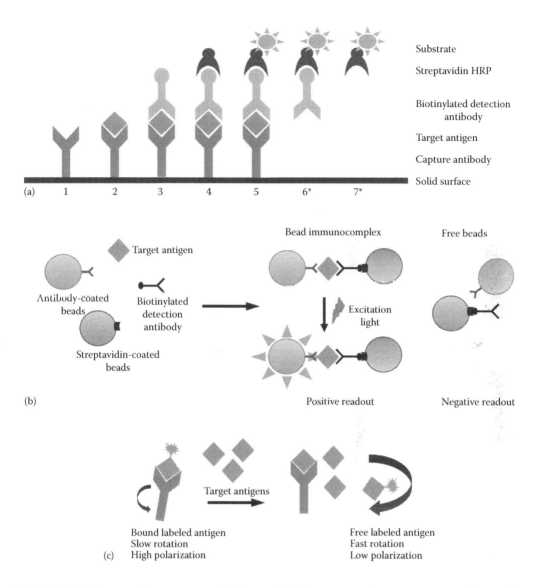

FIGURE 15.2   Common immunoassay platforms. (a) Heterogeneous immunoassay requires washing between each step. (1) Capture antibody binds to solid surface. (2) Target antigen binds to capture antibody. (3) Detection antibody binds to a distinct epitope of target antigen. (4) The streptavidin, horseradish peroxidase (HRP) conjugate binds to biotinylated detection antibody. (5) Substrate is added, eliciting colorimetric, chemiluminescent, or fluorescent readout. (6*) Nonspecific signal generated if unbound detection antibodies are not washed away. (7*) Nonspecific signal results if unbound HRP remains in solution. (b) Homogeneous bead-proximity immunoassay. Target antigen is sandwiched between antibody-conjugated microbeads. Upon laser excitation, chemiluminescent signal is emitted. Unbound beads in solution do not interfere with assay signal. (c) Fluorescence polarization immunoassay. Labeled antigens have slow rotations compared to fluorescence lifetime (high polarization). Unlabeled target antigens displace labeled antigens and the resulting free labeled antigens rotate fast compared to fluorescence lifetime (low polarization).

Although single-wavelength colorimeters are in principle sufficient to measure colorimetric assay signals, dual-wavelength spectrophotometers more efficiently eliminate background noise from assays and improve assay sensitivity. Equipped with reversible motors, dual-wavelength spectrophotometers alternately record sample absorbance at two different wavelengths: one selected wavelength corresponds to the reaction product and the second wavelength is not at all associated with the colored product (Khandur 2006). The instrument automatically subtracts the difference between the absorbance readings, effectively subtracting background interference. In its simplest form, a colorimetric assay only needs the human eye as a measuring instrument. The inherent simplicity of these colorimetric assays makes them very useful in clinical labs and outpatient clinics and as point-of-care diagnostic tools because these simple, portable, low-cost devices provide rapid diagnoses and do not require extensive medical training to use.

15.16.2 Lateral Flow Immunoassays

The most common type of colorimetric assays is the lateral flow immunoassays or dipsticks, which use immunochromatographic techniques to detect antigens and antibodies in patient specimen (Figure 15.3). Although many of these assays can be read qualitatively by the naked eye, instrument-based readings of colorimetric or fluorescent signals can improve accuracy for clinical use. Test results from dipstick assays are strictly qualitative or semiquantitative, providing a negative or preliminary positive outcome. HIV heterogeneous dipstick assays, for example, are designed to be highly sensitive, that is, to miss as few HIV infections as possible. But by engineering highly sensitive HIV antibody assays, assay developers generally lower assay specificity. The inherent disadvantage of HIV dipsticks is that a positive test result may actually correspond to a false-positive measurement, caused by the presence of antibodies to other diseases that the assay wrongly recognizes as antibodies to HIV.

Urine dipsticks are used during routine urinalysis to detect substances like glucose, leukocyte esterase, and nitrite in patient urine samples. Assay sensitivity depends on the chromogenic substrate used. If dipsticks are coated with glucose oxidase, these assays will detect glucose with high specificity. Detection of nitrites and leukocyte esterase in urine identifies urinary tract infections. Nitrite tests are based on the presence of bacteria in tested urine, which reduce nitrates to nitrites. Like most dipstick assays, these tests are mainly qualitative and a positive result is obtained if more than 10 organisms/mL are detected. For leukocyte esterase dipstick assays, esterase molecules released by neutrophils react with esters on the dipstick pad, and the reaction product causes a color change in the chromogenic substrate.

Critical design considerations for lateral flow immunoassay development include membrane composition, capillary flow time, and test line placement (Mansfield 2009). The membrane composition affects capture reagent adsorption at the test and control lines as well as detector particle flow through the system. Often, nitrocellulose is the chosen membrane material. Capture protein reagents initially bind to the membrane via electrostatic interactions between the nitrate ester linkage of the nitrocellulose and peptide bonds of the protein reagent. Later in the manufacturing process, the test strip is dried by lyophilization

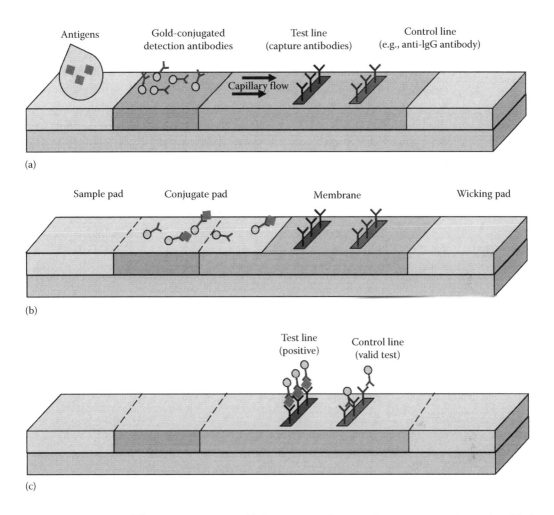

FIGURE 15.3 Lateral flow immunoassay. (a) Patient sample, containing target antigens, is added to the sample pad. (b) Sample wicks through the conjugate pad and antigens conjugate to detector labels. (c) Antigen-label immunocomplexes bind to capture antibodies at the test line. Detector labels also bind to anti-immunoglobin G (anti-IgG) antibodies at the control line.

or forced air. Upon water removal, the proteins unfold and their hydrophobic interior binds strongly to the hydrophobic regions of the nitrocellulose. Detector labels are contained within the conjugate pad. Common detector labels include colloidal gold particles (~40 nm), fluorescent particles, latex or magnetic beads (>0.5 μm), and fluorescent quantum dots (Chun 2009). Since particle size determines particle flow rate (smaller particles flow faster) through the membrane pores, it is necessary to manufacture and use monodisperse particles in order to ensure assay accuracy. To promote particle flow through the membrane, the membrane pore size should be between 3 and 20 μm; below 3 μm, the flow rate is too slow to be practical (Mansfield 2009). Incorporation of a wicking pad into the test trip also promotes in capillary flow by drawing liquid through the membrane.

In lateral flow test strips, sample flow is unidirectional from the location of sample input to the test line. Typically, the sample is placed on the sample pad—an absorbent pad that contains salts, surfactants, and other buffering agents. Assay speed and sensitivity is dependent on the capillary flow time—the time taken for liquid to flow through and completely fill the strip of defined length. Since sample flow through the membrane is affected by the pore size, the membrane composition affects the capillary flow time. For test strips with longer flow times, the antigens and detector particles have more time to form stable immunocomplexes with capture reagents at the test line. Longer flow times therefore improve assay sensitivity. If assay speed is too fast, less interaction occurs between assay reagents.

In multiplexed lateral flow assays, the sample is mixed with detector labels against multiple antigens. Each measured antigen has a different test line, and some test lines will be farther upstream from the sample input location. Since the capillary flow rate changes as a function of distance along the test strip, detector labels that have to travel farther up the strip will have different flow profiles. To account for this difference in test line positioning, optimizing the concentration of each detector label is necessary. Higher concentrations of detector labels are used for the farthest test lines. The width of the capture reagent or test line is also very important. Membrane composition and capillary flow time affect test line width. If large pore sizes or fast flow is used, then wider test lines result. Thinner test lines are desired because they give higher signal-to-background levels and lower limits of detection due to concentration of detector labels within a smaller area (Mansfield 2009).

The lowest reported detection limit for conventional lateral flow immunoassays is 1 ng/mL (Fernandez-Sanchez 2005). However, in one variation of the dipstick assays, dye-loaded liposomes were used for the highly sensitive detection of cholera toxin, and the reported detection limit was 10 fg/mL (Ahn-Yoon et al 2003).

15.16.3 Fluorescence Intensity Reading–Based Assays

Fluorescent molecules or fluorophores can be detected with higher sensitivity and selectivity than colorimetric labels because signals are emitted. As a result, fluorescent immunoassays can have detection limits in the picogram per milliliter levels using optimized detectors (Table 15.3). Other crucial design considerations for fluorescence instrumentation include temperature control within the sample chamber, limiting photobleaching, and the ability to accurately differentiate between signals generated by target antigen fluorescence and background signals.

To eliminate the effect of temperature on emitted fluorescent signals, fluorescence instruments are designed with temperature-controlled sample chambers. As temperature increases, molecular motion and collision frequency increases. Such collisions dissipate energy and decrease quantum yield—the efficiency with which fluorophores convert absorbed light to emitted light. Lower quantum yields result in lower fluorescent signals, and the change in fluorescence is usually 1%/°C (Guilbault 1990).

Fluorescence signals are also greatly affected by photobleaching. Photobleaching is the irreversible loss of fluorescent signal, caused by the degradation of the fluorophores upon illumination with high-intensity light. One method of minimizing photobleaching is by designing fluorescence readers with pulsed light excitation instead of continuous excitation. Application of short exposures of low-intensity excitation light will prevent photobleaching of the sample.

TABLE 15.3   Comparison of Immunoassays

Assay Mechanism	Planar	Lateral Flow	Bead Proximity	Fluorescence Polarization	ECL	SPR	Multiplex Suspension Bead
Signal detection method	Fluorescence, colorimetric, chemiluminescence	Visual, colorimetric, digital camera image, CCD camera	Luminescence	Fluorescence	ECL	Refractive index	Fluorescence, flow cytometer with dual laser
Sensitivity (FDA approved)	<2.0 ng/mL	<20 ng/mL	<0.5 ng/mL	N/A	<20 ng/mL	N/A	N/A
Minimum specimen volume required	<100 µL	<100 µL	<20 µL	<100 µL	<50 µL	<50 µL	<50 µL
Assay time	<5 h	<40 min	<21 min	<1 h	<2.5 h	<90 min	<4 h
Number of wash steps	Up to 4	Up to 1	None	None	Up to 2	Up to 1	0–4
Analytes	Up to 2, up to 9	Up to 4	No	Up to 2[a]	Up to 10[a]	Up to 10	Up to 13, up to 29[a]
Advantages	Good sensitivity	Easy point-of-care diagnostics	Simple *mix-and-read* method	Suitable for small (<10 kDa) analytes	High sensitivity, multiplexing, wide dynamic range	Label-free detection, suitable for small analytes	Multiplexing, high-throughput testing
Disadvantages	Small dynamic range, poor sensitivity due to detection of antibody cross-reactions	Qualitative, poor sensitivity	Requires specialized instrumentation	Requires specialized instrumentation. Poor sensitivity	Requires specialized instrumentation	Inefficient for high-throughput assays	Requires specialized instrumentation, poor sensitivity due to detection of antibody cross-reactions
References	[Federal Drug Administration]	Ahn-Yoon et al 2003; Federal Drug Administration	Federal Drug Administration; McPherson and Pincus 2007	Hong and Choi 2002; Tian et al 2012	Federal Drug Administration; Zhang et al 2012	Krishnamoorthy et al 2010; Krishnan et al 2010	Ellington et al 2010; Federal Drug Administration; McPherson and Pincus 2007

Fluorescence assay sensitivity is reduced by background signals, caused by autofluorescence from sample components, unbound or nonspecifically bound fluorescent probes, or microplate material. To reduce or eliminate background fluorescence, fluorescence instruments are designed with optical filters in addition to excitation and emission monochromators and adjustable gain. Because optical filters restrict the excitation and emission wavelengths more effectively than monochromators, these filters can eliminate interference from stray light, scattered light, and sample interference. Maximum sensitivity is achieved by selecting filters, not necessarily on their ability to transmit particular wavelengths, but more so the ability to reject unwanted interfering wavelengths. In homogeneous assays, background interference from sample components such as hemoglobin, bilirubin, and cellular debris may contribute to emission signals at longer wavelengths. Similarly, inadequate washing in heterogeneous assays as well as autofluorescent material in microplates may also increase background signals. To reduce data corruption by these nonspecific signals, bandpass filters that transmit a selected wavelength of light may be used instead of long pass filters (Lakowicz 2006). Use of black microplates that absorb stray light can further enhance the signal-to-noise ratio of fluorescent measurements. Gain controls are also helpful in enabling the end user to subtract background interference to achieve the highest possible signal-to-noise ratio.

15.16.4 Fluorescence Instrumentation for Singleplexed Immunoassays

Instruments designed for fluorescence measurements usually have four basic components, namely, an excitation light source, monochromators, optical filters, and a photodetector. Polarizers are inserted in the excitation and emission light paths if the reader is compatible with fluorescent polarization assays. The light sources commonly used in commercial fluorescence readers include xenon flash lamps, lasers, laser diodes, and light-emitting diodes (LEDs). Important factors to consider for light sources in clinical fluorescence assays include flexibility in wavelength selection, lifetime, cost, and heat generation during operation. Multiwavelength selection is particularly advantageous for multiplex diagnostics where an array of fluorophores need to be excited at various wavelengths without emission overlap. Xenon lamps do provide a continuous spectrum of light (ultraviolet, visible, and infrared), but these lamps produce excessive heat and usually have short lifetimes (McPherson and Pincus 2007). Lasers and laser diodes have relatively long lifetimes, produce monochromatic radiation but are costly, and can also generate heat. LEDs provide near-monochromatic light that are relatively inexpensive, generate little heat, and have long lifetimes. Given these advantages, there has been increased use of LED-based fluorescence instrumentation.

Monochromators are needed in fluorescence readers if the light source used provides a broad wavelength source, for example, xenon lamps. In these cases, an excitation monochromator selects the desired excitation wavelength (Lakowicz 2006). In spectrofluorometers where the patient specimen is assayed in a quartz cuvette or test tube, the emitted fluorescence is directed at a right angle to the excitation light to avoid interference by the excitation signal. In spectrofluorometers where the sample is stored in a microwell plate, the emitted light cannot be detected by right-angle observation. Instead, the fluorescent signal is directed at a specified angle to optical filters and emission monochromators prior to signal detection via PMTs or photodiodes. The resulting fluorescent signal is stored digitally as an emission spectrum or as an end-point numerical measurement. Durable optical filters that withstand high-intensity light sources are also essential in fluorescence readers. Well-characterized optical filters are particularly beneficial in multicolor assays to prevent optical cross talk and to improve assay sensitivity.

The detection limit of fluorescent clinical immunoassays is heavily dependent upon the sensitivity of the detector in the fluorescence reader. Most commercial fluorescence instruments use carefully selected PMTs as detectors, because these extremely sensitive detectors are able to measure low-intensity fluorescent signals.

One example of a homogeneous, fluorescence-based assay is fluorescence polarization. When fluorescein-labeled antigens (e.g., drugs of abuse) bind to specific antibodies, there is an increase in polarization, that is, the rotational diffusion of the antigens. If the patient specimen contains target antigens, sample antigens compete with the labeled antigen for binding to the antibody and polarization signals decrease. Small molecules, like unbound antigens, rotate quickly in solution and have low polarizations. In contrast, large molecules such as antigen–antibody complexes have high polarization. To detect the fluorescent signals in polarization assays, excitation and emission polarizers are included in the reader. Specifically, the emission polarizer detects the perpendicular and parallel components of the emitted fluorescent signals. Polarization assays are useful for lifetime measurements— the time interval between the absorption and emission of a photon by a fluorophore. The longer the lifetime, the more the fluorophore will rotate and the lower the recorded polarization of emitted light (Jameson and Ross 2010).

15.16.5 Fluorescence Instrumentation for Multiplexed Immunoassays

In clinical laboratories, multiplex immunoassays have several advantages over singleplex assays. First, multiple target antigens can be detected with very small volumes of patient specimen. Second, quantitative data from multiplex antigen analysis often allow clinicians to diagnose and prognose infectious diseases more accurately than decisions based on a single readout. Third, multiplexed assays are economical and reduce total assay turnaround time compared to performing multiple singleplex tests. Currently, flow cytometry–like bead assays are the most widely used multiplexed clinical assays.

In bead-based multiplexed immunoassays, capture antibodies are immobilized to polystyrene microbeads (5.6 μm). To facilitate multiplexed antigen detection, each microbead is coated with reagents that recognize different antigens or autoantibodies in the given patient specimen. Each bead is also loaded with different ratios of red and orange fluorochromes; thus, in theory, up to 100 different microspheres can be distinguished by their fluorescence. In reality, however, only up to 29-plex has been reported for these flow cytometry-based immunoassays because the robustness of multiplexed assays is reduced by antibody cross-reactions (Djoba et al 2008). Cross-reactions arise, because in multiplexed immunoassays, although the primary capture antibodies are individually coated onto different surfaces, the secondary antibodies are all mixed together into a single cocktail that is exposed simultaneously to all the surfaces coated with different primary capture antibodies. As the number of multiplexing increases, so does the total concentration of secondary antibodies in the cocktail, increasing background as well as the possibility of cross-reaction of different secondary antibodies to the same antigen.

To differentiate the multiplexed assay signals between different beads, a two-color flow cytometry–like instrument is used. Using these customized flow cytometers, microspheres are hydrodynamically focused into a sample stream and excited at 635 nm by a red laser. At this wavelength, both the red and orange dyes in the bead are excited, and the resulting emission at 535 nm (orange) and 650 nm (deep red) is used to identify each microsphere and measured antigen. The detection antibody is fluorescently labeled with PE, and a second laser (532 nm) is used to excite this dye, quantifying antigen levels.

As with all heterogeneous assays, a wash step is required prior to addition of the detection antibodies. Since these multiplexed bead-based assays involve freely floating microspheres, washing would involve loss of microspheres and a consequent loss in assay reagents and fluorescent signal. To mitigate the loss of beads, some multiplexed assays use magnetic beads to ensure assay beads remain adhered to the plate surface during washing. *Homogeneous* bead-based multiplexed assay formats are, however, possible because only the intensity of the fluorochrome at the surface of each microsphere is read as it is focused in a narrow stream; therefore, unbound fluorochromes in solution minimally contribute to assay signal. More accurately, this may be considered as in situ washing away or dilution of unbound fluorochromes.

In planar multiplexed immunoassays, capture reagents against different antigens or antibodies are spotted onto microwell plate surfaces or to polymer membranes. To detect fluorescent signals in these planar-based assays, highly sensitive CCD detectors are often needed. Typically multiplexed assays have lower assay sensitivity and lower reliability compared to conventional enzyme-linked immunosorbent assay (ELISA) platforms, particularly if the measured antigen is present in low levels in blood plasma (Liu *et al* 2005). Improvements in detection sensitivity can be achieved by selecting optimum capture/detection antibody pairs and optimizing the coupling of capture antibodies to microbeads.

Although sensitive and reliable multiplexed assays are beneficial to clinical research, validating these complex assays is very challenging. Each multiplexed assay requires extensive cross validation of antibodies, not only to confirm appropriate sensing of the target antigen by the appropriate pair of the capture and detection antibodies but also to confirm lack of cross-reactivity between all other detection antibodies. This extensive cross validation should be performed for each antigen across all antibodies included. Furthermore, as the number of antigens to be simultaneously analyzed increases, so does the total concentration of detection antibodies in solution since they must be simultaneously applied as a cocktail to the capture antibody arrays or beads. This use of higher antibody concentration increases background signal. Given these complexities, it has been difficult to establish a single standard for all immunoassays (Boja *et al* 2011, Liu *et al* 2005). Consequently, the FDA-approved multiplexed diagnostic assays are usually qualitative or semiquantitative tests. In these FDA-approved tests, a cutoff value is set and a positive diagnosis is made when antigen concentrations in patient specimen exceed the cutoff value.

15.16.6 Luminescent Assays

Luminescent assays typically have higher sensitivities than fluorescent assays, as is shown in Table 15.3. Unlike fluorescence assays, luminescence assays do not require a light source. Instead, luminescent assay readers only require a photodetector, simplifying the reader design. In this section, we will briefly discuss three types of luminescent immunoassays, namely, bead proximity, electrochemiluminescent, and chemiluminescent assays.

15.16.7 Bead-Proximity Assay Instrumentation

Bead-based proximity assays are rapid homogeneous assays that can be completed in as little as 10 min. The target antigen is sandwiched between two different types of receptor-coated latex beads (250 nm). One of the assay beads contains a photosensitizing agent, and the second bead contains a chemiluminescent agent. When the photosensitizer bead is excited via 680 nm laser light for 0.1–1 s, singlet oxygens are generated. Since singlet oxygens have a half-life of only 4 μs, these oxygens are only able to diffuse a maximum of 200 nm through the assay buffer. The singlet oxygens react with neighboring chemiluminescent beads, and luminescent signal is emitted at 550–650 nm. To trigger these chemiluminescent signals effectively, a 680 nm laser is needed together with short pass optical filters to selectively collect the emitted signal (Wild 2005).

15.16.8 Electrochemiluminescence Instrumentation

Electrochemiluminescence (ECL) assays are chemiluminescent reactions that are produced from an electrochemical redox reaction. Similar to chemiluminescent reactions, ECL assays are advantageous to clinical labs because of their high sensitivity, short assay times, and large dynamic ranges (5 orders of magnitude). Detection limits of FDA-approved ECL tests are typically less than 20 ng/mL (Table 15.3). Common labels include ruthenium, osmium, and rhenium. In ECL assays, target antigen is sandwiched between a biotinylated ligand and a ruthenium-labeled ligand. Paramagnetic streptavidin-coated beads are added, and these beads conjugate to the biotinylated ligand. These immunocomplexes form at an electrode, and unbound assay reagents are washed away. Once an electrical voltage is applied, chemiluminescent signal is emitted.

15.16.9 Chemiluminescent Instrumentation for Singleplexed and Multiplexed Immunoassays

Singleplexed chromogenic assays that involve the detection of peroxidase activity from horseradish peroxidase–conjugated antibodies can be easily performed as chemiluminescent assays by addition of highly sensitive chemiluminescent substrates to the assay. Chemiluminescent assays have high signal-to-noise ratios, which enable low picogram per milliliter sensitivity for proteins. Multiplexed chemiluminescent immunoassays are planar-based assays, in which capture antibodies against multiple target antigens are spatially spotted onto solid substrates. In sandwich-based ELISA formats, target antigens bind specifically to capture antibodies. Enzyme-linked detection antibodies are later added and incubated with sample antigens forming an immunocomplex. As in all heterogeneous assays, wash steps are employed to remove excess reagents and background levels are lowered. Chemiluminescent substrates are added to the mixture and the enzyme is cleaved, eliciting strong chemiluminescent signals. High-resolution CCD cameras are required to detect emitted chemiluminescent signals from each arrayed spot. Like multiplexed fluorescence assays, the sensitivity of multiplexed chemiluminescent assays is limited by cross-reactions between detection antibodies.

15.16.10 Refractive Index–Based Label-Free Detection Methods

SPR techniques and photonic microring resonators use optical readouts to detect antigen binding–associated refractive index changes at a sensing surface. Although not yet approved for clinical diagnostics, these assays are potentially beneficial to the clinical laboratory because these tests enable label-free antigen detection. When mediated by specific antibody–antigen interactions, refractive index changes are dependent on the quantity of antigen bound to a surface. Since the sample volume used in these assays is kept constant, the detected change in refractive index near the surface reflects a change in antigen concentration within the sample. In some SPR assays, the target antigen binds to capture antibodies that are immobilized to a gold-coated surface. In microring resonators, the capture antibodies are immobilized onto microring structures. Although there are a few examples of highly sensitive SPR assays with reported detection limits of 10 fg/mL, most of these label-free assays have nanogram per milliliter sensitivities particularly when operated in the *label-free* modes where antigen signals are not amplified further by secondary labels (Krishan *et al* 2010, Luchansky and Bailey 2011).

15.17 NUCLEIC ACID ASSAYS

Technical advances are leading to the increased use of nucleic acid–based diagnostic assays in clinical labs. Prenatal and newborn genotyping, cystic fibrosis screening, and chronic leukemia genetic tests are just a few examples of molecular assays that detect genetic mutations. Nucleic acid assays can also complement culture and immunoassays for the detection of pathogens for infectious diseases. Most nucleic acid diagnostic assays involve amplification techniques that generate thousands to billions of copies of the target nucleic acid sequence or amplicons. In these target amplification assays, enzymes are usually required to catalyze amplicon formation. Depending on the assay design, genetic abnormalities can either be detected by the PCR directly, such as with allele-specific PCR, or be detected in a post-PCR step by using other methods such as capillary electrophoresis, nucleic acid sequencing, and microarray-based platforms. In contrast, some diagnostic platforms rely on signal amplification. Therefore, instead of exponentially increasing the number of target molecules, the signal from small number of the target molecule is amplified. In all cases, the assay sensitivity refers to the smallest amount of target nucleic acid in the patient sample that can be reliably measured (quantitative tests) or detected (qualitative tests). If the assay aims to detect genetic mutations, the assay sensitivity will be defined as a percentage (Table 15.4).

In this section, we will briefly describe some key nucleic acid assays and instrumentation used in clinical labs. We will also provide a brief overview of some newly developed technologies that although not yet certified by the Federal Drug Administration (FDA) offer promise to significantly improve detection limits or perform faster assays than currently available platforms.

TABLE 15.4   Comparison of Nucleic Acid Assays

Assay Mechanism	Real-Time PCR	TMA	[a]RCA	PNA-Based FISH	Bead-Based DNA Microarrays
Signal detection method	Fluorescence	Chemiluminescence	Colorimetric Fluorescence	Fluorescence	Fluorescence Flow cytometer with dual laser
Target	RNA	DNA or RNA	DNA or RNA	Ribosomal RNA	DNA or RNA
Sensitivity (FDA approved)	>48 copies/mL	<50 copies/mL	N/A	10^5 colony-forming units (CFUs)/mL	>95%
Minimum specimen volume required	<20 μL	<50 μL	<2 μL	<10 μL	<5 μL
Assay time	<4 h	<4 h	<4 h	<2.5 h	<8 h
Analytes	Up to 4, up to 5[a]	Up to 2[a]	Up to 17	Up to 3[a]	Up to 10[a]
Advantages	High sensitivity High precision High throughput	High sensitivity Does not require thermocyclers Low risk of cross contamination No-wash steps	Simple Does not require thermocyclers Low risk of cross contamination Low costs per reaction Requires little to no assay optimization	Rapid pathogen identification Organism's morphology that offers additional species differentiation	Multiplexing High throughput
Disadvantages	Requires thermocyclers	Needs three different enzymatic steps: transcription, cDNA synthesis, RNA degradation	Requires several hours to obtain a detectable signal	High cost of probes No automated tests available	Risk of false-positives from cross contamination Requires thermocyclers
References	Federal Drug Administration; Sawyer et al 2000	Candotti et al 2003; Federal Drug Administration; Goda et al 2009; Sawyer et al 2000	Ladner et al 2001; Schopf and Chen 2010	Federal Drug Administration; Jang et al 2010	Federal Drug Administration; Millipore Corporation 2002

[a] Not FDA approved.

15.17.1 Enzyme-Mediated Target Amplification Technologies

The numerous copies of target nucleic acid sequences produced by nucleic acid amplification techniques are detected by fluorescence, chemiluminescence, or ECL. The emitted signal is proportional to the amount of generated amplicons. Currently, there are many nucleic acid amplification methods that are certified by the FDA for clinical use. Of these, the PCR technique is viewed as the gold standard. One major problem, common to most amplification methods, is the potential for laboratory contamination by the amplicons. To overcome this drawback, simple, user-friendly, automated systems that facilitate real-time detection of amplicons in a closed instrument are needed. Sealed instruments will minimize cross contamination of samples by amplified products.

15.17.2 Real-Time PCR Amplification Technology

Real-time PCR enables detection of the degree of target amplification as the amplification reactions are carried out. Real-time PCR differs from conventional PCR because it enables end users to monitor fluorescence signals in real time rather than just at the end of all the amplification cycles. This real-time monitoring capability is advantageous because it eliminates time-consuming postanalysis and allows more accurate quantification of starting nucleic acid quantity.

Optimization of assay specificity in PCR-based technologies requires a delicate balance between parameters such as primer length, guanine–cytosine (GC) content, efficiency of thermal cycler, and specificity of fluorescently labeled PCR probes. Assay specificity is improved by designing primers with 15–20 bases and GC content between 40% and 80% of the total sequence. Ideally, primers should not contain complementary bases because this increases undesired primer-dimer formation, which inhibits amplification of the target nucleic acid sequence. Since real-time PCR is not an isothermal process, thermal cyclers that provide uniform heating and cooling of samples within the sample chamber are necessary. If cyclers are able to reach target temperatures rapidly, assay speed improves and fluorescence signals can be quickly analyzed by appropriate software. Examples of fluorescent probes used in real-time PCR include DNA-intercalating agents, fluorescence resonance energy transfer (FRET) probes, and molecular beacons (Frayling et al 2006). One commonly used DNA-intercalating agent is SYBR Green I. During primer extension, SYBR Green I intercalates double-stranded DNA and produces more intense fluorescent signals than the unbound dye. However, the SYBR Green I dye does not bind specifically to the target DNA sequences. Consequently, the dye can also intercalate with nonspecific amplified PCR products such as primer dimers. If assay developers opt to use SYBR Green I in newly developed assays, assay specificity can be improved by constraining system software to analyze fluorescent signals above the primer-dimer melting temperature but below the target temperature. Hydrolysis probes (e.g., Taqman probes), FRET probes (e.g., dual hybridization probes), and molecular beacons overcome the inherent specificity problem of SYBR Green I dye because these probes are designed to hybridize to a specific target sequence within the PCR amplicon (Frayling et al 2006). Taqman probes and molecular beacons have similar mechanisms whereby the fluorescence signal is blocked by a quencher in the unbound state. In the presence of target amplicons, molecular beacons unfold and hybridize to the target sequence. The fluorophore is now far enough away from the quencher, and signal is generated. In contrast, when Taqman probes bind to the target amplicon, DNA polymerase cleaves the probe, thereby physically separating the fluorophore from the quencher to generate fluorescent signal. Consequently, the generated fluorescent signal from Taqman probes is irreversible, whereas the generated signal in molecular beacon-based assays is reversible. The physical separation of fluorophores and quenchers in Taqman probes makes these assays more sensitive than molecular beacon-based assays. If the quencher is unable to effectively suppress fluorescent signal in unbound probes, molecular assays will have high background levels and low assay sensitivity. The most sensitive hepatitis B virus (HBV) DNA tests use Taqman probes and can detect as few as 10 copies/mL.

Dual hybridization probes consist of two oligonucleotide probes, one labeled at the 3′ end with a donor fluorophore and the other labeled at the 5′ end with an acceptor fluorophore. Upon binding of the two independent probes to the target amplicon and fluorescent excitation, the two dyes are brought in close proximity to each other. The donor dye transfers energy to the acceptor dye, eliciting fluorescence energy resonance transfer (FRET). The requirement that both probes hybridize to the target amplicon increases specificity of the reaction.

To improve assay sensitivity in nucleic acid assays involving hydrolysis or molecular beacon probes, fluorophore–quencher pairs need to be carefully chosen for optimal signal. Ideally, fluorophore–quencher pairs should have adequate spectral overlap such that the emission spectrum of the fluorophore overlaps the absorption spectrum of the quencher. For multiplexed real-time PCR assays, spectrofluorometric thermal cyclers that allow simultaneous excitation and detection of multiple fluorophores are needed. These multiplexed real-time PCR assays are, however, limited by the availability of fluorescent dye combinations. Therefore, the maximum reported multiplexing capability of these assays is 5-plex (Peter *et al* 2001).

15.17.3 Digital PCR–Based Amplification Technologies

Unlike real-time PCR, digital PCR is not yet approved for clinical use. Nonetheless, digital PCR offers some unique advantages, including high assay precision of low-abundance targets and single-molecule detection. This high-sensitivity to low-abundance targets is enabled by subdividing one's sample into many small volume compartments. This compartmentalization effectively increases the concentration of target nucleic acids within each chamber, and this physical enrichment step allows for sensitive detection. For these procedures, the target and background nucleic acids are randomly divided and sequestered into hundreds or thousands of separate reaction chambers. These sequential aliquoting steps can be performed via microfluidic circuitry embedded within the digital PCR instrument. Other instruments use water/oil emulsion techniques to divide sample into nanoliter-sized droplets. Each droplet is transported to flow cytometer–like readers for detection. Like most PCR assays, digital PCR technologies require thermal cyclers for target amplification. After PCR amplification, reaction chambers or droplets that contain target nucleic acid sequences are labeled as positive or 1, and chambers that do not contain the target molecules are labeled as negative or 0, thus the term *digital* PCR. Because of compartmentalization, the likelihood that there is more than one target analyte per compartment or droplet is low, allowing this type of quantification. Digital PCR offers single-molecule sensitivity and allows direct counting of the total number of target molecules in a given sample (White *et al* 2009). One disadvantage of digital PCR is the large dead volume; therefore, the more rare the target sequence, the more sample is required to accurately quantify the concentration of target DNA as copies per microliter.

15.17.4 Transcription-Based Amplification Technologies

Two common transcription-based amplification technologies include transcription-mediated amplification (TMA) and nucleic acid–based sequence amplification (NASBA). Unlike real-time PCR, TMA and NASBA are isothermal processes and therefore do not require a thermal cycler to drive reactions to completion. Instead, TMA and NASBA instruments require only a water bath or a heat block to very rapidly produce billions of copies of the target RNA or DNA sequence. In TMA assays, reverse transcriptase creates a double-stranded DNA copy from an RNA or DNA template. Next, RNA polymerase initiates transcription and generates as much as 10 billion RNA amplicons in less than an hour (Hill 2001). Acridinium ester–labeled DNA probes hybridize specifically to the RNA amplicons, eliciting a chemiluminescent signal upon chemical activation. Highly sensitive TMA assays can detect less than 50 RNA copies/mL of hepatitis C virus (HCV) (Sawyer *et al* 2000). In contrast, NASBA assays are usually detected by ECL methods. To facilitate highly sensitive ECL detection, the generated amplicons are typically hybridized with tris (2,2′ bipyridine) ruthenium-labeled probes. The simple, homogeneous, or no-wash assay format of TMA and NASBA assays is very beneficial to the clinical lab. Since reagents are never removed or transferred from reaction tubes, cross contamination and false-positive results are reduced.

15.17.5 Strand Displacement Amplification Technology

Like TMA and NASBA, strand displacement amplification (SDA) is an isothermal process, operating at about 37 °C–55 °C. In SDA-based technologies, nucleic acid amplification is catalyzed by restriction enzymes. Restriction enzymes cleave double-stranded DNA and produce a single-stranded nick, displacing downstream DNA. The amplification process involves serial nick formation, strand displacement, and primer hybridization to displaced strands. When designing new SDA assays, genetic assay developers must choose the most sensitive detection methods to ensure high assay sensitivity. FRET and chemiluminescent-based detection methods are more advantageous than fluorescence polarization techniques, because polarization-based SDA assays have low signal-to-noise ratios. Some only show a twofold increase in signal between dye-labeled probes in double-stranded versus single-stranded DNA. Low signal-to-noise ratio means lower assay sensitivity. In contrast, FRET-based SDA assays can show as much as a 20-fold increase in signal from background fluorescence levels (Hellyer and Nadeau 2004).

15.17.6 Rolling Circle Amplification Technology

Another isothermal enzymatic amplification and detection method is rolling circle amplification (RCA). In RCA assays, one strand of the double-stranded DNA is nicked, and DNA polymerase catalyzes amplification from the 5′ to 3′ end. As the linear replicate of target circular DNA is formed, the 3′ end of the nicked strand is displaced (Jain 2010). As many as 10^{12} copies of the target sequence can be produced in 1 h. RCA assays are highly specific and enable detection of point mutations with attomolar sensitivity (Schopf and Chen 2010). Like TMA and NASBA assays, RCA assays are advantageous because they are contamination resistant and do not require any special instrumentation such as thermal cyclers. Unlike all other isothermal methods, RCA assays are easily adapted to microarray formats because the amplified product remains attached to the DNA primer. This characteristic property of RCA assays enables easy multiplexing because signals are localized at discrete locations on the microarray.

15.17.7 Signal Amplification Technologies

In contrast to target amplification technologies, signal amplification methods do not increase the number of target nucleic acid molecules prior to signal detection. Instead, signal amplification methods improve assay sensitivity by detecting fluorescence or colorimetric signals from highly sensitive and specific reporter probes, which are hybridized to the target nucleic acid molecules. Since the target molecules are not amplified, signal amplification techniques are not susceptible to amplicon contamination. The most common signal amplification methods employ in situ hybridization (ISH) techniques. In this section, we will provide a brief overview of key instrumentation parameters required for these ISH-based clinical assays.

15.17.8 Fluorescence In Situ Hybridization

In FISH assay platforms, target DNA in interphase or metaphase chromosomes hybridizes to highly specific fluorescent probes, facilitating direct detection of chromosomal abnormalities via fluorescence microscopy. The high spatial resolution offered by FISH techniques allows pathologists to precisely identify the chromosomal abnormality. Like heterogeneous fluorescence immunoassays, assay sensitivity of FISH-based platforms depends on the choice of fluorophores, detector sensitivity, and the effectiveness of wash steps to remove nonspecifically bound fluorescent probes.

There are three main types of fluorescent probes used in FISH assays: locus-specific probes, chromosome enumeration probes (CEPs), and whole-chromosome painting probes (Cagel and Allen 2009). Locus-specific probes are most useful for identifying gene deletions (e.g., p. 53) or translocations (e.g., BCR/ABL translocation). These probes hybridize to nonrepetitive nucleic acid sequences that range from 40 to 500 kilobases (kb). If the probe hybridizes to regions less than 40 kb, the fluorescent signals produced are too weak to be effectively detected. If the probe hybridizes to regions greater than 500 kb, the resulting diffuse fluorescent signals are not localized enough to identify the labeled genes. CEPs are used to determine whether an individual has the correct number of chromosomes. In contrast to locus-specific probes, CEP probes hybridize to repetitive nucleic acid sequences that are located near the centromeres of chromosomes. Although these target nucleic acid regions consist of short sequences (~170 bp), they have high copy numbers (up to 5 million bases) (Cagel and Allen 2009). When CEP probes hybridize to these sequences in the compact centromeric regions, the resulting fluorescent signals are very strong. One major limitation of CEP probes is cross hybridization, which occurs because several chromosomes share the same repetitive sequences. For example, because chromosomes 13 and 21 share the same repeat sequence, they cannot be differentiated by CEPs. To overcome this limitation, new probes need to bind specifically to individual chromosomes without interferences by cross hybridization. Whole-chromosome painting probes identify both numerical and structural chromosome abnormalities in metaphase cells. The whole-chromosome painting probes consist of a mixture of smaller probes that each hybridizes to unique sequences along the entire length of one or more chromosomes (Cagel and Allen 2009). Each whole-chromosome painting probe is labeled with a different fluorescent dye, enabling pathologists to *paint* each chromosome in a unique color. The simultaneous painting of all 24 human chromosomes is made possible by two independent FISH techniques: multicolor FISH (M-FISH) and spectral karyotyping (SKY).

The sensitivity and resolution of FISH techniques is also affected by the microscope hardware (microscopes, optical filters, microscope objectives, and CCD cameras). Sensitivity depends on the ability of the microscope to detect the weakest fluorescent signal emitted by probes, hybridized to small target sequences. Resolution depends on the ability of the microscope to adequately delineate two discrete points along the length of the chromosome. In addition to these technical limits, assay developers should also consider the compact conformation of DNA within the chromosome. Compared to naked DNA, interphase chromosomes are roughly 10 times more compacted. In contrast, highly compacted metaphase chromosomes are about 10,000 times more compact than naked DNA.

The degree of multiplexing in M-FISH techniques is also limited by the system's resolution. Low-resolution systems have a high susceptibility to false-positive determinations because what clinicians may interpret as colocalized genetic aberrations may actually be two discrete fluorescent signals in close proximity. In addition, since FISH provides two-dimensional imaging of the three-dimensional nucleus, it is inherently difficult to identify colocalized signals.

15.17.9 Peptide Nucleic Acid–Based FISH

Fluorescently labeled peptide nucleic acid (PNA) probes are also designed for FISH-based assays. PNA probes are DNA mimics that contain a noncharged, synthetic peptide backbone instead of the negatively charged phosphate backbone present in DNA. The hydrophobic characteristic of peptide backbones allows PNA probes to traverse easily through the hydrophobic core of cell membranes and subsequently hybridize with high specificity to complementary target ribosomal RNA (rRNA) strands via hydrogen bonds. Some assays are completed in as few as 90 min. Only a water bath and a fluorescence microscope are required for PNA FISH assays. When designing multiplexed PNA FISH assays, developers must design several PNA probes with different fluorescent labels to ensure accurate signal differentiation between multiple target sequences.

15.17.10 Chromogenic In Situ Hybridization

In contrast to FISH-based platforms that rely on fluorescence detection, chromogenic in situ hybridization (CISH) methods employ colorimetric detection (Cagel and Allen 2009). Additional steps are required in CISH assays because probes are not directly tagged with the chromogenic substrate. Instead, biotinylated or digoxigenin-labeled probes are used.

As in heterogeneous immunoassays, once these probes hybridize to the target nucleic acid sequence, wash steps are required to remove nonspecifically bound probes. Horseradish peroxidase and diaminobenzidine (DAM) are subsequently added for colorimetric detection. CISH assays are advantageous because only a light microscope is necessary for direct visualization of genetic abnormalities. However, chromogenic signals in CISH assays are more diffuse than the fluorescent signals in FISH assays, limiting assay precision and assay sensitivity. Another disadvantage is that CISH assays cannot be multiplexed.

15.17.11 Nucleic Acid Microarray Platforms

Nucleic acid microarrays are also referred to as DNA microarrays or gene chips. DNA microarrays consist of labeled oligonucleotide probes or DNA probes that detect target nucleic acid sequences with high specificity and sensitivity. Although still mainly a research tool, chromosomal microarrays are now being more routinely used for clinical applications, primarily for analysis of development disorders and less frequently in oncology. These microarray-based platforms can be applied to gene expression profiling, single-nucleotide polymorphism (SNP) detection, and comparative genomic hybridization (CGH) (McPherson and Pincus 2007). In planar-based microarrays, the labeled probes are immobilized onto solid substrates, usually glass or silicon. In bead-based microarrays, the probes are immobilized to small beads.

15.17.12 Planar-Based DNA Microarray Technologies

There are two main types of planar-based microarray platforms: oligonucleotide arrays and complementary DNA (cDNA) microarrays. In oligonucleotide arrays, microarray spots consist of oligonucleotides that are synthesized directly onto a coated solid substrate. In some cases, silane-coated quartz substrates are used and probes bind to the available hydroxyl groups on the coated surface. In other cases, the DNA probes are bound via electrostatic interactions between negatively charged phosphate groups on the DNA backbone and positively charged amine-derivatized surfaces. Similarly, in cDNA microarrays, DNA probes that hybridize to target cDNA or mRNA sequences are immobilized to the surface. The sensitivity of microarray-based assays depends on the surface chemistry of the solid substrate and fidelity of the spotted probes.

Microarrays require precise spatial patterning of DNA probes. If the substrate surface promotes nonspecific binding of target or nontarget nucleic acids, not only will background noise levels increase, but also the spatial resolution of the microarray will be compromised. To reduce nonspecific binding to the surface, the chosen substrate is usually made hydrophobic by appropriate coatings. In addition, spacing between printed probes is optimized.

Long probes used in cDNA microarrays (600–5000 bases) may form multiple contacts with the substrate surface and reduce the ability of target nucleic acids to bind along entire length of probe (Coleman 2006). This inherent limitation of cDNA microarrays adversely affects assay specificity. In contrast, oligonucleotide probes are only 25 bases long. Because these probes are so short, it is possible that different gene targets may have complementary sequences to the same probe. Therefore, taken individually, oligonucleotide probes will yield high rates of false-positives. Instead, to improve assay sensitivity, gene targets are identified only if target sequences hybridize to a predefined set of labeled probes. Assay sensitivity is also reduced if microarray probes have erroneous nucleotide sequences. Mismatches in base sequences will prevent target nucleic acid molecules from hybridizing along the entire length of the probe. Poor hybridization of the intended target to probes results in lower detected fluorescent or chemiluminescent signals. In addition, nontarget sequences may hybridize nonspecifically to poorly designed probes and increase background noise levels. To account for background noise, mismatch probes can be included in the microarray as a negative control. Mismatch probes are complementary to the target with the exception of one single-nucleotide substitution. Hybridization of the target nucleic acid to the mismatch probes enables quantification of nonspecific binding.

High-resolution optical systems are required to distinguish fluorescence or chemiluminescence signals from microarray-based assays. In some cases, hybridized targets may be as little as 5–20 μm apart. In some systems, the spot size may vary as a function of signal intensity. For example, saturated signals may have larger spot diameters. Algorithms are required to account for these disparities in acquired data.

15.17.13 Bead-Based DNA Microarrays

Like planar microarrays, bead-based microarrays can perform highly parallel assays to detect nucleic acid mutations. In one format of bead-based microarrays, each microbead is uniquely color coded, enabling each bead to be distinguished from other beads after excitation with a two-color laser beam. First, the target nucleic acid sequence is amplified using PCR-based technologies, and the generated amplicons are labeled with biotin. Second, oligonucleotide probes are conjugated to the microbeads. Since up to 100 beads can be distinguished by their unique spectral emission, up to 100 different probes against different nucleic acid sequences can be simultaneously screened for in one sample tube. Consequently, discrete emission signals from each microbead correlate to a different oligonucleotide probe. Third, biotinylated amplicons hybridize with high specificity to complementary oligonucleotide probes. Thermal cyclers are required to facilitate hybridization. Lastly, biotinylated amplicons bind to streptavidin-PE reporter dyes, enabling fluorescent detection of target nucleic acids. This assay requires custom flow cytometers to identify bead fluorescence. Assay types of this form can be homogeneous (no washing) or heterogeneous. To improve assay sensitivity for heterogeneous bead microarrays, optimizing washing steps is crucial to minimize sample loss. Use of magnetic beads in these heterogeneous assays allows washing to be performed by centrifugation or vacuum filtration. This prevents loss of sample and subsequent reduction in detected fluorescence. Of all the nucleic acid assays, bead-based microarrays achieve the highest degree of multiplexing—as much as 17-plex (Pabbaraju *et al* 2008). However, the highest degree of multiplexing in clinical bead microarrays is 10-plex, as listed in Table 15.4. These multiplexed clinical tests are qualitative, and positive results do not rule out bacterial or viral infections. Conformations from additional laboratory testing are needed before the clinician can make a final diagnosis.

Another example of a heterogeneous bead-based microarray platform is NanoString technology. The assay requires a pair of color-coded capture and reporter probes and enables direct fluorescent detection of gene targets. The capture probe is biotinylated on its 3′ end and contains 35–50 bases, which are complementary to the target mRNA. The reporter probe consists of six color-coded beads on its 5′ end and also contains 35–50 bases complementary to the target mRNA. Since each bead in the reporter probe can be one of four colors, many uniquely coded reporter tags can be used in the same sample tube, enabling highly multi-plexed gene expression, miRNA expression, and copy variant analysis. First, excess probe pairs are mixed with target mRNA to ensure probe-RNA hybridization. Second, unhybrid-ized probes are washed away and the remaining nucleic acid complexes are captured onto a streptavidin-coated surface. Third, an electric field is applied and causes each nucleic acid complex to elongate and orient in the same direction. Immobilized, elongated complexes are then imaged, and target sequences are identified by unique spectral emissions generated by the color-coded reporter probes. Instead of relying on enzymatic amplification like most nucleic acid detection methods, NanoString assays use single-molecule imaging via a highly sensitive CCD camera to directly count target molecules.

15.17.14 Next-Generation DNA Sequencing Technologies

Increasingly, automated sequencing systems are being developed to identify rare genetic mutations and new pathogens. The capability to identify new pathogens makes these DNA sequencing platforms superior to most conventional nucleic acid testing methods, which can only pinpoint known pathogens. Although DNA sequencing has applications in epigenetics, metagenomics, and transcriptomics, these sophisticated tests are still only sparingly used in clinical diagnostics. These systems are, however, heavily used in research settings. Ever-increasing robustness, rapidly decreasing costs, and active research make this a promising area for further development of clinical diagnostics.

15.18 CONCLUSIONS AND FUTURE CHALLENGES

In this chapter, we have provided a basis for understanding the principles and instrumen-tation required for small molecule assays (proteins and nucleic acids) as well as cytometry-based assays. As more and more applications for nucleic acid-, protein-, and cell-based assays are being developed, the use of these diagnostic tests will increase in the clinical laboratory. When designing new diagnostic assays, assay developers should ideally cre-ate user-friendly, low-cost, rapid assays that require small sample/reagent volumes. Less expensive instruments are also advantageous for clinical laboratories because this ensures that all patients, even those in developing countries, have access to optimal diagnostic evaluation to enhance their medical care.

Quantitative and qualitative cellular analyses via cytometry can be performed quite rapidly, at present on the order of 10,000 cells/s. However, accurate detection of rare cell populations, that is, less than 1 target cell in 1000 (0.1%), is still challenging in flow cytom-etry assays. Although SBC assays have improved identification of rare cell populations, SBC assays are currently not available for clinical use. Since many important cells such as circulating tumor cells and fetal blood cells in maternal blood are only present in low quantities, further advances in cytometry and cell sorting are needed. Exciting advances have been made in research labs, and translation to the clinical labs are awaited (Goda *et al* 2009, Nagrath *et al* 2007).

For protein-based clinical immunoassays, highly sensitive single-antigen detection is readily performed via fluorescent and chemiluminescent assays, and low picogram per milliliter concentrations can be measured. Generally, singleplexed chemiluminescent assays are more sensitive than corresponding fluorescent assays due to lower background levels. An obvious area of much need but also much technical difficulty is the development of multiplexed immunoassays. Although many research papers have demonstrated multiplexed immunoassays, translation to the clinical lab has been very difficult due to antibody cross-reactions between detection antibodies introduced into the assay as a cocktail. The challenge for clinical translation and FDA approval arises from the extensive validation that becomes necessary to ensure lack of cross reactivity between antibodies. As the degree of multiplexing increases, the required cross validation increases exponentially. Multiplexed immunoassays, by necessity, also use higher total concentration of detection antibodies, increasing background, and decreasing sensitivity compared to the corresponding single-plexed assays. Thus, clinical multiplexed assays are generally qualitative tests whereby cutoff values are set. Further innovation is eagerly awaited in this area.

Unlike immunoassay diagnostic tests, standards are in place for nucleic acid tests, making multiplexed nucleic acid detection far more robust and more easily validated. Clinical nucleic acid tests encompass amplification or/and detection of target genes or point mutations. Current clinical diagnostic nucleic acid tests are mainly PCR based and encompass amplification and/or detection of target genes or point mutations. The main challenge in nucleic acid tests is the integration of sample preparation with the amplification and detection steps. Improvements in automated systems will help to significantly simplify the workflow of these tests. Nucleic acid tests exhibit high sensitivities and high accuracy, but tests are still costly. Low-cost, nucleic acid array–based technologies, which utilize less expensive equipment, are therefore needed. An exciting area to watch for in nucleic acid detection is whole or targeted genomic DNA sequencing.

REFERENCES

Ahn-Yoon S *et al* 2003 Ganglioside-liposome immunoassay for the ultrasensitive detection of cholera toxin *Anal. Chem.* **75** 2256–61

BD Biosciences. http://www.bdbiosciences.com (Accessed January, 2014)

Boja E *et al* 2011 The journey to regulation of protein-based multiplex quantitative assays *Clin. Chem.* **57** 560–7

Cagel P T and Allen T C 2009 *Basic Concepts of Molecular Pathology* (New York: Springer Science)

Candotti D *et al* 2003 Evaluation of a transcription-mediated amplification-based HCV and HIV-1 RNA duplex assay for screening individual blood donations: a comparison with a minipool testing system *Transfusion* **43** 215–25

Choi *et al* 2008 Sheathless focusing of microbeads and blood cells based on hydrophoresis *Small* **4** 634

Chun P 2009 Colloidal gold and other labels for lateral flow immunoassays in R Wong and H Tse (eds.), *Lateral Flow Immunoassay* (New York: Springer Science) pp 75–93

Di Carlo D 2009 Inertial microfluidics *Lab Chip* **9** 3038–46

Djoba S J *et al* 2008 An evaluation of commercial fluorescent bead-based luminex cytokine assays *PLoS One* **3** e2535

Ellington A *et al* 2010 Antibody-based protein multiplex platforms: technical and operational challenges *Clin. Chem.* **2** 186–93

Federal Drug Administration. http://www.fda.gov/ (Accessed January, 2014)

Fernandez-Sanchez C *et al* 2005 One-step immunostrip test for the simultaneous detection of free and total prostate specific antigen in serum *J. Immunol. Methods* **307** 1–12

Frayling I *et al* 2006 PCR-based methods for mutation detection in W Coleman and G Tsongalis (eds.), *Molecular Diagnostics for the Clinical Laboratorian* (New York: Springer Science) pp 65–74

Goda K *et al* 2009 Serial time-encoded amplified imaging for real-time observation of fast dynamic phenomena *Nature* **458** 1145–9

Goddard G *et al* 2006a Single particle high resolution spectral analysis flow cytometry *Cytometry A* **69** 842–51

Goddard G *et al* 2006b Ultrasonic particle-concentration for sheathless focusing of particles for analysis in flow cytometer *Cytometry A* **69A** 66–74

Gorkin R *et al* 2010 Centrifugal microfluidics for biomedical applications *Lab Chip* **10** 1758–73

Grossman A (ed.) 1998 *Clinical Endocrinology* (London: Wiley-Blackwell)

Hellyer T and Nadeau J 2004 Strand displacement amplification: a versatile tool for molecular diagnostics *Exp. Rev. Mol. Diagn.* **4** 251–61

Henriksen M *et al* 2011 Laser scanning cytometry and its applications: a pioneering technology in the field of quantitative imaging cytometry in Z Darzynkiewicz *et al* (ed.), *Methods in Cell Biology* 102:161–205.

Hong J Y and Choi M J 2002 Development of one-step fluorescence polarization immunoassay for progesterone *Biol. Pharm. Bull.* **10** 1258–62

Jain K 2010 *The Handbook of Biomarkers* (New York: Springer Science)

Jameson D and Ross J 2010 Fluorescence polarization/anisotropy in diagnostics and imaging *Chem. Rev.* **110** 2685–708

Jang H *et al* 2010 Peptide nucleic acid array for detection of point mutations in hepatitis B virus associated with antiviral resistance *J. Clin. Microbiol.* **48** 3127–31

Khandur R S 2006 *Handbook of Analytical Instruments* (New Delhi: Tata McGraw-Hill)

Kost G (ed.) 1996 *Handbook of Clinical Automation, Robotics, and Optimization* (New York: John Wiley & Sons)

Krishnamoorthy G *et al* 2010 Electrokinetic label-free screening chip: a marriage of multiplexing and high throughput analysis using surface plasmon resonance imaging *Lab Chip* **10** 986–90

Krishnan S *et al* 2010 Attomolar detection of a cancer biomarker protein in serum by surface plasmon resonance using superparamagnetic particle labels *Angew. Chem.* **50** 1175–8

Ladner D P *et al* 2001 Multiplex detection of hotspot mutations by rolling circle-enabled universal microarrays *Lab. Invest.* **81** 1079–86

Lakowicz J R 2006 *Principles of Fluorescence Spectroscopy* (New York: Springer Science)

Lee J A *et al* 2008 MIFlowCyt: the minimum information about a flow cytometry experiment *Cytometry Part A* **73A** 926–30

Lerner J *et al* 2010 Approaches to spectral imaging hardware *Curr. Protoc. Cytom.* Chapter 12 Unit 12.20

Liu M *et al* 2005 Multiplexed analysis of biomarkers related to obesity and the metabolic syndrome in human plasma, using the Luminex-100 system *Clin. Chem.* **51** 1102–9

Luchanski M and Bailey R 2011 Rapid, multiparameter profiling of cellular secretion using silicon photonic microring resonator arrays *J. Am. Chem. Soc.* **50** 20500–6

Macey M G 2007 *Flow Cytometry: Principles and Applications* (Totowa: Humana Press Inc.)

Mansfield M 2009 Nitrocellulose membranes for lateral flow immunoassays: a technical treatise in R Wong and H Tse (eds.), *Lateral Flow Immunoassay* (New York: Springer Science) pp 95–113

McPherson R and Pincus M 2007 *Henry's Clinical Diagnosis and Management by Laboratory Methods* (Philadelphia: Saunders)

Millipore Corporation 2002 Rapid lateral flow test strips: considerations for product development Lit. No. TB500EN00 (Bedford, MA)

Nagrath S *et al* 2007 Isolation of rare circulating tumour cells in cancer patients by microchip technology *Nature* **450** 1235–9

Pabbaraju K *et al* 2008 Comparison of the Luminex xTAG respiratory viral panel with in-house nucleic acid amplification tests for diagnosis of respiratory virus infections *J. Clin. Microbiol.* **46** 3056–62

Peter M *et al* 2001 A multiplex real-time PCR Assay for the detection of gene fusions observed in solid tumors *Lab. Invest.* **81** 905–12

Robinson J 2004 Flow cytometry in G E Wnek and G L Bowlin (eds.), *Encyclopedia of Biomaterials and Biomedical Engineering* (New York: Marcel Dekker) pp 630–40

Sawyer L *et al* 2000 Clinical laboratory evaluation of a new sensitive and specific assay for qualitative detection of hepatitis C virus RNA in clinical specimens *J. Hepatol.* **32**(Suppl 2) 116A

Schopf E and Chen Y 2010 Attomole DNA detection assay via rolling circle amplification and single molecule detection *Anal. Biochem.* **397** 115–7

Srivastava S K, Gencoglu A and Minerick A R 2010 DC insulator dielectrophoretic applications in microdevice technology: a review *Anal. Bioanal. Chem.* **399** 301–21

Shi J *et al* 2008 Focusing microparticles in a microfluidic channel with standing surface acoustic waves *Lab Chip* **8** 221–3

Tian J *et al* 2012 Multiplexed detection of tumor markers with multicolor quantum dots based on fluorescence polarization immunoassay *Talanta* **92** 72–7

Tuchin V 2011 *Advanced Optical Flow Cytometry: Methods and Disease Diagnoses* (Weinheim: Wiley-VCH)

Watson D A *et al* 2008 A flow cytometer for the measurement of Raman spectra *Cytometry A* **73** 119–28

White R, Blainey P, Fan H C and Quake S R 2009 Digital PCR provides sensitive and absolute calibration for high throughput sequencing *BMC Genomics* **10** 116

Wild D 2005 *The Immunoassay Handbook* (New York: Nature Publishing Group)

Zhang M *et al* 2012 Ultrasensitive electrochemiluminescence immunoassay for tumor marker detection using functionalized Ru-silica@nanoporous gold composite as labels *Analyst* **137** 680–5

CHAPTER 16

Pediatric Physiological Measurements

Michael R. Neuman

CONTENTS

16.1 Introduction		394
16.2 Well-Child Care		394
16.3 Hospital Care		397
	16.3.1 Cardiac	398
	16.3.2 Breathing Monitoring	399
	16.3.3 Blood Pressure Monitoring	403
	16.3.4 Temperature Monitoring	403
	16.3.5 Blood Gas Monitoring	404
	16.3.6 Partial Pressure of Oxygen	404
	16.3.7 Intravascular Oximetry	406
	16.3.8 pH	407
	16.3.9 Partial Pressure of Carbon Dioxide	408
	16.3.10 Blood Gas Analyzer	409
	16.3.11 Noninvasive Blood Gas Measurements	410
	16.3.11.1 Pulse Oximeter	412
	16.3.11.2 Transcutaneous Bilirubin Measurement	413
	16.3.12 Pediatric Intensive Care Unit	413
	16.3.12.1 Fluid and Electrolyte Therapy	414
	16.3.12.2 Extracorporeal Membrane Oxygenation	414
16.4 Home Care		415
	16.4.1 Temperature Measurement	415
	16.4.2 Apnea Monitoring	416
	16.4.3 Observation Devices for Infants and Children	417
16.5 Summary		418
Bibliography		418

16.1 INTRODUCTION

Physiological measurements in pediatrics are similar to those in any other medical specialty: they involve detecting a physiologic variable from the patient using noninvasive or invasive sensors that are connected to some electronic processing system and an appropriate method for display and storage of the information. Just as in any other medical specialty, the first principle of physical diagnosis in pediatrics is to look at the patient. Often, medical instrumentation can make this difficult to do because so much of the patient's body surface is covered with sensors and instrumentation. This is especially true in the neonate (a newborn infant during the first week of life or a preterm infant) where the patient might be small compared to the sensors being used.

Pediatric physiological measurements can be grouped into three general classifications: measurements associated with well-child care, monitoring of critically ill patients, and monitoring and assessing children at home. In the first case, measurements are carried out in the office of a pediatrician, pediatric clinic, or family practitioner while the second case involves making measurements and monitoring patients in critical care hospital units such as the neonatal intensive care unit (NICU) or the pediatric intensive care unit (PICU). Wherever the measurements are made, a general rule should be followed in the design and application of the instrumentation. This rule states that a pediatric patient should not be considered as just a miniature adult. Pediatric patients, especially the very young ones, in addition to being smaller than adults, have body proportions and physiological functions that can be different from those of an adult. Thus, adult measuring techniques may not always be appropriate for use with children even when appropriately scaled for infants. Size, mobility, ability to communicate, and anxiety are factors that can differ considerably between children and adult patients. Although different modalities of medical imaging will not be considered in this chapter, the pediatric patient presents special problems in this diagnostic area that point out their difference from an adult. These include a strong desire by clinicians to limit children's exposure to ionizing radiation and the fact that some imaging procedures, such as obtaining an MRI image, require the patient to remain still for periods as long as 30 min as the imaging data are obtained. This is often difficult for younger children to do, and the patients need to be sedated, a procedure that in itself has risks. In the remaining paragraphs, we will look at some of the measurement techniques used in the care of pediatric patients.

16.2 WELL-CHILD CARE

Most readers will be familiar with the care they received as a child or the care of their own children and grandchildren. Part of growing up for most children is the periodic visit to the pediatrician or family practitioner for routine care, assessment of growth and development, and administration of vaccines and other preventive measures. Part of the routine examination of a child involves making physiological measurements on the patient. Growth is assessed by measuring the height and weight of the patient and charting it on standard growth charts, an example of one shown in Figure 16.1 (CDC 2014). Height is

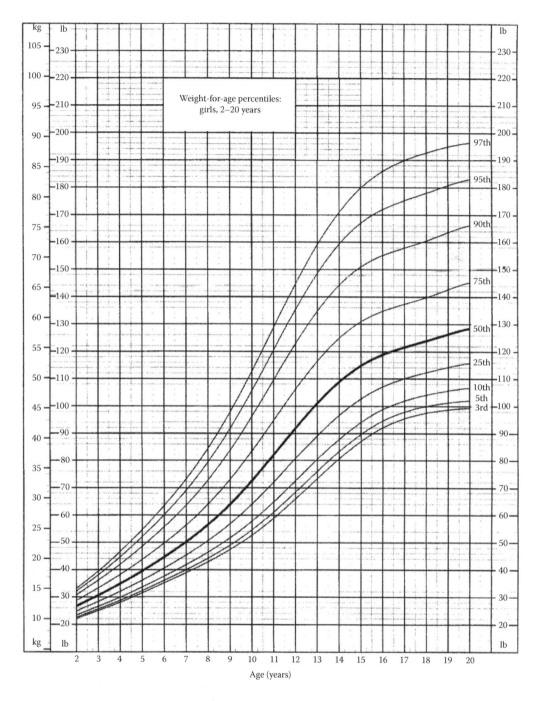

FIGURE 16.1   A standard growth chart for use with children. (Developed by the National Center for Health Statistics in collaboration with the National Center for Chronic Disease Prevention and Health Promotion (CDC 2014).)

measured with the device known as a stadiometer that measures the distance from the foot to the top of the head, or the same measurement can be made on babies lying supine using a tape measure. The stadiometer has a scale that can be read by the operator when a probe attached to the scale touches the head of erectly standing child. Weight is assessed by using either a balancing or electronic scale that provides the display of the weight to the operator. Both of these instruments have been modernized to provide the information in electronic form that can be transferred to a computer file or directly to the patient's electronic medical record at the command of the clinician making the measurement.

Another common measurement made as a part of the well-child care examination is the patient's temperature. Ideally, it is desired to obtain body core temperature, and this can best be done with a temperature sensor placed a few centimeters into the rectum, but patients often object to this method. For this reason, alternative noninvasive or minimally invasive sites are used. Until recently, temperature was measured by mercury-in-glass or alcohol-in-glass clinical thermometers, but the risks of breakage and concerns regarding the introduction of toxic materials as well as the development of inexpensive electronic thermometers have allowed the latter to replace the traditional methods. The electronic thermometers have the advantage of being low mass so that they have a rapid response time, and they provide an easier-to-read display than the traditional thermometer. The electronic thermometer is usually based upon a thermistor temperature sensor that is often protected by a very thin sterile sheath that is replaced after the measurement is made to avoid the necessity of sterilizing the probe between patients. Although this increases the response time, the material of the sheath is very thin so as to add minimal thermal resistance between the sensor and the tissue being measured.

Electronic thermometers are frequently used to estimate core temperature from oral measurements, or very young infants' temperatures can be measured in the axilla (arm pit), but this provides a poorer estimate of body core temperature. Axillary temperatures are, therefore, seldom used for well-child care office visits, yet parents, and their babies, are more comfortable when making this measurement at home. A typical clinical electronic thermometer is illustrated in Figure 16.2a, and a block diagram of its function is shown in Figure 16.2b. The thermistor sensor will change its electric resistance as a function of its temperature in a reproducible way; thus, the strategy of an electronic thermometer is to get this sensor to the body temperature as quickly as possible so that a reading can be made. The temperature of the sensor is determined by measuring its electrical resistance, often a Wheatstone bridge circuit is used, and then converting that resistance to temperature either in degrees Celsius or degrees Fahrenheit. Since the thermistor characteristics are nonlinear, this is often done by a microcontroller with a lookup table in memory that holds the calibration curve for the temperature sensor. Even when a low-mass thermistor and probe assembly are used, the response time of the electronic thermometer to come to body temperature is longer than desired, especially when dealing with small children. For this reason, the temperature sensor–child combination is modeled as a first-order system, yielding an exponential response from the time the sensor comes in contact with the child. It is then possible to use the microcomputer to predict the ultimate temperature before the sensor actually gets there. Thus, the instrument collects data along the rising exponential

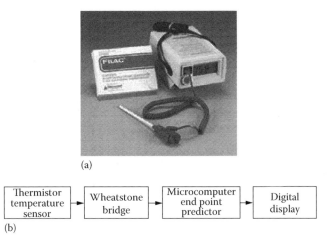

(a)

Thermistor temperature sensor → Wheatstone bridge → Microcomputer end point predictor → Digital display

(b)

FIGURE 16.2   (a) An example of an electronic clinical thermometer for use at home. (b) Functional block diagram of a typical electronic thermometer.

curve after the sensor is placed on or in the body, and when it has sufficient data to make an accurate prediction of the endpoint, it sounds a tone that indicates thermometer can be removed from the subject and that it has estimated the child's temperature.

Other types of temperature sensing instruments can be used in a well-child care clinic. These include radiation temperature measurements from the tympanic membrane (eardrum) and measuring thermal radiation from the temporal artery on the side of the forehead. The clinical versions of these instruments are very similar to those used for measurements in the home with the exception that their packaging is made more robust due to their increased use in the clinic. These instruments will be described in greater detail in the home care section of this chapter.

16.3 HOSPITAL CARE

The care of pediatric patients in the hospital can be classified into the general areas: routine care and critical care. The former involves caring for inpatients on the typical pediatric medical-surgical ward. These patients require routine checking of their vital signs: heart rate, breathing rate, blood pressure, and temperature; but these are sampled readings taken once or a few times during a typical 8 h nursing shift. Although physiological measurement instrumentation can be used to determine these quantities, this level of sophistication is generally not required for routine care. One exception is the use of the electronic thermometer, and this has the advantage of rapid response time so that the clinician collecting the vital signs does not have to spend a lot of time waiting for a conventional thermometer to reach equilibrium.

Critical hospital care, on the other hand, is much more intensive and involves the use of many different types of physiological measurement instrumentation. Often in critical care units, a patient's vital signs are continuously monitored rather than sampled. There are two types of pediatric critical care units in tertiary hospitals: the NICU and the PICU. The former is concerned with premature or newborn infants while the latter covers everyone

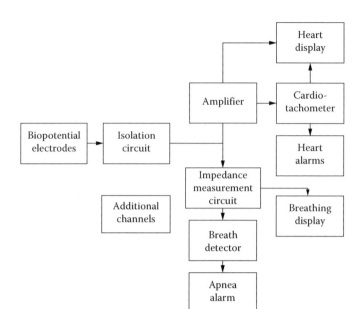

FIGURE 16.3   A functional block diagram of a typical infant cardiopulmonary monitor.

else in the pediatric population who requires intensive care. The central measuring device in both units is the cardiopulmonary monitor. This device monitors several variables related to the cardiovascular and pulmonary systems to continuously assess their status. In the case of the NICU, the cardiopulmonary monitor also watches for periods of apnea (cessation of breathing) and includes some type of alarm when these periods are greater than a preset threshold.

A typical clinical pediatric cardiopulmonary monitor's functional block diagram is shown in Figure 16.3. Although only the cardiopulmonary sections are shown, these monitors often have the capability of monitoring additional physiological variables such as the patient's temperature, oxygen tension or saturation in the blood, expired air or blood carbon dioxide partial pressures, blood pressure, and general patient activity. These functions are indicated by the block labeled additional channels. Each measurement unit is a module that can be added to the monitor when clinically indicated. We concentrate on the cardiopulmonary aspects of the monitor in this section.

16.3.1 Cardiac

The cardiac section of the monitor provides a display and record of the patient's electrocardiogram and heart rate. The former is collected from biopotential electrodes that are usually placed on the patient's chest (Neuman 2010). Although the electrodes might appear to be similar to adult cardiac electrodes, there are important differences. It is obvious that the electrodes should be smaller for infants than they are for adults, and this is indeed the case. The infant electrodes cover approximately one quarter of the area of adult electrodes, yet the infant's mass might be approximately 1/20th of that of an adult male. Thus, although smaller, pediatric electrodes have not been scaled to a size comparable to that of the infant. Because their size is relatively large compared to the infant, when these electrodes are made of materials that are x-ray opaque, they make shadows that interfere with radiographs and have to be removed prior to making the films. This can be quite irritating to infants when conventional electrode adhesive is used to secure the electrodes to the skin, and so pediatric electrodes often use weaker adhesives to hold them in place. A better solution is not to use x-ray-opaque electrode materials in the first place, and this is frequently done with pediatric electrodes so that they do not have to be removed at all when x-rays are taken (Neuman 1979a). Finally, the conductive electrolyte paste used to help the electrode make good electric contact with the skin should also consist of nonirritating components. Thus, the gel used in pediatric electrodes frequently has low chloride ion content and no abrasive materials when compared to adult electrode gels.

The processing of the electrocardiogram is the same as is done in adult cardiac monitors with the exception that children's heart rates are higher than adults', especially in infants and younger children. For this reason, alarms must cover a higher range of heart rates in pediatric monitors; however, they still perform the same function of alerting clinicians when heart rates exceed or fall below preset thresholds. More recent monitors not only display the patient's electrocardiogram, but they also have signal-processing algorithms that recognize arrhythmias that can be life threatening to the pediatric patient, and they display the pattern of the rhythm disturbance as well as alert the caregivers.

16.3.2 Breathing Monitoring

The purpose of breathing monitoring is to record and display breathing rate and pattern as well as to alarm when certain breathing conditions such as intermittent and low-volume breaths or apnea occur. The most critical component of this section of the monitor is that which senses breathing. There are many possible methods to do this, but the most frequently used method in clinical monitors is measurement by transthoracic impedance. In this case, electrodes are placed on the anterior chest at about the level of the fourth interspace and lateral to the midclavicular lines. This is illustrated schematically in Figure 16.4. These electrodes can also be used to obtain the infant's electrocardiogram for purposes

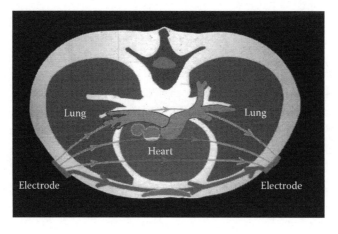

FIGURE 16.4 A cross-sectional view of the thorax showing transthoracic electrical impedance measurement electrodes and approximate current pathways through thoracic structures.

of determining heart rate and rhythm, but the positions are generally not similar to those used for diagnostic electrocardiography. A high-frequency alternating current is passed between the electrodes to determine the breathing signal. This current is generally small, approximately 50 to 100 µA rms, and is at a frequency in the 50 to 100 kHz range. This frequency is used so as to minimize any effects from electrode impedances and to be a sufficiently high frequency so as not to cause electrical stimulation of electrically excitable tissues.

As is seen in Figure 16.4, the current between the chest electrodes passes through most of the thoracic structures. A large portion of it is confined to the anterior thoracic wall due to its relatively low impedance compared to the deeper structures in the thorax, but some also passes through the lungs and heart. Both of these organs have a time-varying electric impedance resulting from their mechanical activity. In the case of the lungs, as they fill with air during inhalation, their impedance increases, while it decreases during exhalation. These impedance changes with breathing produce a voltage change at the electrodes that reflects the breathing pattern. This voltage change, however, is quite small, since the overall impedance change due to breathing seen between the thoracic electrodes is no greater than 0.5% of the baseline impedance and often is much less. To further complicate matters, the impedance between the thoracic electrodes can change due to other causes such as changes in other tissues as a result of patient movement and changes in impedance due to the beating heart. Just as the motion of the lungs changes their impedance, the motion of the heart has a similar effect. Blood is a better conductor of an electric current than solid tissues, and so when the heart contracts, its electric impedance will increase slightly, and the opposite will occur as it relaxes and fills with blood during diastole. These impedance changes are generally smaller than those seen from the lungs, but they can be particularly problematic during periods of apnea when there is no pulmonary air exchange. Both the effects of movement and the beating heart can produce signals that make it difficult for the monitor to determine if the signal was the result of the patient taking a breath or it came from cardiac or other artifactual sources (Geddes and Baker 1989). Figure 16.5 illustrates an example of the breathing waveform determined by measuring transthoracic impedance with an example of normal quiet breathing (top trace) and artifact from patient movement.

There are two types of apnea that are seen primarily in neonates and very young infants: central apnea and obstructive apnea. In the former case, all breathing activity stops, and the patient makes no biomechanical effort to take a breath. In the latter case, the apnea is due to obstruction in the upper airway that prevents the air from reaching the lungs even though breathing efforts are being made. The transthoracic impedance method of detecting breathing activity can usually detect central apneas although motion or cardiogenic artifact can interfere, but this method is unable to detect obstructive apneas. Thus, as can be seen from this description, there are several limitations to the transthoracic impedance method of detecting breathing waveform and apneas. Even with these limitations, the transthoracic impedance method is the most commonly used approach to monitor neonate and infant breathing in hospital and, as we will see later, at home.

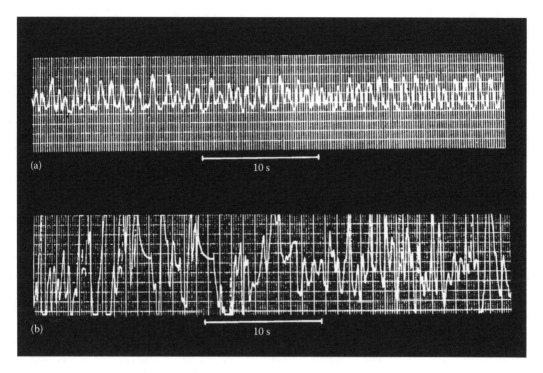

FIGURE 16.5   An example of a breathing waveform pattern generated from transthoracic imped-ance measurements illustrating (a) quiet breathing and (b) motion artifacts.

It is important to point out that there are other methods to detect infant breathing, and some of these are used with clinical monitors even though the dominant technology remains the transthoracic electrical impedance method. These methods include measuring infant breathing air temperature with rapidly responding temperature sensors such as very small thermistors located in the breathing air stream. As the infant inhales, room temperature or incubator temperature air is drawn past the temperature sensor, and this air is at a temperature lower than the infant's core temperature. When the infant exhales, however, the air from its lungs will be at a temperature greater than that of the inhaled air, and the temperature sensor will respond to this elevated temperature, indicating a breath. Temperature sensors have to be very small to respond rapidly enough to accurately indicate infant breaths, and the sensor has to be attached to the infant in such a way that it remains in the stream of breathing air. These constraints make this type of breathing sensor both fragile and inconvenient to use for chronic monitoring.

Another type of breathing sensor that has been used in research studies is a compliant strain gage placed on the infant's abdomen. The strain gage is a liquid metal device that consists of a small diameter, thin-walled elastomeric tube filled with liquid mercury. If this tube is stretched a small amount and taped circumferentially on the infant's abdomen, it will stretch as the infant inhales and relax when it exhales. Electrical connections to the mercury at each end of the tube measure changes in its electrical resistance due to geometric changes as it stretches and contracts, and these changes can be used to indicate breathing activity and apnea. This method has the advantage that there is very little cardiogenic artifact on the signal, and its sensitivity to movement is less than that of the transthoracic electrical impedance signal.

In a study, five different methods of detecting breathing activity were simultaneously monitored on 34 infants for a period of 4 h each. The methods included a nasal thermistor, liquid metal strain gages on the chest and abdomen, transthoracic electrical impedance, and a motion-sensing pad under the infant. The infants' breathing movements were also recorded on videotape, and this served as the gold standard for indicating a breath. The data were separated into periods of quiet sleep and periods when the infant was asleep but indicated movements similar to those associated with the active sleep state. During the periods of quiet sleep, all the sensors behaved similarly and detected 98% of the observed breaths. When movement was associated with the infants' sleep, the sensors behaved quite differently. Figure 16.6 shows the efficacy of each of the sensors studied with regard to detecting breaths as observed directly from the infant or from the videotape. It is interesting to note that under these conditions, transthoracic impedance only detected 43% of the breaths while the nasal thermistor and abdominal liquid metal strain gage were more efficacious. Of course, these two better performing methods have their limitations as well, since in the case of the nasal thermistor, the device has to be attached to the infant's face. The thermistor had to be placed and maintained in the airflow, and the liquid metal strain gages contain mercury, a toxic material. Nevertheless, even though nasal thermistors are not used for intensive care monitoring of neonates and infants, they are used in infant polysomnography (sleep studies). Commercial products are available, utilizing very low

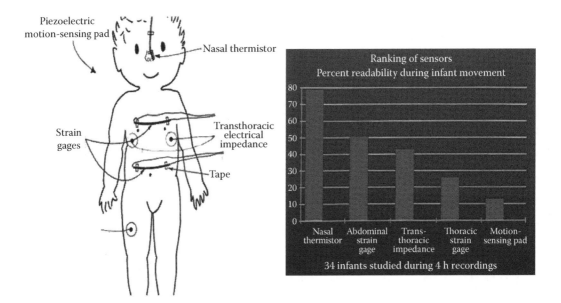

FIGURE 16.6 A comparison of the efficacy of different infant breathing sensors during periods of infant movement while asleep. (a) The placement of the sensors. (b) Their relative efficacy in detecting breaths during periods of movement.

mass thermistors so that they respond quickly to infant breaths or thin-film temperature sensors that have low mass and high surface area to also give a rapid response.

The breathing portion of the cardiopulmonary monitor has important signal-processing functions once the breathing signal is detected. It must determine and indicate the patient's breathing rate in breaths per minute, and it should also give a relative indication of the volume of each breath so that clinicians can see if their patients are breathing at a normal rate, a rate that is too high or too low, or if breath amplitudes are sufficiently small that one questions whether the infant is getting sufficient air exchange (hypopnea). Another function that needs to be carried out by the processing block of the monitor is to determine if an apnea is occurring, and if this apnea exceeds a preset duration, such as 20 s, the processor will provide an alarm to the clinicians so that appropriate actions can be taken. Most of the times monitors tend to err on identifying nonapnea events or nonevents all together as being an alarm-producing apnea.

16.3.3 Blood Pressure Monitoring

Another vital sign that is important to monitor in pediatric critical care is blood pressure. In neonatal intensive care, this is often done using the direct method of blood pressure measurement, whereby a catheter is advanced through one of the still patent umbilical arteries until its distal tip reaches the central circulation. The catheter is coupled to an external pressure sensor through a tube containing a column of normal saline solution. Often, heparin is added to the solution to minimize clotting of blood at the catheter tip and possible subsequent embolization. Provided the catheter is not too long, there are no air bubbles or partial obstructions in it, and the pressure sensor is at the same vertical level as the distal catheter tip in the infant, a high-quality blood pressure waveform as a function of time can be obtained. This is generally displayed on the monitor screen, and values for systolic, diastolic, and mean arterial pressures can be obtained through signal processing and observation.

Older infants no longer have patent umbilical vessels, and so access to the central circulation is more difficult, and generally, blood pressures are taken by the indirect method. In this case, small disposable cuffs are used, and the size is determined by the size of the infant's arm or leg where the cuff is placed. A good rule of thumb is that the cuff width should be approximately 50% of the limb circumference. For this reason, neonatal blood pressure cuffs come in a variety of sizes, and they are usually designated as only being used on a single patient so that the risk of cross infection from one patient to another is minimized.

The oscillometric blood pressure measurement technique is used for indirect pediatric blood pressure measurement. This method is based upon measuring the pulsations in cuff pressure due to the infant's beating heart. With this method, one can obtain systolic, diastolic, and an estimate of mean arterial pressure.

16.3.4 Temperature Monitoring

Another vital sign that is continuously monitored in critical care situations is the patient's core body temperature. As stated earlier, it is undesirable to actually place a sensor in a core location such as the rectum for continuous temperature monitoring, and so core temperature is approximated by monitoring skin temperature over a part of the body where the skin temperature reflects core temperature. This is often done by placing a small thermistor against the skin so that it has good thermal contact to the skin, and this should be done in a central rather than peripheral location. Frequently, the sensor is placed over the right upper quadrant of the patient's abdomen so that it is over the liver, which is highly perfused and can closely approximate core body temperature. An insulating adhesive pad is often used to hold the temperature sensor in place so as to minimize environmental effects on body temperature measurement. Clinical patient monitors often have a module for measuring temperature, and temperature is recorded and displayed on the monitor screen. A noninvasive monitoring of core temperature uses heating a sensor placed on the skin so that the heat flux across the skin is zero. Then the skin temperature as measured by the sensor should be equal to the core temperature.

16.3.5 Blood Gas Monitoring

A very important set of physiologic variables to monitor in neonatal and pediatric intensive care are the blood gases. These include the partial pressures of oxygen and carbon dioxide, hemoglobin oxygen saturation in the blood, and, although it is not technically a gas, blood pH. The traditional way of monitoring these variables involved periodically drawing an arterial or venous blood sample and analyzing it with a blood gas analyzer, an instrument that determines these quantities by electrochemical measurement and calculation. This method has the advantage of the accuracy of a clinical laboratory with its calibration and standardization procedures, but there are delays associated with the transport of the blood sample to the laboratory and the analysis procedure once it gets there. Also, if the drawing of the blood sample required arterial or venous puncture with a needle, this usually causes considerable anxiety and pain for the child that can result in a change in breathing pattern that is likely to affect the values measured. Another limitation is the fact that the measurements can only be made periodically when a blood sample is drawn, but in many intensive care situations, continuous measurements of these quantities are important so that clinical staff can be quickly alerted when significant changes occur.

For these reasons, researchers have been studying ways to continuously monitor blood gases using both invasive and noninvasive techniques. The former has involved developing catheters that can be placed intra-arterially that have either electrochemical or optical sensors for one or more of the blood gases at their distal tip while still retaining a patent lumen so that blood samples can be periodically drawn to validate the continuous measurements and the catheter can be used for other measurements such as direct blood pressure. The electrochemical measurements involve either potentiometric or amperometric determination of the individual blood gases.

16.3.6 Partial Pressure of Oxygen

In the case of oxygen, an amperometric method is used with an electrochemical cell similar to what is shown schematically in Figure 16.7a. Oxygen in the blood diffuses through an oxygen-permeable membrane into an isotonic internal solution of potassium chloride

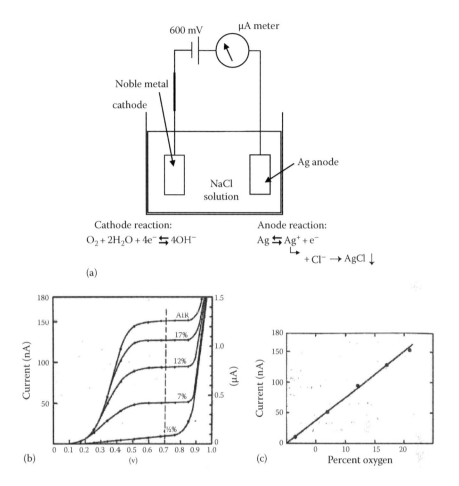

FIGURE 16.7 Measurement of the partial pressure of oxygen. (a) An electrochemical cell that can be used to measure the partial pressure of oxygen in blood and be located at the distal tip of a catheter or in a blood gas analyzer in the clinical laboratory. (b) Current–voltage characteristic for the electrochemical sensor for different oxygen tensions. (c) Current as a function of oxygen tension for the electrochemical sensor.

where it is reduced at the cathode of a two- or three-electrode electrochemical cell. The cell is made up of a small noble metal cathode such as platinum or gold and a silver anode. The fundamental reactions are

$$\text{At the cathode: } O_2 + 2H_2O + 4e^- \rightleftharpoons 4OH^-, \tag{16.1}$$

$$\text{At the anode: } 4Cl^- + 4Ag \rightleftharpoons AgCl \downarrow + 4e^-. \tag{16.2}$$

As can be seen from these reactions, oxygen is reduced at the cathode to produce hydroxyl ions, and four electrons are required for each oxygen molecule that is reduced. Thus, the greater the availability of oxygen (hence its greater partial pressure), the higher the requirement of electrons to reduce it, and thus, the current at the cathode is directly related to the partial pressure of oxygen in the solution.

A second method for determining oxygen partial pressure at the tip of a catheter involves fluorescence quenching. Oxygen is well known as an agent that can interfere with the fluorescence of molecules such as ruthenium complexes. These molecules will normally fluoresce at a wavelength longer than that of the light that excites them, but the presence of oxygen can quench this fluorescence; the amount of quenching being a function of the oxygen partial pressure. Thus, it is possible to entrap some of these molecules at the distal tip of a catheter such that they are protected by an oxygen-permeable membrane and coupled to optical fibers within the wall of the catheter that can provide excitation light and determine the amount of fluorescence. As oxygen diffuses through the membrane, it will cause the fluorescence to be diminished due to this quenching effect. One can then quantitatively calibrate the fluorescence quenching to reflect the partial pressure of oxygen in the blood at the distal tip of the catheter.

Both the optical and electrochemical techniques have been used on indwelling arterial catheters to measure partial pressure of oxygen, but they both have major limitations that must be considered when they are applied. With the optical technique, dye molecules can leach out through the oxygen-permeable membrane or after repeated illuminations with the optical excitation signal, dye molecules can bleach. In either case, this changes the sensitivity of the measurement technique; therefore, frequent calibrations are required. The electrochemical method has its limitations as well. Materials such as serum proteins will deposit on the oxygen-permeable membrane or, in some cases, on the cathode itself. In either case, this interferes with the transport of molecular oxygen, and this will affect the calibration of the sensor. Although researchers have demonstrated the viability of both techniques, these limitations have, at the time of this writing, prevented successful commercial products from appearing and remaining in the marketplace.

16.3.7 Intravascular Oximetry

The spectral properties of oxygenated and reduced hemoglobin have been well established in the red and infrared light regions of the electromagnetic spectrum. The basic principle is that oxygenated hemoglobin has a bright red color while reduced hemoglobin has a deeper red or maroon color. It is also noted that both oxygenated and reduced hemoglobin have the same color at several different wavelengths known as isosbestic points. These optical properties of hemoglobin have been used to measure the oxygen saturation of hemoglobin using optical methods. The basic idea is to illuminate the hemoglobin with each of two different wavelengths of light at different times just a few milliseconds apart. One wavelength is at the isosbestic point in the near-infrared region of the spectrum, 805 nm, and the other in the visible red portion of the spectrum where there is close to the greatest difference between oxygenated and reduced hemoglobin, 660 nm. The ratio of the absorption of light by a blood sample at these two different wavelengths is found to be nearly linearly related to the hemoglobin oxygen saturation. Thus, a laboratory instrument known as an oximeter determines the difference in absorption of light at these two different wavelengths and calculates the hemoglobin oxygen saturation based on an experimentally determined calibration curve.

This technique can also be used invasively by using a catheter-type probe containing fiber optics to illuminate the blood within an artery and collect the reflected light that will have an intensity determined by the incident light minus that which is transmitted or absorbed by the blood, so it will be related to the hemoglobin absorption of light at each of the selected wavelengths. By measuring the intensity of this reflected light, the instrument can calculate the hemoglobin oxygen saturation. Although this invasive technique should work well in principle, there are some limitations to it that can influence the results. Proteins deposited on the surfaces of the illuminating and reflecting fiber optics in contact with the blood or blood clots and the surface can greatly influence the results. If the catheter is oriented such that it illuminates the arterial wall rather than the blood, this can also have a deleterious effect on the measurement. Nevertheless, this technique has been successfully used in human and animal studies.

16.3.8 pH

The other *blood gas* that can be measured directly at the tip of a catheter is the blood pH. As with oxygen, this can be measured using either an electrochemical or an optical technique. Figure 16.8 illustrates an electrochemical sensor that can be used for this purpose. The electrochemical sensor makes a potentiometric measurement. The electrochemical cell consists of an indicating electrode that consists of a reference electrode that makes contact with a pH-buffered solution in an enclosed chamber separated from the external world by an ion-selective glass or a membrane that is made of a polymeric material such as highly plasticized polyvinyl chloride and contains a small amount of a proton ionophore that responds to hydrogen ions (protons). The electrochemical cell is completed by an external reference electrode that makes contact with the fluid being measured, in this case blood. The indicating electrode also makes contact with this blood sample. Thus, we can see that by measuring the potential difference between the two electrodes, one is in

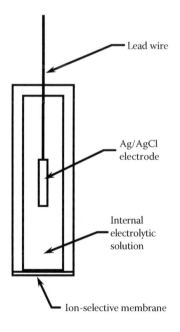

FIGURE 16.8   Measurement of pH using a potentiometric ion-selective electrode.

essence measuring the potential across the ion-selective glass or polymer membrane. This potential will be proportional to the negative logarithm of the concentration differences (or more accurately, activity differences) of hydrogen ions across the membrane. In this case, a potential that is inversely proportional to the pH is generated. This potential is given by the Nernst equation:

$$E = E^0 - \frac{RT}{nF} \ln\left(\frac{[\text{Products}]}{[\text{Reactants}]}\right), \tag{16.3}$$

where
E^0 is the standard potential for the redox reaction
R is the universal gas constant
n is the number of electrons exchanged in the redox reaction
F is the Faraday constant

16.3.9 Partial Pressure of Carbon Dioxide

Carbon dioxide is also measured electrochemically, but the process is more complex than it was for electrochemical oxygen measurements. An intermediate variable, pH, is involved. The sensing principle is based upon the chemical reaction:

$$H_2O + CO_2 \rightleftharpoons H_2CO_3 \rightleftharpoons H^+ + HCO_3^-. \tag{16.4}$$

The relationship between the concentration of CO_2 and the partial pressure of CO_2 is given by

$$[CO_2] = 0.0301(PCO_2). \tag{16.5}$$

The coefficient, 0.0301, is the solubility of CO_2 in blood. Taking Equations 16.4 and 16.5 into account, one sees that the hydrogen ion concentration will be proportional to the partial pressure of carbon dioxide in the blood. One molecule can react to form one hydrogen ion. Thus, by measuring the pH of a bicarbonate-buffered solution, one can find the negative logarithm of the partial pressure of carbon dioxide in that solution. It is important to point out that one still needs to do at least a one-point calibration to be able to determine the PCO_2 concentration of a solution.

This approach is used in the development of a sensor for determining the partial pressure of carbon dioxide in a blood sample. The Stowe–Severinghaus carbon dioxide sensor is illustrated schematically in Figure 16.9. It consists of two chambers: one that contains the blood sample being analyzed, and the other holds a solution containing bicarbonate ions balanced with either sodium or potassium cations that is in contact with a pH electrode. The two chambers are separated by a carbon dioxide–permeable membrane. The volume of the bicarbonate-containing solution should be kept very small so that very little carbon dioxide needs to diffuse across the membrane to bring the PCO_2 of the bicarbonate solution

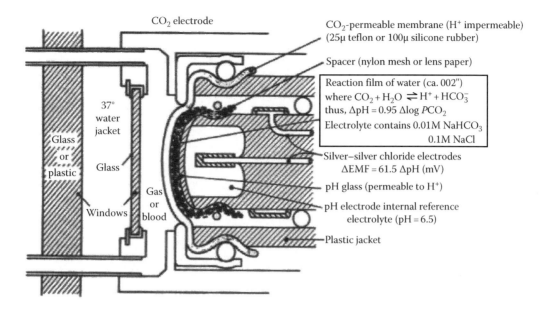

The following labels appear in the figure:

CO$_2$ electrode

CO$_2$-permeable membrane (H$^+$ impermeable)
(25μ teflon or 100μ silicone rubber)

Spacer (nylon mesh or lens paper)

Reaction film of water (ca. 002")
where CO$_2$ + H$_2$O $\rightleftharpoons$ H$^+$ + HCO$_3^-$
thus, ΔpH = 0.95 Δlog PCO_2
Electrolyte contains 0.01M NaHCO$_3$
0.1M NaCl

Silver–silver chloride electrodes
ΔEMF = 61.5 ΔpH (mV)

pH glass (permeable to H$^+$)

pH electrode internal reference
electrolyte (pH = 6.5)

Plastic jacket

37° water jacket

Glass or plastic

Glass

Windows

Gas or blood

Glass

FIGURE 16.9 Measurement of the partial pressure of carbon dioxide using the Stowe–Severinghaus sensor.

into equilibrium with the blood sample. Often, just a thin layer of the solution between the permeable membrane and the pH electrode is all that is needed. The pH electrode and its associated reference electrode can then measure the pH of the inner solution, and the PCO_2 can then be calculated. It is important to note that this sensing system by measuring pH is measuring a quantity proportional to the negative logarithm of the PCO_2. Thus, it gives a quantity that is nonlinearly related to the dissolved carbon dioxide, and since the range of values of PCO_2 in blood is relatively small, the resolution on this measurement has the potential for not being as good as it is for the linearly related oxygen sensor current and partial pressure of oxygen.

16.3.10 Blood Gas Analyzer

The three sensors described in the previous sections can be used intravascularly or in tissue if they are made sufficiently small. They can also be used as part of a blood gas analyzer instrument in the clinical laboratory. A typical instrument is illustrated in Figure 16.10. Since blood gas measurements are often associated with critical care clinical situations, the blood gas analyzer may be located in the intensive care unit rather than the remotely located clinical laboratory. This will help to give the results of the measurements to the clinicians as quickly as possible. The clinical blood gas analyzer consists of a temperature-controlled chamber that will accept the blood sample from the patient, and the blood in this chamber is in direct contact with an oxygen, pH, and PCO_2 sensor. The blood gas analyzer also contains a computer that can calculate other derived variables related to the blood gases such as base deficit, bicarbonate levels, and hemoglobin oxygen saturation. Blood gas analyzers used in pediatric measurements are often optimized to make measurements with very small sample volumes. This is especially important in neonatal intensive

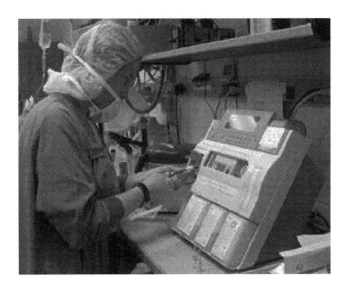

FIGURE 16.10   An example of a blood gas analyzer.

care where the patient is very small with small total blood volumes and frequent blood gas measurements are required.

16.3.11 Noninvasive Blood Gas Measurements

The blood gas measurements described in the previous section involve the collection of an arterial blood sample, which is an invasive procedure. These measurements can be quite accurate when appropriate calibration and standardization procedures are followed, but they only represent the blood gas situation at the time of blood sample collection. Often, in pediatrics, the blood collection involves a painful procedure for the child, and they are frequently upset and crying. This can significantly alter the patient's breathing pattern and hence the measured blood gases. The results can be different from what they were before the blood was drawn and the patient was resting quietly. To follow the time course of the blood gases, multiple samples need to be taken. This can result in significant blood loss for the small premature infant.

Techniques to measure the partial pressure of oxygen and that of carbon dioxide transcutaneously have been developed. These methods take into consideration the fact that oxygen and carbon dioxide can diffuse through the skin. By having the blood in the dermal capillary loops to be as close to arterial blood as possible, one can estimate arterial blood gases from measurements made on the skin surface. This arterialization of the dermal capillary blood can be achieved by heating the skin to temperatures in the range of 42 °C to 44 °C. Figure 16.11 illustrates the basic structure of a transcutaneous oxygen and carbon dioxide sensor. In both cases, the basic sensing mechanism is the same as it was for the sensors in the previous section that came in direct contact with the blood. The difference is that the dermal capillaries now become the blood sample chamber and the skin is part of the gas-permeable membrane for each sensor. In the case of the transcutaneous oxygen sensor (Figure 16.11a), oxygen in the arterialized blood in the dermal capillaries diffuses through the skin and into the oxygen sensor where it is measured. The determined

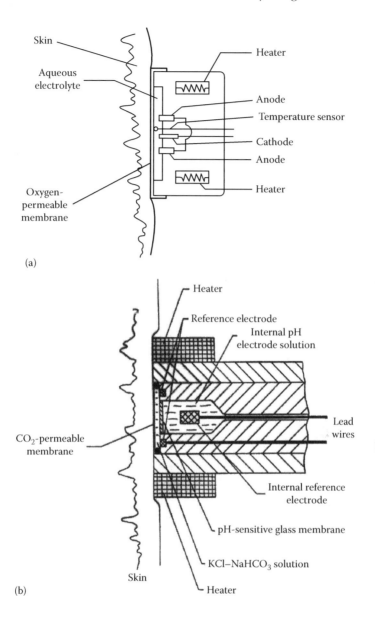

FIGURE 16.11 Schematic illustrations of typical transcutaneous blood gas sensors. (a) A transcutaneous oxygen sensor and (b) a transcutaneous carbon dioxide sensor.

oxygen tension can be close to that of the capillary blood in neonates where the skin thickness is relatively uniform, but in older children and adults, the technique often yields lower oxygen tensions than seen in the arterial blood. The transcutaneous carbon dioxide sensor (Figure 16.11b) gives a better estimate of arterial blood carbon dioxide partial pressure in neonates and older children, and often the sensor can operate at lower temperatures to reduce the possibility of skin irritation or burning.

Transcutaneous blood gas sensors that have both oxygen-sensing and carbon dioxide–sensing elements together in one package have been developed. These devices still heat the skin to arterialize the capillary blood, and they are able to sense the two blood gases simultaneously. As with the other transcutaneous blood gas sensors described in this section, the heating of the skin to temperatures as high as 44 °C carries with it the risk of skin irritation and even the possibility of mild burning. For this reason, it is recommended the transcutaneous sensors be repositioned every 4 h, and this also can lead to skin irritation due to the removal of the adhesive ring that holds the sensor in place. These limitations have reduced the use of transcutaneous blood gas monitors in the pediatric population. Although they were frequently applied in the 1980s, the advent of the pulse oximeter brought a viable new technology to be the method of choice for noninvasive oxygen measurement.

16.3.11.1 Pulse Oximeter

One of the most frequently used medical instruments is the pulse oximeter. This device noninvasively determines the hemoglobin oxygen saturation in blood. It is based on a similar principle to that of the invasive or clinical laboratory whole blood oximeter, but it differs in that it does not directly contact the blood nor does blood have to be withdrawn from the patient and analyzed in a cuvette. Instead, the blood can remain in the capillaries of tissues that can either be transilluminated such as a finger or toe of the pediatric patient or light can be reflected from tissues that cannot be easily transilluminated. As with the oximeter described in the previous section, near-infrared light at a wavelength close to an isosbestic point on the reduced and oxygenated hemoglobin spectra and light in the visible red portion of the spectrum is either passed through the tissue or reflected from it. In conventional oximetry, the steady-state absorption of the blood is determined and processed to give the hemoglobin oxygen saturation, while in pulse oximetry only the pulsatile component of the absorption is processed. This component of the light corresponds to each fresh bolus of arterial blood entering the capillary bed of the tissue being illuminated during cardiac systole. Thus, this blood represents fresh arterial blood entering the tissue. As oxygen is consumed by metabolizing cells close to the capillaries, the hemoglobin becomes less saturated, and this blood ultimately moves to the venous circulation as a fresh bolus of arterial blood enters the capillaries. It is this change in hemoglobin oxygen saturation due to the fresh infusion of arterial blood and the volume increase in systole that causes the pulsations in light absorption seen by the pulse oximeter. By looking at the ratio of the amplitude of the pulsations at the red wavelength to that at the infrared wavelengths, one can determine the percent hemoglobin oxygen saturation in the arterial blood. This ratio, as before, is not linearly related to the hemoglobin oxygen saturation, and so pulse oximeters like their laboratory counterparts need to use a lookup table in memory to determine the hemoglobin oxygen saturation from the ratio of pulse amplitudes.

Most pulse oximeters used in pediatric measurements are based on the absorption of light transmitted through a finger, hand, toe, or foot. In older patients, a spring-loaded clamp containing light-emitting diodes at the two wavelengths on one side of the clamp, and a photodetector such as a photodiode or photo transistor to detect the transmitted light is located on the other side of the clamp. In neonates and young infants, a flexible strip with the light sources at one end and the detectors on the other is wrapped around the hand or foot and secured with tape. In either case, there can be considerable artifact due to motion of the limb that interferes with sensing the pulsatile signal. Thus, one of the major limitations of the pulse oximeter is that it is sensitive to limb movement that alters the optical path length and can cause absorption differences that are much greater than those due to blood volume and saturation changes in the capillary bed. Another limitation of the pulse oximeter, especially when it is used with neonates, is that it cannot indicate *hyperoxia* since its maximum reading is 100% hemoglobin oxygen saturation. Infants on oxygen therapy, however, can have partial pressures of oxygen that can go beyond safe values, but these cannot be indicated by the pulse oximeter. For this reason, clinicians often choose to keep the hemoglobin oxygen saturation values no greater than 94% so that unsafe high PO_2 levels are not reached.

16.3.11.2 Transcutaneous Bilirubin Measurement

Bilirubin is a substance found from the breakdown of hemoglobin in red blood cells that is often elevated in newborn infants. If sufficiently high, significant quantities can cross the blood–brain barrier and result in kernicterus, an irreversible injury to the brain that can lead to various neurologic deficits such as seizures and abnormal reflexes. Elevated bilirubin in infants manifests itself as jaundice or enhanced yellow pigmentation of the skin and sclera of the eyes. The standard clinical test for elevated bilirubin is to draw a blood sample and spectrophotometrically analyze it in the clinical laboratory.

In an effort to have a simpler test that could be used for screening infants who might have elevated serum bilirubin, investigators have developed instrumentation to noninvasively measure bilirubin in the skin. This instrumentation also takes advantage of photometric methods to measure the enhanced yellow color of jaundiced skin. The basic approach is to illuminate the skin with a white light and to determine the spectral distribution of the reflected light. In early instruments, a sensing probe was placed against the skin of the forehead or over the sternum, and a flash of white light was produced using a xenon flash tube such as used in electronic flash photography. The reflected light was analyzed either in terms of its spectral content, especially in the blue and green regions of the spectrum, or in terms of the time taken for various wavelengths to extinguish following exultation by the photo flash unit. Today, microfabricated devices and microcomputers allow the complete spectrum of the reflected light to be analyzed and compared to calibration spectra to determine bilirubin levels. Simple, handheld devices can now be used in the neonatal nurseries as screening devices to determine which infants might need therapeutic intervention to reduce bilirubin levels. Studies comparing these noninvasive methods with traditional blood sampling from infants have demonstrated correlations high enough for pediatric professional organizations to recommend the routine use of these transcutaneous devices without the need for blood sampling.

16.3.12 Pediatric Intensive Care Unit

The PICU cares for patients ranging in age from infants to adolescents. It is similar to the NICU but generally deals with older patients. In this way, it is also similar to the adult intensive care unit. As with the NICU, many physiological measurements are involved in the care of critically ill patients. Most of these have already been discussed in this chapter and include continuous cardiopulmonary monitoring, blood gas monitoring, and temperature monitoring.

16.3.12.1 Fluid and Electrolyte Therapy

As is the case in the NICU, monitoring fluid and electrolyte balance is important in critical care pediatrics. Fluid inputs can be measured by noting the amount of fluid consumed orally and through the use of intravenous fluid administration pumps. Similarly, some aspects of fluid loss can be quantitatively measured by noting the amount of urine passed through catheter or naturally.

Other aspects of fluid loss are more difficult to monitor. This includes fluid and electrolyte loss due to sweating and what is known as insensible fluid loss which includes the fluid loss in exhaled air. Electrolyte balance can also be maintained by monitoring intake and output from the patient. In this case, the insensible loss is only through sweating, and the electrolyte loss through this route is relatively minimal for the hospitalized patient. By knowing the volume of fluid intake and determining its electrolyte content, one can determine the total electrolyte intake, and the same procedure can be used to identify electrolyte loss through measuring urine volume and electrolyte concentration.

16.3.12.2 Extracorporeal Membrane Oxygenation

An aspect of pediatric critical care that is frequently used in PICUs is the application of extracorporeal membrane oxygenation (ECMO) as illustrated in Figure 16.12.

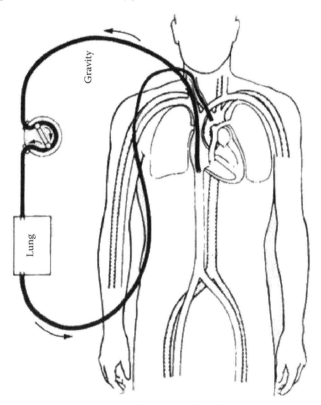

FIGURE 16.12 The fundamental concept of ECMO whereby an external pump and membrane oxygenator is connected to the patient through cannulating a major artery and vein and withdrawing blood from the patient's circulation and reintroducing it at increased pressure and oxygen tension. Illustration courtesy of Wikimedia.

This technology involves circulatory and respiratory support for patients whose hearts and lungs are too severely damaged or diseased to be able to function on their own. By providing an extracorporeal pump and oxygenator for the patient's blood, the patient's own heart and lungs can rest and, hopefully, recover to the point where they can return to their normal functions. Needless to say, when such a system is used, it is important to make measurements to demonstrate that the system is functioning appropriately. These measurements involve hemodynamic assessment, determination of blood gases, determination of the functioning of other organ systems such as the kidneys and brain, the provision of nutrition, and the prevention of hemorrhage or thromboembolism. Thus, many of the measurement techniques described earlier in this chapter must be applied to the patient undergoing ECMO therapy. Although ECMO therapy has been applied to adult patients, it has not to this point been found to be very beneficial. On the other hand, pediatric patients have been shown to benefit from this technology.

16.4 HOME CARE

Up to this point, we have described pediatric physiological measurements that are made by professionals in healthcare institutions, yet there are many cases where pediatric physiological measurements can be made in the home by parents or other nonprofessional caregivers. A few examples of these instruments and the measurements that they can make are given in this section.

16.4.1 Temperature Measurement

Perhaps the most frequent pediatric physiological measurement made in the home is that of body temperature. Traditionally, this has been done with a common mercury or alcohol in glass thermometer placed orally or axillary. Today, better and safer devices are available for this measurement. One of the simplest instruments is a pacifier that contains a thermistor temperature sensor at its tip and has all of the electronics and a temperature readout in the body of the pacifier itself. Oral temperature can be determined as the infant sucks on the pacifier.

A temperature-measuring instrument that is simple to use and records a temperature that is closer to body core temperature is the radiation thermometer that measures thermal radiation from an infant's or child's auditory canal. It has been well established that tympanic membrane (eardrum) temperature reflects the temperature within the brain. The auditory canal with the tympanic membrane at its base can be considered to represent a black body radiator, and so by measuring this radiation, one can determine the temperature deep within the head. Commercial devices have been developed with a disposable ear speculum and a thermal radiation detector. Upon placing the ear speculum in the auditory canal and pressing a button, the sensor detects the radiation from the canal and processes this information to give a readout of the temperature. This measurement can be quickly made, and even though some patients might protest having the speculum placed in their ear, the time to acquire the radiant energy is very short compared to that needed for the glass thermometer to come to equilibrium.

Another approach to measuring deep body temperature in children that has resulted in commercial products is the measurement of the temporal artery temperature (TAT). The instrument for doing this consists of a handheld probe containing a low mass, thermally isolated thermistor that is placed in contact with the temporal skin and slowly scanned from anterior to posterior. The anterior temporal artery is located close to the skin in this region, and because it contains central blood at a relatively high rate of flow, the temperature of the skin over this artery is close to the temperature of the blood and hence body core temperature. The electronic processing in the instrument detects the maximum temperature as the probe scans the skin surface, and this temperature reflects body core temperature.

The principal advantage of this system is similar to that of the auditory canal temperature measurement system in that they both are relatively fast compared to a glass thermometer, and they are particularly useful in making temperature measurements from active children. Both instruments have been shown to detect febrile conditions and have seen some professional use as well as use as a home body temperature measurement instrument. Investigators have compared TAT technology with auditory canal and rectal temperature measurement and found some correlation between all three but also some significant differences in individual cases. Clearly, these methods have important advantages when a quick measurement is needed from an uncooperative patient, yet there are also limitations to the technique that should be considered.

16.4.2 Apnea Monitoring

Monitoring the breathing patterns of infants at home was thought by some to be a way to identify life-threatening conditions in these children early enough that resuscitative measures could be taken. Since the cardiopulmonary monitors used in hospitals were both too expensive and too complex for use at home by untrained individuals, simplified home monitoring devices were developed and marketed. These home infant monitors carried out many of the same functions as done by the hospital instruments, but they required less interaction with the caretaker and provided only fundamental information to them. A typical home infant apnea monitor is shown in Figure 16.13. This device measures

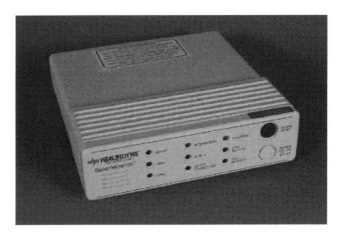

FIGURE 16.13   A home infant apnea monitor.

infant breathing pattern by the same transthoracic impedance method used in the hospital devices, but it does not have the same controls and displays found in the hospital units. Instead, it has a simple indicator light that flashes for each infant breath detected and an alarm that sounds when periods of apnea are sensed that are greater than a preprogrammed threshold. This threshold value can only be inputted by the healthcare provider or vendor supplying the instrument to the home.

Home infant apnea monitors also monitor heart rate in the same way that the hospital monitors do this. The infant's electrocardiogram is sensed by the same electrodes that sense the transthoracic impedance changes associated with breathing. Internal signal-processing circuits in the monitor then determine the infant's heart rate from the electrocardiogram, and a threshold detector determines if the heart rate exceeds a tachycardia threshold or falls below one set for bradycardia. Once again, these threshold levels can only be programmed by the healthcare provider or monitor vendor and not by the family using the monitor at home.

There is controversy as to the value of these electronic devices in reducing risks associated with sudden infant death or other life-threatening conditions. Clinical studies have shown little association between prolonged apnea events and infants who go on to die at home, yet the use of these devices has helped to reassure parents that their infants are being carefully observed. In most cases today, these monitors have a built-in memory loop such that when an alarm condition occurs, recordings of the electrocardiogram and breathing pattern for a brief period up to the alarm condition and continuing until normal patterns are reestablished can be made. These recordings can be downloaded from the monitor by the healthcare provider and examined to determine if any pathological conditions are present.

16.4.3 Observation Devices for Infants and Children

Popular home monitoring devices for observing children at home have been developed based upon wireless and personal computer technologies. The simplest of these consists of a wireless microphone that can be placed in a room with a sleeping infant, and the receiver of the wireless signal is a small device similar to a pager or mobile telephone that can be attached to the clothing of a caregiver. Sounds made by the infant, such as crying, can then be heard by the caregiver even though they are not in the same room. The range of such devices typically covers the region of a home so that the caregiver can go about their normal activities yet still keep an ear open in case the infant wakes up and requires attention.

A more complex version of this device involves the use of a small camera in the room where the infant is located or in rooms where older children are resting, studying, or playing. This wireless system can provide sound and images to the caregiver's computer or over the Internet so that parents can observe their children at home. This latter approach should not be considered an alternative to having a caregiver close to the children, but it does give parents at work or traveling the opportunity to check in on their children. Although these devices cannot really be considered pediatric physiological measurement instruments, they are popular with parents, especially those of infants and young children, and are described here is another way that technology can assist in the care of infants and children.

16.5 SUMMARY

This chapter has surveyed some of the physiological measurement devices that are used in pediatrics. We have seen that many of these devices are related to critical care situations involving children. The development of cardiopulmonary monitors for use in the NICU is an example of how physiological measurement technology can contribute to the care of seriously ill preterm infants. It is thought by many clinicians that such devices were a major contributor to the improved survival of preterm infants, some as small as weighing only 500 g. Of course, technology alone is not much help in improving the care of any patient. It is an adjunct to highly trained personnel who understand the importance of the measurement and the utility of the instrumentation in helping them to assess clinical conditions and provide optimal treatment. Pediatric physiological measurement instrumentation along with highly trained pediatric providers has helped to the improve clinical outcomes in infants and children. The prospects for continuing improvement in pediatric care will come from improved understanding of normal and pathologic conditions of infants and children and the development of technology to diagnose and monitor therapy in these patients.

BIBLIOGRAPHY

Brans Y W and Hay W W Jr. (eds.) 1995 *Physiological Monitoring and Instrument Diagnosis in Perinatal and Neonatal Medicine* (Cambridge: Cambridge University Press)

CDC 2014 Centers for Disease Control and Prevention, Growth charts http://www.cdc.gov/growthcharts/

Geddes L A and Baker L E 1989 *Principles of Applied Biomedical Instrumentation*, 3rd edition (New York: John Wiley & Sons)

Lafeber H N, Aarnoudse J G and Jongsma H W (eds.) 1991 *Fetal and Neonatal Physiological Measurements: Proceedings of the Fourth International Conference on Fetal and Neonatal Physiological Measurements* (Noordwijkerhout: Excerpta Medica)

Neuman M R 1979a The biophysical and bioengineering bases of perinatal monitoring—part V: neonatal cardiac and respiratory monitoring *Perinatol./Neonatol.* 23 17–23

Neuman M R 1979b The biophysical and bioengineering bases of perinatal monitoring—part VI: neonatal temperature, blood-pressure, and blood-gas instrumentation *Perinatol./Neonatol.* 3 25–32 53

Neuman M R 2010 Biopotential electrodes In J G Webster (ed.) *Medical Instrumentation Application and Design* 4th Ed. (Hoboken NJ: John Wiley & Sons)

Ramanathan R, Corwin M F, Hunt C E, Lister G, Tinsley L, Baird L, Silvestri J M *et al* and the Collaborative Home Infant Monitoring Evaluation (CHIME) Study Group 2001 Cardiorespiratory events recorded in the home: comparison of healthy infant with those at increased risk for SIDS *JAMA* **285** 2199–207

Rolfe P (ed.) 1980 *Fetal and Neonatal Physiological Measurements* (London: Pitman Medical)

CHAPTER 17

Measurements
in Pulmonology

Jason H.T. Bates, David W. Kaczka, and G. Kim Prisk

CONTENTS

17.1 Introduction   420
17.2 Sensors   420
    17.2.1 General Theory   420
    17.2.2 Pressure Measurement   423
    17.2.3 Flow and Volume Measurement   425
    17.2.4 Concentrations and Partial Pressures of Gases   427
    17.2.5 Mass Spectrometry in the Gas Phase   429
    17.2.6 Pulse Oximetry   431
    17.2.7 Capnography   432
    17.2.8 Blood-Gas Analysis   433
17.3 Assessing Lung Mechanical Function   433
    17.3.1 Spirometry   433
    17.3.2 Body Plethysmography   435
    17.3.3 Resistance and Elastance   436
    17.3.4 Lung Impedance and the Forced Oscillation Technique 438
17.4 Assessing Pulmonary Gas Exchange   440
    17.4.1 Oxygenation   440
    17.4.2 Shunt and Dead Space   441
    17.4.3 Foreign Gases   443
17.5 Conclusions   445
Glossary   445
References   448

17.1 INTRODUCTION

Pulmonology is a highly quantitative discipline based on physiology. It is primarily concerned with diagnosing and treating the diseases of the respiratory system. Diagnosing these diseases relies on the measurement of physical variables that contain relevant information about the condition of the lung. These variables are then used to derive parameters of clinical significance. The physical variables are pressures, flows, volumes, and concentrations of gases that reflect various aspects of lung function. The physiological parameters may be purely empirical constructs having intuitively obvious relevance to lung function, but often they correspond to the parameters of some descriptive mathematical model of mechanics or gas exchange.

In its broadest sense, monitoring in pulmonology refers to the assessment of gas exchange between the environment and the blood. There are two key events that take place in this sequence: (1) transport of gas between the environment and the alveoli, via the branching airway tree, and (2) diffusion of gases across the physical barrier separating the alveoli from the pulmonary capillaries. There are a variety of methods for assessing the efficiency and effectiveness of each of these two events, all of which involve specialized equipment to measure the relevant physical variables, as well as specialized procedures to ensure that the measured variables accurately reflect the underlying physiological phenomena of interest.

In this chapter, we will summarize the measurement of physical variables associated with the process of respiration. We begin by addressing the instrumentation required to measure pressures, flows, and concentrations of respiratory gases. We then cover some of the various specialized maneuvers and equipment used to diagnose specific aspects of lung function. Where relevant, we show how mathematical models allow the diagnosis of pulmonary disease from the measurements obtained by these procedures.

17.2 SENSORS

17.2.1 General Theory

The instrumentation required to measure respiratory variables, such as gas pressure, flow, and concentration, is subject to various constraints that determine how accurately these measurements can be made.

Quantitative measurements of physical variables that change in time are made with sensors, devices that typically convert the physiological signal of interest into an electrical signal ready for storage and analysis. The electrical signal is digitized prior to storage, and may require an intermediate step of conversion to, for example, an optical signal so that it can be transmitted in an electrically safe and wireless manner to a host computer (Figure 17.1). Such sensors are not perfect recorders of the original signals; however, it is crucial to have a thorough understanding of how recording imperfections arise and the kinds of errors they can cause. The extent to which a sensor accurately represents the physical quantity being measured depends on a number of different factors.

A sensor is said to be *linear* if it produces an output that is *directly proportional* to the physical quantity being measured. Of course, no sensor can ever be perfectly linear, but

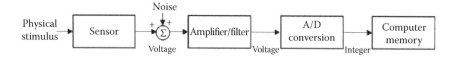

FIGURE 17.1 Typical sensor measurement system, illustrating the sequence of transduction, amplification and filtering, analog-to-digital (A/D) conversion, and computer storage. (Reprinted from Bates, J.H.T., *Lung Mechanics: An Inverse Modeling Approach*, Copyright 2009. With permission from Cambridge University Press.)

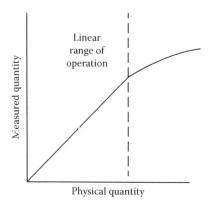

FIGURE 17.2 Relationship between an actual physical quantity and the estimate provided by a sensor. The linear range of operation is at the lower end of values for the quantity being measured.

linearity is a useful approximation for many sensors, particularly over a limited range of their input variable. This range is known as the sensor's *linear range of operation* (Figure 17.2). If the relationship between the input signal and its voltage output is *nonlinear*, then a computer can readily use some suitable nonlinear function to convert output to input provided the relationship is single-valued. Some sensors exhibit *hysteresis*, however; meaning that the value of its voltage output for a given value of the input variable depends on whether that value was approached from above or below. Hysteresis is very difficult to correct for, in general; so it must be kept to a minimum.

The *dynamic response* of a sensor refers to the extent to which it is able to respond to a change in the quantity being measured. One way to assess dynamic response is to subject a sensor to a sudden step change in input signal and observe how quickly its output voltage responds. Ideally, one would like to have the voltage also exhibit a step change in level at exactly the same point in time as the input, but most sensors are low-pass systems that respond with a delayed output (Figure 17.3). The output response to a step input may be sluggish, known as over-damped, or it may be excessively exuberant and exhibit transient oscillations, known as under-damped. The dynamic response of a sensor may also be captured in its *frequency response*, which refers to its ability to accurately track sinusoidally varying changes in the quantity being measured (Figure 17.4). Provided the sensor is linear, an input consisting of a single sine wave of amplitude A_{in} and phase ϕ_{in} will produce

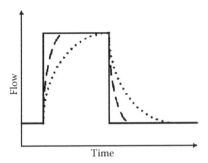

FIGURE 17.3 Step inspiratory flow profile (solid line) with corresponding transduced waveforms from fast-responding (dashed line) and slow-responding (dotted lines) flow sensors.

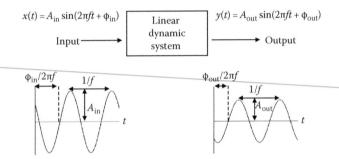

FIGURE 17.4 Schematized representation of the frequency response of a linear system. Input to the system $x(t)$ is a sinusoid with frequency f, amplitude A_{in}, and phase ϕ_{in}. Output from the system $y(t)$ is also a sinusoid of frequency f, but with amplitude A_{out} and phase ϕ_{out}. The output sine wave will differ in amplitude from the input sine wave by a factor $A = A_{out}/A_{in}$ and will be shifted in phase by an angle $\phi = \phi_{out} - \phi_{in}$. These amplitude and phase differences depend only on the excitation frequency f. The dynamic response of a linear sensor can thus be represented completely by two functions of frequency $A(f)$ and $\phi(f)$, which together constitute the *frequency response* of the sensor.

an output that is also a sine wave at the same frequency, although with amplitude A_{out} and phase ϕ_{out}. Thus, the output sine wave will differ in amplitude from the input sine wave by a factor $A = A_{out}/A_{in}$, and will be shifted in phase by an angle $\phi = \phi_{out} - \phi_{in}$ (Figure 17.4). Since these amplitude and phase differences depend only on the excitation frequency f, they will not be affected by the amplitude of the input sine wave if the system under consideration is linear. Accordingly, the dynamic response of a linear sensor can be represented completely by a description of the two functions of frequency $A(f)$ and $\phi(f)$, which together constitute the *frequency response* of the sensor. Ideally, one would like $A(f) = 1$ and $\phi(f) = 0$ for all f, but if this situation pertains at all, it will only ever be over a limited range of f. Outside this range, the output signal produced by the sensor will be a distorted version of the input signal that is being measured, although it is possible to use numerical methods based around deconvolution to improve the situation if the functions $A(f)$ and $\phi(f)$ are known (Bates 2009; Renzi *et al* 1990).

The electrical signal produced by a sensor is converted into digital form by an *analog-to-digital converter* (ADC) (Figure 17.1). Here, the continuous amplitude range of the voltage signal is mapped to a series of equal sized bins numbered consecutively from 1 to 2^n for some integer n. The value of n gives the number of *bits* of the ADC, and must be large enough so that the digitized signal appears smooth. Failure to *amplify* the voltage signal so that it occupies a significant fraction of the input range of the ADC prior to digitization can lead to *discretization error*, producing a choppy appearance in the discretized signal. Also critically important is to low-pass filter the voltage signal prior to digitization in order to eliminate all signal components with frequencies greater than half the digitization frequency. Failure to satisfy the *sampling theorem* in this way risks the appearance of spurious artifacts in the digitized signal, a phenomenon known as *aliasing* (Bates 2009).

17.2.2 Pressure Measurement

The measurement of gas pressure is critical to many assessments of pulmonary function. Examples include *airway pressure* measured at the mouth, nose, trachea, or endotracheal tube; the pressure within a *body plethysmograph*; the pressure within the *esophagus* as a surrogate for *pleural pressure*; and *gastric pressure* within the stomach.

Variable reluctance pressure sensors became popular for pulmonary applications in the mid-twentieth century and consist of a stainless steel diaphragm placed between two induction coils (Bates 2009; Duvivier *et al* 1991). Excitation of one coil by alternating current (typically 3–10 kHz) induces an alternating voltage in the other coil to a degree that is modulated by the pressure-induced bending of the diaphragm. Variable reluctance sensors are sensitive and accurate, but somewhat expensive and bulky. Also, they tend to have a rather limited frequency response (Duvivier *et al* 1991). For these reasons, pulmonary pressures are usually now measured using *piezoresistive* sensors (Bates 2009) that operate by having pressure impinge on a tiny solid-state strain gage that changes its electrical resistance in proportion to its deformation. This change in electrical resistance can be converted to a proportional voltage with a simple Wheatstone bridge circuit. Piezoresistive sensors typically respond well to pressures oscillating at up to hundreds of hertz (Lutchen *et al* 1993; McCall *et al* 1957). However, even a perfect frequency response can be degraded substantially if the sensor is connected to the pressure source by long and/or compliant tubing (Jackson and Vinegar 1979).

The measurement of a single pressure relative to some reference (usually atmospheric pressure) is achieved with a *gage* pressure sensor (Figure 17.5b). By contrast, when two pressures are measured relative to each other, a *differential* pressure sensor is used in which the two pressures of interest are directed to either side of the deformable strain gage (Figure 17.5a). An important consideration for differential sensors is their *common-mode rejection ratio* (CMRR), defined as the capacity of the sensor to produce an output of zero when the two pressures are identical even if the pressures themselves are not zero. CMRR should be as large as possible.

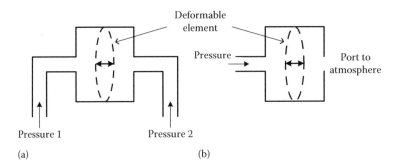

FIGURE 17.5   Pressure sensors arranged in differential (a) or gage (b) configurations. (Reprinted from Bates, J.H.T., *Lung Mechanics: An Inverse Modeling Approach*, Copyright 2009. With permission from Cambridge University Press.)

The measurement of airway pressure is typically achieved with a gage pressure sensor connected via a lateral tap into the main conduit through which air flows. This may provide, for example, the pressure applied to inflate the lungs from an external source, such as a ventilator or constant positive airway pressure (CPAP) machine. In the case of a passively ventilated subject, it is also useful for determining the *compliance* of the respiratory system (i.e., the ratio of gas volume entering the lungs to the pressure applied). However, lateral pressure underestimates the pressure driving gas through the conduit at the site of measurement by an amount given by the *Bernoulli effect*, which is proportional to the square of the gas flow velocity. At low flow rates, this is usually negligible, but at high flows, the Bernoulli effect can lead to a major underestimation of the driving pressure (Bates *et al* 1992). This can be avoided by using a *Pitot tube* that measures pressure facing directly into the oncoming gas stream, but it is directionally dependent.

Transpulmonary pressure (the difference between airway opening and pleural pressures) is a reflection of the distending pressure of the lung, and requires the measurement of pleural pressure. For most applications, esophageal pressure is a suitable surrogate for pleural pressure, and is much more easily (and noninvasively) measured with an esophageal balloon catheter (Loring *et al* 2010). For human subjects, such catheters are typically about 100 cm in length with a thin-walled latex balloon covering the distal tip. The catheter is placed via the mouth or nares into middle-to-distal third of the esophagus, while its proximal end is connected to a pressure sensor (Figure 17.6). A small volume of air is injected through the catheter into the balloon to clear the balloon from the catheter tip, although not so much as to distend the balloon walls. The absolute value of pleural pressure is affected by its position in the esophagus and by mediastinal compression of the balloon in supine patients, as well as by oscillations generated by the beating heart (Hager and Brower 2006, Schuessler *et al* 1998). Balloon position is first checked by the *occlusion test*, in which the subject makes breathing efforts against a closed airway shutter. The excursions in airway opening and esophageal pressures should be equal if properly positioned (Baydur *et al* 1982).

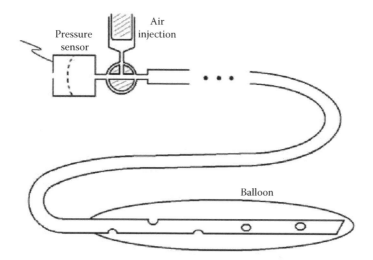

FIGURE 17.6   The esophageal balloon catheter. (Reprinted from Bates, J.H.T., *Lung Mechanics: An Inverse Modeling Approach*, Copyright 2009. With permission from Cambridge University Press.)

Gastric pressure relative to esophageal pressure provides the pressure drop across the diaphragm, and is useful for assessing diaphragm function. Gastric pressure can also be measured with a balloon catheter in the same way as esophageal pressure, except that here the occlusion test for balloon placement is no longer applicable. Instead, one ensures the balloon is in the stomach by observing that its pressure increases during inspiration, in contradistinction to esophageal pressure, which decreases during inspiration.

17.2.3 Flow and Volume Measurement

The earliest measurements of changes in pulmonary lung volume were accomplished by devices that measured volume directly. These devices began as water-filled *spirometers* in which air was introduced beneath inverted containers of known dimensions that were counterbalanced to account for their weights. The containers rose vertically as they filled with air, so the amount of vertical movement multiplied by the container cross-section gave the volume change. Hutchinson was perhaps the first to make significant use of such a device when, in the mid-1800s, he used it to measure the lung capacities of more than 4000 subjects (Spriggs 1977). The concept, however, dates back at least to Lavoisier (Spriggs 1978). Subsequently, other devices were constructed around the same basic principle, such as dry-bellows wedge spirometers and spirometers using dry rolling seals. Today, devices of this nature are often used as gold standard sensors for volume and flow, since the latter can be obtained by differentiating the former with respect to time either electronically or digitally (Schuessler and Bates 1995, Simon and Mitzner 1991).

Flow ($\dot{V}$) can also be measured directly, a simple approach being to measure the differential pressure drop (ΔP) across an element of known flow resistance. This principle forms the basis for one of the oldest and most widely used flow meters, the *pneumotachograph*. One particular design of pneumotachograph, the *Fleisch* type, incorporates a number of small capillary tubes of diameter d parallel to the direction of flow as the resistive element

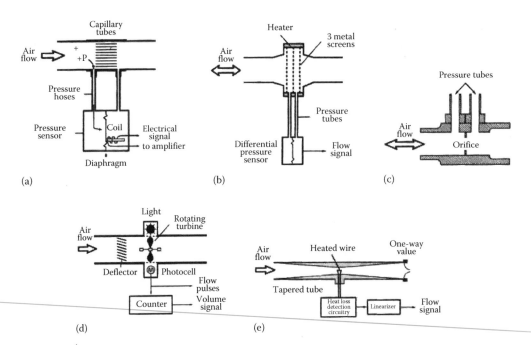

FIGURE 17.7 Various flow sensors typically used in anesthesia and respiratory applications. (a) Fleisch pneumotach, (b) screen pneumotach, (c) orifice flow meter, (d) turbine flow meter, and (e) hot wire anemometer. (Modified from Sullivan, W.J. et al., *Respir. Care*, 29, 736, 1984. With permission.)

(Figure 17.7a). For moderate flow rates, the flow through the element is *laminar,* causing the resulting pressure drop to be proportional to flow according to *Poiseuille's law*:

$$\Delta P = \frac{128\mu l}{\pi N d^4} \dot{V} \tag{17.1}$$

where

l is the distance between the two pressure sensing ports of the pneumotachograph
μ is the viscosity of the flowing gas
N is the number of parallel capillary tubes packed into the flow conduit

An alternative type of resistive element is comprised of one or more fine wire meshes (Figure 17.7b), which provides a linear relationship between ΔP and flow $\dot{V}$ over a limited range of flow, but with the advantage of having a smaller dead space (Sullivan *et al* 1984). To ensure laminar flow through a pneumotachograph, and hence linearity of its $\Delta P - \dot{V}$ relationship, the conduits connecting to both ends of the device should have smooth internal bores and be gradually tapered to match the diameter of the pneumotachograph itself. The lengths of connecting tubing should also, if possible, be at least five times the internal diameter of the pneumotachograph in order to allow for laminar flow to be fully developed before it enters the device (Sullivan *et al* 1984). The resistance of a pneumotachograph should be as small as possible in order to avoid having it affect the system being investigated. Consequently, the differential pressure sensor used to measure the pressure drop across a pneumotachograph needs to be very sensitive, typically having a full-scale range of only a few cmH$_2$O.

Both Fleisch and screen pneumotachs possess frequency responses that are approximately flat out to 20 Hz or so (Finucane *et al* 1972), depending on the geometry of the connecting tubing (Jackson and Vinegar 1979). However, they are very sensitive to changes in temperature, humidity, and gas composition (Sullivan *et al* 1984) and, thus, should be frequently calibrated using various electronic or software-based techniques to ensure accurate measurements (Renzi *et al* 1990, Yeh *et al* 1982). Several different calibrations of a pneumotachograph may be required if it is to be used with varying levels of inspired oxygen (Turner *et al* 1990) or volatile anesthetics (Habre *et al* 2001). Also, while an electrical heating element will limit condensation developing on the capillary tubes or wire screens, the prolonged use of pneumotachographs in ventilated patients frequently results in their becoming clogged with secretions (Grenvik *et al* 1966). This issue, combined with difficulties in cleaning and sterilization, limits their widespread use in clinical environments.

A key goal of the pneumotachograph is to avoid *turbulent* flow, as this will lead to a nonlinear relationship between ΔP and $\dot{V}$. The onset of turbulence depends on gas density (Sullivan *et al* 1984), however, and is exploited via the *Bernoulli* effect (Bates *et al* 1992) for the measurement of flow by *Venturi* and *orifice* flow meters. Orifice flow meters (Figure 17.7c) have the advantage of relatively large internal diameters compared to pneumotachographs, which reduces the problems caused by condensation and secretions. Orifice flow meters can be manufactured using inexpensive plastic injection molding techniques and thus can be disposable, which makes them advantageous for clinical environments. Flow can also be measured by *turbine* flow meters (Figure 17.7d) in which the rotational speed of a vane is measured (Primiano 2010). Turbines are less sensitive to changes in gas composition, humidity, and temperature than other types of flow meters (Ilsley *et al* 1993), and they are not affected by turbulence (Sullivan *et al* 1984), but the inertia of the turbine limits their frequency response (Yeh *et al* 1987). *Hot wire anemometers* (Figure 17.7e) are widely used currently for accurate clinical flow measurement (Plakk *et al* 1998) and exploit the fact that the electrical resistance of a current-carrying wire is modulated by its temperature, which is reflective of the flow of air past it. Hot wire anemometers typically have a better frequency response than pneumotachographs and orifice flow meters (Ligeza 2007), making them suited to measurement of flow in applications such as high-frequency ventilation (Hager *et al* 2006).

Finally, volume can be readily calculated from flow by numerical integration with respect to time, but this often leads to volume drift due to slight zero offsets and asymmetries in the flow calibration. Drift can be removed by high-pass filtering the volume signal in order to maintain it on a stable baseline (Frey *et al* 2000).

Figure 17.8 illustrates a typical arrangement of the instrumentation described earlier for a mechanically ventilated patient. Pressures are measured at the airway opening and in the esophagus using gage sensors, while flow entering the lungs is measured using a Fleisch-type pneumotachograph.

17.2.4 Concentrations and Partial Pressures of Gases

Any mixture of different molecular species in the gas phase will exert a pressure that can be measured using the methods described earlier. Each of the different gas species in the mixture contributes to this pressure, in proportion to its fractional contribution to the total number

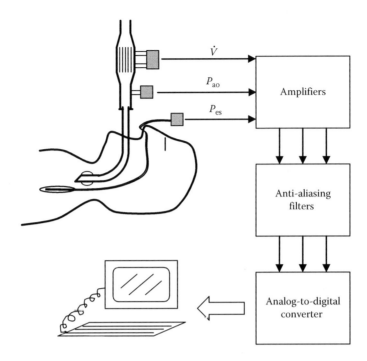

FIGURE 17.8 Measurement setup for assessment of pulmonary mechanics in a mechanically ventilated patient showing gas flow ($\dot{V}$) measured with a pneumotachograph, airway opening pressure (P_{ao}) measured through a lateral tap, and esophageal pressure (P_{es}) measured with a balloon catheter placed in the esophagus. (Reprinted from Bates, J.H.T., *Lung Mechanics: An Inverse Modeling Approach*, Copyright 2009. With permission from Cambridge University Press.)

of gas molecules present. For example, air consists of about 21% oxygen (O_2), 78% nitrogen (N_2), and 1% other gases. At sea level, the pressure of air varies slightly around a value of 1 atmosphere, which is equal to the pressure created by the gravitational weight of a vertical column of mercury (Hg) 760 mm in height. Thus, the *partial pressures* of O_2 and N_2 at sea level are $0.21 \times 760 \, \text{mmHg} = 160 \, \text{mmHg}$ and $0.78 \times 760 \, \text{mmHg} = 593 \, \text{mmHg}$, respectively.

The concept of partial pressure is also relevant when the gases are dissolved in tissues or bodily fluids, such as when O_2 and carbon dioxide (CO_2) are dissolved in blood. This concept can be understood by imagining a closed container partially filled with blood, above which there is a certain partial pressure of a specific gas. As these gas molecules are in continuous motion, some will move randomly from the upper gas phase to become dissolved in the liquid layer in the container. Similarly, some of the dissolved gas molecules will manage to escape the blood and move into the upper gas phase. When these two rates of movement are equal, the system is in dynamic equilibrium, and the partial pressures in the gas and liquid phases are equal by definition. In general, however, the *concentrations* of gas molecules in the two phases (expressed as number of molecules per unit volume of either gas of liquid) will not be equal. A highly soluble gas has a high blood content for a given partial pressure, while the reverse is true for a poorly soluble gas.

Gases generally dissolve in liquids according to *Henry's law*, which assumes a linear relationship between the partial pressure of a gas and its concentration in solution. Henry's law implies a limitless carrying capacity for the liquid, although this assumption breaks down when concentrations become too high. However, for relatively dilute solutions, it works well. For O_2 dissolved in blood, the situation is fundamentally different, since most O_2 is bound to hemoglobin. Each molecule of hemoglobin accommodates up to four O_2 molecules, so when all four of these O_2 receptors are occupied, the hemoglobin molecule is considered fully saturated. At the same time, there are some oxygen molecules that are dissolved in the blood plasma in simple solution according to Henry's law, but at the partial pressures normally encountered in the body, this represents an essentially negligible fraction of the total O_2 content of blood. For all intents and purposes, the O_2 content of blood is expressible in terms of the % *saturation* of the hemoglobin that it contains, denoted S_aO_2. CO_2 is more soluble than O_2, and so a greater percentage of it is carried in the blood in simple solution relative to the amount bound to proteins (principally hemoglobin). Most CO_2 in the blood, however, is carried in the form of *bicarbonate*.

For the reasons outlined earlier, quantification of respiratory gases requires the measurement of either partial pressures (e.g., in terms of mmHg), or content (e.g., in terms of mL of gas per unit volume). These quantities are usually referenced to *standard temperature and pressure* (STP), which is 0 °C and 760 mmHg.

17.2.5 Mass Spectrometry in the Gas Phase

Mass spectrometers analyze respiratory gases on the basis of *atomic mass*. For example, N_2 has 28 atomic mass units (AMU), while O_2 has a mass of 32 AMU. A small quantity of the gas to be analyzed is admitted to a vacuum chamber and immediately *ionized*. This is achieved using an electron source (typically about 70 eV), which is sufficient to strip one electron (but not two) off most molecules. The resulting input gas stream is now singly ionized, and subsequently separated on the basis of the *mass-to-charge ratio*. How this separation occurs varies across devices, but the incoming gas stream is accelerated and focused into an ion beam and then delivered to the mass separation stage (e.g., as illustrated in Figure 17.9). The most common form of respiratory mass spectrometry uses either: (1) a *transverse static magnetic field* to alter the trajectory of the ions based on mass/charge ratio or (2) a *varying quadrupole magnetic field* in which only a specific mass/charge ratio has a stable trajectory capable of reaching a distal amplifier. In the former case, the system can measure multiple gases simultaneously, but the choice of gases is fixed at the time the instrument is constructed. In the latter case, the choice of gases to be sampled is adjustable, but sampling multiple gases requires switching the quadrupole field between them, limiting temporal resolution.

In either case, there is no need for extremely high mass resolution (such as that seen in mass spectrometers used in organic chemistry), as the gases of interest (and thus their masses) are already well known. However in order to be useful, the dynamic response of a respiratory mass spectrometer is an important consideration. This requires careful consideration of the sampling strategy for the gas. Most mass spectrometers are rather bulky, so a catheter system must be used to collect gas samples. However, this introduces delays between sampling and analysis as the gas traverses the catheter, causing a degradation of dynamic response. Acceptable response times are possible with careful design, and additional improvements may be possible with *digital postprocessing* (Bates *et al* 1983).

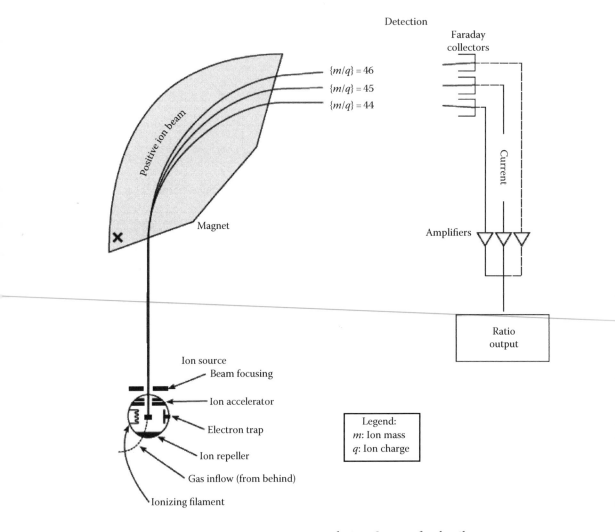

FIGURE 17.9   Magnetic sector mass spectrometer design. See text for details.

Another problem for most respiratory mass spectrometers is the presence of *water vapor* in expired gas, since this will reduce the apparent partial pressures of the other *dry* gases. To deal with this, the sample can be dried by heating, a nominal value for PH_2O can be included in the total gas pressure, or the sum of all the (nonwater) gases can be forced to add to a constant value, in effect ignoring any varying contribution from water vapor.

Specialized analyzers for many gases are now standard in most pulmonary function equipment. CO_2 is typically measured using infrared (IR) absorption (see the following text), while O_2 is often measured using ultraviolet (UV) emission devices. In both cases, it is usually necessary to remove water vapor from the sample as this tends to affect the analyzers. For sampling systems, this has the potential to greatly degrade dynamic response, but the recent advent of water-permeable sampling tubing (Naphion®) has greatly lessened this problem by allowing the water vapor to be extracted passively prior to gas analysis.

Several *foreign gases* are also important in pulmonology, including more soluble gases (e.g., volatile anesthetics) and other gases used for special applications (see the following text).

Mass spectrometry is the standard method for measuring these gases, but some of them, such as methane (CH_4), carbon dioxide (CO_2), and sulfur hexafluoride (SF_6), have an IR signature that can exploited for their quantification.

17.2.6 Pulse Oximetry

As noted earlier, arterial oxygen is carried to the tissues mostly bound to hemoglobin, with a very small additional amount dissolved in the blood plasma. These two contributions sum to give the total content (C_aO_2, expressed in ml of oxygen at STP per 100 mL of blood) approximated as

$$C_aO_2 = \left(1.34 \times S_aO_2 [Hb]\right) + 0.0031 \times P_aO_2, \tag{17.2}$$

where

1.34 mL/g is the nominal oxygen-binding capacity of hemoglobin, although it actually ranges between 1.34 (Boron and Boulpaep 2011) and 1.39 mL/g (West 2008)

[Hb] is the concentration of hemoglobin in the arterial blood (g/100 mL)

0.0031 is the solubility of oxygen in blood (mL/100 mL/mmHg)

P_aO_2 is the arterial partial pressure of oxygen (mmHg)

The relationship between S_aO_2 and P_aO_2 is described by the oxyhemoglobin dissociation curve, which has a markedly sigmoidal shape with very important consequences (Figure 17.10). Under most normal arterial levels of P_aO_2, the relationship approaches 100% saturation and is quite flat. This means that variations in P_AO_2, such as might occur in daily life, have minimal effect on the near-full saturation of blood as it leaves the lungs. Conversely at normal venous levels of P_vO_2, the relationship is steep. This gives the arterial blood the capacity to adjust the amount of O_2 it offloads to the working tissues in response to regional variations in tissue PO_2 levels. For example, local tissue PO_2 decreases when the muscles are exercising, leading to increased O_2 delivery from arterial blood. Second-order effects on the shape of the O_2 dissociation curve for hemoglobin caused by changes in temperature, pH, P_aCO_2, and erythrocyte 2,3-diphosphoglycerate concentration allow the curve to adapt favorably to conditions of exercise (Figure 17.10). It is thus a matter of substantial clinical and physiologic importance to be able to measure both PO_2 and S_aO_2.

S_aO_2 is measured by *oximetry*, a technique based on the absorption of light by hemoglobin at a specified wavelength. The four species of hemoglobin, namely, deoxy-Hb, oxy-Hb, CO-Hb, and met-Hb, absorb most efficiently at slightly different wavelengths, so least four different wavelengths of light must be used to determine their respective concentrations uniquely.

The standard method of measuring S_aO_2 in vivo is by *pulse oximetry*, which produces a continuous assessment (denoted S_pO_2) by measuring the absorption of light through tissue. The tissue concerned is usually a finger or toe, so this technique may not work well in the face of poor peripheral perfusion, nail polish, or in the presence of intravascular dyes. Pulse oximetry exploits the pulsatility of arterial blood flow to provide an estimate of S_aO_2. Because of arterial pulsatility, light absorbed by perfused tissue consists of both pulsatile and non-pulsatile components. By measuring the absorption ratio between these two components at

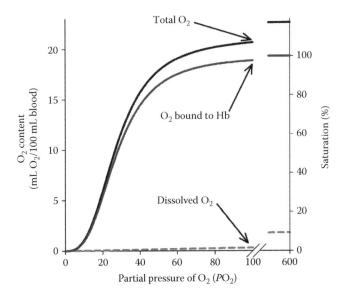

FIGURE 17.10   Oxygen content of the blood as a function of partial pressure, showing the relative amounts bound to hemoglobin and dissolved in the blood plasma.

different wavelengths, the relative amounts of *oxyhemoglobin* and *deoxyhemoglobin* (i.e., the fractional saturation of the blood) are determined. Most manufacturers of pulse oxymeters report accuracies of ±2% to 3% S_pO_2 over the 70%–100% range. However, the accuracy of S_pO_2 is greatly diminished for reported values less than 60%.

17.2.7 Capnography

Capnography refers to the process of measuring CO_2 concentrations in respiratory gases. The actual measurements can be made using mass spectrometry, Raman spectrometry, or gas chromatography (Gravenstein *et al* 1995). Clinically, the most common method uses nondispersive IR absorption (Jaffe 2008) to detect the intensity of a beam of infrared light after it has passed through a gas sample, although filters must be used to avoid spectral overlap with other gases commonly encountered in anesthesia.

Sidestream capnometers use a pump or compressor to extract for analysis a small sample of gas from the main gas stream, and are common in clinical situations. The lateral tubing connecting the gas stream to the site of gas analysis may be 1 m or more in length, which may introduce a significant delay in the resulting CO_2 signal. Characterizing this delay, along with any distortion in the gas concentration profile that results from dispersion during transit along the sampling tube, can in principle be mitigated by numerical correction techniques (Bates *et al* 1983). Gas sampling rates may vary from 30 to 500 mL/min, which are not significant for adult subjects. This sampling volume may be significant for infants, however, in which case returning the sampled gas to the breathing circuit after suitable filtering or scavenging may be necessary. In contrast with mainstream capnographs, the gas analysis hardware is placed directly in the path of the flowing gas, and thus has the advantage of eliminating sampling delay time. However, this has the potential to increase equipment dead space substantially.

17.2.8 Blood-Gas Analysis

It is common clinical practice to draw a sample of arterial blood from a patient in order to measure P_aO_2, P_aCO_2, and pH. The P_aO_2 is usually measured using a *Clark electrode* consisting of a platinum or gold cathode and an anode bathed in electrolytic fluid, all surrounded by a thin membrane that is permeable to O_2. When O_2 diffuses through the membrane, it is reduced by the cathode to produce a current in proportion to its partial pressure (Severinghaus and Astrup 1986). Optical approaches to measure P_aO_2 have also been developed, but these remain experimental (Ganter and Zollinger 2003). P_aCO_2 is quantified from changes in pH that occur when the sample is equilibrated with a standardized bicarbonate solution (Severinghaus 2002). Blood-gas analysis should be performed with the sample at 37 °C. This is important, since the solubilities of gases such as CO_2 and O_2 in blood are inversely related to temperature, so corrections made need to be made for a patient's actual body temperature in a case of, for example, hypothermia (Andritsch *et al* 1981).

17.3 ASSESSING LUNG MECHANICAL FUNCTION

17.3.1 Spirometry

The mainstay of diagnosis in pulmonary disease is termed *spirometry*, which involves measuring volumes of gas inhaled and exhaled from the lungs using either a spirometer to measure volume directly or a flow meter, which provides flow, which is then integrated to volume. The most common clinical use of spirometry involves making a voluntary forced expiration from *total lung capacity* to *residual volume*, a procedure first developed by Robert Tiffeneau in the 1940s (Yernault 1997). Forced expiration can be easily performed in an outpatient setting with simple equipment by any competent subject (Spriggs 1977). The two clinically important physiological parameters derived from spirometry are: (1) *forced vital capacity* (FVC) and (2) the *volume of gas exhaled in the first second of the forced expiration* (FEV_1). FEV_1 is determined not by effort, but by the resistive and elastic properties of the lungs at a specified volume. Our understanding of the mechanistic underpinnings of maximal expiratory flow began with the choke point theory (Fry *et al* 1954, Hyatt 1983) and progressed to include consideration of the Bernoulli effect and *wave speed theory* (Dawson and Elliott 1977).

Because *dynamic airway compression*, the central phenomenon of expiratory flow limitation, is a function of airway stiffness, maximal flow is influenced strongly by the degree to which the airways are tethered open by the parenchyma in which they are embedded, which in turn is strongly affected by lung volume. Accordingly, maximal expiratory flow decreases markedly with decreasing lung volume, giving rise to the approximately linear region of the expiratory flow–volume curve in normal individuals that becomes characteristically distorted in *obstructive diseases* such as asthma and chronic obstructive pulmonary disease (COPD). This gives FEV_1, FVC, and the general shape of the forced expiratory flow–volume curve their diagnostic usefulness. In particular, FVC, FEV_1, and the ratio FEV_1/FVC have well-established normal ranges that are functions of height and gender (Stanojevic *et al* 2008), and so, measured values are typically reported in terms of their % predicted, which

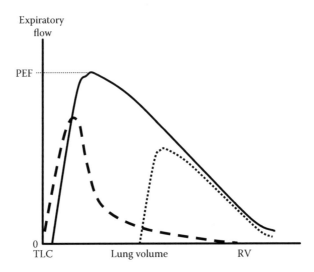

FIGURE 17.11 Representation of forced expiratory flow–volume curves for a normal lung (solid line), an obstructed lung (dashed line), and a restricted lung (dotted line).

gives immediate diagnostic feedback. Figure 17.11 illustrates the usual shape of the forced expiratory flow–volume curve along with idealized pathologic examples.

The measurement of flow, and hence the results of spirometry, is affected by the temperature and humidity of the flowing gas. Gases within the lungs are generally at normal human body temperature of 37 °C, and are also saturated with water vapor at a partial pressure of 6.3 kPa or 47 mmHg. These conditions are referred to as BTPS (body temperature, atmospheric pressure, fully saturated). By contrast, ambient air is typically around 20 °C and is much drier, although the precise temperature and water vapor pressure are obviously quite variable, depending on the location, season, time of day, etc. Thus, the average volume of gas expired during each breath will be larger than the average volume of the preceding inspiration, due to warming and humidification of air in the lungs. With appropriate monitoring of temperature and humidity levels, this can be largely corrected using the *ideal gas law*

$$PV = n\bar{R}T, \qquad (17.3)$$

where
   n is the number of *moles* of gas (which increases in the expired gas due to the addition of water vapor)
   $\bar{R}$ is the *universal gas constant*
   T is temperature in Kelvin

The composition of the flowing gas can also influence flow measurement. For example as noted earlier, the pressure drop across the resistive element of a pneumotachograph for laminar flow is proportional to the apparent gas viscosity, and thus may vary, depending on the relative fractions of oxygen, nitrogen, carbon dioxide, or any other gases that might be present.

17.3.2 Body Plethysmography

Another clinical tool used in assessing lung mechanical function is *whole-body plethysmography* (Gold 2000), which involves having a subject to be enclosed in a rigid-walled air-tight container. For adult human subjects, the volume of the container is about 500 L. *Thoracic gas volume* (V_{TG}) is determined by having the subject attempt to inspire through a completely occluded mouthpiece, while the change in mouth pressure (ΔP_{ao}) behind the occlusion is measured (Figure 17.12a). The inspiratory effort generates a subatmospheric pressure within the lungs that expands the thoracic gas, while generating equal and opposite compression of the plethysmographic gas surrounding the subject. The volume of gas in the plethysmograph, V_{pleth}, is readily determined to be the container volume minus the volume of the subject, and its pressure, P_{pleth}, is easily measured. The amount by which V_{pleth} is compressed, and thus by which V_{TG} is expanded (ΔV_{TG}) during the inspiratory effort is obtained from the change in plethysmographic pressure (ΔP_{pleth}) and the compressibility of the plethysmographic gas. *Boyle's law* is then used to determine the changes in ΔP_{ao}, ΔV_{TG}, and ΔV_{TG}; thus,

$$V_{TG} = \frac{-1000 \Delta V_{TG}}{\Delta P_{ao}}, \tag{17.4}$$

where the factor 1000 corresponds to a pressure of approximately one atmosphere (in units of cmH_2O), and pressures are measured relative to atmospheric in the same units.

Once V_{TG} has been determined as described earlier, body plethysmography can be used to determine *airway resistance* (R_{aw}) by having the subject breathe air from the plethysmograph in a rapid shallow panting manner, while respiratory flow ($\dot{V}$) is measured at the

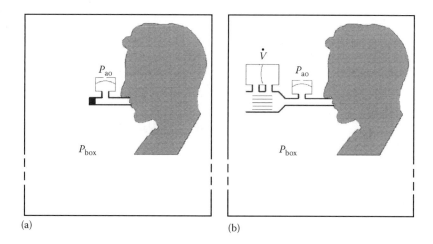

(a) (b)

FIGURE 17.12 Body plethysmograph used to measure (a) thoracic gas volume (V_{TG}) and (b) airway resistance (R_{aw}). P_{ao} is airway opening pressure, P_{box} is pressure inside the plethysmograph, and $\dot{V}$ is mouth flow. (Reprinted from Bates, J.H.T., *Lung Mechanics: An Inverse Modeling Approach*, Copyright 2009. With permission from Cambridge University Press.)

mouth with a pneumotachograph along with P_{box} and P_{ao} (Figure 17.12b). During panting, the alternating negative and positive *alveolar pressures* (P_A) give rise to corresponding changes in V_{TG} according to Boyle's law. As with measurement of V_{TG} described earlier, changes in V_{TG} result in equal and opposite changes in V_{pleth}, with Boyle's law now yielding

$$P_A = \frac{-1000\Delta V_{TG}}{V_{TG}}, \quad (17.5)$$

and thus, airway resistance may be determined as

$$R_{aw} = \frac{P_{ao} - P_A}{\dot{V}}. \quad (17.6)$$

The earlier analysis neglects some complicating factors. For example, inspired gas becomes humidified and heated when entering the lungs, causing V_{TG} to increase more than the volume of the gas passing through the lips. The measurements of V_{TG} and R_{aw} described earlier also assume that the lungs are ventilated *homogeneously*, which is reasonable for normal subjects but may not be for those with severely diseased lungs (Shore *et al* 1982).

17.3.3 Resistance and Elastance

Throughout the majority of the *conducting airways*, gas transport occurs largely via *convective flow*. Since the airway tree comprises a substantial number of segments with widely differing lengths and diameters, the flow rates through its various branches are also highly variable. Nevertheless, the bulk transport of gas through the airway tree may be idealized as flow through an equivalent single cylindrical conduit. This resistance of the conduit is conceptualized as airway resistance (R_{aw}) defined as the ratio of the pressure difference between the two ends of the conduit (i.e., airway vs. alveolar) to the magnitude of flow. Changes in R_{aw} thus become a useful reflection of overall changes in airway caliber (Pedley *et al* 1970). In the case of oscillatory flow, as frequency increases, the pressure drop across the conduit will accrue a significant *inertial* component due to acceleration of the gas column. In normal adults, this gives rise to a *resonant frequency* at about 8 Hz, which tends to increase with lung disease (Verbeken *et al* 1992). As frequencies approach 100 Hz, there can be a significant inertial pressure drop due to acceleration of the respiratory tissues (Mead 1956).

The conducting airway tree ends in tens of thousands of *terminal bronchioles*, each leading into an *acinus* containing the alveoli where gas exchange takes place. Nevertheless, it is still often useful to idealize all these myriad terminal units as if they behave like a single large alveolar compartment. As this compartment increases in size during inspiration, it also stores elastic energy. In its simplest manifestation, the single-compartment model of the lung considers elastic energy to be linearly related to inflation volume, with the constant of proportionality defined as lung *elastance* (E_L). However, because lung tissue is *viscoelastic*, energy is also dissipated as the compartment dynamically changes volume due to tension within the alveolar surface film (Mora *et al* 2000), friction within the pleural space and the various tissue components of the parenchyma and chest wall (Kaczka and Smallwood 2012), and cross-bridge cycling of contractile elements within the airway walls and lung tissues (Fredberg *et al* 1997, Fredberg and Stamenovic 1989). This results in the resistive properties of the tissues being inversely dependent on the frequency at which its volume is cycled (Fredberg and Stamenovic 1989, Hantos *et al* 1992b). In its simplest manifestation, these dissipative properties in the lung parenchyma can be represented as a single constant known as *tissue resistance* (R_{ti}) (Bachofen 1968, McIlroy *et al* 1955). The conventional description of the average distending pressure across the entire lung, known as transpulmonary pressure (P_{tp}), relative to the resistive, inertial, and elastic pressures, is

$$\begin{aligned} P_{tp} &= R_L \dot{V} + E_L V + I_{aw} \ddot{V} + P_0 \\ &= \left(R_{aw} + R_{ti} \right) \dot{V} + E_L V + I_{aw} \ddot{V} + P_0, \end{aligned} \tag{17.7}$$

where P_0 is the value of P_{tp} that maintains the relaxed end-expiratory volume of the lung. Equation 17.7 can also be applied to the mechanical behavior of the total respiratory system, or the chest wall, by replacing P_{tp} by either airway pressure (P_{ao}) or pleural pressure (P_{pl}), both relative to atmospheric pressure (Bates 2009, Kaczka and Smallwood 2012). In the case of P_{ao}, for example, the elastic pressure term across the lung in Equation 17.7 becomes replaced by a term that encompasses the elastic properties of both the lung and the chest wall, which together comprise the elastance of the *respiratory system* (E_{rs}). These structures contribute to the elastic pressure in parallel, and are related as

$$E_{rs} = E_L + E_{cw}. \tag{17.8}$$

The simple linear single-compartment model we have just considered, of course, neglects numerous important details that affect the relationships between pressure and flow in a real lung. For example, gas flow in the conducting airways is largely laminar when flow magnitudes are low, but during exercise or FVC maneuvers, the flow may become turbulent, particularly in the larger airways where *Reynolds numbers* are highest (Pedley *et al* 1970). In this case, the flow resistance of the airways becomes dependent on the magnitude of the flow itself, often being accurately described as the equation of Rohrer (Bates 2009):

$$\Delta P = K_1 \dot{V} + K_2 |\dot{V}| \dot{V}, \tag{17.9}$$

where K_1 and K_2 are empirically determined constants.

The $|\dot{V}|$ denotes the absolute value of $\dot{V}$, which is required, since K_2 is assumed to have an equal contribution to the total resistive pressure drop across the airways for moderate flow rates, regardless of flow direction.

17.3.4 Lung Impedance and the Forced Oscillation Technique

As noted earlier with regard to the dynamic response of a linear sensor (Figure 17.4), if the input to a linear dynamic system is a pure sine wave oscillating at a particular frequency f, then the output will also be a sine wave oscillating at f, although in general, the output sine wave will be scaled in amplitude by factor A and shifted in phase by an amount ϕ. When the input to a system is a variable that expresses rate of flow of material through the system (i.e., velocity, gas flow, electric current, etc.) and the output is a variable that expresses the opposition to this flow (force, pressure, voltage, etc.), then the frequency response relating these two variables is given the special name of *impedance*, usually referred to as Z.

The lungs or total respiratory system can be considered to constitute a linear dynamic system, at least to a useful degree of approximation, if the excursions in P and $\dot{V}$ to which it is subjected are modest. Consequently, a sine wave input in $\dot{V}$ at the tracheal opening (conventionally termed the input) results in a sine wave in P being generated at the same location. The functions $A(f)$ and $\phi(f)$ that relate P to $\dot{V}$ are represented in a single complex function of f known as the *input impedance* of the lung, $Z(f)$; thus,

$$P(f) = Z(f)\dot{V}(f) = \left[R(f) + iX(f)\right]\dot{V}(f), \tag{17.10}$$

where $R(f)$ and $X(f)$ are the *real part* and the *imaginary part* of impedance, respectively. The real part is commonly known as *resistance*, while the imaginary part is commonly known as *reactance*, and i is the positive square root of -1. The impedance of the single-compartment model of the lung given by Equation 17.7 earlier, for example, is (Bates 2009)

$$Z(f) = R + i\left[2\pi fI - \frac{E}{2\pi f}\right]. \tag{17.11}$$

Determining the input impedance of the lung involves applying controlled oscillations in $\dot{V}$ to the mouth or tracheal opening and measuring the variations in P that result. Numerical techniques based around the *fast Fourier transform* are then used to invert Equation 17.10 to yield $Z(f)$ over the range of oscillation frequencies in the $\dot{V}$ signal. The implementation of these various steps constitutes what is known as the *forced oscillation technique* (FOT). A key part of the FOT thus involves measuring pressure and flow with appropriate sensors, as discussed earlier. Also key to the FOT is some means of generating the oscillations in $\dot{V}$. These oscillations must be large enough to create measureable changes in $\dot{V}$ and P at the measurement site, but not so large as to create discomfort for the patient or alter their normal pattern of breathing. In practice, the FOT can generally be implemented with oscillatory changes in V of a few mL, or changes in $\dot{V}$ of a few tens or hundreds of mL/s, depending on the frequency. The devices that have been used to generate these oscillations include loudspeakers (Landser *et al* 1976), piston-cylinder assemblies (Kaczka *et al* 1999, Lutchen *et al* 1993, Schuessler and Bates 1995), and a rapidly varying resistance that chops the mouth flow generated by the subject during normal breathing (Lausted and Johnson 1999).

How $Z(f)$ is interpreted physiologically depends greatly on the range of f over which it is measured. Early investigations with the forced oscillation technique focused on what might be called an intermediate range that typically spanned about 4–30 Hz. In a subject with normal lung function, $R(f)$ is relatively constant over this range, while $X(f)$ is negative for low f and crosses the frequency axis to become positive at the resonant frequency of about 8 Hz (Landser *et al* 1976, Verbeken *et al* 1992). With the development of lung disease, $R(f)$ tends to increase in its mean value and take on a negative dependence on f (Verbeken *et al* 1992), reflecting the development of *regional heterogeneities* of lung mechanical function throughout the lung (Bates 2009). The resonant frequency also tends to increase (Verbeken *et al* 1992) due to stiffening of the lung parenchyma (Bates 2009). More recently, interest has focused on the lower end of the f range that encompasses the frequencies of spontaneous breathing. Here, the viscoelastic properties of the tissues play a major role in determining $Z(f)$, and in particular cause $R(f)$ to decrease markedly as f increases (Bates 2009, Hantos *et al* 1992b, Kaczka *et al* 1999, Lutchen *et al* 1993). At higher frequencies extending into the acoustic range, one encounters *quarter-wave resonances* related to the lengths of the conducting airways (Jackson *et al* 1989).

In animal models, the so-called *constant-phase model* (Hantos *et al* 1992b) is most often used to describe $Z(f)$ over the low f range. This model has an impedance given by

$$Z(f) = R_{\mathrm{N}} + i2\pi f I_{\mathrm{aw}} + \left[\frac{G - iH}{(2\pi f)^{\alpha}} \right], \tag{17.12}$$

where

R_{N} is a *Newtonian resistance* that has been shown to usefully approximate the overall flow resistance of the conducting airway tree (Tomioka *et al* 2002)

I_{aw} is the *airway gas inertance*

G is a parameter characterizing *viscous dissipation of energy* in the tissue (related to tissue resistance)

H reflects *elastic energy storage* in the tissues (related to elastance)

The parameter α is related to G and H via the expression

$$\alpha = \frac{2}{\pi} \tan^{-1} \frac{H}{G}. \tag{17.13}$$

The constant-phase model describes low-frequency $Z(f)$ in animal models extremely accurately despite having only four free parameters (Hantos *et al* 1992a,b, Petak *et al* 1997, Wagers *et al* 2004, 2007).

The FOT has yet to supplant conventional spirometry for the clinical assessment of lung function in human patients. However, it is gaining acceptance in clinical settings (Oostveen *et al* 2003) and has some significant practical advantages over spirometry in that it can be applied in subjects unable to perform forced expiratory maneuvers, such as infants (Frey 2005).

17.4 ASSESSING PULMONARY GAS EXCHANGE

17.4.1 Oxygenation

The overall exchange of gases between the alveolar regions of the lung and the systemic circulation is described by the *Fick principle*, which accounts for the rate at which O_2 is removed from the lungs by the pulmonary blood (the *oxygen consumption* of the body $\dot{V}_{O_2}$) as

$$\dot{V}_{O_2} = \dot{Q}\left(C_aO_2 - C_{\bar{v}}O_2\right), \tag{17.14}$$

where

$\dot{Q}$ is *cardiac output* in mL/min
C_aO_2 is the *arterial* O_2 *content* in mL/mL
$C_{\bar{v}}O_2$ is *mixed venous content* in mL/mL

$\dot{V}_{O_2}$ can be measured as the difference between the volume of O_2 taken up by the lungs and the volume secreted (both measured, e.g., by integrating PO_2 at the mouth during inspiration and expiration, respectively). C_aO_2 can be determined by performing blood-gas analysis on a blood sample drawn from a peripheral artery by relating P_aO_2–C_aO_2 via the oxygen dissociation curve for hemoglobin. $C_{\bar{v}}O_2$ likewise can be determined from a sample of mixed venous blood drawn from the pulmonary artery, but this requires invasive pulmonary artery catheterization. $C_{\bar{v}}O_2$ is the *mean oxygen saturation* of the blood returning to the right heart from various regions of the body, weighted by their respective regional blood flows.

Normal P_aO_2 ranges between 80 and 100 mmHg for a healthy subject, although it decreases progressively with age. *Hypoxemia* is defined as a P_aO_2 less than 80 mmHg, and may result from any of five distinct physiological mechanisms:

1. Low inspired O_2 fraction (F_IO_2)

2. *Hypoventilation* during which increases in the amount of CO_2 in the alveoli displace O_2, leading to an increase in P_aCO_2 and a decrease in P_aO_2

3. *Ventilation-to-perfusion* ($\dot{V}/\dot{Q}$) *mismatching* where areas of lower than optimal ventilation (regions of low $\dot{V}/\dot{Q}$) add poorly oxygenated blood to the arterial circulation

4. *Right-to-left shunting* in which systemic venous blood bypasses the gas-exchanging regions of the lungs

5. *Diffusion limitation* whereby sufficient quantities of O_2 are prevented from being able to move across the blood–gas barrier

These mechanisms of hypoxia can arise for a variety of reasons. For example, reduced F_IO_2 occurs at altitude, while hypoventilation may occur when *respiratory drive* is abnormally low or the *work of breathing* is substantially increased (e.g., asthma exacerbation). Mismatching of $\dot{V}/\dot{Q}$ always occurs as a result of pulmonary embolism or in diseases that affect the architecture of the lung parenchyma such as COPD.

17.4.2 Shunt and Dead Space

The flow of blood through shunting regions of the lung can be quantified by considering the lungs to consist of three separate compartments, as illustrated in Figure 17.13 (Riley and Cournand 1949). One of these compartment represents the fraction of the lung that receives both effective *ventilation* and effective *perfusion*, while the other two compartments represent, respectively, a *shunt compartment* (receiving perfusion but no ventilation) and a *dead space compartment* (receiving ventilation but no perfusion).

Invoking this simple model of the lung with the aid of blood-gas analysis allows the estimation of the fraction of blood that is shunted through the lung as

$$\frac{\dot{Q}_s}{\dot{Q}_t} = \frac{\left(C_cO_2 - C_{\bar{v}}O_2\right)}{\left(C_cO_2 - C_aO_2\right)}, \tag{17.15}$$

where

$\dot{Q}_s$ represents the blood flow shunted past gas-exchanging regions of the lung
$\dot{Q}_t$ is the total *capillary blood flow* (i.e., cardiac output from the right ventricle)
C_cO_2 is the *end-capillary oxygen content*

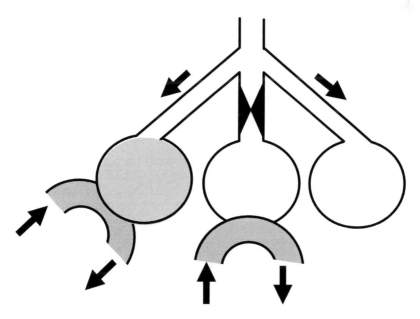

FIGURE 17.13 Three-compartment model of the lung: *Left*, a well-ventilated and perfused compartment; *Middle*, a nonventilated but well-perfused compartment (shunt); *Right*, a well-ventilated but nonperfused compartment (dead space).

The C_cO_2 is usually estimated from the alveolar partial pressure of O_2 (P_AO_2) under the assumption that alveolar gas and end-capillary blood are at equilibrium. P_AO_2 can be determined either from the PO_2 of the gas that is expired toward the end of a breath, or from the alveolar gas equation

$$P_AO_2 = F_IO_2 \left(P_{atm} - PH_2O \right) - \frac{P_aCO_2}{RER}, \qquad (17.16)$$

where

F_IO_2 is the inspired O_2 fraction

P_{atm} is the atmospheric pressure

PH_2O is the partial pressure of water vapor in the lungs (47 mmHg at 37 °C)

RER is the respiratory exchange ratio, which is equal to the ratio of pulmonary CO_2 elimination by the lungs to O_2 uptake by the lungs (at steady state, this is roughly 0.8 under normal diet and rest conditions). In practice, shunt is best estimated when the patient breathes 100% O_2, since in this circumstance, the difference between the end-capillary and arterial PO_2 is maximized due to the low slope of the O_2-Hb dissociation curve at high values of PO_2 (Figure 17.10).

Another important measure of global *gas exchange efficiency* is the difference between alveolar and arterial PO_2. For any given gas exchange unit in a lung, one assumes these two partial pressures to be equal so that the extent to which they differ in the whole lung (the so-called *A-a gradient* equal to $P_AO_2 - P_aO_2$) is a useful indicator of abnormal function (Lilienthal *et al* 1946). The A-a gradient is normally less than 10 mmHg in young adults breathing room air, but it increases with age (Mellemgaard 1966). The A-a gradient increases significantly with F_IO_2 (McCahon *et al* 2008) due to the flatness of the O_2 dissociation curve at high P_aO_2 (Figure 17.10), and also with the development of pathologic abnormalities such as shunt and ventilation-to-perfusion mismatch, which serve to directly lower P_aO_2 (Gowda and Klocke 1997).

Another clinically useful index of gas exchange efficiency is the P_aO_2/F_IO_2 ratio, which becomes decreased as the gas-exchanging ability of the lungs deteriorates. The P_aO_2/F_IO_2 ratio is particularly important in defining the severity of the condition known as *acute respiratory distress syndrome* (ARDS), the most severe cases corresponding to a P_aO_2/F_IO_2 ratio less than 100 (Ranieri *et al* 2012).

The determination of *physiologic dead space*, the effective volume of the lungs that receives no perfusion, is also very important clinically. Some amount of dead space is unavoidable, because the anatomy of the conducting airway tree necessarily means that for an adult human, roughly the last 150 mL of the inspired volume in each breath does not make its way to the gas-exchanging regions of the lung. When significant shunt or low-perfusion regions of the lung exist, however, the effective volume of the dead space increases. The traditional way of measuring the ratio of dead space to *tidal volume* (V_D/V_T) is based on the *Bohr Equation*:

$$\frac{V_D}{V_T} = \frac{P_aCO_2 - P_{\bar{e}}CO_2}{P_aCO_2}, \tag{17.17}$$

where

P_aCO_2 is estimated from the gas leaving the mouth at the end of an expiration (assuming the arterial and alveolar partial pressures of CO_2 are equal)

$P_{\bar{e}}CO_2$ is the *mixed expired* partial pressure of CO_2 measured in the gas expirate averaged over a specified period of time (e.g., the partial pressure of CO_2 in a bag containing all gas expired over, e.g., 1 min)

The three-compartment model of Figure 17.13 is obviously a gross simplification of an actual lung, so more complicated models can be employed for a more precise characterization of gas exchange. This, in turn, requires the collection of additional data. An example is the *multiple inert gas elimination technique* (MIGET) in which an infusion of six gases of differing solubilities is given to a subject, while the steady-state concentrations of these gases in expired air and arterial blood are measured (Wagner *et al* 1974a). A multicompartment model of the lungs then permits distributions of ventilation and perfusion to be identified (Wagner *et al* 1974b). Numerous studies have used MIGET to quantify the manner in which ventilation and perfusion are altered by various interventions and by disease (Gale *et al* 1985, Wagner *et al* 1977, 1978).

17.4.3 Foreign Gases

Measuring *lung volume* requires an assessment of the amount of gas left in the lungs at the end of expiration, since the lungs are never completely emptied of gas. This is typically performed using either body plethysmography (discussed earlier) or by insoluble *gas dilution* using helium. The experiment is simple in concept; the subject re-breathes from a bag of known volume containing a known concentration of helium, until the helium (which is not normally present in the lungs) is equilibrated between the bag and the lungs. Knowing the initial bag volume (V_{bag}) and the initial and final helium concentrations ([He]$_i$ and [He]$_f$, respectively) allows for the calculation of V_{TG} at the time when the test began. This is based on the principle that because helium is completely insoluble, the total amount of helium in the closed bag-lung system remains fixed, leading to the equation:

$$V_{bag}[He]_i = \left(V_{bag} + V_{TG}\right)[He]_f. \tag{17.18}$$

From this, V_{TG} can be estimated as

$$V_{TG} = \frac{[He]_i - [He]_f}{[He]_i} V_{bag}. \tag{17.19}$$

Measuring cardiac output is an extension of the lung volume measurement by helium dilution, and relies on the fact that the lung receives the entire cardiac output from the right heart. In contrast to the measurement of lung volume, which depends on the tracer gas remaining in the lung, measuring cardiac output requires that the tracer gas be soluble in blood. Just like the insoluble tracer gas used for measurement V_{TG}, the soluble gas mixes into the air spaces of the lung, but a small amount is continuously removed as it dissolves in the blood that is pumped through the lungs. If the solubility of the gas in blood (λ) is known, then the rate of gas uptake can be converted to the rate of pulmonary blood flow via the Fick equation (Equation 17.14). When the foreign gas is first inspired, it is assumed to have zero concentration in the mixed venous blood returning to the lungs. This is true for about the first 15 s, until the arterial blood that first acquires some of the foreign gas has had a chance to makes its way back to the lungs via the peripheral circulation. Prior to this recirculation, the Fick equation then simplifies to

$$\dot{V}_{gas} = \dot{Q} \lambda P_{A,gas}, \tag{17.20}$$

where $P_{A,gas}$ is the partial pressure of the foreign gas in alveolar spaces, which can be assessed by analyzing the gas appearing at the mouth toward the end of an expiration. Gases that have been used for this measurement include nitrous oxide (N_2O) (Ayotte *et al* 1970) and acetylene (C_2H_2) (Trebweiser *et al* 1977), the latter being preferable as it has a blood solubility that is close to optimal in terms of the sensitivity of the measurement. Some of the soluble gas also dissolves in the tissues of the lung (comprising mostly water), an effect that can be exploited, in principle, to determine *lung water* (Sackner *et al* 1980, Verbanck and Paiva 1994).

Just as the addition of a soluble gas to the measurement of lung volume allows the measurement of cardiac output, if the physical properties of the gas are different, other inferences about cardiopulmonary function can be made. Carbon monoxide (CO) is poorly soluble in plasma, but once it diffuses into the red cell, it is avidly taken up by hemoglobin. The effect is to remove it rapidly from the plasma, thus maintaining a very low overall partial pressure in the blood. As a direct consequence of this, the uptake of CO in the lung is limited not by the rate of blood flow (as was the case for acetylene), but rather by the ability of CO to diffuse across the blood–gas barrier. Thus, the measurement of CO uptake provides an estimate of the *diffusing capacity* of the lung, which may be clinically useful as it reflects the same physical processes of O_2 diffusion.

Because the uptake of CO is substantially greater compared to gases such as acetylene, it can usually be measured in a single breath-hold (Piiper and Sikand 1966). A single vital capacity inspiration of an insoluble gas (to measure lung volume) with a small concentration of CO (typically 0.3%) is made and the inspiration held at TLC for approximately 10 s. Concentrations of both the insoluble gas and CO are then measured in the subsequent expiration. This test is well standardized (Abboud and Sansores 1996) and is an important component in all pulmonary function laboratories.

17.5 CONCLUSIONS

Modern measurements in pulmonology provide an accurate assessments of lung function. Some of these measurements assess the lung as a mechanical system of air-filled conduits and viscoelastic tissues. Sensors monitor gas pressures and flows during specialized maneuvers, providing the data with which physicians are able to determine if this mechanical system is able to inflate and deflate normally. Abnormalities in mechanical behavior can signal the presence of pathologies such as obstructive and/or restrictive pulmonary disease. Similarly, measurements of the partial pressures and blood contents of various gases, including certain foreign gases, provide clinically important information about the way that these gases distribute throughout the lungs, and also how they pass into or out of the blood that flows through the pulmonary capillaries. Measurements in pulmonology are, for the most part, based on mathematical models of the lung and are thus highly quantitative, making them firmly grounded in the principles of physics and engineering.

GLOSSARY

A	Amplitude
A_{in}, A_{out}	Amplitudes of input and output sine waves, respectively
ADC	Analog-to-digital converter
BTPS	Body temperature and pressure and saturated conditions
C_2H_2	Acetylene
C_aO_2	Arterial oxygen content
C_cO_2	End-capillary oxygen content
$C_{\bar{v}}O_2$	Mixed venous oxygen content
CH_4	Methane
cmH_2O	Centimeters of water pressure
CMRR	Common-mode rejection ratio
CO	Carbon monoxide
CO_2	Carbon dioxide
COPD	Chronic obstructive pulmonary disease
CPAP	Continuous positive airway pressure
d	Diameter capillary tubes comprising a Fleisch pneumotachograph
E	Elastance
E_{cw}	Chest wall elastance
E_L	Lung elastance
E_{rs}	Respiratory system elastance
f	Frequency in Hz (i.e., s^{-1})
F_IO_2	Fraction of inspired oxygen
FVC	Forced vital capacity
FEV_1	Volume of gas expired in first second of a maximal force expiration from TLC
G	Energy dissipation parameter of the constant-phase model of impedance
H	Elastic energy storage parameter of the constant-phase model of impedance

Hb	Hemoglobin
[Hb]	Concentration of hemoglobin
He	Helium
$[He]_i$	Initial concentration of helium
$[He]_f$	Final concentration of helium
i	Unit imaginary number (i.e., $\sqrt{-1}$)
I	Inertance
I_{aw}	Airway inertance
IR	Infrared
K_1, K_2	Two parameters of the Rohrer equation for flow resistance
l	Distance between two pressure sensing ports on a Fleisch pneumotachograph
MIGET	Multiple inert gas elimination technique
mmHg	Millimeters of mercury pressure
n	Number of moles of gas
N	Number of capillary tubes in a Fleisch pneumotachograph
N_2	Nitrogen
N_2O	Nitrous oxide
O_2	Oxygen
P	Gas pressure
P_A	Alveolar pressure
P_{atm}	Atmospheric pressure
pH	Negative logarithm of hydrogen ion concentration
PH_2O	Partial pressure of water vapor
P_{ao}	Airway opening pressure
P_aCO_2	Partial pressure of arterial carbon dioxide
P_aO_2	Partial pressure of arterial oxygen
P_A	Alveolar pressure
$P_{A,gas}$	Alveolar partial pressure of a specific gas
P_AO_2	Partial pressure of alveolar oxygen
P_{box}	Pressure inside plethysmograph
$P_{\bar{e}}CO_2$	Partial pressure of mixed expired carbon dioxide
PEF	Peak expired flow
P_{es}	Esophageal pressure
P_0	Value of transpulmonary pressure at functional residual capacity
PO_2	Partial pressure of oxygen
P_{pl}	Pleural pressure
P_{pleth}	Plethysmographic pressure
P_{tp}	Transpulmonary pressure
ΔP	Differential pressure drop
ΔP_{ao}	Change in airway opening pressure
ΔP_{pleth}	Change in plethysmographic pressure

$\dot{Q}$	Cardiac output
$\dot{Q}_s$	Pulmonary blood flow shunted past gas-exchanging regions of the lung
$\dot{Q}_t$	Total pulmonary blood flow
R	Resistance
$\bar{R}$	Ideal gas constant
R_{aw}	Airway resistance
R_{cw}	Chest wall resistance
R_L	Lung resistance
R_N	Newtonian resistance of the constant-phase model of impedance
R_{rs}	Respiratory system resistance
R_{ti}	Tissue resistance
RER	Respiratory exchange ratio
RV	Residual volume
SF_6	Sulfur hexafluoride
S_aO_2	Arterial oxygen saturation
S_pO_2	Arterial oxygen saturation as estimated using pulse oximetry
STP	Standard temperature and pressure (760 mmHg, 0°C)
T	Temperature
TLC	Total lung capacity
UV	Ultraviolet
V	Volume
V_{bag}	Volume of re-breathing bag
V_D	Dead space volume
V_{gas}	Volume of a specific gas
V_{pleth}	Volume of plethysmograph
V_T	Tidal volume
V_{TG}	Thoracic gas volume
ΔV_{pleth}	Change in plethysmographic gas volume
ΔV_{TG}	Change in thoracic gas volume
$\dot{V}$	Gas flow
$\dot{V}/\dot{Q}$	Ventilation-to-perfusion ratio
$\dot{V}_{O_2}$	Oxygen consumption
X	Reactance
Z	Impedance
α	Exponent of the constant-phase model of impedance relating G and H
ϕ	Phase angle
ϕ_{in}, ϕ_{out}	Phase angles of input and output sine waves, respectively
λ	Solubility of gas in the blood
μ	Gas viscosity
π	Radians in a semicircle

REFERENCES

Abboud R T and Sansores R 1996 ATS recommendations for DLCO *Am. J. Respir. Crit. Care. Med.* **154** 263

Andritsch R F, Muravchick S and Gold M I 1981 Temperature correction of arterial blood-gas parameters: a comparative review of methodology *Anesthesiology* **55** 311–6

Ayotte B, Seymour J and McIlroy M B 1970 A new method for measurement of cardiac output with nitrous oxide *J. Appl. Physiol. (1985)* **28** 863–6

Bachofen H 1968 Lung tissue resistance and pulmonary hysteresis *J. Appl. Physiol. (1985)* **24** 296–301

Bates J H, Prisk G K, Tanner T E and McKinnon A E 1983 Correcting for the dynamic response of a respiratory mass spectrometer *J. Appl. Physiol. Respir. Environ. Exerc. Physiol.* **55** 1015–22

Bates J H, Sly P D, Sato J, Davey B L and Suki B 1992 Correcting for the Bernoulli effect in lateral pressure measurements *Pediatr. Pulmonol.* **12** 251–6

Bates J H T 2009 *Lung Mechanics: An Inverse Modeling Approach* (Cambridge: Cambridge University Press)

Baydur A, Behrakis P K, Zin W A, Jaeger M and Milic-Emili J 1982 A simple method for assessing the validity of the esophageal balloon technique *Am. Rev. Respir. Dis.* **126** 788–91

Boron W F and Boulpaep E L 2011 *Medical Physiology*, 2nd edition (Amsterdam: Elsevier Saunders)

Dawson S V and Elliott E A 1977 Wave-speed limitation on expiratory flow—a unifying concept *J. Appl. Physiol. Respir. Environ. Exerc. Physiol.* **43** 498–515

Duvivier C, Rotger M, Felicio da Silva J, Peslin R and Navajas D 1991 Static and dynamic performances of variable reluctance and piezoresistive pressure transducers for forced oscillation measurements *Eur. Respir. Rev.* **1** 146–50

Finucane K E, Egan B A and Dawson S V 1972 Linearity and frequency response of pneumotachographs *J. Appl. Physiol. (1985)* **32** 121–6

Fredberg J J, Inouye D, Miller B, Nathan M, Jafari S, Raboudi S H, Butler J P and Shore S A 1997 Airway smooth muscle, tidal stretches, and dynamically determined contractile states *Am. J. Respir. Crit. Care Med.* **156** 1752–9

Fredberg J J and Stamenovic D 1989 On the imperfect elasticity of lung tissue *J. Appl. Physiol. (1985)* **67** 2408–19

Frey U 2005 Forced oscillation technique in infants and young children *Paediatr. Respir. Rev.* **6** 246–54

Frey U, Stocks J, Sly P and Bates J 2000 Specification for signal processing and data handling used for infant pulmonary function testing. ERS/ATS Task Force on Standards for Infant Respiratory Function Testing. European Respiratory Society/American Thoracic Society. *Eur. Respir. J.* **16** 1016–22.

Fry D L, Ebert R V, Stead W W and Brown C C 1954 The mechanics of pulmonary ventilation in normal subjects and in patients with emphysema *Am. J. Med.* **16** 80–97

Gale G E, Torre-Bueno J R, Moon R E, Saltzman H A and Wagner P D 1985 Ventilation-perfusion inequality in normal humans during exercise at sea level and simulated altitude *J. Appl. Physiol. (1985)* **58** 978–88

Ganter M and Zollinger A 2003 Continuous intravascular blood gas monitoring: development, current techniques, and clinical use of a commercial device *Br. J. Anaesth.* **91** 397–407

Gold W 2000 Pulmonary function testing in J Murray and J Nadel (eds.), *Textbook of Respiratory Medicine* (Philadelphia: Saunders) pp 793–9

Gowda M S and Klocke R A 1997 Variability of indices of hypoxemia in adult respiratory distress syndrome *Crit. Care Med.* **25** 41–5

Gravenstein J S, Paulus D A and Hayes T J 1995 *Gas Monitoring in Clinical Practice* (Boston: Butterworth-Heinemann)

Grenvik A, Hedstrand U and Sjogren H 1966 Problems in pneumotachography *Acta Anaesthesiol. Scand.* **10** 147–55

Habre W, Asztalos T, Sly P D and Petak F 2001 Viscosity and density of common anaesthetic gases: implications for flow measurements *Br. J. Anaesth.* **87** 602–7

Hager D N and Brower R 2006 Customizing lung-protective mechanical ventilation strategies *Crit. Care Med.* **34** 1554–5

Hager D N, Fuld M, Kaczka D W, Fessler H E, Brower R G and Simon B A 2006 Four methods of measuring tidal volume during high-frequency oscillatory ventilation *Crit. Care Med.* **34** 751–7

Hantos Z, Adamicza A, Govaerts E and Daroczy B 1992a Mechanical impedances of lungs and chest wall in the cat *J. Appl. Physiol.* (*1985*) **73** 427–33

Hantos Z, Daroczy B, Suki B, Nagy S and Fredberg J J 1992b Input impedance and peripheral inhomogeneity of dog lungs *J. Appl. Physiol.* (*1985*) **72** 168–78

Hyatt R E 1983 Expiratory flow limitation *J. Appl. Physiol. Respir. Environ. Exerc. Physiol.* **55** 1–7

Ilsley A H, Hart J D, Withers R T and Roberts J G 1993 Evaluation of five small turbine-type respirometers used in adult anesthesia *J. Clin. Monit.* **9** 196–201

Jackson A C, Giurdanella C A and Dorkin H L 1989 Density dependence of respiratory system impedances between 5 and 320 Hz in humans *J. Appl. Physiol.* (*1985*) **67** 2323–30.

Jackson A C and Vinegar A 1979 A technique for measuring frequency response of pressure, volume, and flow transducers *J. Appl. Physiol. Respir. Environ. Exerc. Physiol.* **47** 462–7

Jaffe M B 2008 Infrared measurement of carbon dioxide in the human breath: "breathe-through" devices from Tyndall to the present day *Anesth. Analg.* **107** 890–904

Kaczka D W, Ingenito E P, Israel E and Lutchen K R 1999 Airway and lung tissue mechanics in asthma. Effects of albuterol *Am. J. Respir. Crit. Care. Med.* **159** 169–78

Kaczka D W and Smallwood J L 2012 Constant-phase descriptions of canine lung, chest wall, and total respiratory system viscoelasticity: effects of distending pressure *Respir. Physiol. Neurobiol.* **183** 75–84

Landser F J, Nagles J, Demedts M, Billiet L and van de Woestijne K P 1976 A new method to determine frequency characteristics of the respiratory system *J. Appl. Physiol.* (*1985*) **41** 101–6.

Lausted C G and Johnson A T 1999 Respiratory resistance measured by an airflow perturbation device *Physiol. Meas.* **20** 21–35

Ligeza P 2007 Constant-bandwidth constant-temperature hot-wire anemometer *Rev. Sci. Instrum.* **78** 075104

Lilienthal J L Jr, Riley R L *et al* 1946 An experimental analysis in man of the oxygen pressure gradient from alveolar air to arterial blood during rest and exercise at sea level and at altitude *Am. J. Physiol.* **147** 199–216

Loring S H, O'Donnell C R, Behazin N, Malhotra A, Sarge T, Ritz R, Novack V and Talmor D 2010 Esophageal pressures in acute lung injury: do they represent artifact or useful information about transpulmonary pressure, chest wall mechanics, and lung stress? *J. Appl. Physiol.* (*1985*) **108** 515–22

Lutchen K R, Kaczka D W, Suki B, Barnas G, Cevenini G and Barbini P 1993 Low-frequency respiratory mechanics using ventilator-driven forced oscillations *J. Appl. Physiol.* (*1985*) **75** 2549–60

McIlroy M B, Mead J, Selverstone N J and Radford E P 1955 Measurement of lung tissue viscous resistance using gases of equal kinematic viscosity *J. Appl. Physiol.* (*1985*) **7** 485–90

McCahon R A, Columb M O, Mahajan R P and Hardman J G 2008 Validation and application of a high-fidelity, computational model of acute respiratory distress syndrome to the examination of the indices of oxygenation at constant lung-state *Br. J. Anaesth.* **101** 358–65

McCall C B, Hyatt R E, Noble F W and Fry D L 1957 Harmonic content of certain respiratory flow phenomena of normal individuals *J. Appl. Physiol.* (*1985*) **10** 215–8

Mead J 1956 Measurement of inertia of the lungs at increased ambient pressure *J. Appl. Physiol.* (*1985*) **9** 208–12

Mellemgaard K 1966 The alveolar-arterial oxygen difference: its size and components in normal man *Acta Physiol. Scand.* **67** 10–20

Mora R, Arold S, Marzan Y, Suki B and Ingenito E P 2000 Determinants of surfactant function in acute lung injury and early recovery *Am. J. Physiol. Lung. Cell. Mol. Physiol.* **279** L342–9

Oostveen E, MacLeod D, Lorino H, Farre R, Hantos Z, Desager K and Marchal F 2003 The forced oscillation technique in clinical practice: methodology, recommendations and future developments *Eur. Respir. J.* **22** 1026–41

Pedley T J, Schroter R C and Sudlow M F 1970 The prediction of pressure drop and variation of resistance within the human bronchial airways *Respir. Physiol.* **9** 387–405

Petak F, Hantos Z, Adamicza A, Asztalos T and Sly P D 1997 Methacholine-induced bronchoconstriction in rats: effects of intravenous vs. aerosol delivery *J. Appl. Physiol. (1985)* **82** 1479–87

Piiper J and Sikand R S 1966 Determination of D-CO by the single breath method in inhomogeneous lungs: theory *Respir. Physiol.* **1** 75–87

Plakk P, Liik P and Kingisepp P H 1998 Hot-wire anemometer for spirography *Med. Biol. Eng. Comput.* **36** 17–21

Primiano F P Jr 2010 Measurements of the respiratory system in J G Webster (ed.), *Medical Instrumentation: Application and Design*, 4th edition (Hoboken: John Wiley & Sons) pp 377–448

Ranieri V M, Rubenfeld G D, Thompson B T, Ferguson N D, Caldwell E, Fan E, Camporota L and Slutsky A S 2012 Acute respiratory distress syndrome: the Berlin definition *JAMA* **307** 2526–33

Renzi P E, Giurdanella C A and Jackson A C 1990 Improved frequency response of pneumotachometers by digital compensation *J. Appl. Physiol. (1985)* **68** 382–6

Riley R L and Cournand A 1949 Ideal alveolar air and the analysis of ventilation-perfusion relationships in the lungs *J. Appl. Physiol. (1985)* **1** 825–47

Sackner M A, Markwell G, Atkins N, Birch S J and Fernandez R J 1980 Rebreathing techniques for pulmonary capillary blood flow and tissue volume *J. Appl. Physiol. Respir. Environ. Exerc. Physiol.* **49** 910–5

Schuessler T F and Bates J H 1995 A computer-controlled research ventilator for small animals: design and evaluation *IEEE Trans. Biomed. Eng.* **42** 860–6

Schuessler T F, Gottfried S B, Goldberg P, Kearney R E and Bates J H 1998 An adaptive filter to reduce cardiogenic oscillations on esophageal pressure signals *Ann. Biomed. Eng.* **26** 260–7

Severinghaus J W 2002 The invention and development of blood gas analysis apparatus *Anesthesiology* **97** 253–6

Severinghaus J W and Astrup P B 1986 History of blood gas analysis IV Leland Clark's oxygen electrode *J. Clin. Monit.* **2** 125–39

Shore S, Milic-Emili J and Martin J G 1982 Reassessment of body plethysmographic technique for the measurement of thoracic gas volume in asthmatics *Am. Rev. Respir. Dis.* **126** 515–20

Simon B A and Mitzner W 1991 Design and calibration of a high-frequency oscillatory ventilator *IEEE Trans. Biomed. Eng.* **38** 214–8

Spriggs E A 1977 John Hutchinson, the inventor of the spirometer—his north country background, life in London, and scientific achievements *Med. Hist.* **21** 357–64

Spriggs E A 1978 The history of spirometry *Br. J. Dis. Chest* **72** 165–80

Stanojevic S, Wade A, Stocks J, Hankinson J, Coates A L, Pan H, Rosenthal M, Corey M, Lebecque P and Cole T J 2008 Reference ranges for spirometry across all ages: a new approach *Am. J. Respir. Crit. Care. Med.* **177** 253–60

Sullivan W J, Peters G M and Enright P L 1984 Pneumotachographs: theory and clinical application *Respir. Care* **29** 736–49

Tomioka S, Bates J H and Irvin C G 2002 Airway and tissue mechanics in a murine model of asthma: alveolar capsule vs. forced oscillations *J. Appl. Physiol. (1985)* **93** 263–70

Trebweiser J H, Johnson Jr R L, Burpo R P, Campbell J C, Reardon W C and Blomqvist C G 1977 Noninvasive determination of cardiac output by a modified acetylene rebreathing procedure utilizing mass spectrometer measurements *Aviat. Space Environ. Med.* **76** 445–54

Turner M J, MacLeod I M and Rothberg A D 1990 Calibration of fleisch and screen pneumotacho-graphs for use with various oxygen concentrations *Med. Biol. Eng. Comput.* **28** 200–4

Verbanck S and Paiva M 1994 Theoretical basis for time 0 correction in the rebreathing analysis *J. Appl. Physiol. (1985)* **76** 445–54

Verbeken E K, Cauberghs M, Mertens I, Lauweryns J M and Van de Woestijn K P 1992 Tissue and airway impedance of excised normal, senile, and emphysematous lungs *J. Appl. Physiol. (1985)* **72** 2343–53

Wagers S, Lundblad L K, Ekman M, Irvin C G and Bates J H 2004 The allergic mouse model of asthma: normal smooth muscle in an abnormal lung? *J. Appl. Physiol. (1985)* **96** 2019–27

Wagers S S, Haverkamp H C, Bates J H, Norton R J, Thompson-Figueroa J A, Sullivan M J and Irvin C G 2007 Intrinsic and antigen-induced airway hyperresponsiveness are the result of diverse physiological mechanisms *J. Appl. Physiol. (1985)* **102** 221–30

Wagner P D, Dantzker D R, Dueck R, Clausen J L and West J B 1977 Ventilation-perfusion inequality in chronic obstructive pulmonary disease *J. Clin. Invest.* **59** 203–16

Wagner P D, Dantzker D R, Iacovoni V E, Tomlin W C and West J B 1978 Ventilation-perfusion inequality in asymptomatic asthma *Am. Rev. Respir. Dis.* **118** 511–24

Wagner P D, Naumann P F and Laravuso R B 1974a Simultaneous measurement of eight foreign gases in blood by gas chromatography *J. Appl. Physiol. (1985)* **36** 600–5

Wagner P D, Saltzman H A and West J B 1974b Measurement of continuous distributions of ventilation-perfusion ratios: theory *J. Appl. Physiol. (1985)* **36** 588–99

West J B 2008 *Respiratory Physiology: The Essentials*, 8th edition (Philadelphia: Lippincott Williams & Wilkins)

Yeh M P, Adams T D, Gardner R M and Yanowitz F G 1987 Turbine flowmeter vs. Fleisch pneumota-chometer: a comparative study for exercise testing *J. Appl. Physiol. (1985)* **63** 1289–95

Yeh M P, Gardner R M, Adams T D and Yanowitz F G 1982 Computerized determination of pneu-motachometer characteristics using a calibrated syringe *J. Appl. Physiol. Respir. Environ. Exerc. Physiol.* **53** 280–5

Yernault J C 1997 The birth and development of the forced expiratory manoeuvre: a tribute to Robert Tiffeneau (1910–1961) *Eur. Respir. J.* **10** 2704–10

CHAPTER 18

Radiology

Melvin P. Siedband and John G. Webster

CONTENTS

18.1 X-Ray and Gamma Radiation · · · · · 454
    18.1.1 Radiation Defined · · · · · 454
    18.1.2 Radiation Measurement Units · · · · · 455
    18.1.3 X-Ray Generation · · · · · 455
    18.1.4 X-Ray Spectra · · · · · 456
    18.1.5 Radiation Absorption · · · · · 457
    18.1.6 Patient Absorption and Shielding · · · · · 458
    18.1.7 Unintended Radiation Sources · · · · · 458
18.2 X-Ray Tubes · · · · · 458
    18.2.1 Fixed Anode Tube · · · · · 458
    18.2.2 Rotating Anode Tube · · · · · 459
    18.2.3 Collimator and Beam Filter · · · · · 459
    18.2.4 Linear Accelerator · · · · · 459
18.3 Information and Noise · · · · · 460
    18.3.1 Modulation Transfer Function and N_e · · · · · 460
    18.3.2 Noise · · · · · 460
    18.3.3 Image Compression · · · · · 460
18.4 Image Detection · · · · · 461
    18.4.1 Plain Film · · · · · 461
    18.4.2 Computed Radiography · · · · · 461
    18.4.3 Image Intensifier · · · · · 462
    18.4.4 Crystal Diode Linear Arrays · · · · · 462
    18.4.5 Scintillator Plate/PMT Detector · · · · · 462
    18.4.6 Grids · · · · · 462
18.5 Test Instruments · · · · · 463
    18.5.1 Ion Chamber · · · · · 463
        18.5.1.1 Gas-Filled Detector · · · · · 463
    18.5.2 Geiger Tube · · · · · 464
    18.5.3 Semiconductor Diode · · · · · 465

18.5.4   Scintillator Diode   465
18.5.5   Thermoluminescent Detector   466
18.6   Basic Clinical X-Ray System   466
18.6.1   X-Ray-Generating Systems   466
18.6.2   Mechanical Components   466
18.6.3   Automatic Exposure Control   467
18.7   Fluoroscopy   467
18.7.1   Intensifier/TV   467
18.7.2   Image Storage: Digital Subtraction Angiography   468
18.7.3   Image and Information Communication Systems   468
18.8   Computed Tomography   468
18.8.1   Simple Tomography   468
18.8.2   Picture of the Universe   469
18.8.3   Fan Beams and Detector Arrays   469
18.8.4   Spot Filming   469
18.8.5   Contrast Agents   470
18.9   Magnetic Resonance Imaging   471
18.9.1   Nuclear Magnetic Resonance   471
18.9.2   Image Generation   471
18.10 Ultrasound   472
18.10.1 Acoustic Impedance and Reflection   472
18.10.2 Ultrasonic Imaging   473
18.11 Quality Assurance   473
18.11.1 Need for Quality Assurance   473
18.11.2 Phantoms and Test Objects   473
18.11.3 Test Instruments and Test Tools   474
18.11.4 Test Procedures   474
References   475

18.1 X-RAY AND GAMMA RADIATION

18.1.1 Radiation Defined

X-rays and gamma rays are carried by wave packets called photons and differ only by the source of radiation. X-rays are created by the interaction of an energetic electron with a target surface, and gamma rays are created by certain types of nuclear disintegrations. They are defined as ionizing radiation because they can cause a breaking of chemical bonds of molecules or the loss of electrons in single atoms. They are a form of electromagnetic radiation, similar to light and radio waves except that they are of shorter wavelength and higher energy than light photons or radio wave photons. The energy of a photon is related to wavelength by $v\lambda = c$, where v is the frequency in Hz, λ is the wavelength in meters, and c is the velocity of light in m/s. The energy of a single photon is $E = hv$, where E is the energy in joules, J, and h is Planck's constant of 6.63×10^{-34} Js, where s is the time in seconds. The charge of an electron is 1.60×10^{-19} C, where C is in coulombs. Combining these relationships, we find that the energy of a photon accelerated through an electric field of e volts can also be defined in electron volts, eV, where $1\ eV = 1.60 \times 10^{-19}$ J and the energy and wavelength are related, $E = hc/\lambda$ J or $E = hc/\lambda$ $(1.60 \times 10^{-19}$ eV), or, more simply, in the units usually encountered, $E = 1.240$ keV/λ where λ is in nm $(10^{-9}$ m).

18.1.2 Radiation Measurement Units

Ionizing radiation can be defined in several ways: the number of photons/s emitted by the radiation source, the ionization produced in a volume of dry air, or the energy deposited/kg at a distance from the source.

Incident ionizing radiation can deposit energy per unit mass as absorbed dose defined as 1 Gy (gray) = 1 J/kg. The older unit, the rad, is defined as 1 Gy = 100 rad. When radiation is absorbed by a living tissue, a weighting factor, relative biological effectiveness (RBE) is applied. In the midrange of diagnostic x-rays, the RBE is about 1.0 but is higher for higher photon energies such as used in radiotherapy. Weighted units of human exposure are given in Sieverts (Sv) or rem (1 Sv = 1 Gy in the range of energies used in diagnostic radiology and 1 rem = 1 rad). Incident ionizing radiation may also be measured in terms of its ability to ionize dry air. The roentgen, R, was originally defined as 1 esu (electrostatic unit) of charge deposited in 1 mL of dry air (0.001293 g), 1 esu = 3.34×10^{-10} C, which is very close to modern value of 1 R = 2.58×10^{-4} C/kg. On average, about 33.7 eV energy is needed to create one ion pair in dry air so that 1 R = 2.58×10^{-4} C/kg = $2.58 \times 10^{-4} \times 33.7$ J/kg = 87×10^{-4} J/kg. A rad = 100×10^{-4} J/kg so that 1 R = 0.87 rad approximately.

The old unit of radioactivity was the Curie originally defined as the number of radioactive disintegrations/s of 1 g of radium. This was not a precise unit because of the variations of radium isotopes. The Curie was later defined as 3.700×10^{10} disintegrations/s. The new unit, the Becquerel (Bq) is simply just the number of disintegrations/s. If the energy of the gamma photon produced by a particular isotope, its activity (Bq), and the absorption characteristics are known, it is possible to calculate dose equivalence versus distance. However, measurement data are usually tabulated for most isotopes and their interactions with many materials and tissue types.

Summary of Radiation Units	
Curie (Ci)	3.7×10^{10} disintegrations/s (or photons of a particular energy/s)
Becquerel (Bq)	1 disintegration/s
Roentgen (R)	2.58×10^{-4} C/kg (usually measured in dry air)
Gray (Gy)	1 J/kg (energy deposited)
Sievert (Sv)	1 J/kg × RBE (relative biological effect)
Old units	1 rad (r) = 0.01 Gy, 1 rem = 0.01 Sv), 1 R = 0.87 r (approx.)

18.1.3 X-Ray Generation

Figure 18.1 shows that free electrons are emitted from a hot filamentary cathode of a vacuum tube following the smaller of the Richardson–Dushman or Langmuir equations and that of thermal heating and dissipation: $J_1 = AT^2 e^{-u/kT}$ (Richardson–Dushman), $J_2 = BV^{3/2}$ (Langmuir), and $\sigma T^4 = I^2 R$ (thermal dissipation), where J_1 and J_2 are current density, T is

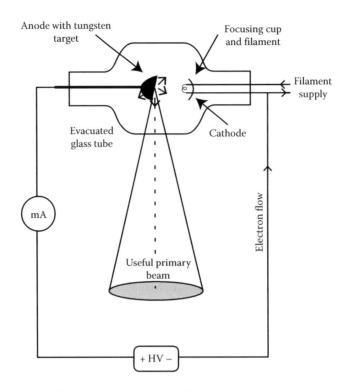

FIGURE 18.1 An x-ray tube has an anode and cathode mounted inside an evacuated glass tube. The hot cathode filament boils off electrons, which are accelerated by 100 kV to hit the tungsten anode to generate x-rays. (From Wikibooks Basic Physics of Digital Radiography.)

the filament temperature, u is the work function of the filament, k is Boltzmann's constant, σ is the Stefan–Boltzmann constant, A and B are constants determined by experiment (functions of the geometry of the system), I is the filament heating current, and R is the resistance of the filament. Electron emission is very sensitive to filament heating current; a 1% change of I yields approximately a 10% change of J_1.

The emitted electrons are accelerated by a strong electric field and strike a metal target, usually tungsten embedded in a molybdenum or copper substrate. Most of the electrons dissipate their energy as heat in the target. Some electrons are decelerated by the charge of the target nuclei, and that deceleration energy is radiated as a continuous spectrum of x-rays (ranging from photon energies equal to that of the arriving electrons down to zero). Deceleration radiation is also called braking radiation, in German, bremsstrahlung. Some arriving electrons remove various shell electrons (those that orbit the nuclei), and these are replaced by other orbiting electrons with the differences of electron energies radiated as photons of discrete energies related to the particular target element. Since these discrete energies are characteristic of each element, they are called characteristic radiation.

18.1.4 X-Ray Spectra

An x-ray photon produced by bremsstrahlung processes cannot have an energy, which exceeds that of the arriving electron that caused it. Because of the probabilities of producing

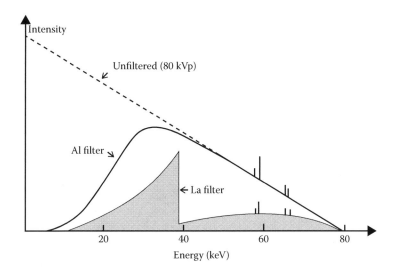

FIGURE 18.2 The x-ray energy spectrum filtered by aluminum and lanthanum. This hardens the spectrum to decrease low-energy x-rays, which would not travel through the body to yield an image. (From Wikibooks Basic Physics of Digital Radiography.)

an x-ray photon versus anode penetration by the electrons, the energy distribution of bremsstrahlung photons is continuous, as shown in Figure 18.2. Characteristic radiation is created when the arriving photon removes an inner electron from an atom of the anode material. When that electron space is refilled by an outer electron, the difference of binding energies between inner and outer atomic electrons is radiated as an x-ray photon. Because the binding energies are characteristic of each particular element, this radiation is called characteristic radiation and the photon energies are determined by whether the replacement electron comes from an adjacent orbit or from electron orbits farther away from the inner orbit or electron shell. X-rays are also produced when electrons are removed from other than the inner shell but have energies too low to be classified as useful x-rays. The anode material and the glass wall of the x-ray tube absorb lower-energy x-ray photons more readily than the higher energies so that the spectrum of x-rays is further shaped, as shown by Figure 18.2.

18.1.5 Radiation Absorption

X-ray photons are absorbed by a variety of mechanisms. Two are most important: photoelectric absorption, where the arriving x-ray photon causes the emission of electrons from the absorber, and Compton scatter, where the arriving photon produces a scattered electron and another x-ray photon of lower energy. Absorption of monochromatic photons is similar to the absorption of light by seawater and follows Beer's law, $I = I_0 e^{-ux}$, where I_0 is the initial beam intensity, u is the attenuation coefficient, and x is the thickness of the absorbing layer. However, u is a function of the energy of the arriving photon, roughly proportional to $1/E^2$, where E is the energy of an arriving x-ray photon. This means that the lower energies of the broad spectrum of x-rays are more readily absorbed. One useful, although not very precise, measure of beam energy distribution is the half-value layer (HVL), the thickness of aluminum foil needed to attenuate the beam 50%. Electrons cannot interact with arriving photons unless their energy exceeds the binding energy of the electron. Since the binding energy is characteristic of each element, the absorption of x-rays versus photon energy shows a sharp increase at the binding energy of, say, the K-shell electrons also called the K-edge. This is a useful characteristic when designing x-ray beam filters for particular applications.

18.1.6 Patient Absorption and Shielding

Ionizing radiation is harmful. The mean lethal whole body exposure to humans is about 0.6 Gy, which means that of a population so exposed, about half will die in 1 month. Because the lower energies in a broad distribution of x-ray photon energies cannot penetrate a patient to contribute to an image, x-ray beam filters are used to absorb those lower energies, which could harm the patient. X-ray sources are enclosed in metal shields to limit the exposed area the smallest required. X-ray examining rooms are designed with shielding to reduce the hazard to operators.

18.1.7 Unintended Radiation Sources

Electron tubes that operate at high voltages can produce x-rays. A high voltage radio or radar transmitting tube can produce x-rays as part of normal operation. While a high voltage rectifier or thyratron should not produce x-rays in normal operation, certain failure modes can cause the generation of x-rays. In normal operation, a high voltage rectifier has a low forward voltage and no reverse current so that no x-rays should be produced. If the tube becomes gassy with age, the ionized gas will conduct electrons, which are accelerated by the high inverse voltage to produce x-rays. During the initial gas ionization time of a thyratron, the lightly ionized gas will conduct energetic electrons to the anode before the gas fully ionizes and the anode voltage falls to a low value. Traveling wave tubes, power klystrons, and magnetrons operating at high voltage will also produce x-rays as part of their normal operation and must be shielded. In general, the operation of any vacuum tube at voltages above 15 kV should prompt an examination of performance and a failure analysis to be certain that no stray radiation can be emitted.

18.2 X-RAY TUBES

18.2.1 Fixed Anode Tube

Fixed anode tubes are used in most dental and cabinet x-ray machines. The cathode uses a tungsten helix, the filament, in a stainless steel focusing structure. The high voltage applied to the anode creates an electric field in the focusing structure, which causes the electrons to be emitted as a sheet of electrons as wide as the hot part of the filament length and just a bit thicker than the filament helix diameter. This line-focused beam strikes the angled anode so that the beam area appears foreshortened to the target (the patient or test object). The dimensions of the foreshortened area define the focal spot size. Oxide-coated filaments are not used as positive ion bombardment caused by gas ions striking the filament would quickly erode any coating. The anode usually consists of a small square of tungsten alloy embedded in a slug of copper. The limitation to performance of the tube is the thermal dissipation of the anode. Since most of the beam energy is dissipated as heat, a small tube operating at 70 kV and 20 mA would dissipate 1.4 kW during the exposure. The tube design would select the smallest focal spot to permit maximum exposure time without melting the anode.

18.2.2 Rotating Anode Tube

While dental x-ray examinations can be performed with a small fixed anode tube, head or abdominal or multiple exposures during a cardiac examination or the long exposures of a computer tomography (CT) machine require a tube of far greater capacity. One approach is to embed a circular tungsten track in molybdenum disk. This disk is part of the rotor of an induction motor contained within the vacuum tube with an external stator. Just prior to exposure, the motor is energized so that cooler anode material is brought under the beam during the exposure. The instantaneous power rating of the tube can be increased to levels approaching 250 kW, especially if the motor is operated at high speeds (inverter powered at 180 Hz). The new limit becomes the melting point of the entire anode disk, related to the total anode heat capacity. Tube ratings are described in tube data sheets for each x-ray tube.

18.2.3 Collimator and Beam Filter

An optical collimator focuses a light source into a beam of parallel light rays. Certain telescopes are also called collimators when used to align optical instruments. An x-ray collimator, unlike an optical collimator, does not alter the direction of the beam. It is used to define the limits of the exposed x-ray field and to limit the radiation source area to a small area of the tube anode. While most x-rays are generated by the impact of focused beam electrons on the anode, electron scatter and other anomalies also produce undesired off-focus radiation which will reduce the quality of the radiographic image. Off-focus radiation can be reduced by using a small lead cone fitted close to x-ray tube. The collimator may also have a radiolucent mirror and lamp equidistant from the focal spot. The light field should be congruent with the x-ray field and defines the x-ray field before the exposure. The collimator may also contain a metal filter to remove the lower x-ray photon energies to reduce patient exposure. The tube is mounted in an enclosure lined with lead with a hole for the exiting x-rays, the mounting for the collimator (or a field defining metal cone).

18.2.4 Linear Accelerator

This device uses a microwave source to set up standing waves in a succession of cavities of increasing length. An electron beam source (hot cathode and focusing electrodes) injects electrons, which are accelerated by microwave electric fields of the cavities. As the electrons move faster, the cavities can become longer for the same transit time until the electrons acquire energies of, say, 500 keV when they collide with a thin metal target to produce forward-projected x-rays. The series of cavities may also be folded and the electron beam bent by a strong magnetic field to change the shape of the apparatus. Linear accelerators are made with energies up to 30 MeV for radiotherapy.

18.3 INFORMATION AND NOISE

18.3.1 Modulation Transfer Function and N_e

The information capacity of the field of a digital image can be defined by the number of picture elements, pixels, and the maximum level in bits/pixel. An analog image, such as an ordinary photograph, can be defined by the image resolution and the modulation depth. Resolution and modulation depth are best defined by plotting the amplitude response versus spatial frequency to show the modulation transfer function (MTF). A film image of a line resolution pattern can be scanned with a slit microscope and phototube with the results plotted. The scanned signal is normalized, that is, the maximum measured signal at the lowest frequency is set equal to 1.00. For x-ray images, an ideal sine-function slit test pattern is difficult to obtain so that a simple pattern of open slits in a thin lead strip is used. The spacing of the slits is varied to have a range of spatial frequencies. The *square wave* scanned signal can be converted to a sine wave response by a matrix inversion of a set of Fourier equations of sine wave expansions of square waves:

$$S(f) = \pi/4 \; [M(f) + M(3f)/3 - M(5f)/5 + M(7f)/7 - \ldots],$$

where
    $S(f)$ is the sine wave amplitude at frequency f
    $M(f)$ is the measured square wave amplitude at that frequency
    $3f$ is the measured value at $3f$, etc.

The sine wave response is plotted as MTF. The MTF can be converted to a useful expression, the noise-equivalent resolution, N_e, which is very close to the spatial frequency of a digital (discrete element) detector. It is as if N_e is defined an equivalent mosaic of discrete detectors of N_p pixels/cm. The noise-equivalent resolution, N_e, is obtained by integrating the square of the MTF amplitudes:

$$N_e = \int_0^\infty S^2(f)df.$$

18.3.2 Noise

Each detected x-ray photon is statistically independent, and an apparently uniform field of detected x-rays will have a Poisson standard deviation (SD) of $N^{1/2}$ photons, where N is the number of photons/pixel. For a maximum signal-to-noise ratio (SNR), where the signal is caused by modulating the number of photons/pixel, the number of photons/pixel is simply SNR^2. An SNR of about 4 is required to just see an object (not very precise as the image background, object shape, and other factors are involved) so that a minimum exposure is required for x-ray imaging. Since an excessive exposure to x-rays is harmful, effort must be taken to use the lowest exposure to produce an image of useful SNR.

18.3.3 Image Compression

Much of the information in a digital image is either redundant or of lesser importance. Just as in normal photography, a digital image can be compressed to reduce image bandwidth for transmission or storage. Two methods dominate the field: fast Fourier transformation (FFT) or block truncation (including Joint Photographic Experts Group (JPEG)). The FFT is based on the conversion of a amplitudes of a time-varying function to Fourier coefficients, which require less bandwidth for the same information. The FFT can achieve compression ratios of 5 or so with little loss of image quality or 10 or more with some loss. Block truncation considers a small block of the digital image field, say 4 × 4 pixels, calculates the mean value, and sets a value of 1 for each pixel above the mean and 0 if below. The mean value, SD, and the pattern of 1s and 0s are used to reconstruct the image requiring only 3.5 bytes rather than the original 12 bytes. Each value 1 is given a value of the mean plus the SD; each 0 is given the value of the mean minus the SD. Splining (matching the edges of adjacent blocks) and filtration smooth the reconstructed image.

Image compression can be evaluated by amplifying and displaying the pixel differences of the original and reconstructed images and examining that difference image to determine whether anything recognizable can be seen. A lossless process should show no differences while a modest and random noise problem would simply require a small increase of x-ray exposure to compensate for the loss of image quality. A strong visible difference is not acceptable.

18.4 IMAGE DETECTION

18.4.1 Plain Film

The most basic x-ray image detector is a sheet of photographic film placed in a light-tight cardboard envelope. Such detectors are still used for industrial radiography and some dental images. The film may also be made with double emulsion coatings for increased sensitivity. However, the silver of the film emulsion is a poor absorber of x-rays, about 1% or less than that of the same mass of tungsten, so that fluorescent screens are often used with film to increase the x-ray absorption. The simplest screens were cardboard sheets coated with phosphor powders. Modern screens still use phosphor powders with selected granule size in a plastic sheet. Experiments have shown that two thin fluorescent screens with double emulsion film will produce higher resolution images than one thick screen for the same total x-ray absorption. Metal x-ray film cassettes with a thin black plastic face are commonly used, which contain two screens with the film reloaded in the darkroom after exposure. The thickness of the fluorescent screens is a compromise between maximum absorption of x-rays and image resolution: thicker screen yields higher absorption/lower resolution. For medical applications, thin screens are often used for radiography of the bony extremities and thick screens are used for imaging the abdomen.

18.4.2 Computed Radiography

Some screens had the unfortunate property of producing a faint residual or shadow image of a previous exposure. Heating the screen would release electrons from crystalline traps for complete erasure. Research exploited this *memory* property so that new screens were developed, which could be scanned after exposure with an infrared laser, the stimulated light of each scanned pixel was proportional to the x-ray exposure, and a computer was used to synthesize the image. The convenience of such digital images is leading to the rapid replacement of film by computed radiography as images can be transmitted, recalled, archived, and even analyzed by common computer techniques. Computer image processing for noise reduction, edge enhancement, and shape recognition is easily accomplished. Specialized screens of varying thickness and energy sensitivity have been developed for particular applications.

18.4.3 Image Intensifier

The earliest fluoroscopic systems comprised a thick calcium tungstate screen covered by a viewing hood. The radiologist, with dark-adapted vision, studied the faint image on the screen to make his diagnosis. Unfortunately, such dark-adapted images were of low resolution and the radiologist was also exposed to high levels of scattered radiation. Very fast camera lenses were used to couple the screen image to a television camera, but it was realized that even the best optical coupling lost image statistical information and so required higher exposures. The first practical image intensifier used a thin clock glass window coated on one side with an x-ray-sensitive phosphor and the other side coated with photocathode similar to the sensitive surface of a photoelectric tube. The thin glass window did have some lateral light spread but was still useful for many imaging applications. The photoelectrons from, say, a 13 cm diameter photocathode were accelerated by a 20 kV electric field toward a, say, 2 cm diameter, small fluorescent screen. This produced a smaller but brighter image, about 10,000–100,000 times brighter than that of the original. The smaller image is easy to couple to a vidicon or charge-coupled diode television camera. While earlier fluoroscopic imaging systems had mirror beam splitters to also share images with various 70 mm or 100 m roll film cameras, such film cameras have been replaced by digital image archiving schemes.

18.4.4 Crystal Diode Linear Arrays

A simple radiation detector for radiation monitoring or for use in many radiation measuring instruments comprises a small scintillator crystal cemented to a silicon photovoltaic diode. Arrays of such detector diodes are used in CT machines.

18.4.5 Scintillator Plate/PMT Detector

A large diameter single crystal plate can be coupled to a combination of, say, 19 end-on photomultiplier tubes (PMTs). A gamma ray photon striking the plate and captured by the crystal will produce thousands of light photons, depending on the energy of the arriving photon. The PMT circuit will analyze the sum of the PMT signals to determine the source location of the detected photon as well as the mean energy range of that photon. Energy discrimination circuits are helpful so that undesired signals (from unwanted photons) can be rejected. Summing these location signals permits the synthesis of an isotope (nuclear) image.

18.4.6 Grids

Scattered x-rays will degrade image contrast. One scheme is to use a grid device that resembles a thin lead venetian blind. For example, an array of lead strips, $350 \times 2.5 \times 0.10$ mm with strips of thin plastic of $350 \times 2.5 \times 0.9$ mm between the lead strips, can be mounted in a frame and placed just in front of the x-ray detector. The thin plastic strips can be slightly tapered so that the 2.5 mm axis of the strips points toward the x-ray source (the x-ray tube). Photons that arrive at the grid will be transmitted through the plastic space between the lead strips. Those outside the incident angle window of about 5 degrees will be intercepted by the lead strips. To keep the lead grid lines from being imaged, a motor drive is often used to cause a lateral oscillation of the grid. A moving grid is called a Bucky grid after its inventor.

18.5 TEST INSTRUMENTS

18.5.1 Ion Chamber

A very simple form of ion chamber was used by Roentgen. This was the gold leaf electrometer where a gold leaf was suspended on an insulated cross-bar within a glass Leyden jar. When charged, the two halves of the gold leaf would repel each other. When irradiated by ionizing radiation, the charge would be reduced by conduction of the ionized air and the leaves would sag in proportion to their mass, size, shape, and remaining charge. Because of the complexity of reading the angle of the leaves, charge, etc., and the calculations needed, this method is rarely used. Rather, an ionization chamber is connected, with one side (cathode) connected to a polarizing voltage source, typically −300 V, and the other side (anode) connected to an electrometer arranged as a coulometer (to accumulate current over time to yield charge). Some ion chambers are integrated with a small capacitor, charged, and the voltage drop measured to determine radiation exposure of the chamber. Figure 18.3 shows ionization chambers may be configured as small cylinders or parallel plates.

18.5.1.1 Gas-Filled Detector

The chamber wall material (inner surface) should have the same atomic number as dry air for the chamber to have *air equivalence,* and it will attenuate and filter the input

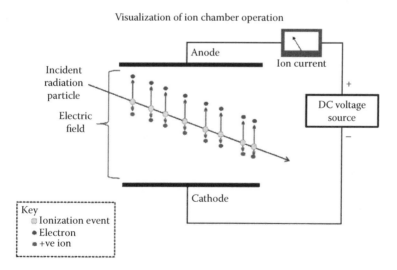

FIGURE 18.3   In an ionization chamber, radiation ionizes gas, which causes current to flow to the ammeter. (From Wikipedia Ion chamber operation.)

radiation as well. The construction, particularly of the smaller chambers, must have wall material and thickness so that it responds as a simple air cavity when placed within a phantom for *charge buildup equivalence* for the energy range of each range of applications. Ion chambers may be cylindrical or planar and may also comprise several smaller chambers as an assembly to measure x-ray field uniformity.

From Section 18.1.2, 1 R will cause the conduction of 2.58×10^{-4} C/kg. For a 100 mL chamber, the conducted charge will be 2.58×10^{-4} C/kg $\times$ (1.293×10^{-4} kg (100 mL dry air) = 3.336×10^{-8} C. For a small 10 mL chamber exposed to 1 mR, the conducted charge would be 3.336×10^{-12} C, equivalent to a 3.336 V discharge of a 10 pF capacitor. An electrometer is a very sensitive current meter capable of measuring currents of less than 10^{-9} A. A coulombmeter is an integrating electrometer capable of integrating electrometer currents and displaying the result directly as R or mR when calibrated in combination with particular ion chambers. Ion chamber currents and cumulative charges are very small so that electrometers or coulombmeters are used to display signals.

Several forms of ion chamber are in common use:

1. Small *thumb* chambers

2. Large cylindrical chamber

3. Small planar or *pancake* chambers

4. Large planar chambers

5. Transparent planar chambers

6. Combined ion chamber-electrometer

Figure 18.3 shows the general construction of ion chambers. One electrode is connected to a voltage source so that the internal voltage gradient is sufficient to sweep out and collect ions, the chamber current, but not so high that the gradient is high enough to ionize the air within the chamber or energize the ions so that multiple ion–air collisions result in per photon interaction. The voltage source electrode is connected to the voltage source and bypassed by a capacitor and the inner electrode connected to the electrometer. For planar chambers, the inner electrode may be a thin plastic sheet coated with a conducting material, often graphite, placed midway between two voltage source surfaces. For some portable combined ion chamber-electrometers (direct reading R-meters), the outer surfaces may be made of graphite fiber sheets at ground potential and the collecting electrode and electrometer circuit floating at the ion collecting potential. This self-shielding construction makes the instrument far less sensitive to stray electric fields.

18.5.2 Geiger Tube

A thin-walled chamber can be filled with a low pressure gas and connected to a voltage source just below the point where gas will ionize. If the voltage source is a capacitor fed by a high value resistor, gas ionization initiated by an x-ray photon or gamma ray will discharge the capacitor to a very low value and the gas ions will recombine, that is, deionize. The *RC* combination must have a time constant τ longer than the recombination time of the gas. If the ground return of the capacitor is connected to a small loudspeaker, a click will be heard for each ionization cycle, the familiar Geiger counter series of clicks. A simple rate meter circuit can display the cycle of ionizations as approximate exposure in R/s or mR/s. This is not very accurate but small and inexpensive. Accuracy is poor because the gas can be ionized by any photon of any energy sufficient to just ionize the gas. During the time the capacitor voltage is less than the ionization voltage of the gas, the device is insensitive to radiation, the *dead time*. Thus, Geiger tubes will read low for pulsed radiation. Because of low cost and simplicity, the Geiger counter is useful for detecting ionizing radiation but not for measurement purposes.

18.5.3 Semiconductor Diode

Diode detectors can operate in the photovoltaic mode where incoming radiation produces a small current, which can be amplified by an electrometer. This mode is not practical in many applications as the active junction thickness at zero bias is quite small. The diode may also be operated reverse-biased to deplete the semiconductor junction, and this junction is much larger than that of the unbiased diode. Detected photons create hole–electron pairs in the solid, which serve the same role as ionized gas ions. While a common silicon rectifier diode will detect radiation, specially designed diodes are far more sensitive. The required reverse bias is only a few volts with little change in performance versus voltage. Very large diodes can be constructed of ultrapure germanium or silicon and operated at higher voltages to sweep out hole–electron pairs and increase the probability of detection. Thermal electrons can produce thermal dark currents (background current when not exposed to radiation) in excess of that produced by low exposure to ionizing radiation requiring that such large diodes (often a single crystal of 1 mL or larger) must be cooled to liquid nitrogen temperature (–77 K) to keep dark current to reasonable levels. Such large and cooled diodes can produce hole–electron pairs where N = photon energy/band gap (approx.), where N is the number of hole–electron pairs produced per interaction. This means that maximum SNR can approach $N^{1/2}$; for example, a 50 keV photon impinging a 0.7 eV Ge crystal can have an SNR of 267, equivalent to an energy resolution of 187 eV. Energy resolution is an important factor in discriminating energies produced by different isotopes.

18.5.4 Scintillator Diode

A small diode operated in the photovoltaic mode can have improved sensitivity by cementing a small scintillator crystal of, say, $CaWO_4$ on the surface. For example, a small photovoltaic light sensor of 1 cm² can have a scintillator of $1 \times 1 \times 0.5$ cm glued in place with optical cement. The diode operates in the photovoltaic mode detecting the crystal glow. A photolytic diode may also be coated with a layer of radiation-sensitive phosphor (scintillator powder) to produce a practical radiation sensor. Such scintillator diodes are used for personal radiation detectors or in x-ray beam energy measuring instruments.

18.5.5 Thermoluminescent Detector

Heating certain crystalline materials drives some electrons from trapped states to ground level. Exposure to radiation drives these electrons back from the ground state to the trap levels. Subsequent heating returns the electrons to the ground state with the emission of light, the total light emitted in proportion to the radiation exposure. Small quantities of these materials, such as LiF, CaF_2, $CaSO_4$, and other compounds, can be incorporated into radiation monitoring badges. This technique is called thermoluminescent dosimetry (TLD).

18.6 BASIC CLINICAL X-RAY SYSTEM

18.6.1 X-Ray-Generating Systems

The x-ray generator comprises a filament control, a high voltage power supply, a timer, and other controls for automatic exposure termination or for coupling to other parts of the system. The high voltage applied to the tube may be rectified but unfiltered direct current with a high ripple, a series of half-sine waves. By custom, the voltage is described as kilovolts peak (kVp), rather as a simple DC value. Because x-ray generation is voltage sensitive and because the energy distribution of x-ray photons is affected by the time-varying tube anode voltage, tube performance can be improved by making the voltage more constant. Because of the high tube currents and the need for sharp termination of exposure times, capacitor filtration is impractical. Polyphase systems or high frequency systems, where the high voltage is obtained from an inverter operating at 25–200 kHz, produce high voltage with lower ripple. A clinical x-ray system will operate the x-ray at anode voltages from 25 (mammography) to 150 kVp (dense abdomen). Anode current is controlled by the adjustment of the filament temperature. The anode current is very sensitive to changes of filament current: typically, a 1% change of filament current results in a 10% change of anode current. To maximize tube life, the filament is usually operated at a lower current until just prior to the exposure and then *boosted* to the value required for the anode current needed for the exposure. For systems using rotating anode tubes, the anode motor is not energized until just prior to the exposure and then about two times the usual motor voltage is applied for a short period to bring the anode to proper rotational speed for the exposure and then reduced to about half voltage to allow the motor to coast (for multiple exposures at lower motor dissipation). Some systems may operate the motor at higher than usual speeds by using an inverter to energize the motor at higher than the power line frequency to rotate faster. Such ultraspeeds of 10,800 rpm (the normal line-operated speed is 3,600 rpm) permit higher instantaneous exposure dissipation but also require that the motor be braked to prevent high frequency resonance vibrations from damaging the delicate bearings. Braking is often accomplished by direct current through the motor windings for a short period after the exposure (Figure 18.4).

18.6.2 Mechanical Components

A typical radiographic-fluoroscopic (R&F) x-ray examining room will have an x-ray table with a spot film device, an overhead x-ray tube, an x-ray tube within the table, and an optional chest board (a device holding a film or screen cassette, antiscatter grid, and exposure sensor) for taking chest exposures. The spot film device will also hold the fluoroscopic image

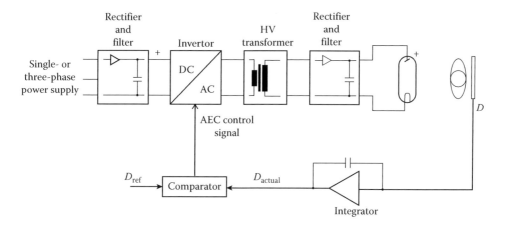

FIGURE 18.4   Block diagram of a high frequency HV generator equipped with automatic exposure control (AEC). Radiation detector(s) are placed at the image receptor to measure the exposure and feed a signal back to the HV generator when termination is required. (From Wikibooks Basic Physics of Digital Radiography.)

intensifier, which usually has a television camera feeding a monitor on a cart or on an additional ceiling suspension. The overhead x-ray tube may be aimed toward the table for taking pictures using a cassette holder in the table or aimed toward the chest board. The spot film device may contain a cassette holder for taking exposures during fluoroscopic procedures.

Specialized tables and assemblies are used for some procedures. Mammography-ray machines will have devices for breast compression during exposure. Genitourinary (GU) tables will position the patient for image-guided surgical procedures.

18.6.3 Automatic Exposure Control

Just as in modern photographic cameras, AEC eliminates the need for precise setting of the exposure factors of kVp, anode current, and exposure time. The AEC sensor is often a three-chambered ion chamber (to measure in areas of the two lung fields in either anterior–posterior (AP, i.e., front back) or lateral (side view) positions placed between the grid and the detector screen (film-screen for film systems or storage screen for computed radiographic systems). The ion chamber current is integrated until a preset charge value is reached, and the circuit causes the control timer to terminate the exposure. Other sensors may be scintillator/photodiodes after the screen or thin plastic paddles with small amounts of fluorescent phosphor coupled to a PMT. The principle of integrating the current, etc., is the same as the ion chamber method.

18.7 FLUOROSCOPY

18.7.1 Intensifier/TV

Fluoroscopy is real-time x-ray imaging. This includes various forms of angiography, especially cardiac imaging, swallowing studies, and image-guided procedures, where the therapy device (a very hot or cold device or precision cutting tool) is guided directly to the site for local treatment. To minimize radiation exposure of the operator, the imaging device is coupled to a TV camera. The imaging device is usually an image intensifier although solid-state panel imagers are under development, which rival the image intensifier. The TV camera is often used with an image storage device for last image hold (LIH), which stores the last TV image as the x-ray control foot switch is released. Rather than a continuous display, the operator may depress the foot switch for a series of *still* images to reduce patient exposure to very low levels.

18.7.2 Image Storage: Digital Subtraction Angiography

In addition to making possible a series of stills or LIH for better examination of the image without further exposure to the patient, stored images permit the display of image differences. For example, one image may be stored just prior to the injection of a contrast agent (e.g., iodine-bearing compound injected in a blood vessel, iodine is a good absorber of x-rays) and subtracted from the image taken with iodine. The difference between the images may be amplified to enhance that part of the image with iodine. This is a key feature of digital subtraction angiography (DSA). These improved iodine images are in common use for determining the patency of blood vessels. Other applications involve taking a series of stills for the estimation of blood velocity or perfusion. Difference images taken at different-ray beam energies (or with different x-ray beam filters) can reveal subtle differences of tissue density or composition, helpful in diagnosis of cancer. High-resolution fluoroscopic images often rival computerized radiographic images and can be stored (archived as part of a patient record) or transmitted to remote locations for diagnosis or comparison as part of patient treatment. If clinics or hospitals are linked by computer communication systems, it is possible for the patient to visit any participating clinic and for the local physician to have access to all medical records and pertinent medical images.

18.7.3 Image and Information Communication Systems

There are dozens of information and image communication systems used on the Internet. Medical image system parameters cannot be standardized for resolution, contrast, and dynamic range (related to best SNR) because different image modalities have different requirements. However, an information data header format has been standardized so that electronic conversion of images can be accomplished so that most images can be displayed, transmitted, archived, and displayed on the same hardware. Combining imaging and patient reporting requirements in a single system with ready access by physicians is causing a revolution in medical record keeping. Various medical imaging systems use different image formats, that is, resolution, field size, and orientation. Converting images to a standard format permits comparison and differencing as ways to assist in the diagnostic process. Progress of a disease may be tracked by comparing present to earlier images. Expert remote consultation is possible by image communication systems.

18.8 COMPUTED TOMOGRAPHY

18.8.1 Simple Tomography

While an ordinary radiograph is a *shadow image* of the body orthogonal to the direction of the x-ray beam, a tomography is a cut or image of a section of the body in the plane of the axis of the x-ray beam. The original tomography systems used a linkage to move the film plane and the x-ray tube around a fulcrum in the plane of the body part of interest. Objects outside of this plane would be blurred, and the image of, say, the bones of the inner ear or the nerves of the spine would be clear. Simple linear tomo attachments as well as systems having complex spiral or trefoil motions were used to blur objects of little interest without the generation of false images or artifacts.

18.8.2 Picture of the Universe

It is interesting to note that the abstract science of astronomy is used in medical imaging. Astronomers were searching to discover the shape of the Milky Way when certain parallax discrepancies prompted research into synthesizing an image of the plane described by the axes of several parallax images. The first computerized tomographic images were made by a step-and-repeat process and then a rotate process. The x-ray tube and detector scanned a set of 128 parallel lines about 2 mm apart through the skull, and then the entire device was rotated about 3° and the process repeated a total of 60 times over a 5 min period. A computer processed the resulting signal, and, after about 30 min, a 128 × 128 pixel image of a section or cut of the brain was displayed. For the first time, it was possible to see brain structure, voids, and tumors without surgery. Of course, these systems, elegant at that time, have been improved a few orders of magnitude (Figure 18.5).

18.8.3 Fan Beams and Detector Arrays

While the first CT systems used a single source and single detector, progress was quick to reduce scan time by using a fan beam and detector array and a rotating gantry. Scan time went from 5 min to 5 s and then to less than 1 s with image acquisition synchronized to heart motion. Computer data processing was also improved to the point where images appear almost as soon as the scan is completed. In order to examine several image planes in the body, the patient table is translated through the gantry in a series of steps. Newer systems operate the table smoothly so that the scanning process is that of a continuous spiral. Computer processing also permits the reconstruction of any image plane for selective examination of any organ or structure.

18.8.4 Spot Filming

Older systems had the image intensifier assembly mounted on a film cassette device arranged so that the radiologist could take a film radiograph with a small interruption during fluoroscopy. A switch on the spot film device caused a film cassette to move in front of the image intensifier for the exposure and then returned the cassette to a shielded enclosure at the rear of the assembly. Modern systems use the image intensifier and television camera for spot filming using an increased x-ray tube beam current to compensate for the increased image noise, which would otherwise occur because of the short spot filming exposure time. It should be noted that the optical lag of the eye is about 0.2 s so that image noise during fluoroscopy corresponds to an exposure of about 0.2 s. A spot film exposure of 0.05 s (for stop motion) would require 10 times more tube current for the same noise level as the fluoro image. For even less image noise,

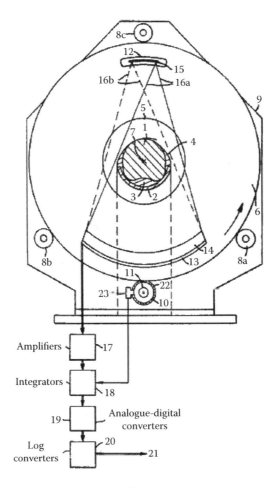

Amplifiers — 17

Integrators — 18

19 — Analogue-digital converters

Log converters — 20 — 21

FIGURE 18.5   X-ray CT produces a volume of data that can be manipulated in order to demonstrate various bodily structures based on their ability to block the x-ray beam. Images are in the axial or transverse plane, perpendicular to the long axis of the body. (From Wikipedia x-ray CT.)

the exposure current may be increased by a factor of 20 or more, say, 100 mA spot film current during a 2 mA fluoroscopy procedure.

18.8.5 Contrast Agents

To improve the visibility of objects, contrast agents can be injected into body cavities (air), swallowed to outline parts of the digestive system (barium based), or injected into blood vessels (iodine based). The energy sensitivity of a contrast agent depends on the binding energy of its innermost electrons, the K-edge. For barium, this energy is 37.4 keV, and for iodine, it is 33.2 keV so that maximum contrast is achieved at radiation photon levels just above these values. Air is almost transparent at all energies and shows no energy dependence. Iodine-based and barium-based contrast agents are compounded to be tolerated by the body; plain iodine and plain barium and many of their compounds are poisonous. Many imaging techniques are not possible without contrast agents: imaging of the urinary system, small bowel, heart vessels, etc.

18.9 MAGNETIC RESONANCE IMAGING

18.9.1 Nuclear Magnetic Resonance

A small magnetic field is generated by spinning nuclei of certain elements. When atoms of these elements are placed within an external field, the spinning nuclei tend to align with the field, the spin-up or normal state. Some particles can be excited to align themselves in a reverse state (N-N or S-S rather than N-S), the excited or spin-down state. The ratio of normal to excited particles is $N_n/N_e = e^{hfkT}$, where h is Planck's constant, f is the frequency, k is Boltzmann's constant, and T is the temperature. The spin is not perfect and has a wobble or nutation determined by the charge of the nucleus and other characteristics of the atoms. The nutation has a precessional frequency determined by the Larmor relationship, $\omega = \gamma B$, where ω is 2π times the precessional frequency, γ is a property called the gyromagnetic ratio, and B is the magnetic field. Adding radio frequency energy at the nutational frequency (magnetic resonance) may cause the amplitude of the wobble to increase and the atom to change state (spin up to spin-down). Turning off the source of radio frequency energy will cause excited atoms to return to their normal state and return radio frequency energy to the system (the antenna or exciting coil). The Larmor frequency or resonance frequency is specific to each polar element. As examples, hydrogen resonance is 42.57 MHz/T and carbon resonance is 10.70 MHz/T (magnetic field in T, Tesla). The resonance of hydrogen is used in most medical MRI imagers.

18.9.2 Image Generation

The object of interest is placed in a strong magnetic field and a pulse of RF energy is applied at the Larmor frequency by an antenna placed so that the RF field is orthogonal to the magnetic field. Following the pulse, the nuclei will return to their random energy distribution state and return the excitation energy either to the RF transmitting antenna or to a second receiving antenna as nuclei return to their normal or rest state at the Larmor frequency. If the magnetic field has changed, the Larmor frequency will also change. If the magnetic field is arranged so that there is a slight gradient across an area of interest, if the exciting RF energy is swept through a narrow band of frequencies during a succession of pulses, or if the magnetic field is altered from a uniform field to one with a small gradient, the relaxation frequency will vary across the area of interest as will lines of atoms across the area of interest returning energy at slightly different frequencies. A multichannel receiver can produce signals for each line of excited atoms at each frequency. The system can be rotated and the process repeated until many sets of signal lines are obtained and sent to a computer for processing using techniques similar to those used for CT. The system does not have to be physically rotated but equivalently rotated by adjusting the direction of the magnetic field gradient.

Only one plane of atoms is excited at a time and then that plane of atoms is examined line by line to construct the image. For example, if the patient is in a magnetic field, which varies from head to toe and then a radio frequency field is applied, only those hydrogen atoms of the patient in a plane where the radio frequency and the magnetic field exactly match the Larmor frequency of hydrogen can accept radio frequency energy and become excited. That plane is orthogonal to the head–toe axis. Other patient atoms will not be excited. When the RF is turned off, exited atoms will return to their rest (unexcited) by radiating the RF energy of excitation at the Larmor frequency, which may be detected by a radio receiver. If the magnetic field is altered by having a small gradient from, say, right to left side of the excited plane, just as the RF source is turned off, then the radiated Larmor frequencies will also vary. A radio receiver can be made with detectors for the expected frequencies. This is a variant of a multichannel analyzer used for pulse height analysis of nuclear decay. For MRI, the output of each channel corresponds to the signal strength produced by just one line of the excited plane. Those signals can be processed in the same way as the detector signals of a CT system to generate the image.

The decay and energy return has two time constants: T_1, the exponential time constant to return to equilibrium with the applied magnetic field (also called spin-lattice relaxation time), and T_2, the exponential time constant to lose signal phase coherence with excitation RF (also called spin–spin relaxation time). This process of coded pulsing of the applied RF pulses and gating the received signals permits the generation of signals favoring T_1 or T_2 signals, which have different sensitivities or contrast for different tissue types. In addition, if the RF source is pulsed with time durations related to specific nuclear decay rates, other useful information can be derived to identify tissue types, blood velocity, etc.

18.10 ULTRASOUND

18.10.1 Acoustic Impedance and Reflection

Just as sound waves in water can be used to detect submarines, sound waves can be used for tissue imaging. While audible frequencies work for submarines, one wavelength at, say, 10 kHz, would be $\lambda = u/f$, where λ is the wavelength and u is the velocity of sound in water (1480 m/s) showing a wavelength of 14.8 cm, too large to image anatomic structures. To resolve structures down to 0.5 mm, the frequency would have to increase to about 3.0 MHz, an ultrasound frequency.

The acoustic impedance, Z, of a medium is a fundamental property related to the velocity of sound, u, and the density, ρ, of the material: $Z = \rho u$. The fraction of energy, R, reflected at the normal interface of two different tissue types is related to their impedances:

$$R = [(Z_2 - Z_1)/(Z_2 + Z_1)]^2.$$

Ultrasound signals decrease as functions of distance, geometry, and attenuation by the medium. The signal intensity, I, at distance, r, from the source, where α is the coefficient of attenuation is

$$I = I_0 = I_0\, e^{-\alpha r}/r^2,$$

where I_0 is the initial signal intensity.

18.10.2 Ultrasonic Imaging

An A-mode scan uses a piezoelectric ceramic transducer to transmit a 1 to 15 MHz pulse of ultrasound into the body. The reflected ultrasound returns to the transducer and generates a voltage amplitude that can be displayed versus transit time.

A Time–Motion (T–M) ultrasonic scan can display movements of the mitral valve of the heart versus time.

A B-mode scan can use spinning head transducers to scan the body in different directions, then display a composite two-dimensional cross sectional image of the body.

A phased array ultrasonic duplex scanner combines real-time two-dimensional imaging with the pulsed-Doppler method to measure directional blood velocity noninvasively (Siedband 2010).

It is often helpful to characterize a medium by its HVL, the thickness of the material required to diminish the signal by half.

18.11 QUALITY ASSURANCE

18.11.1 Need for Quality Assurance

Every machine will wear out eventually or drift away from proper settings. The apparent quality of an image is no assurance that the system is working properly as anatomical structures, not visible in the image, may not be capable of being resolved by the system: small blood vessels close to the image detector may be clear while those farther away may not be clear. Operation at the incorrect anode voltage of the x-ray tube may not produce high contrast images of soft tissue. Filters may have been removed by mistake during repairs or replacement, resulting in excessive exposure to patients. Normal wear and tear of an x-ray tube may cause the anode current to drift away from preset values, causing a change in anode voltage. The effective resistance of the high voltage transformer can be as high as 50 kΩ, causing a voltage drop of 5 kV for 100 mA current. This high resistance is an excellent limiter of short circuit current if the tube arcs (a normal failure mode) but also means that a tube miscalibrated by 200 mA will also be miscalibrated by 10 kV. Three principal methods are used to test medical imaging systems: patient simulating phantoms, test objects, and instruments. A phantom is any object used in place of a patient to produce a specific, expected image. A phantom may also incorporate test objects as resolution wedges, meshes of different holes/cm, and other objects useful for diagnosing system performance.

18.11.2 Phantoms and Test Objects

Plastic materials, which have x-ray attenuation characteristics close to water, materials that are similar to bone, air chambers, and various small parts to simulate blood vessels can be assembled to simulate the body for the evaluation of x-ray and CT systems. Specialized materials have also been developed for testing MRI machines. Gels of varying acoustic properties are used to evaluate ultrasound imagers, and ultrasound phantoms will also contain nylon fibers of various diameters and plastic spheres of various diameters and densities. CT phantoms are similar to x-ray test phantoms but will also contain small aluminum sheets set at angles so that their images can be used to determine slice thickness.

MRI phantoms are similar to CT phantoms except that phantom materials are chosen to simulate tissues of various hydrogen concentrations with plastic sheets set at angles to determine slice thickness. Some phantoms will also contain motorized objects to measure dynamic performance. Other test objects, for example, are plastic tubes filled with iodine compounds to test blood vessel imaging with iodine-bearing contrast agents. Other phantoms may contain other test objects to determine special performance characteristics. The important point is that phantoms are a substitute for patients, which eliminate concern for patient exposure to x-rays and provide constant, reference test images. Frequently used test objects are resolution stars, resolution bars and wedges, wires or fibers of varying diameter and density, plastic balls, metal plates with varying size holes, and metal or plastic absorbers (attenuators).

18.11.3 Test Instruments and Test Tools

In addition to the instruments described in Section 18.5, special instruments for calibrating x-ray systems have been developed. These include the following:

1. The mAs meter, an integrating milliammeter for reading the product of x-ray tube current and exposure time. This is required because exposure times are too short to be read on a milliammeter alone.

2. The kVp meter, an instrument that uses at least one pair of x-ray detectors preceded by different beam filters. The meter circuit determines the ratios of the detected signals to estimate the effective kV applied to the x-ray tube.

3. The timer test tool, an instrument for measuring x-ray exposure time.

4. Dosimeters of various forms, small devices containing either a small film or thermoluminescent material, which glows when heated in proportion to x-ray exposure. These are used to measure occupational exposure of equipment operators or exposure to patients during certain procedures.

5. Multiple mesh patterns are often made with eight pie-slice sections of various mesh resolutions (i.e., mesh lines/cm). These are useful refocusing or testing imaging systems for center-to-edge focus.

18.11.4 Test Procedures

Different test procedures are used for different purposes. There may be a requirement for hospital or other organization for periodic testing to standards to maintain institutional qualifications. There are government and professional society standards or other standards for certification or insurance purposes. There are manufacturer test procedures for calibration and adjustment. There may be special test procedures to verify that a machine performs as needed without exposing patients to radiation. An important point is that test procedures should be done on a regular, scheduled basis with written records for comparison to spot failure trends. Frequent failures or frequent requirements for calibration should cause an adjustment of routine quality assurance schedule and procedure.

REFERENCES

Brown M A and Semelka R C 2010 *MRI: Basic Principles and Applications*, 4th edition (Hoboken: Wiley-Blackwell)

Chen M, Pope T and Ott D 2010 *Basic Radiology*, 2nd edition (New York: McGraw Hill)

Dhawan P A 2003 *Medical Imaging Analysis* (Hoboken: Wiley-Interscience)

Gore R M and Levine M S 2008 *Textbook of Gastrointestinal Radiology*, 3rd edition (Amsterdam: Saunders)

Hajnal J V, Hawkes D J and Hill D L 2001 *Medical Image Registration* (Boca Raton: CRC Press)

Herring W 2012 *Learning Radiology*, 2nd edition (Amsterdam: Saunders)

Hornak J P 2014 *The Basics of MRI* https://www.cis.rit.edu/htbooks/mri/index.html

James A P and Dasarathy B V 2014 Medical image fusion: a survey of state of the art *Information Fusion* arXiv:1401.0166

Roobottom C A, Mitchell G and Morgan-Hughes G 2010 Radiation-reduction strategies in cardiac computed tomographic angiography *Clin. Radiol.* **65** 859–67

Shrestha R B 2011 Imaging on the cloud *Appl. Radiol.* **40** 8–12

Siedband M P 1968 X-ray image storage, reproduction and comparison system US patent 3,582,651.

Siedband M P 2010 Medical imaging systems in J G Webster (ed.), *Medical Instrumentation Application and Design*, 4th edition (Hoboken: John Wiley & Sons)

Udupa J K and Herman G T 2000 *3D Imaging in Medicine*, 2nd edition (Boca Raton: CRC Press)

Weiss R E, Egorov V, Ayrapetyan S, Sarvazyan N and Sarvazyan A 2008 Prostate mechanical imaging: a new method for prostate assessment *Urology* **71** 425–9

Yochum T R 2004 *Essentials of Skeletal Radiology* (Philadelphia: Lippincott Williams & Wilkins)

CHAPTER 19

Rehabilitation

John G. Webster

CONTENTS

19.1 Wearable Sensors for Rehabilitation   477
    19.1.1 Health and Wellness Monitoring   478
    19.1.2 Safety Monitoring   478
    19.1.3 Home Rehabilitation   478
    19.1.4 Assessment of Treatment Efficacy   478
    19.1.5 Early Detection of Disorders   478
    19.1.6 Dementia   479
19.2 Stroke   479
19.3 Fall Risk   480
19.4 Parkinson's Disease   480
19.5 Carpal Tunnel Syndrome   481
19.6 Gait Rehabilitation   481
19.7 Vascular Rehabilitation   482
19.8 Diabetic Foot   482
References   482

19.1 WEARABLE SENSORS FOR REHABILITATION

Patel et al. (2012) summarize wearable sensors and systems for rehabilitation. They note the need for monitoring the elderly and transmitting diagnostic information to their mobile phone or personal computer, then to the Internet, and then to the hospital. Wearable systems for patients' remote monitoring consist of (1) the sensing and data collection hardware to collect physiological and movement data, (2) the communication hardware and software to relay data to a remote center, and (3) the data analysis techniques to extract clinically relevant information from physiological and movement data. Sensors can measure heart rate, respiratory rate, blood pressure, blood oxygen saturation, and muscle activity. Sensors can be embedded in or attached to a garment to collect electrocardiographic and electromyographic signals by weaving electrodes into the fabric and to gather movement data by printing conductive elastomer-based sensors on the fabric and then measuring changes in their resistance associated with stretching of the garment due to a subject's movements.

Home ambient instrumented environments can include sensors and motion detectors on doors that detect opening of, for instance, a medicine cabinet, refrigerator, or the home front door. Feedback could be provided on the performance of rehabilitation exercises and checking whether the stove or the coffee machine has been switched off.

19.1.1 Health and Wellness Monitoring

An in-shoe pressure and acceleration sensor system can classify activities including sitting, standing, and walking with the ability of detecting whether subjects were simultaneously performing arm reaching movements. An accelerometer-based device can count steps in patients with Parkinson's disease (PD). Sensors can monitor heart rate, blood pressure, oxygen saturation, respiratory rate, body temperature, and galvanic skin response.

19.1.2 Safety Monitoring

Systems can detect falls and relay alarm messages to a caregiver or an emergency response team. Advanced microprocessors and accelerometers can monitor the body's position. Systems can detect falls as distinct events from normal movements, and automatically relay a message to the designated response center or nurse call station. Smartphones can provide localization of the person who fell via a global positioning system (GPS)-based method.

19.1.3 Home Rehabilitation

A medical back training device improves patient's compliance to achieve increased motivation by real-time augmented feedback based on trunk movements. It transfers trunk movements from two wireless sensors into a motivating game environment and guides the patient through exercises specifically designed for low back pain therapy. To facilitate challenging the patient and achieving efficient training, the exercises can be adjusted according to the patient's specific needs.

19.1.4 Assessment of Treatment Efficacy

Portable triaxial accelerometers placed on the shoulder can monitor the severity of dyskinesias in PD patients. Dyskinesias are a side effect of medication intake, and they can cause significant discomfort to patients. There is a correlation between accelerometer output and severity of dyskinesia in patients with PD. A biaxial accelerometer mounted on the lower back can measure gait features such as stride frequency, step symmetry, and stride regularity. Accelerometer data can provide objective information about real-world arm activity in stroke survivors. Two wireless accelerometers are placed on each leg to monitor walking in stroke survivors. Results showed that the system was able to monitor the quantity, symmetry, and major biomechanical characteristics of walking.

19.1.5 Early Detection of Disorders

Measurement of human movement in three dimensions over 3 days showed that the magnitude of the acceleration vector recorded in patients with chronic obstructive pulmonary disease (COPD) was correlated with measures of patients' status such as the 6 min walk distance, the forced expiratory volume in 1 s (FEV1), the severity of dyspnea, and the physical function domain of health-related quality of life scale. A multisensor system, which measured galvanic skin response, heat flow, and skin temperature in addition to motion, provided accurate estimates of energy expenditure.

19.1.6 Dementia

Remote monitoring can play an important role in the management of patients with dementia. Systems that can assist these patients with remembering daily activities and monitoring daily behavior for early signs of deterioration can allow patients to live independently longer. Such systems range from monitoring activities of daily living to tracking medication compliance to monitoring changes in social behavior.

19.2 STROKE

Pennycott et al. (2012) note that for stroke rehabilitation, neurodevelopmental training (NDT) is particularly prevalent, with the best well-known stream being the Bobath concept. This therapy attempts a holistic approach where emotional, social, and functional problems are targeted in addition to the main sensory-motor deficits. The general aim is to suppress abnormal movement synergies and move toward normal motor patterns. Despite the acceptance of NDT and other conventional rehabilitation techniques, evidence demonstrating their efficacy is lacking. Better outcomes in gait rehabilitation have been elicited from the more direct approach of body weight–supported treadmill training where the patient walks on a treadmill with his or her body weight partially supported, and one or more therapists support the patient and guide their limbs where required. This type of therapy has the advantages of being task specific and repetitive but is often very physically intensive. As a result, the training duration can be limited by the fitness of the therapists themselves. Robot-driven gait therapy, where the patient is aided by robotic actuators rather than a therapist, is becoming an increasingly prominent feature of rehabilitation worldwide. In addition to alleviating the physical load on therapists, a robot can accurately and objectively measure a patient's output, for example, in terms of joint kinematics and kinetics. Stationary walking systems, the Lokomat and the AutoAmbulator, are commercially available devices, which support patients to perform walking on a treadmill. The Lokomat consists of an exoskeleton in combination with a body weight support system and controls the joint angles at the knee and hip by means of linear actuators. The AutoAmbulator (HealthSouth, Birmingham, United States) employs robotic arms attached laterally to the patient for the control of the lower limbs. An alternative approach is to use foot plates to guide the feet and thereby reproduce gait trajectories. The Gait Trainer (Reha-Stim, Berlin, Germany) uses a crank and gear system to guide the feet, simulating stance and swing phases. The system can provide a varying degree of support to the patient, as can the Haptic Walker, which is designed for more arbitrary movements of the feet to simulate walking on different surfaces. Some robotic devices are mobile and thus offer over-ground walking. The KineAssist (Kinea Design, Evanston, United States) has a mobile base and provides partial body weight support and assistance for movements of the pelvis and torso while leaving the patient's legs unobstructed in order to allow therapist assistance.

19.3 FALL RISK

Howcroft et al. (2013) notes that approximately one third of over 65 year old will fall each year. Sensors that measure whole body motion, ground reaction forces, and electromyographic signals provide objective, quantitative measures for fall risk assessment. However, the associated equipment is typically located in a gait laboratory and requires a time-consuming setup that is difficult to practically integrate into typical clinic schedules. Small wearable sensors, gyroscopes (angular velocity) and accelerometers (linear acceleration), are inertial sensors. Successful single-variable models used: the mean squared modulus ratio for postural sway derived from lower back angular velocity, sit-to-stand transition duration determined from sternum acceleration and angular velocity, sit-to-stand fractal dimension derived from sternum acceleration and angular velocity, and sit-to-stand lower back jerk (derivative of acceleration). The mean squared modulus ratio for postural sway is the ratio of the rotational kinematic energy during eyes-open or closed postural sway on a foam surface to the rotational kinematic energy during eyes-open postural sway on a firm surface. The sit-to-stand fractal dimension represents the regularity of the sit-to-stand movement, with larger values associated with greater movement irregularity and fall risk. The five best performing models, in terms of overall accuracy, specificity, and sensitivity, used neural networks, naive Bayesian classifier, Mahalanobis cluster analysis, and a decision tree. In the majority of papers, sensors were placed on the lower back. However, the justification for this location is limited to an intention to approximate the body center of mass and the location's unobtrusive nature for long-term use. High subject acceptance for long-term lower back sensor placement was found in a 20-day case study. While the lower back is a promising sensor location, the upper back, hip, and thigh have potential as long-term sensor locations.

19.4 PARKINSON'S DISEASE

Moore et al. (2013) assessed freezing of gait (FOG), a paroxysmal block of movement when initiating gait. FOG entails an increase in high frequency (3 to 8 Hz) leg movement (*trembling*) in the relative absence of lower frequency (0 to 3 Hz) locomotor activity. Using seven sensors attached to the lumbar back, thighs, shanks, and feet and signal-processing parameters, patients performed 134 timed up-and-go (TUG) tasks. PD patients performed the TUG task, in which subjects start from a seated position and walk 5 m to a target box marked on the floor with tape, execute a turn, and return to the seated position. Subjects were instrumented with seven inertial measurement units (IMUs: Xsens MTx, Enschede, Netherlands) secured to the lumbar region of the back (approximately L2), the lateral aspect of each thigh and shank, and the superior aspect of each mid-foot, with elasticized straps such that the sensors were affixed to the segment as a rigid body. The IMUs were small ($38 \times 53 \times 21$ mm; 30 g) and did not interfere with natural movement. During testing, each IMU acquired triaxial linear acceleration (only longitudinal acceleration of each body segment was considered in this study), transmitted wirelessly to a computer at a sample rate of 50 Hz. Synchronization of the video and accelerometer recordings was performed prior to data collection by the alignment of the video camera and data-acquisition computer clocks. Three parameters must be considered: the location and number of sensors, the width of the sampling window for determining the ratio of freeze (3 to 8 Hz) to locomotor (0 to 3 Hz) band power at each point in time, and the threshold above which this ratio indicates an FOG event. Evaluation of FOG in clinical practice is notoriously difficult due to its paroxysmal nature, and current subjective approaches lack utility. There is a clear need for the development of objective standardized measures for the assessment of future pharmacological and nonpharmacological interventions, and the results presented here provide a practical basis for the implementation of accelerometry-based FOG monitoring. The seven-sensor approach was both the most complex and most robust system, scoring highly on all measures of performance, and is better suited to demanding research applications. A single shank sensor provided FOG measures comparable to the seven-sensor approach but is considerably simpler in execution, facilitating clinical use. A single lumbar sensor exhibited only moderate agreement with the clinical raters in terms of absolute number and duration of FOG events but had high sensitivity and specificity as a binary test of FOG and lent itself readily to ambulatory monitoring given the ubiquitous nature of accelerometers in mobile telephones and other belt-worn devices.

19.5 CARPAL TUNNEL SYNDROME

Heuser et al. (2007) describe the carpal tunnel as a narrow rigid passageway of ligaments and bones at the base of the hand by which finger flexor tendons are allowed to translate during grasp. The carpal tunnel syndrome occurs when swelling causes the median nerve to become compressed within the carpal tunnel. The resulting symptoms are numbness and tingling of the thumb, index, middle, and ring fingers. The Rutgers Masters II haptic glove was tested on five subjects, who were two weeks post–hand surgery. Patients performed a fixed sequence of virtual telerehabilitation (therapist was remote from the patient), ball squeezing grasps, power putty, and DigiKey, and then the difficulty was increased for later sessions. Improvement in their hand function was also observed (a 38% reduction in virtual pegboard errors, and 70% fewer virtual hand ball errors). Clinical strength measures showed increases in grip (by up to 150%) and key pinch (up to 46%) strength in three of the subjects, while two subjects had decreased strength following the study. However, all five subjects improved in their tip pinch strength of their affected hand (between 20% and 267%). When asked whether they would recommend the virtual reality exercises to others, four subjects very strongly agreed and one strongly agreed that they would.

19.6 GAIT REHABILITATION

Luu et al. (2014) developed a robotic-assisted gait rehabilitation system for functional recovery of walking in spinal cord injury and stroke patients. It is designed as a pair of anthropomorphic legs. Two robotic arms hold the subject at both sides of pelvis, which provides unrestricted pelvic movements to promote body weight shifting. The robotic orthosis has a total of six degrees of freedom (DoFs) with three DoFs on each side. The system provides the following: (1) *Pelvic motion*—Unrestricted pelvic movements play an important role in normal locomotion, and thus, it is included as one of the essential features in this work. It promotes body weight shifting during gait rehabilitation. (2) *Body weight support*— Unique body weight support approach, without restricting pelvic movements. (3) *Overground mobility* provides gait practicing in functional over-ground walking, as opposed to treadmills. (4) *Reciprocal stepping* assists motion of hip, knee, and ankle in sagittal plane and provides actual functional over-ground walking. The results show a 10 m walking trial yielded a reduction in manpower. The task-specific repetitive training approach and natural walking gait patterns were also successfully achieved.

19.7 VASCULAR REHABILITATION

Andrews et al. (2005) describe that vascular diagnostic testing is typically performed to confirm a clinical diagnosis and document the severity of disease. The ankle-brachial index (ABI) is measured using blood pressure cuffs around the lower calves or ankles and arm. A handheld Doppler detects systolic blood movement. When ABI < 0.4, the 5-year mortality rate approaches 50%. Photoplethysmography can be used to show distal disease in the hands or feet. Continuous wave Doppler can confirm arterial patency or identify occlusive lesions. Transcutaneous oximetry ($TcPO_2$) can assess skin perfusion and the potential for healing. Duplex scanning using B-mode imaging combined with directional Doppler can visualize and assess arterial aneurysms and detect flow velocity changes as sites of localized stenosis or occlusion. Magnetic resonance angiography (MRA) can identify arteriostenoses and occlusions. Contrast arteriography is employed to plan vascular reconstruction. Rehabilitation consists of self-care—monitoring extremities carefully for redness or skin breakdown and proper footware. Exercise augments limb flow. Intermittent venous occlusion with an externally applied inflatable cuff increases skin blood flow.

19.8 DIABETIC FOOT

Shah and Patil (2005) developed an optical pedobarograph with a glass–plastic interface to measure the foot pressure distribution for subjects standing or walking. Scattering of light at the glass and plastic sheet interface is converted to a high-resolution intensity image. The intensity is proportional to the applied pressure. The light image is converted to an analog electrical signal. This is connected to an International Business Machines Personal Computer-Advanced Technology (IBM PC-AT), which grabs the image and loads data as the subject stands or walks over a glass platform mounted on the wooden box standing on load cells. Foot areas of diabetic subjects are scanned in 10 specified areas using Semmes–Weinstein nylon monofilaments to determine quantitatively the degree of neuropathy as per the following classification (1) Normal: sensation level of 3 g force exerted by the nylon monofilament. (2) Diminished light touch: sensation level of 4.5 g force exerted by the nylon monofilament. (3) Diminished protective sensation: sensation level of 7.5 g force exerted by the nylon monofilament. (4) Loss of protective sensation: sensation level of 10 g force exerted by the nylon monofilament.

REFERENCES

Andrews K L, Gamble G L, Strick D M, Gloviczki P and Rooke T W 2005 Vascular diseases in J A DeLisa (ed.), *Physical Medicine and Rehabilitation*, 4th edition, Vol 1 (Philadelphia: Lippincott Williams & Wilkins)

Heuser A, Kourtev H, Winter S, Fensterheim D, Burdea G, Hentz V and Forducey P 2007 Telerehabilitation using the Rutgers Master II Glove following carpal tunnel release surgery: proof-of-concept *IEEE Trans. Neural Syst. Rehabil. Eng.* **15** 43–9

Howcroft J, Kofman J and Lemaire E D 2013 Review of fall risk assessment in geriatric populations using inertial sensors *J. Neuroeng. Rehabil.* **10** 91

Luu T P, Low K H, Qu X, Lim H B and Hoon K H 2014 Hardware development and locomotion control strategy for an over-ground gait trainer NaTUre-Gaits *IEEE J. Transl. Eng. Health Med.* **2** 1–9

Moore S T, Yungher D A, Morris T R, Dilda V, MacDougall H G, Shine J M, Naismith S L and Lewis S J G 2013 Autonomous identification of freezing of gait in Parkinson's disease from lower-body segmental accelerometry *J. Neuroeng. Rehabil.* **10** 19

Patel S, Park H, Bonato P, Chan L and Rodgers M 2012 A review of wearable sensors and systems with application in rehabilitation *J. Neuroeng. Rehabil.* **9** 21

Pennycott A, Wyss D, Vallery H, Klamroth-Marganska V and Riener R 2012 Towards more effective robotic gait training for stroke rehabilitation: a review *J. Neuroeng. Rehabil.* **9** 65

Shah S R and Patil K M 2005 Processing of foot pressure images and display of an advanced clinical parameter PR in diabetic neuropathy *Proceedings of the 2005 IEEE Ninth International Conference on Rehabilitation Robotics* Chicago

CHAPTER 20

Urology

Michael Drinnan and Clive Griffiths

CONTENTS

20.1	Introduction		487
	20.1.1	What Is Urodynamics?	487
		20.1.1.1 Disorders of Storage	489
		20.1.1.2 Disorders of Voiding	489
	20.1.2	Physiological Parameters of Interest	490
	20.1.3	Conventional Urodynamics	490
20.2	Urine Flow and Voided Volume		490
	20.2.1	Short History of Uroflowmetry	492
	20.2.2	Dipstick Flowmeter	492
		20.2.2.1 Implementation	492
		20.2.2.2 Practicalities	494
	20.2.3	Load-Cell Flowmeter	495
		20.2.3.1 Implementation	495
		20.2.3.2 Practicalities	495
	20.2.4	Spinning Disk Flowmeter	496
		20.2.4.1 Implementation	496
		20.2.4.2 Practicalities	497
	20.2.5	Summary of Attributes of Flowmeters	497
20.3	Uroflow Signal Conditioning and Processing		498
	20.3.1	Analog Signal Processing	498
	20.3.2	Digital Signal Processing	498
20.4	Flowmeter Calibration		499
20.5	Measuring Flowmeter Frequency Response		500
20.6	Uroflowmetry and Natural (Physiological) Filling		500
20.7	Recording and Display of Flow and Volume Data		500
	20.7.1	Accuracy and Sampling Precision	500
		20.7.1.1 Voided Volume	500
		20.7.1.2 Flow Rate	501
	20.7.2	Bandwidth and Sampling Rate	501
	20.7.3	Visual Presentation	501

20.8   Interpretation of Flow and Volume Data   501
         20.8.1   Summary Uroflowmetry Statistics   501
                  20.8.1.1 Total Voided Volume V_{void} 501
                  20.8.1.2 Maximum Flow Rate Q_{max} 502
         20.8.2   Flow Artifacts   502
                  20.8.2.1 Instrumental Artifacts   503
                  20.8.2.2 Artifacts of Patient Origin   503
         20.8.3   Automated Analysis of Uroflowmetry   505
         20.8.4   Clinical Value of Uroflowmetry   505
20.9   Bladder Pressure during Filling and Voiding   506
         20.9.1   Short History of Cystometry   506
         20.9.2   Detrusor Pressure versus Vesical Pressure   508
20.10   Measurement of Urodynamic Pressure   509
         20.10.1 Pressure Sensors   509
         20.10.2 Water-Filled Catheter and External Sensor   510
         20.10.3 Air-Filled Catheter and External Sensor   510
         20.10.4 Catheter-Tip Sensors   511
         20.10.5 Measuring Pressure in the Rectum or Other Dry Spaces   512
         20.10.6 Pressure References   512
         20.10.7 Summary of Attributes of Sensor Arrangements   512
20.11   Pressure Signal Conditioning and Processing   513
         20.11.1 Analog Signal Processing   513
         20.11.2 Digital Signal Processing   513
20.12   Pressure Sensor Calibration   513
20.13   Measuring Pressure Sensor Frequency Response   515
20.14   Cystometry and Artificial (Non-Physiological) Filling   515
         20.14.1 Measuring Filled Volume   516
         20.14.2 Filling Catheters   517
20.15   Recording and Display of Pressure Data   517
         20.15.1 Accuracy and Sampling Precision   517
         20.15.2 Bandwidth and Sampling Rate   517
         20.15.3 Visual Presentation   518
20.16   Interpretation of Pressure Recordings   518
         20.16.1 Quality Control   518
         20.16.2 Normal Cystometry   519
         20.16.3 Detrusor Overactivity   519
         20.16.4 Stress Incontinence (Women)   520
         20.16.5 Outlet Obstruction (Men)   520
                  20.16.5.1 ICS Nomogram   521
20.17   Pressure Artifacts   521
         20.17.1 Filling Artifacts   521
         20.17.2 Dislodged Catheter   522

20.17.3	Bubbles	522
20.17.4	Rectal Activity	523
20.18	Complete Urodynamics System	523
20.18.1	Load Cell and Preamplifier	523
20.18.2	Volume and Flow Circuitry	523
20.18.3	Pressure Circuitry	526
20.19	Adjuncts to Conventional Urodynamics	526
20.19.1	In Vitro Testing	526
20.19.1.1	Urine Dipstick Testing	526
20.19.1.2	Prostate-Specific Antigen	526
20.19.2	Simple Devices for Home Use	526
20.19.2.1	Frequency–Volume Chart	526
20.19.2.2	Leakage Diaper	526
20.19.3	Ultrasound	527
20.19.4	Ambulatory Urodynamics	527
20.19.5	Electromyography	529
20.19.6	Video Urodynamics (Cystography)	530
20.19.7	Cystoscopy	530
20.19.8	Urethral Pressure Profilometry	530
20.20	The Future	531
20.20.1	Home Uroflowmetry	532
20.20.2	Noninvasive Measurement of Bladder Contractility	532
20.20.3	Indwelling Sensors	532
20.21	Equipment Suppliers	533
Acknowledgments		533
References		533
Bibliography		534

20.1 INTRODUCTION

20.1.1 What Is Urodynamics?

The urological symptoms illustrated in Figure 20.1 affect more than half of us at some time in our lives. Unfortunately, many symptoms are not specific to a particular disease process. Cancers are naturally the most feared of the diseases and are always ruled out first, but thankfully they are responsible for symptoms in a minority of cases. Less than 0.1% of the population every year present with a new case of prostate or bladder cancer. For the majority of patients, some further investigations are needed. In common with many physiological measurements, urodynamics has two aims: first, to reproduce and make an objective measurement of bothersome symptoms and, second, to make a rapid and specific diagnosis to inform appropriate treatment.

In order to understand the physiological measurements that play a part in urodynamics, it is helpful first to consider the function of the healthy urinary tract. The urinary system has two distinct functions: the storage of urine for typically a period of hours

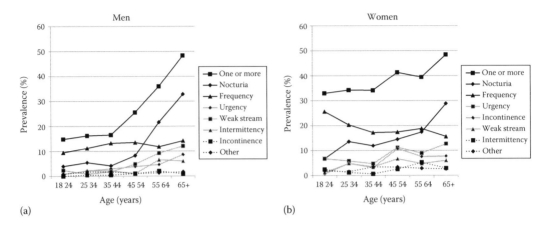

FIGURE 20.1   The rising prevalence of common urological symptoms with age, for (a) men and (b) women. (From Pinnock, C.B. and Marshall, V.R., *Med. J. Aust.*, 167, 72, 1997.)

and the efficient and complete passage or voiding of urine at a socially convenient time. Symptoms of disrupted storage or voiding go under the generic term of lower urinary tract symptoms (LUTS). Many of the symptoms are common to a range of disease processes. Nevertheless, the literature makes a clear distinction between the symptoms and disease processes affecting storage and those that affect voiding.

In simple terms, urine storage and voiding is mediated by just two factors: pressure and opposition to flow. The driving pressure to expel urine comes from the bladder detrusor muscle and from abdominal pressure. To a close approximation, these sum to give the vesical (bladder) pressure p_{ves}; note that the contribution from abdominal pressure is measured even when the bladder is not actively contracting.

The opposition to flow is presented by the bladder outlet, which includes active elements such as the sphincters and passive obstructions posed by the prostate gland in men or surgical interventions in some women. It is convenient to think of this in terms of a so-called *opening pressure*. Flow takes place, whether or not it is convenient, when the driving pressure p_{ves} exceeds the opening pressure of the outlet. Many disorders of the lower urinary tract can be understood in terms of this simple hydrodynamic model (Figure 20.2).

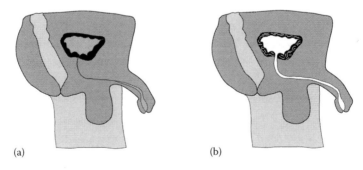

FIGURE 20.2   Under most circumstances, the patient is continent (a). Flow takes place, whether desired or not, when the pressure in the bladder is high enough to overcome the opening pressure of the urethra and sphincters (b).

20.1.1.1 Disorders of Storage

The bladder is the organ responsible for urine storage. Urine is produced by the kidneys at a rate of typically 1 to 4 mL min^{-1}, and so a functional bladder capacity of 300 to 500 mL is sufficient to store some hours of urine production. During filling and storage, continence is maintained primarily by the sphincters and pelvic floor muscles. In order to remain continent, the opening pressure of the sphincters must exceed the pressure p_{ves} in the bladder, including any contribution from abdominal pressure. Incontinence by leakage of urine is the result when the opening pressure of the outlet is not high enough to prevent flow.

For women, their shorter urethra, weaker muscles, and disruption of the anatomy typically due to childbirth make them more susceptible to incontinence. The day-to-day symptoms include leakage of urine during any maneuver that increases abdominal pressure. Typically, patients with this form of incontinence (genuine stress incontinence) will report leakage on coughing or during exercise.

Incontinence may also be due to detrusor overactivity, where the bladder contracts inappropriately and overcomes the outlet opening pressure. This can affect men or women and may be partly psychological in origin; latchkey incontinence where a patient gets a strong urge or leak in anticipation of a nearby toilet, or similar urges when hearing the sound of running water, are common. The symptom of overactive bladder may be actual incontinence or a poor working bladder capacity adopted by the patient as a compensation strategy.

20.1.1.2 Disorders of Voiding

Just as storage disorders are more common in women, voiding problems are more common in men. Difficulties in voiding are reported when the bladder contraction is insufficient to open the outlet fully for the whole void period. Women have a short and relatively straight urethra that provides little opposition to flow, and so voiding difficulties are uncommon in women with normal anatomy and physiology. Indeed when the sphincters are relaxed, women sometimes void by gravity alone with no detectable contraction of the detrusor muscle. However, women can report difficulty emptying their bladder after surgery for the treatment of incontinence, which creates a passive urethral obstruction.

The primary cause of bladder outlet obstruction in men is benign prostatic hyperplasia (BPH), a natural and benign change of the prostate gland that is diagnosed histologically. Benign prostatic enlargement (BPE) describes an increase in prostate size, often secondary to BPH. In about one-third of older men, BPE progresses to the point where it causes benign prostatic obstruction. Signs and symptoms include: weak, hesitant, or intermittent urine flow, incomplete emptying, and as a consequence a poor functional bladder capacity. The most bothersome symptoms are reported to be frequency (frequent visits to the toilet) and nocturia (night-time visits), which have a big impact on the quality of life for active older men.

As with storage, there are voiding disorders that compromise the contraction of the bladder detrusor muscle and therefore can affect men and women alike. Detrusor muscle failure can be the result of old age or neurological injury. Less commonly, sphincter dyssynergia where the bladder contracts but the sphincters do not relax has a similar effect to outlet obstruction. It is seen in neurological conditions such as multiple sclerosis.

20.1.2 Physiological Parameters of Interest

In order to assess LUTS, it follows that we are interested in the following physiological parameters:

- The volumes of urine stored in the bladder and passed from the bladder during voiding

- Urine flow, either of leaked urine during storage or the flow passed during voiding

- Bladder pressure, both during storage and during voiding

These are the principal parameters of interest, and they are measured routinely in urodynamic investigations. While other useful measurements have their place later, we will consider volume, flow, and pressure first.

20.1.3 Conventional Urodynamics

Figure 20.3 shows a urodynamic system and a conceptual view of the components: a load-cell flowmeter with stand, funnel, and jug; a load cell to measure infused volume; two external pressure sensors; amplifiers, filters, and signal conditioning electronics; and a computer for recording, display, and analysis of the urodynamic signals. This suite of physiological measurements would be included in a full conventional urodynamic system. This full urodynamic investigation might be cited as cystometry, cystometrogram (CMG), or pressure-flow studies (PFS).

We will now consider the components of the cystometry system, before considering how the measurements are combined to assess symptoms and make a specific diagnosis.

20.2 URINE FLOW AND VOIDED VOLUME

We will begin with uroflowmetry. The urine flowmeter forms part of the full cystometry system, but in addition, it is the single most-used stand-alone urodynamic test. In some simple tests that are suitable for home use, either volume or flow may be measured in isolation. In the clinic, flow (Q) and voided volume (V_{void}) are almost invariably measured together in the same instrument. Flow rate (Q) and volume (V) are related by simple calculus:

$$\text{Flow rate is the rate of change of volume } Q = \frac{dV}{dt}$$

$$\text{Volume is the time integral of flow rate } V = \int Q \, dt$$

The terminology and practice for urodynamics have been defined largely by the International Continence Society (ICS), particularly in their 2002 standardisation documents on Good Urodynamic Practices and Standardization of Terminology. According to Good Urodynamic Practices, the recommended units are mL for volume and mL s^{-1} for flow rate, because for most patients, they give manageable numbers.

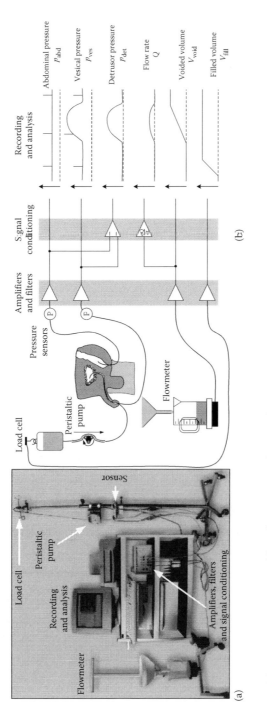

FIGURE 20.3 A cystometry system (a) and a conceptual view of its components (b).

In most instruments, the volume is measured first, and then flow is calculated by differentiation. The notable exception is the spinning disk, which measures flow rate and then calculates volume by integration. We will now describe briefly the history leading to the three most-used implementations of the urine flowmeter.

20.2.1 Short History of Uroflowmetry

In a very readable summary, Perez and Webster (1992) have provided a history of urodynamics to the present date. It is clear that until the end of the nineteenth century, the urologist's armoratorium consisted of the physical examination and very little else. The ancestors of what we would recognize today as uroflowmetry were first described in 1912, where Ramon Guitaras described his patients voiding into glass vessels:

> The size, shape and force of the stream is noted … a normal man will pass a fairly large stream, projecting from his body from three to five feet.

Edgar Ballenger reinforced the value of *cast uroflowmetry* in 1932, opining that men over 45 years of age should:

> Make the test from time to time, preferably when alone or in the country or out by the barn.

Importantly, Guitaras suggested the clinical value of uroflowmetry, which undoubtedly helped its development. Note that he has recognized the two principal causes of voiding disorders. But importantly, he has recognized that uroflowmetry alone is not adequate to make a differential diagnosis:

> A stream which slowly becomes smaller and less forcible points to either some obstruction…or a lack of tone in the bladder.

The first true uroflowmeter was described by Willard Drake in 1948 (Figure 20.4). It was a simple mechanical apparatus where the weight of urine in a collecting vessel was transferred to a stylus that wrote a permanent record on a revolving drum. In that respect, it might perhaps be more accurately described as urovolumetry. Then in 1956, von Garrelts used an electronic sensor in place of the mechanical linkage, and the first modern uroflowmeter was born. All our modern flowmeters owe a lot to these early designs.

20.2.2 Dipstick Flowmeter

In the simplest type of electronic flowmeter, a dipstick is arranged such that some electrical property varies with urine height. The height can be converted to volume simply by multiplying by the cross-sectional area of the vessel. Note that in this context, the term *dipstick* should not be confused with the urine dipstick test, a well-used microbiology test to detect abnormalities such as infection in the urine.

20.2.2.1 Implementation

Whereas a resistance dipstick has been proposed, it is quite difficult to realize in practice. Therefore, in most practical devices, capacitance is the measurand of choice.

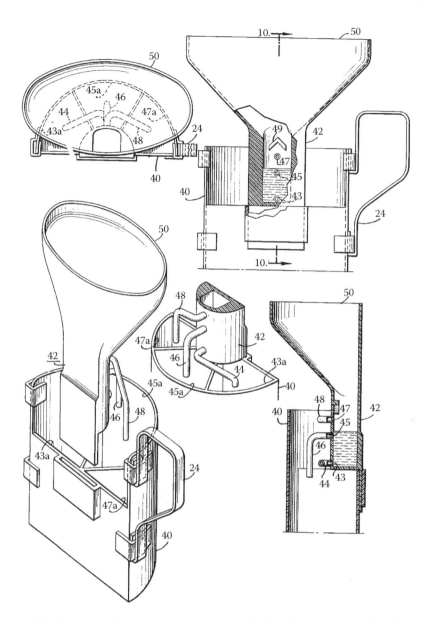

FIGURE 20.4   The Drake uroflowmeter, dating from 1948. This illustration is taken from US patent 2 648 981.

The capacitance of a parallel plate capacitor is directly proportional to its cross-sectional area. It is easy to construct a parallel plate capacitor such that the dipstick forms one plate and the second plate is the urine under test (Figure 20.5). Typically, a length of copper-clad printed circuit board with a water-impervious dielectric coating is used. If the dipstick is now stood on edge in the collecting vessel, the capacitance will be directly proportional to the contact area of urine with the dipstick and hence the height of urine in the vessel.

Using a lacquer or self-adhesive film as the dielectric, the capacitance is typically in the range 50 pF to 1 nF. This is large enough to be measured by standard capacitance

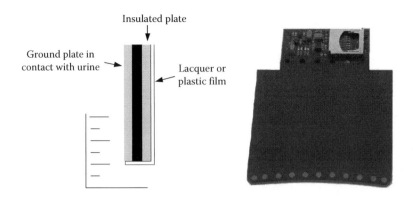

FIGURE 20.5 Construction detail of sensor for dipstick flowmeter. On the right is a working design—contact with urine is made by the conducting pads visible along the bottom of the dipstick.

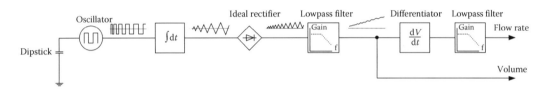

FIGURE 20.6 Block diagram for a dipstick uroflowmeter.

measurement techniques. A block diagram for a working implementation is given in Figure 20.6. The capacitor is built into an oscillator, whose time constant is proportional to C and hence to urine volume. The period-to-voltage conversion is achieved by an integrator, which generates a triangle wave whose amplitude is proportional to the timer period of the oscillator. This is rectified to give a DC level and then filtered to give volume.

20.2.2.2 Practicalities
So long as the cross-sectional area of the vessel is known, the capacitance dipstick gives a direct indication of urine volume. It will therefore work with any conductive liquid, including contrast medium. Though distilled water is not a good conductor, the mineral content of tap water is in practice sufficient to make a measurement. Nevertheless, the dipstick has significant practical problems. The vessel needs preloading with a small volume of saline to avoid a dead spot at the start of voiding and produces a noisy signal because of ripples in the urine. There are also difficulties with measuring reliably the relatively small capacitances involved. The dipstick can be susceptible to interference, for example, due to the presence of nearby body parts. Most surface dielectric coatings are to some extent hydroscopic, meaning that the dipstick capacitance slowly increases when immersed in water. And finally, a dipstick quickly goes out of calibration if not cleaned regularly, because crystals of urea and other solutes precipitate on the dielectric. These problems mean that in practice, the dipstick flowmeter has been superseded in most clinical applications.

20.2.3 Load-Cell Flowmeter

The load-cell flowmeter is currently the most common implementation, being used by the majority of stand-alone flowmeters and full urodynamic systems on the market. This device is also known as the gravimetric flowmeter, giving a hint as to its principle of operation; it is in essence a urine weighing scale. Since the density of urine is within 3% of water, the weight of the urine can be converted with good accuracy to volume and then to flow rate.

20.2.3.1 Implementation

The term *load cell* is used to mean an electromechanical component that transduces a force to a proportional electrical quantity. Many designs are available, typically using a billet of aluminum such as shown in Figure 20.7. The transduction takes place in one, two, or four strain gages mounted in the cantilevers. In the design shown, there are four strain gages connected as a Wheatstone bridge. When force is applied to the load cell, the electrical resistance of two diagonally opposite arms of the bridge increases, and that of the other two arms decreases.

The voltage output of this arrangement is proportional not only to the applied force but also to the bridge excitation voltage. Sensitivity is expressed in millivolts of output, per volt of excitation voltage, per kg force applied. A typical load cell for uroflowmetry would have a full-scale range of 2 kg with a sensitivity of 1 mV V^{-1} kg^{-1}. With an excitation of 5 V, this is equivalent to 5 $\mu V mL^{-1}$ of urine. Further amplification is required using a differential instrumentation amplifier before the usual signal conditioning circuits.

20.2.3.2 Practicalities

The load-cell flowmeter is easy to clean, since only the collecting vessel comes into contact with urine. The load cell measures weight and unlike the dipstick will work with any shape and size of collecting vessel. However, it needs calibration if using contrast medium for video urodynamics, which can be 10% more dense than urine. Similarly, there will

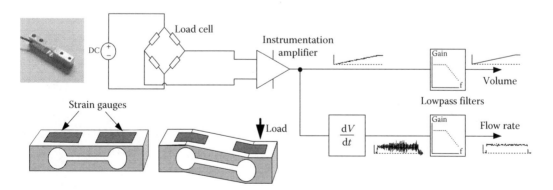

FIGURE 20.7   Block diagram for a strain gage flowmeter. At the bottom left is an exaggerated view of the cantilever strain gage mechanism: one strain gage stretches and the other shortens. Strain gages can also be fitted under the load beam in an arrangement with 4 strain gages. The photograph shows a load cell from the authors' homegrown flowmeter.

be errors due to hydration status, since urine from a dehydrated patient can be up to 3% denser than it would be normally. This error cannot easily be corrected, but is clinically not important.

Load cells are precision mechanical components. Even when fitted with a mechanical end stop, they are inherently susceptible to mechanical damage. A dropped load cell may not survive the experience and should be treated with suspicion. It is recommended to check the calibration of a load-cell flowmeter regularly, preferably before each session. In practice, a modern strain gage is far more susceptible to errors in its zero offset than in its sensitivity. So long as the flowmeter can be set to read zero volume before the study starts, it is probably not clinically important if a calibration check is occasionally forgotten.

20.2.4 Spinning Disk Flowmeter

The spinning disk or momentum-flux flowmeter is currently made by just one manufacturer (formerly Dantec, now Mediwatch) but has been very well used in the clinic and particularly the research setting. The instrument is arranged such that the urine stream falls on a rapidly spinning disk. The urine is quickly thrown off the disk but gains kinetic energy from the disk that tends to slow the disk down. The amount of energy gained by the urine is directly proportional to its mass and hence to its volume.

20.2.4.1 Implementation

The practical implementation of the spinning disk flowmeter is rather complicated. A servomotor is used to keep the disk spinning at constant speed. Therefore, the urine is flung off the rim of the disk at a constant speed v. The kinetic energy of a mass m of urine is $1/2\ mv^2$, and the servomotor must supply this energy. The engineer is faced with the technical problem of measuring the power demanded by the servomotor, which is directly proportional to the mass flow rate.

In one implementation (Figure 20.8), we can use a variation on the phase-locked loop. Conceptually, a servomotor is a voltage-controlled oscillator; feedback from a shaft encoder tachometer is the output of the oscillator. The phase-locked loop will adjust the

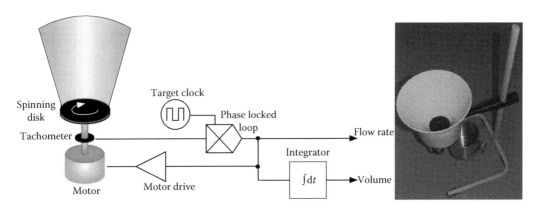

FIGURE 20.8 Block diagram of a spinning disk flowmeter and a picture looking into the commercial implementation.

control voltage in order to maintain the spinning disk at fixed speed set by the target clock. The control voltage relates to the power being supplied to the motor and with some additional processing gives a measurement of flow.

Note that this a much-simplified assessment of the design. There are technical problems related to the measurement of drive power. In addition, integrators are particularly sensitive to small input offset voltages that equate to a low but long-term constant flow into the device. The authors do not know the commercial solution to this problem but are aware of reports that a dribbling flow can be missed by the commercial device.

20.2.4.2 Practicalities

Uniquely, the spinning disk flowmeter does not require a collecting vessel and can be fitted directly over a toilet. Nevertheless, it is not easy to clean and has become less popular due to hygiene and infection control; fecal incontinence poses its own special risks with the spinning disk.

The spinning disk flowmeter measures mass flow and not volume flow. Therefore, as with the load cell, it needs calibration for contrast medium. Unlike the load cell, the primary measurand is flow, which must be integrated to give volume. For reasons explained later, this makes it less prone to some artifacts but at the expense of errors in voided volume, which can be clinically significant at very low flow rates.

20.2.5 Summary of Attributes of Flowmeters

Table 20.1 summarizes the key attributes of flowmeters.

TABLE 20.1   Summary of Key Attributes of Urine Flowmeters

	Capacitive or Resistive Dipstick	Load Cell (Gravimetric)	Spinning Disk (Momentum Flux)
One-Off Cost	Very cheap.	Medium—load cell element can be expensive.	High—technically difficult to engineer.
Measurand	Urine height. Requires vessel of known cross section. Differentiate to get flow.	Mass. Differentiate to get flow.	Mass flow. Integrate to get volume.
Re-calibration	Required if vessel changes.	Required for contrast medium.	Required for contrast medium.
Hygiene	Moderately easy, but must be cleaned well for correct calibration.	Easy—rinse a jug.	Can be hard, particularly in cases of fecal incontinence.
Response Time	Typically 1 s, due to mechanical delays in funnel and low-pass filtering.	As for dipstick.	Best response, because of minimal funnel and direct measurement of flow.
Artifacts	Start up artifact unless vessel is precharged with water. Wagging.	Sensitive to being knocked and to wagging. Best with baffle.	Good immunity to knocking and wagging.
Typical Use	Not widely used today.	General-purpose, clinical, and home use.	In-clinic urodynamics and research, particularly where good flow resolution is needed.

20.3 UROFLOW SIGNAL CONDITIONING AND PROCESSING

Since the load cell is the basis of most modern flowmeters, we will consider it first. Nevertheless, the principles apply equally to the dipstick and any flowmeter where flow is derived from volume. Typically, there are two steps in the signal processing—differentiation and low-pass filtering. These can be achieved in analog hardware but since the signal bandwidths are relatively low, it is possible to implement them using digital signal processing algorithms.

20.3.1 Analog Signal Processing

Flow rate is the time derivative of voided volume V_{void}. Since V_{void} is normally represented by an analog voltage, then the textbook operational amplifier differentiator circuit can be used to obtain flow rate. This flow rate signal is too noisy to be useful. This reflects the fact that flow is not continuous; by the time the stream reaches the meter, it has broken into a series of droplets. Therefore, a stage of low-pass filtering with a typical cutoff frequency of 1 Hz is added, as shown in Figure 20.9.

20.3.2 Digital Signal Processing

In theory, it is possible to digitize the voided volume directly using an analog-to-digital converter and to implement the differentiation and filtering algorithms in software. Digital filters have many advantages, for example, the potential to create filter characteristics that cannot be achieved in analog hardware. However, a brief consideration of the practicalities indicates that it is more difficult than it might first seem. Differentiation is in theory and in practice a noisy process, because the gain of a differentiator increases proportionally to

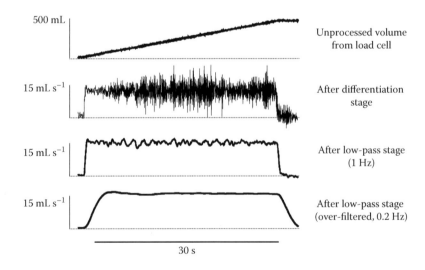

FIGURE 20.9   The effects of differentiating, low-pass filtering, and overfiltering (0.2 Hz) the volume signal from a load cell flowmeter. Flow input was from a 500 mL calibrated milk bottle as described later.

frequency, and noise tends to be prevalent at high frequencies. In uroflowmetry, the differentiator interacts with the quantization noise of the analog-to-digital converter in an unexpected way. Consider a flowmeter with 2 L capacity and a 10-bit analog-to-digital converter. The volume resolution is approximately 2 mL, being 2 L divided by 2^{10} quantization intervals. This is more than adequate for reporting volume. However, we can only detect flow when the volume has changed by at least 2 mL from its previous value. This is inadequate; practical experience tells us that >12 bits of resolution is necessary to achieve the subjective impression of an analog signal. The situation can be improved by oversampling and signal processing, but this is beyond the intended scope of this chapter.

20.4 FLOWMETER CALIBRATION

It is good practice to check the calibration of a flowmeter regularly, preferably at the start of every session. It is easy to generate a known volume, but not a known flow rate; when water flows from an open vessel such as a funnel, the flow rate decreases as the vessel empties. A useful alternative is shown in Figure 20.10, comprising a bottle with known volume, rubber bung to fit the bottle, short length of rigid tube, and longer length of flexible neoprene or silicone rubber tubing. In use, the inside face of the bung is fixed at atmospheric pressure because it is vented at this level by the rubber tubing. In principle, it is possible to calculate the flow rate:

$$Q = A\sqrt{2gl}$$

where
    Q is the constant flow rate
    A is the cross-sectional area of the pipe
    l is the length of the pipe
    g is the gravitational constant

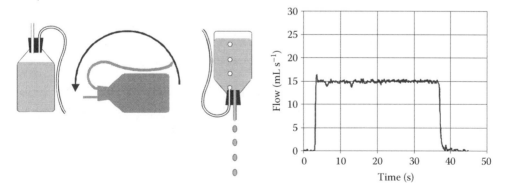

FIGURE 20.10 The *milk bottle* constant flow test. An example flow rate produced by the device is shown. (Griffiths, C.J., Murray, A., and Ramsden, P.D., *Brit. J. Urol.*, 55, 21, 1983.)

Thus, the flow rate depends only on the length and diameter of the metal pipe. In practice, there are significant resistive and turbulent losses; it is easier to make the device and measure the flow, which for a constant flow is the total volume divided by total time to empty.

20.5 MEASURING FLOWMETER FREQUENCY RESPONSE

The frequency response of a uroflowmeter can be estimated by measuring its step response (Gammie et al. 2014). The simplest way to do so is to establish a steady flow using the milk bottle described earlier and then interrupt the flow. To measure the response of the sensor alone, the funnel should not be used. With a load-cell flowmeter, the jug can be filled with sponge or some other absorbent material to help avoid the momentum artifact described later. Under the assumption of a first-order system, the measured flow will have dropped to 37% of its starting value in one time constant τ. Alternatively, if flow rate Q is plotted on a logarithmic axis, then the gradient of the best straight line can be used to estimate more accurately the time constant and the cutoff frequency f_c:

$$\tau = -\frac{\Delta t}{\Delta \log_e Q} \qquad f_c = \frac{1}{2\pi\tau}$$

For second- and higher-order systems, a more sophisticated analysis is required. For example, an underdamped step response will exhibit an overshoot from which a natural frequency and damping ratio can be estimated.

20.6 UROFLOWMETRY AND NATURAL (PHYSIOLOGICAL) FILLING

For flow tests and ambulatory urodynamics as described later, it is usual to let the patient fill naturally. Filling is normally helped through provision of juice, but nevertheless, with a filling rate of 4 mL min^{-1}, it will take at least an hour to fill a typical bladder. While this is more natural than filling via a catheter, the clinical environment probably doesn't reflect normal circumstances, and the voided volume may be substantially higher or lower than the habitual volume. There has been recent emphasis on home testing as a more representative assessment of flow.

20.7 RECORDING AND DISPLAY OF FLOW AND VOLUME DATA

20.7.1 Accuracy and Sampling Precision

For a crude assessment of the desired accuracy in uroflowmetry, we will use the *rule of 10*. That is, the resolution of the instrument should be a factor of 10 better than the smallest effect it supposes to measure. We will adopt test–retest variability as the effect we wish to measure. Since any diagnostic decision is subject to this variability, it offers a pragmatic yardstick of what might constitute a clinically significant effect.

20.7.1.1 Voided Volume

There are many reports in the literature of test–retest variability for uroflowmetry parameters (Rosier *et al* 1995, Sonke *et al* 1999, 2000, Tammela *et al* 1999). Within the individual, the test–retest agreement quantified by the standard deviation of consecutive measurements is typically 25–50 mL. Therefore, we might hope for an instrumental error of 5 mL or better, in a full-scale measurement of up to 2 L. Quantization noise is equally important as a source of error. In digital systems, this requirement equates to an absolute minimum of 9-bit resolution. Note however, this is assuming that an analog flow rate signal is available. If flow rate is to be calculated from voided volume, refer to the earlier discussion of digital differentiation.

20.7.1.2 Flow Rate

For maximum flow rate Q_{max}, the standard deviation of test–retest agreement is typically 2.5 mL s^{-1}. Therefore, we would hope for accuracy of 0.25 mL s^{-1}, in a full scale of typically 50 mL s^{-1}. In digital systems, this equates to an absolute minimum 8-bit resolution.

20.7.2 Bandwidth and Sampling Rate

Urine flow is a result of detrusor muscle contraction. Since the detrusor is smooth muscle, rapid changes in contraction strength are not physiologically possible (van Mastrigt *et al* 1986). In addition, the elastic compliance of the urethra and mechanical delays in the flowmeter funnel introduce low-pass filtering and delay. This means that in practice, any changes in urine flow rate on a timescale much shorter than a second are artifactual. An analog bandwidth of 5 Hz or a sample rate of 10 Hz is usually considered adequate.

20.7.3 Visual Presentation

Though computerized reporting has become the norm, the data are nevertheless presented in chart recorder style as flow rate versus time. Typical maximum scales are as follows:

- *Flow* 0–50 mL s^{-1}, in divisions of 5 mL s^{-1}.

- *Volume* 0–2 L, in divisions of 100 mL. In practice, 1 L is almost always adequate.

- *Time* divisions of 5 s, typically to 2 min though longer may be required.

20.8 INTERPRETATION OF FLOW AND VOLUME DATA

20.8.1 Summary Uroflowmetry Statistics

Many summary statistics for uroflowmetry data have been proposed. These include voiding time, flow time, time to maximum flow, mean flow, maximum flow, and total volume voided. However, voided volume (V_{void}) and maximum flow rate (Q_{max}) have been recorded since the early days of urodynamics, and a wealth of knowledge has accumulated about technical measurement issues, normative ranges, and response to treatment. Therefore, Q_{max} and V_{void} have become *de facto* the most important summary statistics.

20.8.1.1 Total Voided Volume V_{void}

Voided volume changes relatively slowly through the void and should be stable when the void has finished; V_{void} is easy to read and present as a summary measurement. Typically, a normal bladder capacity is 300–400 mL, with the first desire to void between

200 and 300 mL. A functional bladder capacity substantially less than 200 mL would adversely affect day-to-day life. Note that unless the patient has emptied completely, V_{void} will not be the same as the bladder capacity. Therefore, it is best practice and diagnostically useful to measure postvoid residual (PVR); high PVR (>50 mL) is one potential explanation for a poor functional bladder capacity, often leading to the symptom of frequency. The residual volume is typically measured by ultrasound as described later, though during full urodynamics, the bladder can be drained through the filling catheter into a measuring vessel.

20.8.1.2 Maximum Flow Rate Q_{max}

Q_{max} is very well used in clinical and research practice, particularly for the assessment of men with LUTS. Figure 20.11 shows an unambiguous flow trace, typical of a normal man. In men, a Q_{max} above 15 mL s^{-1} would be considered normal by most authorities, whereas a Q_{max} below 10 mL s^{-1} would typically be considered for active treatment (Reynard *et al* 1998). In women, the short urethra and lack of prostate means that a normal Q_{max} may be as high as 50 mL s^{-1}.

The shape of the flow curve is considered important by some workers. The bell shape of Figure 20.11 with a rapid rise and slower fall is normal. A low Q_{max} with the same bell shape may indicate obstruction, whereas a low, steady flow with an abrupt finish to the void may be indicative of urethral stricture. The evidence is equivocal and it would be risky to base a diagnosis on the shape of a flow curve. Nevertheless, the eye becomes adept at recognizing patterns; the authors feel strongly that the plotting system should not autoscale in either the horizontal (time) or the vertical (flow) axis.

20.8.2 Flow Artifacts

There are a number of common artifacts when making flow measurements. We divide them broadly into two groups: instrumental artifacts that are a result of inadequacies in the flowmeter design and the remainder that are introduced by the patient's voiding habits.

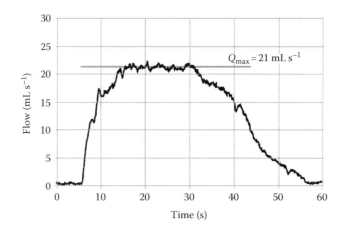

FIGURE 20.11   A normal male flow. Maximum flow rate Q_{max} is maintained for 15–20 s and would be recorded as 21 mL s^{-1}.

20.8.2.1 Instrumental Artifacts

20.8.2.1.1 Mass Flow and Contrast Medium   Load cell and spinning disk flowmeters measure mass flow. The volume flow rate is calculated by assuming the density of urine is approximately 1 g mL^{-1}. If using a denser contrast medium or if the patient is particularly dehydrated, the indicated flow rate will be proportionally high.

20.8.2.1.2 Momentum Artifact   The stream of urine falling on a load-cell flowmeter has momentum, which is registered as a force by the load cell, force being the rate of change of momentum. This is registered as an abrupt change in volume at the start of flow, or equally, a brief surge in flow. The size of the effect will depend on the amount and velocity of urine hitting the load cell and the filtering in the electronics. Momentum artifact is reduced by fitting a baffle and by a funnel spout that reaches into the jug. These slow the urine flow at the impact with the load cell, at the expense of increasing the time delay in the flowmeter.

20.8.2.1.3 Time Delay   There is inevitably a delay between a change in bladder pressure and the corresponding change in flow rate being detected. This is caused by mechanical delays due to urethral compliance and then as the urine flows down the funnel and into the collecting vessel (Kranse *et al* 1995). The low-pass filter in the flowmeter electronics will introduce a further delay. Figure 20.12 shows the response of a flowmeter to a step change in flow at time zero. The flowmeter has no response whatever for almost half a second; in clinical practice, there are additional delays due to flow through the distensible urethra and flight from the external meatus to the funnel. Therefore, a total delay of one second is a pragmatic estimate. This delay is of no importance for plain uroflowmetry but might be relevant during a simultaneous high-resolution pressure measurement.

20.8.2.2 Artifacts of Patient Origin

20.8.2.2.1 Straining   Straining increases the abdominal pressure and therefore the pressure in the bladder; in some patients with voiding difficulties, straining will improve the

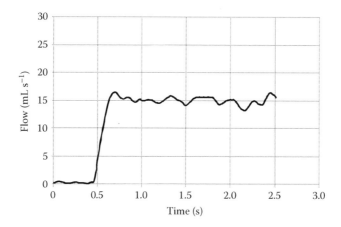

FIGURE 20.12   Estimation of the time delay in a flowmeter. Flow into the funnel began at $t = 0$, but was not registered for approximately 0.6 s.

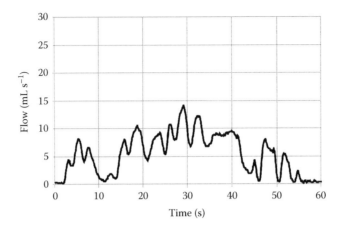

FIGURE 20.13 Uroflowmetry for a man who is intermittently straining to void. There would be a clear difficulty in measuring Q_{max} from this trace.

flow rate. Figure 20.13 shows the uroflowmetry for a man who was straining to void, hence the uneven flow. Without straining, his Q_{max} would probably be 9–10 mL s^{-1}.

20.8.2.2.2 Pinching the Penis   Some men develop the habit of repeatedly pinching the penis while they void. The flow trace thus produced will be intermittent as the maneuver is repeated, with a burst of flow following each occlusion. The benefit of penis pinching may be partly psychological but it is possible that it helps to stretch and dilate the urethra by exposing it to the full bladder pressure. Pinching causes an artifactual surge of urine, as is clear in Figure 20.14. Nevertheless, the size of the surge should in theory relate to the bladder contraction pressure, an effect that has been used to assess bladder obstruction noninvasively (Sullivan and Yalla 2000).

20.8.2.2.3 Wagging   Wagging is a second talent restricted to men, who alternately direct the flow down the spout, then into the funnel rim. When directed down the spout,

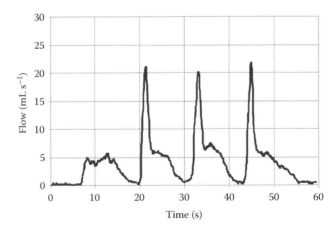

FIGURE 20.14 Uroflowmetry from penis pinching. Q_{max} is probably closer to 7 mL s^{-1} than the 22 mL s^{-1} recorded value.

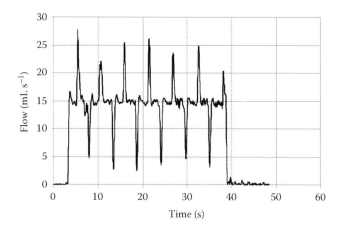

FIGURE 20.15 The effect of wagging on uroflowmetry. Notice that the upward and downward spikes are approximately symmetrical. This reflects the fact that the long-term flow rate is uniform.

the measured flow increases briefly. When the flow is directed back into the rim, the opposite happens. In Figure 20.15, maximum flow rate Q_{max} is 15 mL s^{-1} and not 28 mL s^{-1}. Wagging can appear similar to straining. In the absence of a full confession, one should suspect wagging when the area of the peak and the area of the trough in the flow recording are equal. Wagging affects all types of flowmeter, but less so the spinning disk. A funnel fitted with a baffle will greatly reduce wagging artifacts. More inventive solutions such as a plastic target fly on the funnel have also been found effective.

20.8.2.2.4 Low Voided Volume Q_{max} has a known relationship with V_{void}, particularly for $V_{void} < 200$ mL. While a low voided volume in itself is diagnostically important and may arguably not be an artifact, many clinics would repeat a flow test if the voided volume is less than typically 150 mL. If the patient is absolutely unable to produce an adequate volume, one can use the Siroky or Liverpool nomograms to estimate what Q_{max} would have been with a normal bladder capacity, or one might consider a home uroflowmetry device.

20.8.3 Automated Analysis of Uroflowmetry

Most computer uroflowmetry systems provide automated measurement of V_{void} and Q_{max}. The authors have a deep suspicion of automated analysis, particularly when applied indiscriminately. As we said earlier, V_{void} changes slowly and should be stable at the end of a void; it is relatively easy to measure automatically. Q_{max} does not and is not. To illustrate the issue, we refer the readers to the earlier figures showing straining, pinching, and wagging. At least one very widely used system records the absolute maximum flow rate. This is clearly wrong for all the examples presented earlier and can have clinical consequences as shown in Figure 20.16.

20.8.4 Clinical Value of Uroflowmetry

Stand-alone uroflowmetry is most useful for documenting voiding symptoms as a first-line assessment in men. Typically, $Q_{max} < 10$ mL s^{-1} is considered abnormally low and,

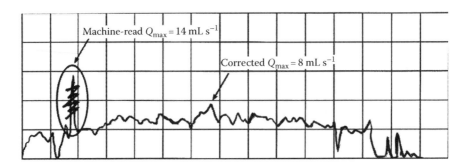

FIGURE 20.16 The one that got away. A flow trace from our clinic that was misreported and shows the dangers of automated Q_{max} detection. Scale divisions are 5 s and 5 mL s^{-1}.

in conjunction with other clinical tests, would be used to inform treatment. Nevertheless, there is considerable overlap of uroflowmetry parameters between pathological and normal populations. Given this overlap, the best possible diagnostic accuracy *with respect to a normal population* is around 75%. This does not constitute a differential diagnosis. It is not possible to separate bladder outlet obstruction from detrusor failure on the basis of flow rate alone, and up to 70% of raters choose a different diagnosis when presented with the same uroflowmetry trace on two occasions (van de Beek *et al* 1997). In conclusion, while uroflowmetry alone is useful to document symptoms, the evidence suggests that it is inadequate for differential diagnosis. For this, uroflowmetry must be used in conjunction with bladder pressure measurements.

20.9 BLADDER PRESSURE DURING FILLING AND VOIDING

Whereas uroflowmetry is more used in men, bladder pressure measurements by invasive cystometry are very important in both men and women. Used properly, a bladder pressure measurement can confirm the normal filling and voiding behavior of the bladder or conversely can indicate a number of pathologies that include:

- Unstable contractions during filling, also called detrusor overactivity

- Abnormally high contraction pressure during voiding

- Abdominal contractions due to straining or physical exertion

- A weak, absent, or unsustained contraction during voiding

When considered in the presence of the bothersome signs and symptoms such as incontinence or low flow rate, the pressure measurement is extremely helpful in providing a specific diagnosis.

20.9.1 Short History of Cystometry

The liquid manometer was described by Torricelli in 1643, and by a long-adopted definition, the units of cm H_2O express the pressure required to support a column of water of the given height. Since urine is mostly water, it is convenient and *de facto* the standard that

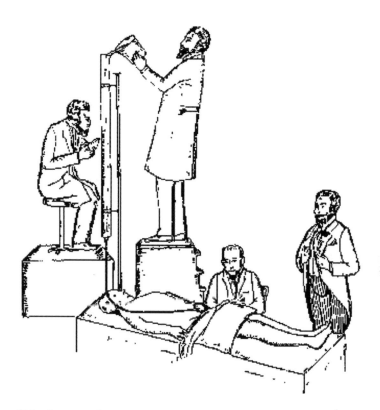

FIGURE 20.17 *"The drawing shows an early attempt at cystometry. In modern times, the staffing level can be halved and the dress code is slightly less formal."* (Courtesy of Doug Small. From Anders, E.K. and Bradley, W.E., *Urology*, 22, 335, 1983.)

urodynamic pressures are measured in cm H_2O, by analogy with the units of clinical blood pressure in mm Hg. These units are adopted by the ICS for use in clinical urodynamics.

Torricelli and Figure 20.17 suggest a crude but instructive and perfectly viable way to measure bladder pressure. One can simply arrange for the bladder to be connected to a vertical tube, open at the top, and ask the patient to void. The height in cm reached by the urine gives the bladder pressure in cm H_2O. Precisely this instrument was described as early as 1882, considerably before the uroflowmeter. This is not as surprising as it seems; in 1733, the Reverend Stephen Hales measured a horse's arterial blood pressure using the same method.

Without the benefit of electrical sensors the early workers used all manner of ingenious mechanical contraptions to make a permanent record of pressure. For example, an apparatus described by Rehfish in 1897 used a very similar mechanism to the first flowmeters to write the pressure to a rotating drum. The manometer tube was also connected to a syringe that could be used to fill the bladder. This arrangement was adopted by the physician D. K. Rose who in the 1920s drove forward the clinical application of cystometry. In fact, the development of electrical sensors was relatively slow. Some devices made use of early strain gages but were beset with technical problems because of the very small strains to be measured. Therefore, liquid manometers still found some use until the 1970s when capacitive sensors became available.

It is implicit in the liquid manometer that it is not absolute pressure but *gage pressure* that is measured—relative to the atmospheric pressure pushing the liquid back down the tube. To put it another way, the zero reference for bladder pressure is atmospheric pressure. This makes physiological sense because blood pressure, bladder pressure, and all other physiological pressures are regulated relative to the ambient pressure. We can still pass water while at the bottom of a swimming pool, despite it being at a pressure in excess of 100 cm H_2O—or so these authors are led to believe.

In Figure 20.17, unless the patient has an exceptionally strong bladder so that urine can reach the top of the tube, the urine has nowhere to go. The system will reach equilibrium and the bladder stops emptying when the pressure from the standing column of urine exactly opposes the contraction pressure of the bladder. The patient is maintaining a bladder contraction without any flow—*isovolumetric contraction*. While isovolumetric bladder pressure might be a useful measure of the bladder's ability to generate a contraction pressure (Griffiths and van Mastrigt 1985), it isn't particularly representative of a normal voiding cycle. By preference, we would measure pressure while the patient is in the act of voiding.

20.9.2 Detrusor Pressure versus Vesical Pressure

Pressure measured by the manometer tube in Figure 20.17 would be termed vesical pressure, p_{ves}. There is always a positive vesical pressure, even when the bladder is not contracting. Consider a simplified torso composed of water as depicted in Figure 20.18. Since the lungs and diaphragm are at approximately atmospheric (zero) pressure, the pressure descending down through the abdomen must increase by 1 cm H_2O per cm descended.

The measured pressure p_{ves} in the bladder is the sum of contributions from abdominal pressure and from the bladder detrusor muscle. Even without a bladder contraction, there is a resting bladder pressure, typically 35 ± 10 cm H_2O in a standing patient but lower when seated or lying. Therefore, much as the detrusor contraction pressure p_{det} is generally the parameter of interest, it cannot be measured directly. In order to determine p_{det},

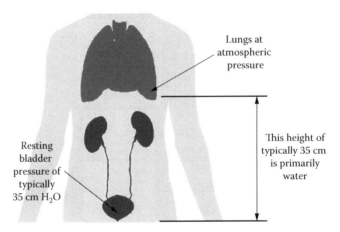

Lungs at atmospheric pressure

This height of typically 35 cm is primarily water

Resting bladder pressure of typically 35 cm H_2O

FIGURE 20.18 The origin of the resting bladder pressure is the hydrostatic pressure from a standing column of (mostly) water in the torso. The resting pressure will be considerably less when lying since the height of the torso vertically above the bladder is much lower.

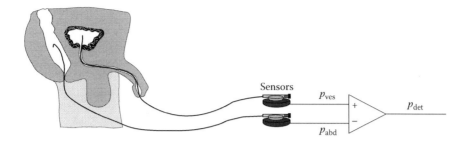

FIGURE 20.19   Obtaining detrusor pressure p_{det} from abdominal and vesical pressures, where $p_{det} = p_{ves} - p_{abd}$.

the abdominal contribution must be subtracted from the vesical pressure p_{ves}, as shown in Figure 20.19. Abdominal pressure p_{abd} is most commonly measured in the rectum but can also be measured in the vagina or via a stoma. The means of measuring pressure in a non-liquid-filled space are debated later.

20.10 MEASUREMENT OF URODYNAMIC PRESSURE

Urodynamic pressure measurements unfortunately need to be made at sites where most patients would prefer they weren't. And if measurements must be made, patients would express a strong preference that the entire instrumentation rack is not taken to the site. The manometer of Figure 20.17 has a certain simplicity that explains its longevity; helpfully, only a fine tube need be taken to the measurement site, but to the dismay of these authors, it has fallen into disuse. This poses some difficulties. First, we will consider the sensor technology that has replaced the manometer and then the solutions to the measurement problem.

20.10.1 Pressure Sensors

Modern pressure sensors used in urodynamics contain a flexible diaphragm; one side is exposed to the liquid or gas under measurement, with the opposite side exposed to the atmosphere to establish the zero pressure reference. When the pressure in the fluid increases, it displaces the diaphragm a little way (Figure 20.20); the diaphragm is arranged so that the applied pressure changes the properties of some electrical components. For example, the diaphragm may form one plate of the capacitor, a measurement being made much as for the dipstick flowmeter. However, recent devices tend to be resistive, with an electrical arrangement very similar to a load cell.

(a)   (b)

FIGURE 20.20   A stylized and exaggerated view of a pressure sensor. Under no pressure (a), the strain gages lie flat along the diaphragm. With pressure applied to the diaphragm, the strain gages are bent downwards (b).

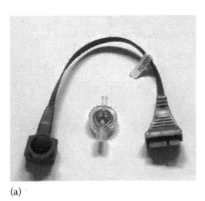
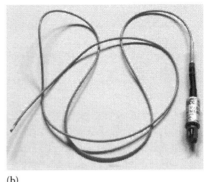

(a)   (b)

FIGURE 20.21   Sensors for urodynamics: external sensor with dome (a) and a catheter-tip sensor (b).

As with the load cell, the traditional circuit is the Wheatstone bridge. A typical sensitivity is 1 mV per V per 100 cm H_2O, and so an instrumentation amplifier is inevitably required. The amplifier is built into some recent devices, which therefore provide a higher signal level.

20.10.2 Water-Filled Catheter and External Sensor

This is the most-used arrangement to the measurement problem and is recommended by ICS *Good Urodynamic Practices*. A saline-filled hollow catheter is passed to the measurement site, and the pressure is transmitted along the catheter to a sensor that is external to the patient. This is a cost-effective arrangement, where sterility is assured by using a disposable catheter. The entire sensor may be disposable but more typically has a disposable dome that avoids direct contact of urine with the diaphragm, thus to preserve the relatively expensive sensor element (Figure 20.21a).

This arrangement requires practice to set up properly. Since the catheter is part of the measurement system, it will not work reliably if there is any air in the catheter, and large bubbles cause dramatic artifacts. Even when set up correctly, a water-filled catheter is sensitive to being jostled.

Since there is a continuous column of saline along the catheter, the pressure at the sensor is the same as that in the bladder *at the vertical level of the sensor*, irrespective of where the catheter tip is resting (Figure 20.22a). By convention, the sensor is levelled to the pubic symphysis, an anatomical landmark for the bladder. This can mean moving the sensors in the middle of the study as the patient is moved between standing and sitting, and in practice it is often forgotten in the heat of battle. This will affect bladder and abdominal pressures equally and so should not affect the detrusor pressure.

20.10.3 Air-Filled Catheter and External Sensor

This arrangement is similar to the water-filled system, except that the catheter is filled with air. In order to prevent the air leaking away into the bladder, there is a small balloon covering the catheter tip. The density of air can be neglected, and so the pressure is transmitted

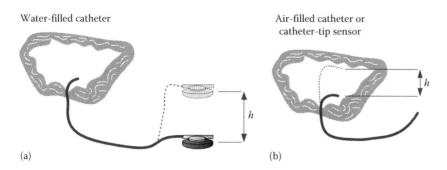

FIGURE 20.22 The effect of height changes with the different catheter arrangements. At (a), only the height of the external sensor is important. At (b), only the height of the catheter tip in the bladder is important. In either case, raising the height by h cm will reduce the pressure measurement by h cm H_2O.

directly from the balloon at the catheter tip to the sensor. Therefore, unlike the water-filled system, the position of the catheter tip will affect the measurement (Figure 20.22b). For example, at the top of the bladder, the measured pressure will be typically 10 cm H_2O lower than at the bottom; there is no easy way to correct for this error. However, the height of the external sensor does not affect the measurement, and this is a significant benefit in clinical practice.

This system is not trivial to set up. Air is compliant and an increase in pressure is associated with a reduction in volume. Therefore, the balloon must be part filled with air for normal use; when it has been squeezed flat by the applied pressure, any further pressure will not be transmitted and so there is an upper limit to the measurement range. However, if the balloon is filled with air beyond its natural volume such that it is under tension, it will cause an artifactual increase in the measured pressure. Since air is compliant, there is little or no artifact due to knocking the line, but at the expense of a poorer and in our experience less predictable frequency response (Cooper *et al* 2011); some workers argue that the frequency response is inadequate for clinical urodynamics.

20.10.4 Catheter-Tip Sensors

Catheter-tip sensors are the simplest to understand (Figure 20.21b). The sensor is mounted at the tip of the catheter and is passed directly to the site of measurement. Since the sensor needs to be small, it is more difficult and therefore expensive to manufacture. In addition, it is exposed directly to the urine and must be sterilized or discarded after use. Taking the sensor to the site avoids many of the drawbacks of water- or air-filled systems. A high measurement bandwidth can be achieved with little or no artifact. As with the air-filled system, it is clear with catheter tip sensors that the pressure is measured at the catheter-tip (Figure 20.22). Therefore, this arrangement suffers similarly from the uncertainty in positioning of the catheter tip in the bladder.

There are other measurement technologies suitable for catheter-tip sensors. For example, fiber optic sensors have been developed. In one design, a Bragg grating takes the place of the electrical strain gage; this acts as a tuned optical filter whose wavelength varies with pressure. The catheter can be relatively simple and robust, at the expense of the complex instrumentation required to perform a spectral analysis of the reflected light.

20.10.5 Measuring Pressure in the Rectum or Other Dry Spaces

A pressure is defined only in a liquid or gas. Since the bladder is liquid filled, a pressure measurement within the bladder is meaningful. However, the rectum and other sites for abdominal measurements are not reliably wet. The air-filled catheter in particular addresses this problem by giving the rectum walls something to press against. Intuitively, the balloon takes up a shape such as to be in equilibrium with the rectal wall pressing inwards, and the pressure recorded is such as to maintain the balloon's size and shape. For other types of rectal catheter, the catheter tip is terminated in a water-filled balloon with a small vent hole to create a liquid medium. The physics of how this pressure measurement happens is not well documented, but pragmatically it works well. In other situations, it is desirable to measure pressure while pulling the catheter through a dry rectum or urethra, where the balloon would detach or obstruct the movement. This is described later as the *urethral pressure profile*.

20.10.6 Pressure References

As for most physiological measurements, *zero pressure* is atmospheric pressure. Therefore, for most designs, the rear of the sensor is vented to atmosphere. This is simple to achieve for the external sensor but requires an air-filled hollow lumen to a catheter-tip sensor.

20.10.7 Summary of Attributes of Sensor Arrangements

Table 20.2 summarizes the key attributes of the main sensor arrangements.

TABLE 20.2　Summary of Key Attributes of Sensor Arrangements

	Water-Filled Catheter with External Sensor	Air-Filled Catheter with External Sensor	Catheter-Tip Sensor
One-off cost	Medium—the sensor has no particular size constraints.	Medium—the sensor has no particular size constraints.	High—technically difficult to engineer.
Disposable	The catheter and dome or the entire sensors are disposable.	The catheter is disposable.	Not normally any disposable components.
Sterilization	The catheter and sensor dome are supplied sterile.	The catheter and sensor dome are supplied sterile.	The catheter is sterilized before every study.
External equipment	Sensor, dome, tap, and syringe for flushing.	Sensor.	None.
Bandwidth	Typically <30 Hz but adequate for most purposes.	Low, typically <10 Hz, and may vary. Does not respond to rapid pressure changes.	High, >500 Hz. Can respond to rapid pressure changes.
Artifact	Artifacts from infusion pump or line tapping.	Relatively immune.	The sensor is in the bladder; almost immune.
Error due to height	None—if the sensor height is set correctly at pubic symphysis.	Random error of typically ±5–10 cm H_2O relative to the center of the bladder.	Random error of typically ±5–10 cm H_2O relative to the center of the bladder.
Typical use	ICS standard for in-clinic cystometric measurements.	General-purpose, laboratory, and ambulatory use.	Ambulatory urodynamics.

20.11 PRESSURE SIGNAL CONDITIONING AND PROCESSING

As with the flowmeter, the signal conditioning has historically been achieved electronically but can now be implemented in software.

20.11.1 Analog Signal Processing

The signal conditioning would normally consist of two phases: low-pass filtering at typically 1 Hz to remove high-frequency components and then subtraction of p_{abd} from p_{ves} to give p_{det}. In theory, the two processes are linear and can be applied in either order. However, these authors would expect filtering before subtraction; the separate p_{abd} and p_{ves} traces are of interest, and therefore it is beneficial to filter them separately. In addition, commercial pressure amplifier designs are likely to include a *per channel* filter. Of course, it requires that the filters in the two channels be well matched.

20.11.2 Digital Signal Processing

As with any signal processing, it is possible to reproduce the filtering and subtraction algorithms in software. This has a particular advantage in pressure measurement, because one can arrange that the abdominal and vesical pressures be filtered identically. Note that an abdominal contraction such as a cough can contain components at relatively high frequency, and indeed there are some investigations where a rapid frequency response is required for pressure measurements. By comparison with uroflowmetry, a substantially higher sampling rate would be required to avoid aliasing, which will have consequences for the computing power required for signal processing.

20.12 PRESSURE SENSOR CALIBRATION

To calibrate a pressure sensor, the normal procedure is to set a zero point at atmospheric pressure and then calibrate the gain using a known pressure, often 100 cm H_2O. This is not easy to get right. Received wisdom is to connect a standard manometer line to the sensor, fill it with distilled water using a syringe, and raise the catheter tip to 100 cm above the sensor height. The hydrostatic pressure of the water will create a pressure of 100 cm H_2O at the sensor. This method is good for a quick check of calibration, but in our experience, it is easy to introduce air to the catheter and inadvertently establish a reference pressure lower than desired.

Catheter dimension is normally cited in French (F) units, 1 F being 1/3 mm. For a typical 4 F or 1.33 mm diameter lumen, each meter of a catheter has a volume of around 1 mL. A single drop of water lost from a catheter might have a volume of 0.1 mL, introducing an error of up to 10 cm H_2O. All pressure measurements made with this badly calibrated system will be correspondingly high. An alternative arrangement for calibration is in Figure 20.23a, which helps avoid air bubbles but may compromise sterility. In fact, modern sensors are stable enough that some centers would be better off not calibrating their system at all.

We would recommend less frequent calibration, using an arrangement similar to that in Figure 20.23b. A reference pressure is set using a standard sphygmomanometer bulb connected to the sensor. The pressure is checked using a good-quality air pressure meter that can be bought and should be calibrated yearly. If the setup is too sensitive to set the pressure accurately,

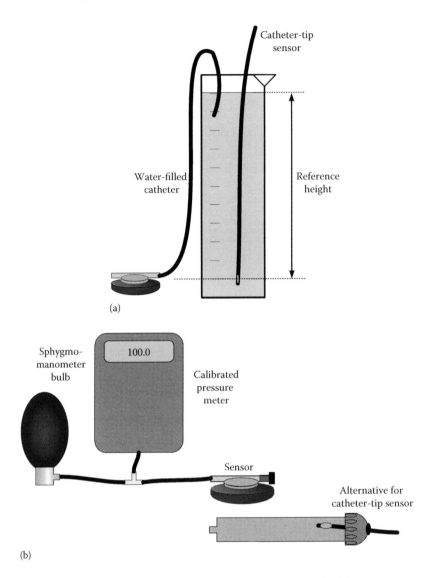

FIGURE 20.23 Calibrating a pressure sensor. At (a) is a conventional calibration setup. For a water-filled catheter, the water level must be at the reference height above the sensing element. For a water-filled catheter, the catheter tip is at the required depth. At (b) the reference pressure is set using the sphygmomanometer bulb or angioplasty balloon inflator. For a catheter-tip sensor, an airtight chamber can be made using a syringe and an airtight cap with a rubber gland through the center.

compliance can be added by using a long length of wide tube. We replace the sphygmomanometer bulb with an angioplasty syringe with a threaded barrel for fine pressure control.

For catheter-tip sensors, the easiest way to generate a known pressure is to immerse the sensor tip in a measuring cylinder of water to a known depth. Though it relies on a static column of water, this is markedly less prone to serious errors than raising the water-filled catheter. Alternatively, an airtight chamber can be constructed using a syringe with an airtight entry port also shown in Figure 20.23b, but there is a less obviously clear case for doing so. Neither procedure is sterile and the catheter must be sterilized after calibration.

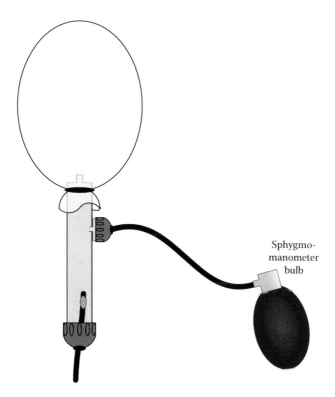

FIGURE 20.24 The pop test to assess frequency response of a pressure sensor and catheter arrangement. The tip of the catheter is held in an airtight compartment. The balloon is inflated and then burst with a lighted match to generate a step change in pressure. The apparatus must be held rigidly in order that its vibrations do not contribute to the measured step response.

20.13 MEASURING PRESSURE SENSOR FREQUENCY RESPONSE

The frequency response of a pressure sensing system can be measured using specialized instrumentation to generate swept pressure waveforms across the frequency range of interest, typically up to a few 10s of Hz. However, it is far more readily measured using the so-called *balloon pop* test (Figure 20.24). In this test, the catheter is first exposed to the pressure generated by an inflated balloon. The balloon is then burst most effectively with a lit match, generating a step change in pressure that can be analyzed as for the step response of a flowmeter (Hok 1976, Nichols and O'Rourke 2005). In our experience, this test is difficult to conduct reliably. In particular, vibrations in the support can interfere with the measurement.

20.14 CYSTOMETRY AND ARTIFICIAL (NON-PHYSIOLOGICAL) FILLING

At a natural filling rate of 4 mL min^{-1}, it will take at least an hour to fill the typical bladder. In order that the urodynamic study can be completed in a reasonable time, the patient is usually filled with saline via a catheter connected to an infusion pump. The infusion pump is normally of the peristaltic type where a series of rollers compress a flexible tube to drive the saline (Figure 20.25). Since saline never comes in direct contact, the pump need not be sterilized.

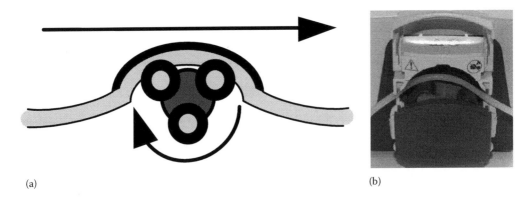

(a)   (b)

FIGURE 20.25   A stylized view of a peristaltic pump (a) and a pump head (b). The rollers in the rotating pump head drive saline through the compressible tube. A crude estimate of flow rate can be made from the product of pump speed and tube internal cross-sectional area.

Good practice would suggest that the patient be filled at the upper limit of the physiological rate for their body size. The ICS has proposed that fill rate is given by the following formula:

$$\text{Filling rate in mL min}^{-1} = \text{Body mass in kg}/4$$

This suggests perhaps 20 mL min^{-1}, well beyond the typical natural fill rate, but even this generous rule of thumb is routinely exceeded in clinical practice. Nevertheless, there is tacit consensus that not more than 100 mL min^{-1} be infused; this is often reduced where the patient shows signs of detrusor overactivity.

It is known that filling with cooled saline can promote overactivity. Following this thinking to its conclusion, some centers warm the infused saline to body temperature, though there is no conclusive evidence that this is useful. It is also worth noting that historically, CO_2 gas has been used in place of saline to fill the bladder. Simultaneous pressure measurements are possible and indeed carbon dioxide is also reported to provoke overactivity during filling, but it is not easy to measure flow rate or voided volume. The practice has been abandoned.

20.14.1 Measuring Filled Volume

The infusion pump is typically driven by a stepper or servomotor to give a controlled rate. It is possible in principle to determine the filled volume by counting revolutions of the rollers, but it is easy to show in a bench model that this is susceptible to errors of typically 10% due in particular to variations in tube cross section and downstream resistance. Many peristaltic pumps will turn and indicate flow even when the downstream tube is completely blocked. It is therefore common to use a load-cell arrangement to measure infused volume V_{fill} by weighing the infusion bag. This is essentially identical to the load-cell measurement of voided volume described earlier, except the bag grows lighter and not heavier as the patient is filled. As with the flowmeter, contrast medium is denser than saline and will lead to overindication of the filled volume if its density is not taken into account.

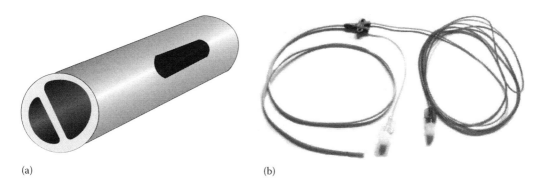

(a)   (b)

FIGURE 20.26   Double-lumen catheters. At (a), a cross section through the tip of a double-lumen catheter, showing the wider filling lumen, the smaller measurement lumen, and the eyelet where saline exits the catheter for filling. There is a similar eyelet for pressure measurement on the opposite side of the catheter. Pulsations in the thin wall between the two lumens can give rise to artifact at high fill rates. At (b), a double-lumen catheter. An eyelet is just visible near the catheter tip.

20.14.2 Filling Catheters

Historically, a dedicated filling line was passed alongside the pressure-measuring catheter. The filling catheter is substantially wider than the measuring line because it needs to support a relatively high saline flow rate. For ease of catheterization, the measuring line can be piggybacked into the eyelet of the filling catheter. Now, it is more common that a dual-lumen catheter is used. It has completely separate channels for filling and for measurement—Figure 20.26. This makes for easier catheterization but is more expensive than the dedicated filling catheter.

20.15 RECORDING AND DISPLAY OF PRESSURE DATA

20.15.1 Accuracy and Sampling Precision

Test–retest agreement for maximum bladder pressure, quantified by the standard deviation of consecutive measurements in the same patient, is typically 10 cm H_2O (Rosier *et al* 1995, Sonke *et al* 2000, Tammela *et al* 1999). Using the *rule of 10* as for uroflowmetry, we might hope for a resolution of 1 cm H_2O, in a full scale of −100 to +400 cm H_2O. This equates to a minimum of 9-bit resolution.

20.15.2 Bandwidth and Sampling Rate

Since detrusor is smooth muscle, rapid changes in contraction pressure are not physiologically possible. However, some physiological pressure events have a different origin; for example, a cough will have frequency components well above 10 Hz (Kim *et al* 2001). A sample rate of 10 Hz is considered adequate when the only concern is the detrusor contraction, but note that a low-pass filter would be required to avoid aliasing of the higher-frequency components. In some specialist urodynamic investigations, the cough pressure profile is important, particularly when correlated with other physiological measurements such as leakage or urethral pressure measurement. In that case, a higher sampling rate in the region of 100 Hz would be appropriate.

20.15.3 Visual Presentation

As with flow rate, pressure data are inevitably presented in chart recorder (pressure versus time) format. Typical settings might be as follows:

- *Pressure* −100 to 300 cm H_2O, in graduations of 25 or 50 cm H_2O

- *Time* Divisions of minutes, typically 5 or 10 min to view though longer may be required

Note that the filling phase would normally be recorded as well as the voiding phase and the entire test may well be repeated. Therefore, the total time of the study may be half an hour or longer.

20.16 INTERPRETATION OF PRESSURE RECORDINGS

We now consider the interpretation of pressure recordings, and present some of the most important clinical conditions that can be diagnosed on cystometry.

20.16.1 Quality Control

An important tool to avoid artifacts in urodynamic pressure measurements is quality control, using each channel as a benchmark for the other. At rest, the two channels should be nonzero but within a few cm H_2O of each other; the difference may be larger with air-filled or catheter-tip sensors given the uncertainty in position of the sensor. Resting pressures while seated or standing are typically 35 cm H_2O, but supine pressures are lower.

Before the study starts, throughout the measurement, and even during voiding, coughing is a useful quality control measure—Figure 20.27. The sharp spikes in the recording are coughs, which cause a rapid rise in intra-abdominal pressure recorded in p_{abd} and p_{ves}.

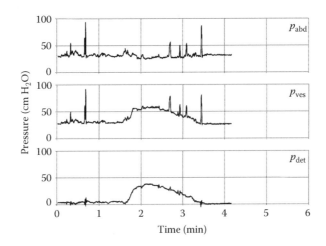

FIGURE 20.27 Coughing during urodynamics, a useful quality control measure. Note the good cancellation even during the detrusor contraction from 1 min 40 s to 3 min 20 s. From top, traces are abdominal pressure, vesical pressure, and detrusor pressure.

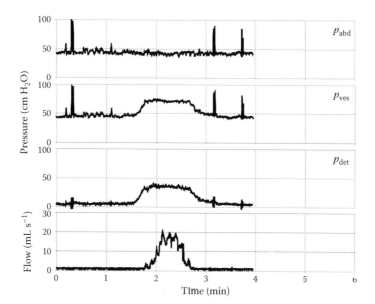

FIGURE 20.28   Normal voiding cystometry. Note good cancellation of coughs throughout. The fill was uneventful and is not shown.

Since the pressure rises in p_{abd} and p_{ves} are similar, the net effect on detrusor pressure p_{det} is small. If the p_{abd} and p_{ves} pressures do not respond equally, the smaller and/or slower response normally indicates the source of the problem. It is not unusual to observe a small biphasic fluctuation in p_{det}, because the pressure wave due to the cough reaches the two sensors at slightly different times. Note that even during the detrusor contraction, the effect of the coughs cancels out.

20.16.2 Normal Cystometry

An example of normal voiding cystometry is shown in Figure 20.28. Note that there is little activity in the detrusor until the voiding phase. Flow commences just after the p_{det} rise and is quite regular. There is just a hint of wagging on the flow trace; we can rule out abdominal straining because there is no activity on the abdominal pressure trace. The pressures in this trace are typical for a normal void; the relatively low detrusor pressure of less than 50 cm H_2O is consistent with the normal flow rate of 20 mL s^{-1}.

20.16.3 Detrusor Overactivity

Detrusor overactivity, formerly bladder instability, is the involuntary contraction of the detrusor muscle during the filling phase. Note in Figure 20.29 the strong detrusor contractions during the filling phase in the first 3 min of the trace. These involuntary bladder contractions are associated with the strong urge to void. Note that the bladder can contract with increased pressure against a closed outlet and contractions in excess of 200 cm H_2O are not uncommon. Abdominal pressure p_{abd} is stable, ruling out abdominal contractions as the source of the bladder pressure rise. The large contraction at the end of the trace is associated with flow, with a low voided volume of less than 200 mL;

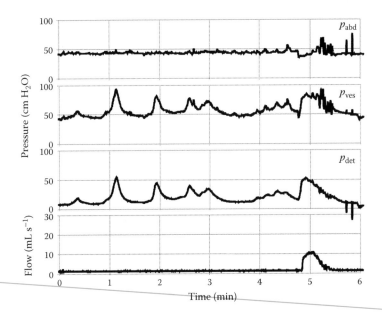

FIGURE 20.29 Detrusor overactivity on cystometry. Following the void, note that p_{ves} no longer responds to the cough test; the line was dislodged during the void.

the patient was unable to hold any more. The low functional bladder capacity is a key symptom of detrusor overactivity.

20.16.4 Stress Incontinence (Women)

Stress incontinence is the leakage of urine, normally in women, as a result of a raised abdominal pressure acting on the bladder. In day-to-day life, the trigger can be any vigorous activity where the abdominal muscles are contracted. Similarly, stress incontinence can be provoked in the clinic by coughing, bending, or making a Valsalva maneuver. The investigation of stress incontinence requires the measurement of pressure in the bladder, with a simultaneous indication of leakage as sensed by patient, seen by the clinician or recorded in the flowmeter. The pressure at which the leak took place is known as the abdominal leak point pressure. Typically, a pressure below 60 cm H_2O is presumed to be diagnostic of stress incontinence. An instructive example is shown later in the section on *ambulatory urodynamics*.

20.16.5 Outlet Obstruction (Men)

Bladder outlet obstruction (Schaefer 1985) is associated with a low flow rate, and so the patient takes a long time to void. In Figure 20.30, note the frequent abdominal straining, nevertheless having little effect on the detrusor pressure; this indicates the importance of the separate display of abdominal pressure. Whether abdominal straining has much effect on flow rate depends on the patient. In women, a raised abdominal pressure is the origin of stress incontinence and clearly therefore does effect flow. Likewise, straining will be helpful to men in the early stages of obstruction and may become a learned behavior. However, as the obstruction progresses the raised abdominal pressure is also brought to bear on

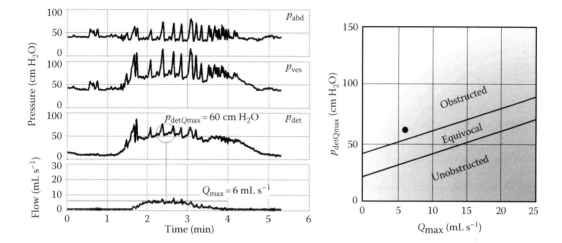

FIGURE 20.30   Obstruction on cystometry. Note the low flow rate and strong abdominal contractions due to straining. To assess obstruction, Q_{max} is read from the flow trace, and $p_{det}Q_{max}$ is the bladder pressure at the time when Q_{max} was recorded. These measurements are plotted on the ICS nomogram (*right*). Note also that abdominal pressure drops during the void, an incidental observation in many patients. We speculate the contracting bladder becomes more rigid and helps support the abdominal contents, thereby lessening the pressure on the rectum.

the enlarged prostate, so straining also increases the obstruction to flow. In Figure 20.30, straining has become relatively ineffective.

20.16.5.1 ICS Nomogram

Bladder outlet obstruction is a particularly common complaint, to the extent that the ICS standardization committee (1997) published separate guidelines specifically for the assessment and diagnosis of the condition. The ICS nomogram is similar to earlier nomograms produced by Paul Abrams and Derek Griffiths and by Werner Schäfer. Broadly, a high contraction pressure with low flow rate indicates an obstruction.

According to the ICS procedure, first locate the point of maximum flow Q_{max}. Then measure the bladder pressure $p_{det}Q_{max}$ at the time when maximum flow is recorded. The point defined by Q_{max} and $p_{det}Q_{max}$ is plotted on the nomogram (Figure 20.30), which indicates *unobstructed*, *equivocal*, or *obstructed*. In the example, the point defined by 6 mL s^{-1} and 60 cm H$_2$O lies in the *obstructed* region.

20.17 PRESSURE ARTIFACTS

Since there is redundancy in having two pressure channels, artifacts in the pressure measurement are typically recognized and dealt with through proper quality control. Here, we describe the most common artifacts.

20.17.1 Filling Artifacts

With separate filling and measurement lines, during filling there will be a dramatic and constant positive pressure offset in p_{ves} if the measuring catheter isn't disengaged from its piggyback position before filling commences. Dual-lumen catheters are susceptible to a different filling artifact where the pressure generated by the infusion pump affects the pressure in the measuring line, particularly at high filling rates. The effect is due to peristalsis of the pump interacting with the compliance of the thin catheter wall and is normally manifested as a rhythmic signal from the pump rollers superimposed on the p_{ves} signal. Filling artifacts disappear if the infusion pump is stopped.

20.17.2 Dislodged Catheter

A dislodged catheter is relatively common but easy to spot with good-quality control—the measurement in the affected catheter will show a reduced or absent response to coughs. If the catheter has moved significantly, the measurement may also show a dramatic offset from its baseline value.

20.17.3 Bubbles

Air bubbles introduce two issues with water-filled catheters. First, the nonuniform density of fluids in the catheter will introduce an offset to pressure measurements. The size of the offset depends on the difference in height between the two ends of the bubble, which changes as the catheter is moved and as the measured pressure changes. Second, water is incompressible and pressure changes are transmitted *without flow of water*. Air bubbles are compressible; a change in pressure requires flow to compress the air bubble. The bubble is a physical analog to a capacitor, and the water–air system becomes a low-pass filter that affects the frequency response of the catheter dramatically. Note that this problem doesn't affect air-filled catheters to the same degree, because the opposition to flow (equivalent to resistance) and inertia (equivalent to inductance) offered by air is very low. The effect of a bubble in the abdominal line is shown in Figure 20.31.

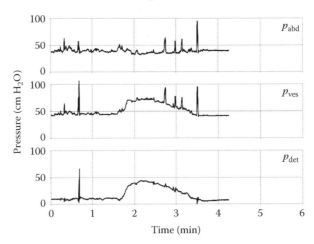

FIGURE 20.31 Pressure recordings with a bubble in the abdominal line. At 40s, the rise in p_{ves} is present but without an accompanying change in p_{abd}. The net effect is therefore visible in the subtracted detrusor pressure, p_{det}. The abdominal line would normally be flushed with saline to clear the bubble, but there is no sign of the abrupt pressure rise that normally accompanies flushing; the problem appears to resolve spontaneously.

20.17.4 Rectal Activity

Good cancellation of abdominal activity makes the assumption that p_{abd} is a good measure of resting abdominal pressure. However, if the rectum is loaded, the rectal sensor will often measure peristaltic waves. These will be manifested as negative-going waves on the resting detrusor pressure (Figure 20.32), which may sometimes appear to be below zero pressure (e.g., around time 01:34). For good physiological reasons, a substantially negative value of p_{det} is unlikely if not impossible.

20.18 COMPLETE URODYNAMICS SYSTEM

We have described the essential components of modern urodynamics. To complete the picture, we now present a complete circuit diagram for a urodynamic system. The circuit in Figure 20.33 is taken from instrumentation built in our own lab; given the recent rapid improvements in instrumentation, we do not claim it is the best or even a particularly good way to implement a new system, but it illustrates well the principles we have described.

20.18.1 Load Cell and Preamplifier

The load cell is of course situated close to the patient, and therefore a preamplifier is provided in the board with the load cell. The entire load-cell circuit is shaded darker gray in Figure 20.33. The preamplifier provides some immunity to electrical interference and in addition allows for gain and zero adjustment so that the load-cell unit provides a standard output of 50 mV L^{-1}. Note also the liberal use of decoupling capacitors; this is particularly important for EMC immunity where signals join or leave a board.

Being active, the load-cell unit needs power. In our case, the unit is powered from a single 5 V rail, for compatibility with off-the-shelf load-cell amplifiers having a standard 5 V excitation output. An onboard 3.3 V regulator not shown in Figure 20.33 is used to drive the strain bridge and provide the amplifier reference. This particular load cell has a half bridge, with the second amplifier input from a potential divider that allows for zero adjustment. These resistors must be of a high-stability type.

20.18.2 Volume and Flow Circuitry

This comprises an instrumentation amplifier, zero and gain adjust, and a second order low-pass filter. All these are standard circuits and are well described elsewhere. In modern systems, there is a move to avoid the use of trimmer potentiometers and provide calibration features in software.

Not shown here, the same circuit is used to measure filled volume. Much as the weight on the sensor decreases with filled volume, this is easily addressed by reversing the polarity of the supply to the strain bridge.

To derive flow rate, the output from the volume circuitry passes through a differentiator and a further gain-zero stage. Differentiation is particularly sensitive to interference because the gain of a differentiator increases linearly with frequency. Therefore, it is usual to roll off the gain of the differentiator above the frequencies of interest. In our case, this purpose is served by the 33 nF capacitor in the feedback path that gives a time constant in the region of 20 ms or a corner frequency around 10 Hz.

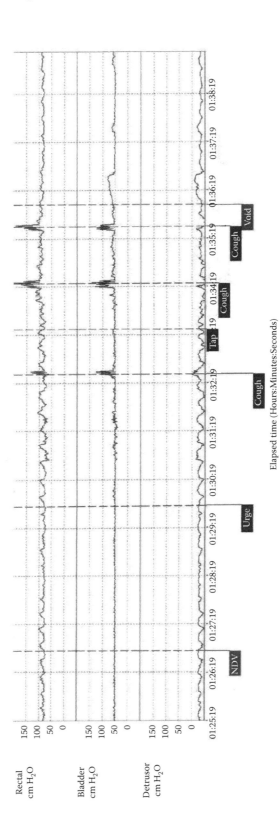

FIGURE 20.32 Ambulatory urodynamics from a lady with a loaded rectum. Note the peristaltic waves of rectal activity, which incorrectly might be taken for detrusor contractions. The vesical pressure trace acts as quality control, indicating the true origin of the contractions. (NDV = normal desire to void).

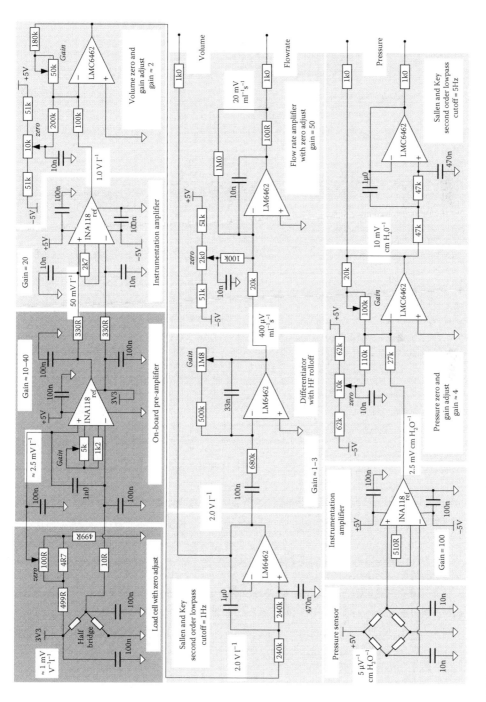

FIGURE 20.33   Circuit diagram for a urodynamic system.

20.18.3 Pressure Circuitry

The circuitry for the pressure channel is almost identical to that for volume. With no separate preamplifier, the pressure sensor uses the normal Wheatstone bridge configuration; the quoted sensitivity of 5 μV V^{-1} cm H$_2$O^{-1} is standard for urodynamic sensors. Two pressure outputs are provided: One is filtered to 5 Hz that provides an anti-aliasing function compatible with the *de facto* standard sampling rate of 10 Hz. The other is not filtered at all and is provided for the rare occasions when a higher bandwidth is required.

The same circuit is used equally for vesical and abdominal pressures, and so only one channel is shown. In the interests of space, we have not shown any further filtering or the subtraction unit to give p_{det}; these functions are increasingly provided in software, but if not, the circuits are very similar to what is shown.

20.19 ADJUNCTS TO CONVENTIONAL URODYNAMICS

We have described the essential components of modern urodynamics. There is a range of physiological measurements used to augment these techniques. We will describe them in order of increasing complexity.

20.19.1 In Vitro Testing

20.19.1.1 Urine Dipstick Testing

The urine dipstick is a first-line test for a number of conditions. The stick is dipped in a midstream urine sample and a series of reagents on the stick change color to indicate the presence of pathological substances. These can include blood, protein, sugar, and ketones. Some of the tests (e.g., hematuria or proteinuria) are nonspecific while others (e.g., glucose) are specific to particular diseases.

20.19.1.2 Prostate-Specific Antigen

The serum prostate specific antigen (PSA) measurement is a first-line test for prostate cancer. It is rather encumbered in that many men with raised PSA do not have cancer (poor specificity); more worryingly, many men with cancer do not have raised PSA (poor sensitivity). Nevertheless, it is a very simple test with some predictive value; in some countries, a do-it-yourself PSA test can be bought over the counter as a healthy man screening tool.

20.19.2 Simple Devices for Home Use

20.19.2.1 Frequency–Volume Chart

The frequency–volume chart gives a semi-objective documentation of symptoms as they affect a patient's home life. He or she will keep a diary of the time of each toilet visit for a period of 4 to 7 days. Where possible, a jug is used to capture the volume. The diary is traditionally kept in paper form, but at the time of writing, the benefits of handheld PDA devices are making themselves felt.

20.19.2.2 Leakage Diaper

The leakage diaper can be as simple as an incontinence pad, which is weighed before and after the test period in an investigation of incontinence. The weight gain gives a semiquantitative indication of total leakage, though of course it will be subject to inaccuracies due to sweating and evaporation.

20.19.3 Ultrasound

Ultrasound is widely used in combination with uroflowmetry to estimate the volume of urine in the bladder. If we assume that natural filling during uroflowmetry is insignificant, the volumes of interest have a simple relationship:

$$V_{starting} = V_{void} + V_{residual}$$

V_{void} is known from uroflowmetry, and therefore only $V_{starting}$ or $V_{residual}$ need be measured to calculate the other. Typically, the bladder is measured in two or three planes to make the volume estimate. However, there are known inaccuracies in estimation of volume due to the limitations of the spherical or ovoid model, plus uncertainties in placing the bladder wall. Therefore, in practice, it is better and more usual to measure $V_{residual}$ because the error on an ultrasound measurement is smaller when the volume is smaller.

In its wider context, transrectal ultrasound can be used to assess sphincter anatomy in women and prostate size in men. There is also evidence that bladder wall thickness relates to bladder hypertrophy and can be used as an index of obstruction, much as cardiac hypertrophy indicates high blood pressure (Manieri *et al* 1998). The value of this test remains to be proven in routine clinical practice.

20.19.4 Ambulatory Urodynamics

Ambulatory urodynamics (Brown and Hilton 1997) is used where conventional urodynamics has not reproduced the bothersome symptoms of the patient; this is typically in relation to storage symptoms, and so ambulatory urodynamics is most used in women. The instrumentation and rationale for ambulatory urodynamics is essentially identical in principle to conventional urodynamics as described earlier. It is convenient to use catheter-tip sensors because they are relatively immune to common artifacts and they need little external equipment.

Ambulatory urodynamics invariably uses natural filling, part for practicalities and part to reproduce the natural situation. This is in keeping with the need to document transient storage symptoms such as overactivity or leakage. In fact, there is some evidence that benign contractions of the detrusor are detected and reported on ambulatory urodynamics, and these may lead to the overdiagnosis of detrusor overactivity (Brown and Hilton 1997).

An electric version of the leakage diaper can be used in conjunction with ambulatory urodynamics. In implementation, this uses the electrical properties of urine. For example, conductance or capacitance between metal strips embedded in an absorbent pad gives a semiquantitative indication of leakage.

An instructive example of ambulatory urodynamics for the assessment of stress incontinence is given in Figure 20.34, using solid-state catheter-tip sensors and an electric diaper for semiquantitative leak detection. Note the maneuvers used to provoke incontinence (02:32–02:37) and the corresponding rises in rectal and bladder pressure. There is excellent

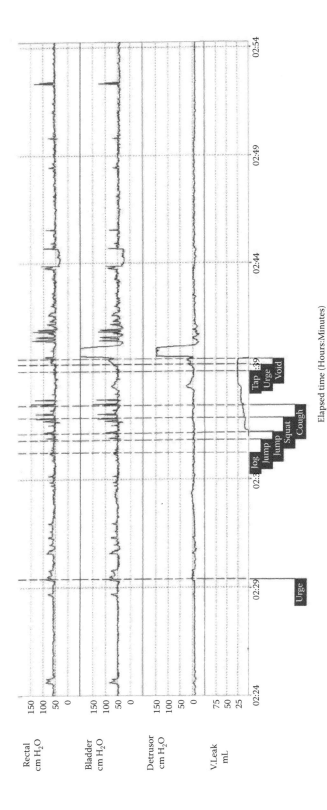

FIGURE 20.34 Stress incontinence on ambulatory cystometry. Note the maneuvers used to provoke leaking, beginning around 02:37.

cancellation of pressures in the detrusor trace, a feature of the high bandwidth of the catheter-tip sensors. The estimated leak volume of 25 mL would be highly inconvenient and embarrassing for daily living. In addition, note the urge at 02:39, provoked by a running tap, which led to a full void; flow rate was not recorded. The total study length of 2 h 54 min is far longer than would be considered acceptable for conventional urodynamics and gives adequate time to reproduce intermittent symptoms.

20.19.5 Electromyography

Electromyography (EMG) measurements are particularly associated with the investigation of neuropathic disorders of the lower urinary tract. In the past, needle electrodes have been used to investigate individual muscle action potentials, but these are technically difficult to record, and are not pleasant for the patient. Therefore, in most centers, the use of EMG is limited to surface electrodes measuring the activation of the pelvic floor muscles as shown in Figure 20.35.

The pelvic floor is important in maintaining continence, and so healthy pelvic floor muscles become increasingly active during the filling phase, from 0 to 12 s in the recording. Nerve or muscle damage during childbirth or due to spinal injury is a risk factor for incontinence and can be diagnosed from the lack of EMG activity during filling.

Conversely, the pelvic floor should relax during the normal void cycle, as from 12 to 46 s. Note the contractions from 47 to 50 s, used by this author to expel the final drops of urine from the bulbar urethra. If the pelvic floor does not relax during the void, the bladder will contract against a closed outlet, with an outcome similar to bladder outflow obstruction, that is a high bladder contraction pressure with little or no flow. This is termed bladder–sphincter dyssynergia and may be the result of neurological injury or disease. Note however that neuropathic bladder disorders are often associated with a whole constellation of symptoms and sequelae and pelvic floor EMG is only one of a number of different ways to diagnose them.

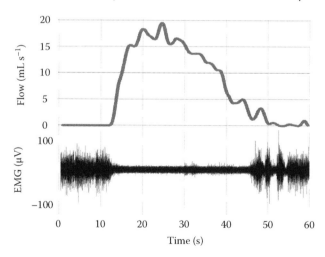

FIGURE 20.35 Uroflowmetry with simultaneous surface pelvic floor EMG. Data recorded from one of the authors using a home flowmeter, a PC with a sound card, and *Audacity*, the open-source sound editing program.

20.19.6 Video Urodynamics (Cystography)

In video urodynamics or cystography, the bladder is filled with a radiopaque contrast medium and imaged at intervals by fluoroscopy during the fill and void phases. Otherwise, the instrumentation is identical to conventional cystometry. Cystography will indicate some pathology, but the added benefit for urodynamics is to show anatomical events such as the descent of the bladder base and the bladder neck opening. Cystography can be unnatural and awkward, with the added risk of abdominal x-ray exposure and the need for staff with radiographic training. Therefore, as with ambulatory urodynamics, it is used in difficult cases. When viewed in conjunction with the pressure waveforms, cystography is useful for diagnosis of neurological difficulties and may also be used to assess surgical failure.

20.19.7 Cystoscopy

Cystoscopy is the endoscopic investigation of the urethra and bladder. The cystoscope is introduced *per urethra,* and the bladder can be filled to expand it and allow for a better view of the bladder wall. Cystoscopy can be used to investigate a range of symptoms but is particularly used with symptoms that are suggestive of serious pathology such as bladder cancer.

20.19.8 Urethral Pressure Profilometry

In some circumstances, it may be beneficial to quantify the pressure along the length of a dry urethra or less commonly a dry rectum. In the Brown and Wickham (1969) method, a water-filled catheter is passed to the bladder and then withdrawn using a catheter puller, typically at 2 to 5 mm s^{-1} (Figure 20.36). Meanwhile, continuous pressure measurements are made as for standard cystometry. Since the distal urethra is dry, the line must be perfused with saline, typically at 2 to 5 mL min^{-1}. In effect, the sensor measures the minimum pressure required to cause the infused saline to leak past the mucous membrane pressing against the catheter.

Under complete occlusion, the pressure in the catheter can only increase at a limited rate, the slew rate. The slew rate is a function of the catheter's volume compliance, because the catheter and other saline-filled components will expand slightly under the higher pressure.

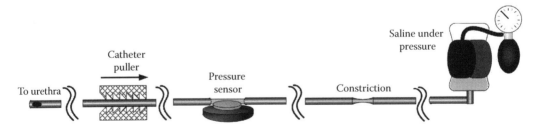

FIGURE 20.36 The Brown and Wickham technique for urethral pressure profile measurements. The catheter is pulled at 2 to 5 mm s^{-1} while being perfused with saline. Saline flow is maintained by a bag at typically 300 cm H_2O pressure, perfusing the catheter via a narrow constriction; this acts as a constant flow system. The pressure required to cause saline flow is measured at the sensor.

Faster infusion takes up compliance faster and improves the slew rate but may cause an offset in pressure due to resistive losses caused by the flow.

The urethral pressure profile has been most used to assess stress incontinence in women, making a measurement of closure pressure in the urethra where a low maximum value is suggestive of problems. Urethral pressure profilometry has also been used in a research context to measure the closure pressure under a prostatic obstruction in men. In the male urethra, a high pressure would suggest problems, but this method is not widely used for clinical assessment of obstruction.

Urethral pressure profilometry is falling from favor, since the measurement is reported to be poorly reproducible; the meaning and value of such measurements has been hotly debated. While this may admittedly be due to the lack of standardization and good clinical technique, it has not been helped by the relative complexity of the instrumentation. It is not widely cited but in our experience, similar measurements can be made using a solid-state catheter-tip sensor coated in an aqueous gel. Once again, the exact physics at the measurement site is not well understood, but in practice, it seems to work (Lose *et al* 2002).

20.20 THE FUTURE

With some exceptions, the urological complaints we describe are seldom fatal but are socially extremely embarrassing and debilitating. Therefore, there has been a substantial change of emphasis from objective measures of urological dysfunction to patient-oriented measures of bother. For example, a patient's self-report of her leakage symptoms would be held to be more important than a change in her urethral closure pressure, and a patient's view would be held in high regard when considering the appropriate treatment. This trend manifests itself also in the choice of investigation; we observe that the more esoteric, complex, or unpleasant tests (e.g., sphincter needle EMG) are going out of favor unless they have a direct bearing on diagnosis and treatment. The corollary is that patient-oriented investigations to reproduce the patient's experience in their own day-to-day life are becoming more important. It is widely said that *the bladder is an unreliable witness*, and it is understood that the urodynamics clinic is not a convivial environment—Figure 20.37.

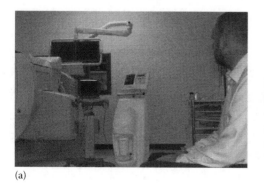
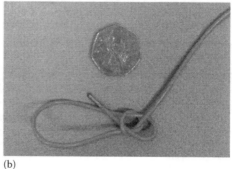

(a)   (b)

FIGURE 20.37   Two of the more unpleasant aspects of urodynamics:(a), waiting for the clinic; (b), this catheter was withdrawn in this state from a lady after ambulatory urodynamics. Thankfully, this is a rare outcome.

Therefore, while the core of urodynamics has remained similar for 30 years or more, a new breed of simpler investigations is emerging.

20.20.1 Home Uroflowmetry

Uroflowmetry is widely used to document symptoms, but there is a lot of evidence that an in-clinic flow test is not representative of home behavior. For example, *bashful bladder*, where patients are inhibited or sometimes completely unable to perform at the allotted time, affects over 10% of the population. It normally manifests itself in a public urinal but, if present, will almost certainly manifest itself in the urology clinic. The funnel uroflowmeter is a recent adjunct to the frequency–volume chart (Figure 20.38). A range of implementations have been described, but each is a funnel with one or more fixed outlets (Pel and van Mastrigt 2002). The height of urine in the funnel rises until outflow equals inflow, and so the height of urine in the funnel gives a crude indication of maximum flow rate. The devices have a poor resolution of perhaps 5 mL s^{-1}, but nevertheless, there is evidence that multiple measurements using these crude devices are more representative of a patient's true flow than a single in-clinic measurement (Caffarel *et al* 2007).

20.20.2 Noninvasive Measurement of Bladder Contractility

In keeping with the move away from complex investigations, there has been recent interest in noninvasive measurement of bladder pressure (van Mastrigt and Pel 1999). By analogy with a sphygmomanometer, the principle is to measure the pressure required to stop urine flow. A number of methods have been described using a condom catheter or occlusive penile cuff (Figure 20.38). However, all provide a pressure measurement under conditions of low or zero urine flow, that is, isovolumetric contraction.

20.20.3 Indwelling Sensors

Inevitably, improvements in technology are also a strong driver for development. It is now feasible to make a pill-like pressure sensor device that dwells in the bladder for hours or days, using telemetry to recover the data. This is technically more difficult than might be apparent, because an indwelling sensor cannot take its zero pressure reference from the atmosphere. Recent designs of sensor can measure relative to a sealed pressure reference

FIGURE 20.38 Three devices that improve the assessment of LUTS. From the left, the Uflow home funnel uroflowmeter (*MDTI Ltd*), the PeePod, and the noninvasive cuff test (*MediPlus Ltd*).

encapsulated with the sensor. This reference will need to be factory calibrated at body temperature. However, since all urodynamic measurements are relative to atmospheric pressure, which may vary even during the test, a separate external measurement of atmospheric pressure may still be required.

We recognize the technical difficulties that must be solved to produce a sensor that can offer stable measurements over a period of days while operating on next to no power; the opportunities for implantable pressure instrumentation are huge. Yet for urodynamics, it is still necessary to recover the pill after the study, which suggests either an umbilical cord or endoscopy. It is not clear to these authors that this is a step forward for clinical urodynamics, and no commercial devices have appeared to date.

20.21 EQUIPMENT SUPPLIERS

At the time of writing, costs of urodynamic instrumentation vary greatly. One might expect to pay around $2,000 for the most basic electronic flowmeter or in excess of $30,000 for a full urodynamic system with EMG and video capability. Among the biggest suppliers (in alphabetical order) are

- Andromeda   www.andromeda-ms.de

- Laborie   www.laborie.com

- MMS   www.mmsinternational.com

ACKNOWLEDGMENTS

The authors thank the following for contributing ideas, inspiration, materials, and proof-reading (in alphabetical order): Alison Bray, Jennifer Caffarel, Becky Clarkson, Andrew Gammie, Derek Griffiths, Chris Harding, Ron van Mastrigt, Rob Pickard, Peter Ramsden, Wendy Robson, Werner Schaefer and Doug Small; Mike Whitaker.

REFERENCES

Anders E K and Bradley W E 1983 History of cystometry *Urology* **22** 335–50

Andromeda 2014 http://www.andromeda-ms.de

Brown K and Hilton P 1997 Ambulatory monitoring *Int. Urogynecol. J.* **8** 369–76

Brown M and Wickham J E A 1969 The urethral pressure profile *Brit. J. Urol.* **41** 211–17

Caffarel J, Robson W, Pickard R, Griffiths C and Drinnan M 2007 Flow measurements: Can several wrongs make a right? *Neurourol. Urodyn.* **26** 474–80

Cooper M A, Fletter P C, Zaszczurynski P J and Damaser M S 2011 Comparison of air-charged and water-filled urodynamic pressure measurement catheters *Neurourol. Urodyn.* **30** 329–34

Gammie A, Clarkson B, Constantinou, C *et al* 2014 The International Continence Society urodynamic equipment working group: International Continence Society guidelines on urodynamic equipment performance *Neurourol. Urodynam.* **33** 370–9

Griffiths C J, Murray A and Ramsden P D 1983 A simple uroflowmeter tester *Brit. J. Urol.* **55** 21–4

Griffiths D J and van Mastrigt R 1985 The routine assessment of detrusor contraction strength *Neurourol. Urodyn.* **4** 77–87

Hok B 1976 Dynamic calibration of manometer systems *Med. Biol. Eng.* **14** 193–8

Kim K, Jurnalov C D, Ham S, Webb M J and An K 2001 Mechanisms of female urinary continence under stress: Frequency spectrum analysis *J. Biomech.* **34** 687–91

Kranse R, van Mastrigt R and Bosch R 1995 Estimation of the lag time between detrusor pressure- and flow rate-signals *Neurourol. Urodyn.* **14** 217–29

Laborie 2014 www.laborie.com

Lose G, Griffiths D, Hosker G *et al* 2002 Standardisation of urethral pressure measurement: Report from the standardisation sub-committee of the international continence society *Neurourol. Urodyn.* **21** 258–60

Manieri C, Carter S S, Romano G, Trucchi A, Valenti M and Tubaro A 1998 The diagnosis of bladder outlet obstruction in men by ultrasound measurement of bladder wall thickness *J. Urol.* **159** 761–5

Medical Measurement Systems MMS 2014 www.mmsinternational.com

Nichols W W and O'Rourke M F 2005 *McDonald's Blood Flow in Arteries*, 5th edition, (London: Hodder Arnold) pp 110–5.

Pel J J and van Mastrigt R 2002 Development of a low-cost flowmeter to grade the maximum flow rate *Neurourol. Urodyn.* **21** 48–54

Perez L M and Webster G D The history of urodynamics *Neurourol. Urodyn.* **11** 1–21

Pinnock C B and Marshall V R 1997 Troublesome lower urinary tract symptoms in the community: A prevalence study *Med. J. Aust.* **167** 72–5

Reynard J M, Yang Q, Donovan J L *et al* 1998 The ICS-'BPH' study: Uroflowmetry, lower urinary tract symptoms and bladder outlet obstruction *Brit. J. Urol.* **82** 619–23

Rosier P F, de la Rosette J J, Koldewijn E L, Debruyne F M and Wijkstra, H 1995 Variability of pressure-flow analysis parameters in repeated cystometry in patients with benign prostatic hyperplasia *J. Urol.* **153** 1520–5

Schaefer W 1985 Urethral resistance? Urodynamic concepts of physiological and pathological bladder outlet function during voiding *Neurourol. Urodyn.* **4** 161–201

Sonke G S, Kiemeney L A, Verbeek A L, Kortmann B B, Debruyne F M and de la Rosette J J 1999 Low reproducibility of maximum urinary flow rate determined by portable flowmetry *Neurourol. Urodyn.* **18** 183–91

Sonke G S, Kortmann B B, Verbeek A L, Kiemeney L A, Debruyne F M and de La Rosette J J Variability of pressure-flow studies in men with lower urinary tract symptoms *Neurourol. Urodyn.* **19** 637–51

Sullivan M P and Yalla S V 2000 Penile urethral compression-release maneuver as a noninvasive screening test for diagnosing prostatic obstruction *Neurourol. Urodyn.* **19** 657–69

Tammela T L J, Schaefer W, Barrett D M *et al* Repeated pressure-flow studies in the evaluation of bladder outlet obstruction due to benign prostatic enlargement *Neurourol. Urodyn.* **18** 17–24

van de Beek C, Stoevelaar H J, McDonnell J, Nijs H G, Casparie A F and Janknegt R A 1997 Interpretation of uroflowmetry curves by urologists *J. Urol.* **157** 164–8

van Mastrigt R, Koopal J W, Hak J and van de Wetering J 1986 Modelling the contractility of urinary bladder smooth muscle using isometric contractions *Am. J. Physiol.* **251**(5 pt. 2) R978–83

van Mastrigt R and Pel J J M 1999 Towards a noninvasive urodynamic diagnosis of infravesical obstruction *BJU Int.* **84** 195–203

BIBLIOGRAPHY

Abrams P. 2006 *Urodynamics*, 3rd edition (London: Springer).

Abrams P, Cardozo L, Fall M *et al* 2002 The standardisation of terminology of lower urinary tract function: Report from the Standardisation Sub-committee of the International Continence Society *Neurourol. Urodyn.* **21** 167–78

Chapple C R, McDiarmid S A and Patel A (2009) *Urodynamics Made Easy*, 3rd edition (London: Churchill Livingstone).

Griffiths D J 1982 *Urodynamics* (Bristol: Adam Hilger Ltd).

Griffiths D J, Hofner K, von Mastrigt R, Rollema H J, Spangberg A and Gleason D 1997 International Continence Society standardisation committee: Standardization of terminology of lower urinary tract function: pressure-flow studies of voiding, urethral resistance, and urethral obstruction. International Continence Society Subcommittee on Standardization of Terminology of Pressure-Flow Studies *Neurourol. Urodyn.* **16** 1–18

Mundy A, Stephenson T and Wein A 1994 *Urodynamics: Principles, Practice and Application* (Edinburgh: Churchill Livingston).

Schaefer W, Abrams P, Liao L *et al.* Good urodynamic practices: Uroflowmetry, filling cystometry, and pressure-flow studies *Neurourol. Urodyn.* **21** 261–74

Small D A urodynamics home page 2014 http://www.gla.ac.uk/departments/clinicalphysics/urodynamics/.

CHAPTER 21

Data Processing, Analysis, and Statistics

Gordon Silverman and Malcolm S. Woolfson

CONTENTS

21.1	Role of Data Processing in Physiological Measurement	538
21.2	Information	539
21.3	Information Limitations	540
21.4	System Capacity and Sampling Rate	540
21.5	Information Coding	541
21.6	Efficient Coding of Information	542
21.7	Huffman Coding	542
21.8	Huffman Decoding	544
21.9	Signal Processing and Computer Interface	544
21.10	Analog Filters	545
	21.10.1 Amplitude and Phase Response	545
	21.10.2 Distortion Caused by Filtering	549
21.11	Sampling	550
21.12	Analog-to-Digital Conversion	553
	21.12.1 Dual-Slope ADC	554
	21.12.2 Successive Approximation ADC	555
	21.12.3 Flash ADC	556
21.13	Digital Signal Preprocessing	556
	21.13.1 Fast Fourier Transform	557
	21.13.2 Digital Filters	558
	21.13.3 Emerging Preprocessing Developments	559
	21.13.4 Embedded Systems	559
	21.13.5 Software Tools	562
	21.13.6 Software Architectures	562
21.14	Statistics and Hypothesis Testing	564
	21.14.1 Overview	565
	21.14.2 Descriptive Statistics	565

21.14.3 Graphic Descriptions 567
21.14.4 Hypothesis Testing; Statistical Inference; Statistical Significance 567
21.14.5 Statistical Inference: Validity of Assumptions and Prediction 570
21.14.6 Bayesian Analysis: Statistical Inference: Looking Backward 570
21.14.7 BBN Considerations 573
21.14.8 Markov Chains and Analysis: Looking Forward 574
21.14.9 Computerized Markov Analysis Tools 577
21.15 Emerging Developments in Physiological Instrumentation 577
21.15.1 Knowledge-Based Techniques in Instrumentation 577
21.15.2 Neural Nets 578
21.15.3 Application of NN to Physiology and Medicine 581
21.15.4 Backpropagation (Training) 582
21.15.5 Summary of Knowledge-Based Technology 584
21.15.6 Medical Record Keeping and the Paperless Medical Office 584
Bibliography 585
Recommended Reading List 587

21.1 ROLE OF DATA PROCESSING IN PHYSIOLOGICAL MEASUREMENT

Data processing, analysis, and statistics play key roles in the measurement and interpretation of physiological information. These areas of data processing can be expanded to include the role of the computer in automated instrumentation, the nature of information and measurement, and various associated hardware and software components. These will form the focus of this chapter.

The word *physiology* is derived from the Greek *physis* meaning *nature* and *logos* meaning *word*. In its modern context, it refers to the study of mechanical, physical, and biochemical functions of living organisms embracing both animal and plant models. There may not be clear origins for such systematic observations, but Herman Boerhaave can be identified as a key historical figure. His work entitled *Institutiones medicae* (1708) might give him the right to the title of *father of physiology* (Lindeboom 2007).

While physiology is a broad label, it has become highly specialized and has numerous subdivisions: electrophysiology is concerned with nerves and muscles; neurophysiology addresses the functioning of individual cells and therein the brain. (Other subdisciplines include plant physiology, ecophysiology, genetics, and others as well.) Written observations have been replaced by mechanical and electrical instrumentation methods so that the information content of contemporary records is an ever-increasing record of evolution with the potential for greater understanding of our origins and the evolutionary process as well as our biological makeup. With such data, it becomes possible to extend life and improve the delivery of health care as well as the quality of living. Modern instruments that process information aid the user in recording, classifying, and summarizing information and subsume a number of tasks: data handling, instrumental control, experimental development, operator control, documentation, and report generation, and its newest purposes include automated interpretation of data augmented through methods of *artificial*

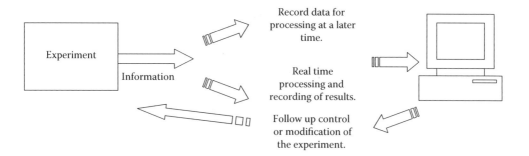

FIGURE 21.1   Functional diagram of information processing in instrumentation applications.

intelligence (Silverman and Silver 1995). Figure 21.1 summarizes the physiological data processing paradigm.

As shown, data handling includes distinct operations with well-defined purposes:

- *Data acquisition*: detection, classification, coding, and representation.

- Data compression to improve recording efficiency and system capacity.

- *Data reduction*: eliminate spurious and/or irrelevant information.

- *Data interpretation*: deciding whether the data satisfy the constraints of a theory or conform to the anticipated outcome of the experiment. (As will be discussed subsequently, *artificial intelligence* (e.g., Neural Nets) is rapidly augmenting human interpretation of data.)

- *Documentation*: journal maintenance.

21.2 INFORMATION

Experiments are characterized by the amount of *information* contained in the process, and information is intimately related to the number of possible outcomes and their associated probabilities (Abramson 1963, Hamming 1980). The basic unit of information is the *bit*—not to be confused with a single unit in the binary number systems—and can be understood in the following context:

Consider an experiment with n equally likely outcomes; the amount of information is derived from the following relation:

$$2^H = n.$$

In this formula, H is the amount of information contained in the n outcomes. By taking the logarithm of both sides of the equation, this equation is normally written as

$$H = \log_2 n,$$

where the log is taken to the base 2. Units of H are *bits* (of information).

However, most experiments do not have equally likely outcomes. In such cases, each outcome is associated with an amount of information (h_i) determined by its probability:

$$h_i = \log_2 \frac{1}{p_i} = -\log_2 p_i.$$

In this formula, p_i is the probability associated with an experimental outcome. The total information can now be computed by averaging over all outcomes:

$$H = -\Sigma_i p_i \log_2 p_i.$$

Experiments with equally likely outcomes contain the maximum information content.

21.3 INFORMATION LIMITATIONS

Noise as well as inherent instrumentation limitations confines representation of information to numbers or codes using discrete formats with a fixed (finite) number of digits or symbols. These representations (i.e., numbers) are defined by the resolution and accuracy of the overall system in the form of *resolution* and *accuracy*.

- Resolution concerns the ability to repeat the measurement result each time the same outcome is presented to the system.

- Accuracy is a measure of the system's ability to repeat the *correct* measurement.

Hence, resolution is reported by the number of digits (significant figures) in the measurement, while accuracy is an error number (e.g., percentage) defining the quality of the measurement.

21.4 SYSTEM CAPACITY AND SAMPLING RATE

In addition to describing the amount of information, the speed at which a system must process the information is also important. This is included in the *capacity* requirement, which is defined as follows:

$$\text{Capacity} = \frac{H}{T},$$

where
   H is the information
   T is the time allotted to process the information

This leads to a rate—the *sampling rate* (Unser 2000)—at which the information must be tested (*sampled*). It is constrained by the Nyquist sampling rate—also referred to as the Shannon sampling rate—which requires that the dependent information be sampled at more than twice the highest frequency contained in the information source. If it is not sampled at least at this rate, the original information cannot be recovered. Sampling is the process of converting a signal (e.g., a function of a continuous independent or dependent variable) into a numeric or coded sequence (see discussion later). The Nyquist–Shannon formula summarizes the relationship between system capacity (C), bandwidth (B), and number of allowable levels per signal element in the message.

$$C = 2B \log_2 M$$

(If two levels per signal element is used (0 and 1), then the formula reduces to $C = 2B$. The concepts associated with system bandwidth are explored later in this chapter.)

21.5 INFORMATION CODING

The discrete events generated within an experiment must be converted to a form for the processing indicated in Figure 21.1. The transformation of information (Avudaiammal 2009, Hamming 1973) has differing objectives, which leads to a variety of coding methodologies, which fall into one of two general groups:

1. Compressive coding: representation of the data in an efficient form while reducing any redundancies.

   An example of such coding is used in many Internet applications where the common *zip* data compression algorithm reduces network traffic and also results in smaller file sizes. In a more popular vein, people using the (social) Internet might use *IMHO* to represent the phrase *in my humble opinion*. Such shorthand coding reduces file size and speeds delivery of the message.

2. Reliability coding: reliability is achieved by adding *meta-information* to the source (code), which provides information that is descriptive of the code itself. Codes like these can correct errors as well as provide barriers to distortions (dirt, scratches, and noise) on CDs and can enhance the quality of musical reproduction. A sketch of the concept is shown in Figure 21.2.

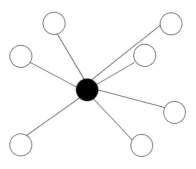

FIGURE 21.2   Representation of experimental outcomes, both correct and with errors.

In Figure 21.2, the black node represents the correct code for an experimental outcome; white nodes are those produced by an error generated in the code during processing. By adding additional information to the underlying code, these erroneous states (errors) can be detected and in some circumstances can be corrected.

21.6 EFFICIENT CODING OF INFORMATION

The probabilities of the experimental outcomes are closely linked to the efficiency of the coding process. Notice, for example, that in the Morse code—whose fundamental coding elements consist of two symbols, a *dot* and a *dash*—the code for the letter *e* in the English language is a *dot*. Such a simple code takes advantage of the fact that the letter *e* is the most frequently used letter in the language and therefore has a high probability of appearing in a string of characters. This results in a shorter coded sequence for a string of characters, because code positions are not *wasted* on highly likely outcomes. This concept can be generalized as shown in the Huffman code.

21.7 HUFFMAN CODING

As previously noted, the number of bits of information generated by an experiment depends in part on the probabilities of the underlying events (Huffman 1952, Silverman and Silver 1995, http://cs.pitt.edu/~kirk/CS1501/animation). In 1951, David Huffman, then an MIT student of information theory, was challenged by his professor (Robert Fano) to come up with finding the most efficient binary code for representing information. Just before he gave up, Huffman struck upon the idea of a frequency-based binary tree algorithm for such a code. He subsequently demonstrated that this method was the most efficient code for such representation. The following example illustrates the algorithm.

The defining elements of the code are limited to 0 and 1; thus, the coded outcome representing a sequence of experimental outcomes will consist of a sequence of such characters. Table 21.1 consists of five experimental outcomes identified only by a number together with its associated frequency of occurrence (or probability) listed in decreasing order. These outcomes must be coded for subsequent processing, as shown in Figure 21.3.

We need to establish a variable-length sequence of 0s and 1s that represents a sequence of experimental outcomes shown in Table 21.1. The algorithm shown in Figure 21.3 describes the Huffman methodology.

If this algorithm is applied to the results depicted in Table 21.1, the following binary tree is produced (Figure 21.4).

TABLE 21.1   Experimental Outcomes

Experimental Outcome (Number)	Frequency of Occurrence
1	15
2	7
3	6
4	6
5	5

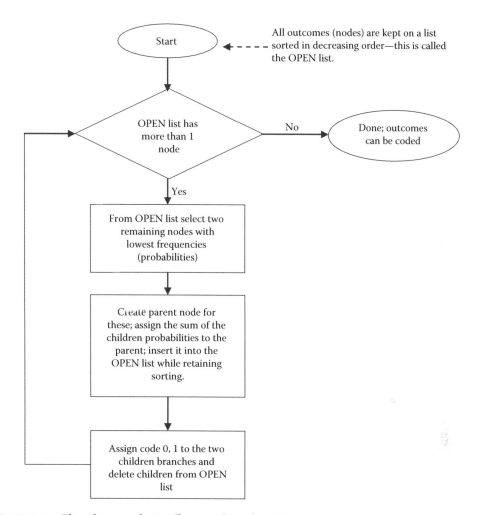

FIGURE 21.3   Flow diagram for Huffman coding algorithm.

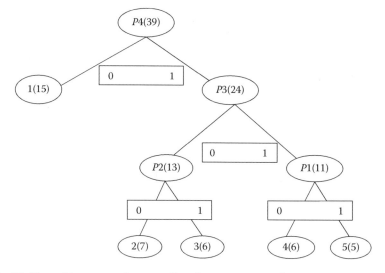

FIGURE 21.4   Huffman binary tree for encoding five experimental outcomes.

TABLE 21.2   Huffman Code for Example

Outcome	#Occurrences	Code	# Bits in Msg
1	15	0	15
2	7	100	21
3	6	101	18
4	6	110	18
5	5	111	15

The resulting Huffman code for each of the five outcomes and the number of times that they are produced in the experiment (and sent to a computer-based data processing system) is summarized in Table 21.2.

The total number of bits in the entire message is 87—add the last column in Table 21.2. The experiment itself consisted of 39 outcomes—add column 2 in Table 21.2. The average number of bits/message is thus 2.23 (bits), which is close to the entire information content of the underlying experiment and therefore represents highly efficient coding.

21.8 HUFFMAN DECODING

How is the message to be decoded? The sequence is summarized as follows:

- Start with first bit in the stream.

- Use successive bit to determine whether to go *left* or *right* in a binary decoding tree.

- If a leaf is reached, the experimental outcome is decoded and it can be placed in the uncompressed output stream.

- The next bit that is encountered is then the first bit of the next character.

21.9 SIGNAL PROCESSING AND COMPUTER INTERFACE

As implied in Figure 21.1, digital signal processing (DSP) techniques are used extensively in engineering to extract information about the underlying information (signal) (Balmer 1997, Horowitz and Hill 1989, Kudeki and Muson 2009, McClellan *et al* 2003, Mitra 2011). However, it is important that the preprocessing steps carried out on the signal introduce the least distortion and loss of information as possible. Often, the experimental sources may produce continuous analog information and the general processing steps that are carried out on such signals are shown in Figure 21.5.

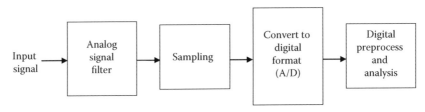

FIGURE 21.5   Block diagram of analog signal processing steps.

The analog filter reduces the effect of noise and other unwanted information. The resulting filtered signal is then sampled and digitized using an analog-to-digital converter (A/D). The digitized signal is further processed (using DSP) to further filter out noise and to extract the *meaningful* parts of the input signal. Finally, the data are analyzed in a computer using digital signal analysis and statistical methods to obtain information about the signal. Each of the processing steps may distort or destroy information in the original signal. Such distortions may be minimized with well-designed instrumentation.

21.10 ANALOG FILTERS

21.10.1 Amplitude and Phase Response

Electrical signals may be thought of as composed of a *pure* signal plus an unwanted component, which could be noise or interference. The noise could come from various sources. For example, when processing an electrocardiogram (ECG), noise could come from extraneous signals generated in muscle (electromyogram (EMG)); the instrumentation itself, or the power lines. Initially, the effects of noise and interference may be reduced before converting the signal-to-digital form. To achieve this, an *analog filter* is used (as shown in Figure 21.5).

To understand the general effects of filters on signals, consider a (co)sinusoidal input to such a filter:

$$v_{in}(t) = \cos(2\pi f t). \tag{21.1}$$

The output signal from such a filter will be given by

$$v_{out}(t) = A(f)\cos(2\pi f t + \phi(f)), \tag{21.2}$$

where the amplitude, $A(f)$, of the output signal could be less than, equal to, or greater than the signal defined in Equation 21.1 where its amplitude is 1. $A(f)$ and $\phi(f)$ (phase shift) depend on the *frequency* of the input signal; $A(f)$ as a function of f is termed the *amplitude response* of the filter and $\phi(f)$ is the *phase response* of the filter.

Suppose that we design a band-pass filter, which passes cosines with frequencies in the range

$$f_l < f < f_h. \tag{21.3}$$

As an example, the *ideal amplitude response* for a band-pass filter that only passes frequencies between 1 and 2 Hz would look like the dashed curve in Figure 21.6.

Noise and interference effects outside the range of frequencies f_l to f_h would then be eliminated. However, we do not wish to eliminate useful frequency components of the

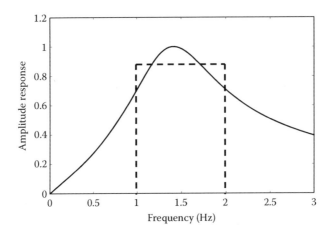

FIGURE 21.6   Frequency response of a band-pass filter with the dotted curve representing an ideal response. The solid curve is representative of a practical band-pass filter.

signal itself by inappropriate choice of f_l and f_h; otherwise, signal information is lost and the resultant signal is distorted. When designing an appropriate filter, we model the input signal as a set of cosines to see what frequencies of the signal are significant. A powerful analytic technique follows from the fact that signals can often be decomposed into a series (sum) of sinusoids of varying frequency. This permits us to determine the important informational frequencies and hence f_l and f_h for the filter. The series representation is referred to as a *Fourier series* and the general method to perform the conversion is called the *Fourier transform*. The form of such signals is given by a linear combination of underlying cosine function as follows:

$$s(t) = \sum_{p=1}^{N} a[p]\cos\left(2\pi f_p\, t + \phi_p\right), \tag{21.4}$$

where $a[p]$ is the amplitude of the cosine with frequency f_p and the corresponding phase is designated by ϕ_p. The bandwidth of the signal is determined by the range of frequencies $\{f_p\}$ such that the amplitudes $\{a[p]\}$ contain significant information.

   For example, one way to determine how significant a spectral frequency is for a signal is to use *correlation*. Suppose that we take the signal of interest to be a cosine with $f_p = 5$ Hz, $a = 10$ and $\phi = 1.1$, as illustrated in Figure 21.7.

   One strategy for determining the potential informational content of this cosine is to multiply this signal by a *test* cosine signal with frequency f—a test frequency—and then to integrate the product from $t = -\infty$ to $t = +\infty$. If, for example, a test frequency of 8.5 Hz is selected, then the product of this with the original cosine shown in Figure 21.7 will result in a signal depicted in Figure 21.8. The integral or area under this resultant signal will be extremely small because positive areas will cancel with negative areas, and thus, this signal contains little informational content.

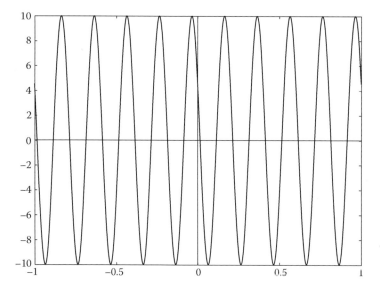

FIGURE 21.7   Experimental cosinusoidal signal.

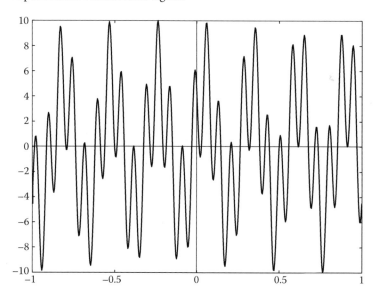

FIGURE 21.8   Product of signal of interest ($f = 5$) and test signal ($f = 8.5$).

Alternatively, if a test signal having a frequency = 5 Hz is used, the product (integrand) of this with the signal of interest has a very different character, which is depicted in Figure 21.9. The area under the curve, $S(f)$, is represented by the following equation:

$$S(f) = \int_{-\infty}^{\infty} A\cos(2\pi f_0 t + \phi) \cdot \cos(2\pi f t)dt,$$
(21.5)

where

f is the test frequency

f_0 is the frequency that we wish to find

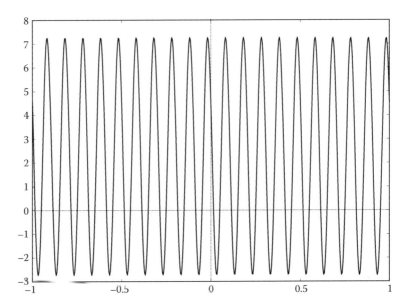

FIGURE 21.9   Integrand of signal of interest ($f = 5$) and test signal ($f = 5$).

The integral in Equation 21.5, represented by the area under the curve, is now nonzero. The area is nonzero because the test cosine is *synchronized* to some extent with the original signal of interest. The two signals *reinforce* each other when multiplied together, resulting in a net nonzero area under the curve, as shown in Figure 21.9. The integral in Equation 21.5 is termed a correlation of the original signal with a test cosine with frequency f. Hence, correlation with a test cosine with the cosine of frequency f is a maximum when the test cosine has the same frequency as the original signal. For a general signal, the correlation is carried out for various test frequencies f.

Complete information about the pth term in Equation 21.4 ($A[p]$ and $\phi[p]$) can be obtained by also correlating the underlying signal with a test *sine* with frequency f and correlation with a resultant cosine and sine signal can be combined to form the Fourier transform:

$$S(f) = \int_{-\infty}^{\infty} s(t)\exp(-j2\pi ft)dt. \qquad (21.6)$$

The amplitude spectrum is found by plotting $|S(f)|$ versus f. The range of frequencies for which the amplitude spectrum is significant gives the *bandwidth* of the signal and can be used to choose the *cutoff frequencies* for the analog filter. In practice, the Fourier transform is calculated digitally by using the *fast Fourier transform* (FFT). (We note later that prefiltering of the analog signal is used to reduce the sampling frequency when digitizing the signal. See also the discussion on information theory.)

21.10.2 Distortion Caused by Filtering

Prefiltering of the analog signal introduces distortion to the underlying signal. There are two reasons for this:

1. Distortion through the Amplitude Response

   The amplitude response of a filter describes those frequencies that are passed by the filter; an example is given for a band-pass filter in Figure 21.6. The dotted curve is the response for an *ideal* filter. This filter *passes* frequencies of the input signal between 1 and 2 Hz and *stops* (rejects) frequency components outside this range. The response is flat within the *passband*—the filter does not favor any frequencies within the passband. A filter with this type of idealized response cannot be constructed in practice. A typical response of a practical filter is shown as a solid line in Figure 21.6. It can be seen that there is not a sharp response at the cutoffs and that not all frequencies within the passband are given equal weight.

2. Distortion through the Phase Response

   The phase response of a filter describes the effect of the filter on the phase of each of the frequency components of the input signal. To see this, consider an example of a simple signal consisting of two frequencies (4 and 7 Hz):

$$v_{in}(t) = \cos(2\pi(4)t) + \cos(2\pi(7)t). \tag{21.7}$$

Now suppose that the phase response of the analog filter is given by

$$\phi(f) = -0.1f^2. \tag{21.8}$$

(Suppose that the amplitude response is unity for all frequencies.)

The output signal according to Silverman and Silver (1995) will then be given by

$$v_{out}(t) = \cos(2\pi(4)t - 1.6) + \cos(2\pi(7)t - 4.9) \tag{21.9}$$

and can be rewritten as

$$v_{out}(t) = \cos\left(2\pi(4)\left(t - \frac{0.2}{\pi}\right)\right) + \cos\left(2\pi(7)\left(t - \frac{0.35}{\pi}\right)\right). \tag{21.10}$$

Comparing Equations 21.7 and 21.10, we see that the 10 Hz component of $v_{in}(t)$ has been delayed in time by $0.2/\pi$ s and the 7 Hz component has been delayed by $0.35/\pi$ s. *Hence, different frequency components have been delayed by different times, resulting in distortion of the output signal.*

For the output signal to be an undistorted version of the input signal, the phase response should be proportional to the frequency

$$\phi(f) = -\alpha f, \tag{21.11}$$

where α is a constant. We refer to this type of filter as a *linear phase filter*.

Suppose this linear phase filter has an amplitude response that is unity for all frequencies. In this case, if the input signal consists of two frequencies f_1 and f_2 as shown in Equation (21.12):

$$s(t) = \cos(2\pi f_1 t) + \cos(2\pi f_2 t) \tag{21.12}$$

then the output signal is given by

$$s(t) = \cos(2\pi f_1 t - \alpha f_1) + \cos(2\pi f_2 t - \alpha f_2) = \cos\left(2\pi f_1\left(t - \frac{\alpha}{2\pi}\right)\right) + \cos\left(2\pi f_2\left(t - \frac{\alpha}{2\pi}\right)\right)$$

$$\tag{21.13}$$

(i.e., the same time shift to both frequency components).

In summary, if the phase response of a filter is not linear with frequency, then distortion will occur, no matter how sharp is the amplitude response. Unfortunately, no analog filter has an exactly linear phase response. Several filters have been used extensively:

- *Chebychev filters:* relatively sharp amplitude response; nonlinear phase response

- *Bessel filters:* nearly linear phase response; *gentle* amplitude response

- *Butterworth filters:* a compromise between Chebychev and Bessel filters

As a consequence, one has to accept that there will be distortion of some kind to the underlying clean signal and selection of an appropriate filter is an *art* in which experience plays a significant role. One would also need to take into account whether any distortion caused to the underlying signal would lead to features on the filtered signal, which could cause misinterpretation of physiological signals.

21.11 SAMPLING

Signal filtering is followed by digitizing or conversion, as shown in Figure 21.5. Digital signals are obtained from analog signals by two operations: *sampling* and *binary coding*. When we *sample* an analog signal, we collect (sample) information from this signal at constant intervals of time. This process produces a discrete series of values resulting in the sampled signal, $x_s(t)$. The time between adjacent samples = $\Delta\tau$. The *sampling frequency* is defined as

$$f_s = \frac{1}{\Delta\tau}\,\text{Hz}. \tag{21.14}$$

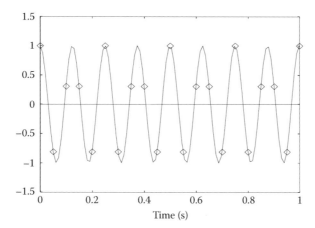

FIGURE 21.10   Sinusoidal signal with diamonds indicating the sampled results.

Sampling of any signal means that we are ignoring information between samples and hence we are losing this information. In particular, there would be ambiguities in determining the frequency of the cosine. To see this, suppose that an 8 Hz cosine signal is sampled with a sampling frequency $f_s = 20$ Hz. The original and sampled signals are shown in Figure 21.10.

However, a sinusoidal wave with a frequency of 12 Hz would also pass through the samples that are shown in Figure 21.10. Thus, the original 8 Hz signal could be misread as a 12 Hz signal. In fact, it can be shown that it is possible to put an infinite number of cosines with the following frequencies through the samples: 8, 12, 28, 32, 48 Hz, etc.

To explain what is happening, note that the original cosine signal with frequency f is given by

$$s(t) = \cos(2\pi f t). \tag{21.15}$$

For a signal of unit amplitude, the sample values from Equation 21.15 are

$$1, \cos\left(2\pi \frac{f}{f_s}\right), \cos\left(2\pi \frac{2f}{f_s}\right), \ldots, \cos\left(2\pi \frac{mf}{f_s}\right). \tag{21.16}$$

However, another frequency that would give rise to the same sample values is $f + f_s$, replacing this frequency into the mth sample value:

$$\cos\left\{2\pi \frac{m(f + f_s)}{f_s}\right\} = \cos\left\{2\pi \frac{mf}{f_s} + 2\pi m\right\} = \cos\left(2\pi \frac{mf}{f_s}\right). \tag{21.17}$$

Yet, another frequency that gives the same sample values is $f - f_s$:

$$\cos\left\{2\pi \frac{m(f - f_s)}{f_s}\right\} = \cos\left\{2\pi \frac{mf}{f_s} - 2\pi m\right\} = \cos\left(2\pi \frac{mf}{f_s}\right). \tag{21.18}$$

In the general case, it can be shown that cosines with the following frequencies pass through the same sample values:

$$f, f_s - f, f_s + f, 2f_s - f, 2f_s + f, \dots. \tag{21.19}$$

This ambiguity in determining the frequency is termed aliasing; a consequence of this is that the sampled signal can be confused with any of the aforementioned frequency components. When sampling a signal, bear in mind that eventually we would like to recover information about the original analog signal from the samples. It can be seen from Equation 21.19 that f is one of the frequencies that can be put through the sample points. Now, if the following condition holds

$$f \leq f_s - f, \tag{21.20}$$

then this will always be the *lowest* frequency that can be fitted to the sample points. This means that we can recover the original frequency, f, by application of an analog low-pass filter to the samples, with cutoff between f and $f_s - f$.

This condition Equation 21.20 can be rewritten as

$$f_s \geq 2f.$$

In general, an analog signal will have frequency components as follows:

$$f_{min} \leq f \leq f_{max}. \tag{21.21}$$

Generalizing Equation 21.21, it can be seen that if the sampling frequency satisfies

$$f_s \geq 2f_{max}, \tag{21.22}$$

then the original analog signal, represented by the range of frequencies

$$f_{min} \leq f \leq f_{max},$$

can be recovered from the samples by applying an analog band-pass filter to the samples, with lower and upper cutoffs of f_{min} and f_{max}, respectively. The minimum sampling frequency, sometimes called the *Nyquist frequency*, that can be used is given by

$$f_s^{min} = 2f_{max}, \tag{21.23}$$

a result attributed to Claude Shannon (Unser 2000).

However, in practice, one always loses information about the original signal through sampling. One reason for this is that all analog signals of finite duration have infinite bandwidth, that is, $f_{max} \to \infty$. Hence, according to the sampling theorem, no matter how high the sampling frequency, loss of information about the underlying signal will always occur. Additionally, noise accompanying the input signal has a wide range of spectral frequencies and produces distortions. In practice, one can define an effective maximum significant frequency of the input analog signal, f_{max}. The analog filter that is used in the first stage has a cutoff frequency equal to f_{max} that serves to filter off some of the noise and to filter *insignificant* frequency components of the underlying clean signal. (An estimate of which frequencies contribute to the information content helps to make this determination.) Given that the amplitude response of practical analog filters is nonideal, one generally uses sampling frequencies that are larger than f_s^{min}. The analog low-pass filter that is used prior to sampling is called an *anti-aliasing filter*; this filter reduces the effects of noise and *unimportant* frequency components of the original continuous signal that would contribute to aliasing in the sampled signal frequency spectrum if retained.

21.12 ANALOG-TO-DIGITAL CONVERSION

After sampling the analog signal, each sample is binary coded; this step is called *quantization* (Mitra 2011). When quantizing, each analog sample is converted to one of 2^N values, where N is the number of bits used for quantization. Suppose that a sample of an analog signal has a value of 2.5 V. As a very simple example, consider a 2-bit quantization of this sample:

In this example, the sample is coded as 10. However, it can be seen that another sample with close analog value, say 2.3 V (shown dotted in Figure 21.11), will also result in a quantized value equal to 10. This leads to what is known as *quantization error* representing a loss of information due to the quantization.

If the underlying analog signal is a sine wave, which has a peak-to-peak value equal to the dynamic range of the *analog-to-digital converter* (ADC), it can be shown that the signal-to-noise ratio associated with quantization is given by

$$SNR_{quant} = 6N + 1.76 \text{ dB}, \tag{21.24}$$

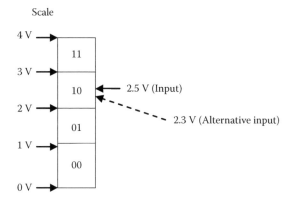

FIGURE 21.11   Simple quantization scheme.

where N is the number of bits used in the quantization. Hence, the larger the N, the less significant is the quantization error. However, increasing N leads to a larger bit rate in the system, given by Nf_s, which would slow down the hardware—see limits to system capacity. It should also be pointed out that the aforementioned result (Zarzoso and Nandi 1999) is for a sine wave that covers all the binary levels from $00...000$ to $11...111$, using up all the dynamic range. Smaller amplitude features will have a lower SNR_{quant} and if these features are of interest medically, then this must be borne in mind when choosing how many bits to use in the quantization process. In addition, thermal noise will reduce the overall signal-to-noise ratio. There are various methods of converting an analog sample to binary coded form; three methods are described here, each with its own advantages and disadvantages.

21.12.1 Dual-Slope ADC

The principle behind this ADC is illustrated in Figure 21.12. The first step is to integrate the sample of the analog voltage, v_A, for a fixed time T_1. The output from the integrator, v_A, after time T_1 will be proportional to v_A. A second integration follows this time in a negative potential direction with a known reference voltage V_r; its magnitude V_r is chosen to represent the *maximum* analog voltage that will be digitized—in this case, the digital word representing V_r would be $111...111$. The time it takes to complete the integration (to zero) is proportional to v_A. Figure 21.12 shows how this is accomplished, where it is also shown that smaller input analog voltage, v_B, leads to a smaller down integration time.

From Figure 21.12, we note that the time to complete the second integration can be used as a measure of the value of the input voltage. If a binary counter is used to determine this time, then the reading of the binary counter when the output from the integrator is zero will give the binary coded value of the input voltage. The dual-slope converter has relatively simple hardware and is accurate because component tolerances do not contribute to error in the integration process. The conversion time is relatively large because the total conversion time is $2^{N+1} \times T$ (where T is the clock period).

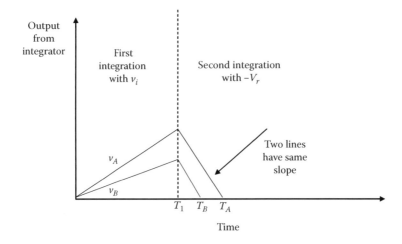

FIGURE 21.12   Graphical representation of the dual-slope A/D integrator.

21.12.2 Successive Approximation ADC

This ADC successively determines each bit of the final result. Consider a 3-bit ADC and an analog voltage V_{max} with a maximum value of 8 V. We construct a table showing how the analog voltage is converted to a 3-bit word (Table 21.3).

The bits in the code have the corresponding equivalent weight:

Most significant bit: half the scale or 4 V

Next most significant bit: ¼ of the scale or 2 V

Least significant bit: ⅛ of the scale or 1 V.

A block diagram for achieving successive approximation is shown in Figure 21.13.

The operational procedure successively divides the range of possible voltage levels in half at each iteration of the clock. The ADC converts this binary number into an equivalent analog quantity that is compared against the input. The bit will be either retained or reset (to 0) as determined by the comparator. Thus, complete conversion is accomplished in N clock cycles, where N is the number of bits of the digital result. Such converters are faster than dual slope but require higher tolerance components (and thus more costly).

TABLE 21.3   Successive Approximation ADC

Analog Voltage v_i	Binary Code
$v_i \geq 7$ V	111
$6 \leq v_i < 7$	110
$5 \leq v_i < 6$	101
$4 \leq v_i < 5$	100
$3 \leq v_i < 4$	011
$2 \leq v_i < 3$	010
$1 \leq v_i < 2$	001
$0 \leq v_i < 1$	000

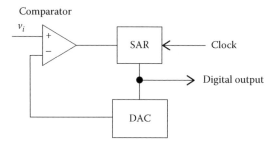

FIGURE 21.13   Block diagram for successive approximations. SAR is the successive approximations register; DAC is a digital-to-analog converter, which takes a digital input and generates an equivalent analog output.

TABLE 21.4   Equivalent Analog and Binary Values for a Flash Converter

Analog Voltage v_i	Binary Code
$3 \leq v_i < 4$	11
$2 \leq v_i < 3$	10
$1 \leq v_i < 2$	01
$0 \leq v_i < 1$	00

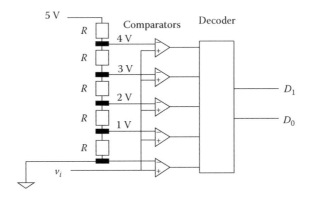

FIGURE 21.14   Block diagram of a flash converter.

21.12.3 Flash ADC

The principle behind this ADC is the simplest of the three; the input signal is compared against all possible digital equivalents at the same time. It is thus the fastest because this can be accomplished in one clock interval. Consider, for example, that an analog signal is to be converted into a 2-bit digital result and that the maximum voltage for the input is 4 V. The input levels and binary equivalent are shown in Table 21.4.

A block diagram for a 2-bit flash ADC is shown in Figure 21.14.

The resistive divider network produces five distinct voltage outputs: 0, 1, 2, 3, and 4 V. The input is compared against each of these voltages simultaneously, and those comparators with reference inputs that exceed the input will be 0 while those that are less than the input can assume a value of (say) 5 V. These comparators present their results to the decoder whose output is the binary equivalent of the comparator outputs, thus producing a binary code for each possible value of the analog input. Flash converters are the fastest type of converter but for high bit values create tolerance problems for the resistor chain as well as the comparators. In addition, the decoder has greater logical complications, resulting in large (chip) switch count. The limit for such converters is generally 8-bit quantization; however, for larger number of bits, pipelining techniques can be used for this ADC.

21.13 DIGITAL SIGNAL PREPROCESSING

With the developments in the speed and power of digital processing hardware, there has been an increase in the last few decades in the range of DSP methods that have been applied to the data; most of these methods cannot be implemented using analog means. For example, one can design digital filters to have frequency responses that *cannot* be obtained with analog filters—hence, a wider range of filtering operations may be obtained with digital rather than analog filters.

The application of DSP methods to biomedical engineering and physiology can be divided into two categories as required by the application:

1. *Preprocessing*: Further processing is carried out on the samples after digitizing— filter out any noise or interference; separate sources from component mixtures; or to enhance features of a signal and then to detect those features automatically.

2. *Extraction*: Extraction of information from the signal.

Information about the signal that is of medical interest, for example, heart rate if an ECG signal, can be extracted by the application of the appropriate DSP method.

In this material, we will be concerned with (1), namely, the further processing of the digital signal that is required prior to extracting information. The processing that is carried out will depend on the information that one wishes to extract from the signal. For example, if one wishes to extract just the R-waves of an ECG signal, then distortion of the underlying signal is acceptable if that would make it easier to extract this information. This would not be an acceptable procedure, however, if one wishes to preserve the shape of the signal. What follows is a brief survey of some of the methods that are used.

21.13.1 Fast Fourier Transform

The Fourier Transform is used to estimate the lowest and highest frequency components of the signal, and this information may then be used to design the analog filter. In practice, this Fourier transform is worked out digitally using the discrete Fourier transform (DFT). (It can be found on Excel spreadsheets.) The DFT relates samples of the signal $\{x[n]\}$ with samples of the Fourier transform $\{X[p]\}$:

$$X[p] = \sum_{n=0}^{N-1} x[n] \cdot \exp\left(-j\frac{2\pi np}{N}\right) \quad p = 0, 1, \ldots, N-1, \quad (21.25)$$

where N is the number of samples in the data and the frequency index p represents a frequency

$$f_p = \frac{pf_s}{N} \quad (21.26)$$

and f_s is the sampling frequency. The amplitude spectrum $|X[p]|$ is plotted as a function of p. The DFT assumes that the underlying signal is periodic, with period equal to the duration, T, of the data.

The FFT is an improved version of the DFT wherein the number of calculations (multiplications) shown in Equation 21.25 is greatly reduced (Cooley and Tukey 1965).

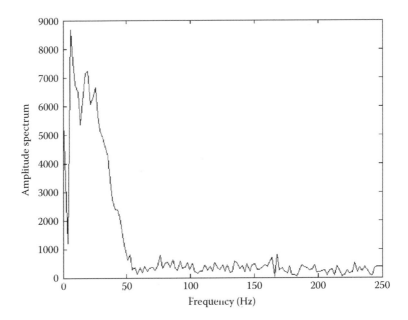

FIGURE 21.15 FFT of fetal ECG.

Consider an application of the FFT as shown in Figure 21.15, which depicts a typical amplitude spectrum for a fetal ECG signal.

The FFT is very defensive against noise and has the effect of spreading the noise across a wide range of frequencies. This contributes to the background in Figure 21.15. From Figure 21.15, one could estimate the bandwidth as approximately 50 Hz. However, features associated with the R-wave will extend to higher frequencies than this in the amplitude spectrum and a bandwidth of 120 Hz may be more realistic.

21.13.2 Digital Filters

A representative sampling of various digital filters is shown in Table 21.5 together with appropriate comments.

Adaptive and other digital filters have been used in many other biomedical applications and can be found in the following references: Bozic (1994), Database for the Identification of Systems (2011), de Weerd (1981), Ferrara and Widrow (1982), Haykin (1996), Hyvärinen (2113), Pan and Tompkins (1985), Smith (2011), Strang and Nguyen (1996), Woolfson *et al* (1995), and Zarzoso and Nandi (1999, 2001). In Zarzoso and Nandi (2001), for example, the authors compare the performances of adaptive filters and a blind source separation (BSS) method for separating maternal and fetal sources for an 8-lead system and conclude that BSS works better for the dataset analyzed. Many biomedical signals are repetitive in nature, where it is of interest to enhance the repeating feature above noise. One example is in ECG monitoring where it may be required to study the morphology of the ECG complex. Another application is in EEG-evoked potential analysis where one applies repeated stimuli to a subject to obtain an estimate of the evoked response. In these two examples, signal averaging is carried out to detect the relevant data.

TABLE 21.5   Examples of Digital Filters

Filter Type	Key Form	Comments
Finite impulse response (FIR) (constant coefficient filters)	$$y[n] = \sum_{p=-M/2}^{M/2} h[p]x[n-p]$$	$x[n]$ are inputs; $y[n]$ outputs; $h[p]$ are impulse response coefficients. MATLAB has toolboxes for design of such filters
Matched filter (Gibson 1995)	As above with $h[n] = s[n]$, the underlying signal. $s[n]$ may be approximated since it is not known a priori	Enhancement and detection of signal features. Estimate features as opposed to preservation of signal shape (e.g., heart rate from ECG).
Adaptive filters	$$y[n] = \sum_{p=-M/2}^{M/2} h[p]x[n-p]$$	$x[n]$ are inputs; $y[n]$ outputs; $h[p]$ are impulse response coefficients. In its simplest form, $h[p]$ is determined from difference between $y[n]$ and a reference signal, which is correlated with $x[n]$; $h[p]$ changes with time according to changes in the characteristics of the underlying signal.
Wavelets (Addison 2005)	$$\psi_{m,n}(t) = \frac{1}{\sqrt{a_0^m}} \Psi\left(\frac{t - nb_0 a_0^m}{a_0^m}\right)$$ (discrete transform). a_0, b_0 usually 2, 1.	Good where physiological signals are not good matches for sinusoidal analysis (FFT). Windows for sampling vary: small windows for high frequencies, longer windows for lower frequencies.

21.13.3 Emerging Preprocessing Developments

Noted earlier are various analog and digital processing steps that are carried out on biomedical and physiological signals. Each step involves either a loss or distortion of information. There is a need to evaluate whether such losses and distortions could lead to wrong clinical diagnosis, stressing the requirement for biomedical engineers to work closely with clinicians. In Figure 21.5, the analog filtering, sampling, and A/D steps have been well established; the only differences between different systems being, for example, choice of cutoff frequencies for the analog filter, sampling frequency, and number of bits used for A/D. There is a much greater variety in the *digital* preprocessing techniques that are used after the A/D, and new methods of digital processing are continually being reported in the journals in this area; a few of these preprocessing methods are noted earlier. Some of the methods used, for example, BSS and adaptive filtering, require the use of algorithms that have several input parameters. Other methods are much simpler in concept, employing the use of simple linear and nonlinear operations on the samples; these have sometimes been shown to work just as well as the more sophisticated algorithms. Many opportunities exist to evaluate the sensitivity of these algorithms to changes in the input parameters. In addition, there is a need to evaluate the robustness of these algorithms to noise.

21.13.4 Embedded Systems

The increased density that characterizes digital technology has led to new architectures as well as applications for computer-based information processing. Such systems continue to develop and will do so in the coming decades. Of keen interest is the development of totally integrated devices that will detect neural signals in brain, convert these and process them in accordance with the information-processing techniques discussed earlier, and communicate the resultant data to the external world—possibly a computer. An article in the *New York Times* describes some of these developments (Kennedy 2011). As noted in the article:

> For years, computers have been creeping ever nearer to our neurons. Thousands of people have become cyborgs, of a sort, for medical reasons: cochlear implants augment hearing and deep-brain stimulators treat Parkinson's. But within the next decade, we are likely to see a new kind of implant, designed for healthy people who want to merge with machines. With several competing technologies in development, scientists squabble over which device works best; no one wants theirs to end up looking like the Betamax of brain wear. Schalk is a champion of the ECoG (electrocorticographic) implant because, unlike other devices, it does not pierce brain tissue; instead it can ride on top of the brain-blood barrier, sensing the activity of populations of neurons and passing their chatter to the outside world, like a radio signal. Schalk says this is the brain implant most likely to evolve into a consumer product that could send signals to a prosthetic hand, an iPhone, a computer or a car.

There are numerous definitions of the term *embedded systems* but one that resonates most with the concept is the following: "*an embedded system is a microprocessor based system that is built to control a function or a range of functions*" (Heath 2003). Such systems are the result of evolutionary changes in integrated technology, which is characterized by high connectivity (circuit densities). Such devices are considered to be *computer-on-a-chip* devices dedicated to real-time computing environments. (This, in contrast to a personal computer (PC) that is configured for flexibility with an application focus on end-user needs.) Examples of such embedded system devices abound in industrial, commercial, and consumer environments: digital watches, factory controllers, nuclear power control, GPS receivers, MP3 players, and numerous others. In biomedical environments, such systems are found in vital signs monitoring, electronic stethoscopes, and medical imaging environments (PET, SPECT, CT, and MRI) or noninvasive inspections. Embedded controllers—a parallel designation—can trace its origins to the 1960s, with perhaps the first such dedicated system employed at the MIT Instrumentation Laboratory (Hall 1996).

The explanation provided earlier is not an *unequivocal* definition of an embedded device. Rather, they may be characterized as dedicated *controllers*. Many systems so described have some element of extensibility or programmability. Devices such as hand-held computers have operating systems and include microprocessors and may allow different application downloads (and peripheral connections) much as those found in *general purpose* computers. The ultimate distinction is thus the number of *dedicated* functions that embedded systems perform.

A summary of embedded system characteristics includes the following:

- Designed to some specific task (or limited number of tasks).

- Real-time constraints (e.g., for safety or reliability).

- Hardware simplification to reduce costs (replacements for integrated circuit logic devices).

- May consist of computerized parts within larger devices that serve more general purposes (e.g., microcontrollers in cars for the automatic braking systems (ABSs) or air conditioning systems).

- Program instructions are referred to as *firmware,* stored in read-only memory (or flash drives); the software is normally proprietary.

- Run with limited computer hardware resources: little memory, small, or nonexistent keyboard, and rarely have visual (screen) output resources.

- User interfaces may be limited to buttons, LEDs, graphic or character-based LCDs and may feature a simple menu system for user input. Also found are touch screens, or joysticks as *pointing* facilities.

- When processors are not intimately available to users, remote communications are possible (e.g., RS 232, USB, and network (Ethernet)).

There are numerous examples of computational element types or *processors.* These can be roughly divided into two categories: *ordinary* microprocessors (µP) and *microcontrollers* (µC). In the case of the µC, additional on-chip peripherals (I/O facilities) can be found, resulting in reduced cost and size. Software architectures can include Von Neumann or RISC systems. While there is no one overarching architecture for such systems, Figure 21.16 is representative of a simplified embedded system device.

A common architecture found in high-volume applications is the *system-on-a-chip* (SoC), which includes a complete system consisting of multiple processors, multipliers, cache memory, and interfaces on a single chip. They can be designed for application-specific-integrated-circuit (ASIC) uses or using a field-programmable gate array (FPGA) where the firmware defines the application.

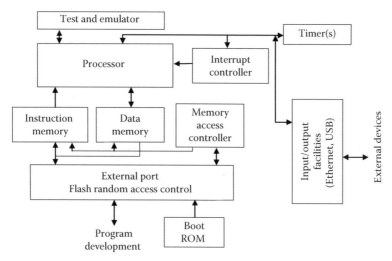

FIGURE 21.16   Simplified embedded system architecture.

21.13.5 Software Tools

Embedded systems require appropriate software facilities to program the on-board processors and peripherals. A number of these are noted in the following:

- Debuggers or emulators

- Embedded system error detection (e.g., checksum or CRC).

- As embedded systems are often organized around DSP tasks, a math workbench may be available to simulate the mathematics. Additionally, libraries for DSP simplification may also be available.

- Custom compilers and linkers to improve optimization for the hardware.

- May have its own special high level (programming) language (HLL) or design tool or enhancements to existing languages.

- A real-time operating system (RTOS).

- Modeling and code generating tools based on state machine models.

21.13.6 Software Architectures

The software architectures can be classified in one of some seven categories. Block diagram descriptions of a sampling of these are depicted in Figure 21.17. Table 21.6 clarifies the software architecture.

However, embedded systems are increasingly found to have unique or proprietary architectures and may not resemble those defined within the set of characteristics noted earlier. One of the most recent examples of such devices has been produced by IBM (http://www-03.ibm.com/press/us/en/pressrelease/35251.wss 2011). It is a result of the SyNAPSE project, which is funded by the Defense Advanced Research Project Agency (DARPA) as part of the Systems of Neuromorphic Adaptive Plastic Scalable Electronics (SyNAPSE) initiative. As described, this new generation of experimental computer chips will emulate the brain's abilities for cognitive capabilities (perception, action, and cognition). These so-called *cognitive computers* represent sharp departures from traditional concepts in designing and building computers. (See the discussion in this chapter on artificial intelligence as it applies to information processing in physiological environments.) The IBM device will:

- Learn by experience

- Find informational correlations

- Create hypotheses

- And provide recall (memory)

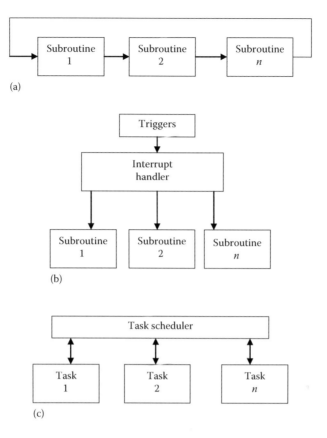

FIGURE 21.17 Software architectures of embedded systems. (a) Simple loop; (b) interrupt driven; (c) cooperative multitasking.

TABLE 21.6 Embedded System Software Configurations

Type of Architecture	Description
Simple control loop	Overall design consists of a repeating loop of subroutines. Each subroutine manages a part of the hardware (or software).
Interrupt controlled systems	Tasks to be performed by the system are triggered by associated events. The processor manages the priorities and execution of these subroutines.
Cooperative multitasking	Similar to the simple control loop except that the loop is *hidden* in an *application programming interface* (API) or a particular set of rules (*code*) and specifications that software programs can follow to communicate with each other.
Preemptive multitasking (multithreading)	A *low-level* code segment switches between tasks or threads based on a timer (interrupt controlled); system is controlled by an *operating system* kernel.
Microkernel	An operating system allocates memory and switches the processor to different threads of execution. The microkernel works best when task switching and intertask communication is fast. It is not well suited when switching and communications are slow.
Monolithic kernel	A relatively large (operating system) kernel with sophisticated facilities is adapted to suit the embedded environment.
Custom operating systems	When the embedded system needs to meet specific targets associated with things such as safety, timeliness, reliable, or efficient operation and cannot be satisfied by other software architectures, custom architectures may be required.

With such characteristics, they will *mimic the brain* both in structure and in synaptic plasticity. The potential for use in physiology and emerging instrumentation for analyzing complex experimental information from multiple sensory modalities is evident. They will be able to dynamically *rewire* themselves as they interact with the environment. Consider, for example, how it could be employed as aids in rehabilitation where instrumental conditioning techniques can be used to manage behavioral deficits resulting from cognitive insults (stroke) for ambulatory patients (Silverman 2005a,b, 2003).

21.14 STATISTICS AND HYPOTHESIS TESTING

Statistics and its applications trace its roots to a number of sources. In 1663, Jaohn Graunt published *Natural and Political Observations upon the Bills of Mortality* (http://www.britannica.com/EBchecked/topic/2011, Wilcox 1938). Graunt was a prosperous merchant and public official. He collected causes of death and began noticing certain patterns; this led to the work cited earlier. He ended up publishing several volumes, one of which was sponsored by the Royal Society. His observations led to a number of conclusions about populations:

- The male birth rate was higher than the female: but there was greater mortality rate for males, leading to a population that was evenly divided between the sexes.

- He produced the important concept of a *life table*; it viewed mortality in terms of survivorship. From this, he was able to predict the percentage of individuals who will live each successive age as well as their life expectancy. One could extract estimates of community economic losses caused by deaths.

These computational tools have provided a basis for much economic and commercial activities. From a linguistic perspective, the word statistics (in English) is traceable to German wherein a translation of the book *Bielfield's Elementary Universal Education* includes reference to the following:

> the science that is called statistics teaches us what is the political arrangement of all the modern states in the known world (David 2001).

Importance of the word let to inclusion in Webster's American Dictionary—in 1828—of its definition as

> a collection of facts respecting the state of society, the condition of the people in a nation or country, their health, longevity, domestic economy, arts, property and political strength, the state of their country, etc.

Three names can be identified with the mathematical developments in probability, which underscores statistics:

- Blaise Pascal and Pierre de Fermat (seventeenth century), who both dealt with the theory of games and gaming as it is related to chance

- Friedrich Gauss (1794), who made contributions to our understanding of the method of least squares

In the late nineteenth century, understanding and quantification of random error led to mathematical statistics, which provided tools to interpret the data from observation and experiment (Mlodinow 2009). *Statistics* provides tools to interpret the data from observations and experiment. It can provide important tools for addressing *real-world* issues such as the effectiveness of drugs and other physiological phenomena.

As related to statistics, *mathematical statistics* concerns itself with the theoretical basis of the subject. The word "statistics" might be referred to as *Statistics is an art*, thereby providing a basis by which to distinguish it from the word *statistic* that refers to a quantity (e.g., mean and variance), which is calculated from a dataset—the plural of this word is *statistics*—and constitutes a *discipline* wherein a data-based measure (*statistic*) is a descriptive measure.

Statistics—as it relates to this chapter—has come to be interpreted with all aspects of the science of the collection, organization, and interpretation of data, including (especially) data collection planning with regard to experiments.

Statistical methods taken together comprise *applied statistics* and include

- *Descriptive*: summarize or describe a collection of data and is used in research to communicate results (among other purposes)

- *Inferential*: patterns of data modeled to account for randomness and uncertainty and draw inferential conclusions about the process or populations

21.14.1 Overview

An experimental (or other) process or population is the starting point of statistical studies. Populations can include observations of experimental outcomes. A picture of this organization is shown in Figure 21.18.

The subset consisting of individual outcomes from within the population constitutes a *sample*; when considering the entire population, the applicable term is the *census*. The representative sample is analyzed with respect to its (statistical) properties, and conclusions are then drawn regarding the population or dataset as a whole. There are two important elements of such analyses: the *descriptive statistics* and the *inferential statistics*.

21.14.2 Descriptive Statistics

Numerical uncertainty characterizes all experimental data (unless there is an immutable deterministic relationship among the variables). These uncertainties result from the presence

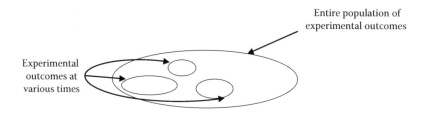

FIGURE 21.18   Schematic representation of an experimental population and its potential variable outcomes.

TABLE 21.7   Definitions of Central Tendencies of Data

Term	Definition
Mean	Average value of the observations.
Median	Midpoint between the lowest and highest values in the sample.
Mode	Most frequently occurring value.
Span	As described by the range (i.e., min and max values) of values within the sample.
Standard deviation (strictly a variation about the central tendency)	The standard deviation gives an idea of how close the entire set of data is to the average value: small standard deviations have tightly grouped, precise data; large standard deviations have data spread out over a wide range of values.

of noise in the experimental system (both internal and external); the inherently discrete nature of independent and dependent variable quantities; and instrumental limitations in resolution and accuracy. It is therefore desirable to represent such data such that the valuations can impart understanding of the underlying experiment. Two important elements describe such statistics:

- Central tendencies of the sample

- Variations about the central tendency—*data distributions*.

Central tendencies include mean, median, mode, standard deviation, min, and max values of a sample provide information about the central tendencies. These terms are defined in Table 21.7 (https://controls.engin.umich.edu/wiki/index.php).

For example, Figure 21.19 shows two histograms from representative data samples.

The samples have very different appearances related to other characteristics of the sample even though the sample averages are the same; in this case, it is in reference to a statistic referred to as the *skew*. Therefore, additional descriptors are desirable for a more complete determination of the nature of the sample data.

Descriptive data include variance (standard deviation as noted earlier); skew, kurtosis (and higher-order *moments* about the mean).

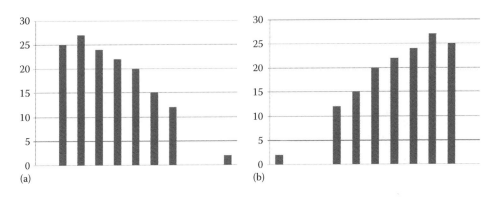

FIGURE 21.19   Histogram of sample data (a) with positive skew and (b) with negative skew

Skewness is a measure of symmetry, or more precisely, the lack of symmetry. A distribution, or dataset, is symmetric if it looks the same to the left and right of the center point. The formula for the *skewness* parameter is given by

$$\text{Skewness} = \frac{\sum_{i=1}^{N} \left(Y_i - \bar{Y}\right)^3}{(N-1)s^3},$$

(21.27)

where

   Y_i are the sample points
   N is the number of data points
   $\bar{Y}$ is the sample mean
   s is the standard deviation

The corresponding formula for *kurtosis* is

$$\text{Kurtosis} = \frac{\sum_{i=1}^{N} \left(Y_i - \bar{Y}\right)^4}{(N-1)s^4}.$$

(21.28)

Additional *moments* may be calculated, but for many applications, skew and kurtosis provide the majority of description.

21.14.3 Graphic Descriptions

Pictorial or alternative characterizations of sample data can provide additional insights into the experimental samples and using modern computer-based tools, relationships between the independent and dependent variables become evident. (In the latter instance, Excel is very useful for experiments where data are restricted to a small sample set.)

For example, data from an experiment may be better expressed by using a logarithmic scale for the abscissa and is readily accomplished using an Excel spreadsheet (see Figure 21.20). The redrawn data greatly enhance the distribution provided by the sample.

A number of graphic presentations are possible, with those listed in Table 21.8 being the most commonly found in statistical analysis. (The plots noted in Table 21.8 are not exhaustive.)

21.14.4 Hypothesis Testing; Statistical Inference; Statistical Significance

Using the sample observations, the goal of *statistical inference* is to extract conclusions about the underlying population as well as the confidence in the probability of various alternative outcomes. In its simplest form, a *hypothesis* is formed about the population (or observations) and the *validity* of the hypothesis is tested; it is the central concept that underscores uncertainty. Even if a test confers *validity* on one hypothesis or another, it does not grant truth (or falsity) on such a hypothesis. A second or subsequent experiment could produce an opposite outcome. However, inferential statistics can provide a measure

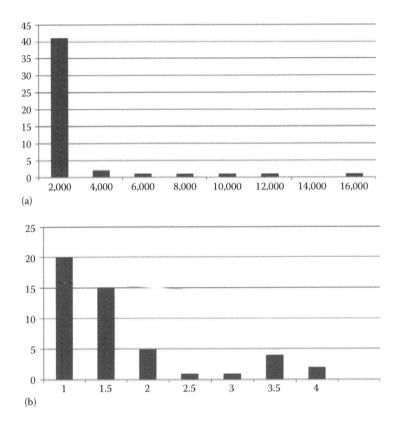

FIGURE 21.20   Graphic presentation of experimental data. (a) Data on a linear scale. (b) Data redrawn on a logarithmic scale.

of belief or confidence in the results. The kinds of results that can be obtained include experimental associations (effect of temperature on the experimental preparation) in the form of data *correlations* using *regression analysis*. (It is to be noted that correlation does not imply any necessary causation.)

An example will serve to illustrate the basic concept. To the extent possible, a population of individuals with similar characteristics (e.g., diet and age distribution) is selected and subsequently studied to determine if there is any effect on their serum cholesterol if they follow two different diet plans (Schefler 1979). The serum cholesterol levels for the two diets are shown in the following:

A	B
177	142
200	155
251	141
239	205
190	147
180	171
210	213
185	164

TABLE 21.8   Statistical Plots and Their Characteristics

Chart Type	Description	Comments
Histogram	A bar graph of the frequency or proportion of times that a variable takes specific values, or a range of values for continuous data within a sample.	Easy to apply to continuous, discrete, unordered data. Use lots of ink and space for small amounts of information. Difficult to display multiple graphs concurrently for comparison.
Stem-and-leaf (created by John Tukey)	Similar to a horizontal bar graph. The value of all elements associated with each is also shown, however, and constitute the bar.	Useful for getting distribution information on the fly. Display the full sample set and waste very little ink. Back-to-back samples for comparison. More descriptive than bar graph. Busier than a bar chart.
Scatter	Display complete dataset, one point per dot.	Trend, curve fitting easy. Ordering of dots and labels on axes provide additional information. High data density; difficult to read.
Box plots	A chart where the median, and first and third quartiles are represented as a box with a line through it. The min and max values are then represented as a bar at the end of a line extending out from the box. These lines cannot be longer than 1.5 times the interquartile range (i.e., $Q3-Q1$) from the box. Beyond bars are *outliers*, and are represented as circles. Box width is often the sample size. The mean is shown as a star.	This type of display contains a large amount of information. Do not use for small numbers of observations, (plot the data). For data with lots and lots of observations, omit the outliers plotting if you get so many of them that you do not see the points.
Kernel density estimates	Kernel density estimates are essentially more modern versions of histograms providing density estimates for continuous data. Observations are weighted according to a *kernel*—Gaussian.	*Bandwidth* of the kernel plays the role of the bin size for the histogram. Too low of a bandwidth yields a too variable (jagged) measure of the density. High bandwidth generates *smoothing*.
Q–Q plots	These are *quantile–quantile* plots in which empirical quantiles are plotted against theoretical quantiles of a distribution.	Most useful for diagnosing *normality* or adherence to a normal distribution.

The theory to be tested rested on the *null hypothesis*, which stated that there is no difference in reduction of serum cholesterol for the individuals following diet A or B. (As a matter of *philosophy*, the null hypothesis is proposed such that it will end up being *rejected*, which means that we *cannot* prove there is no difference.). An *alternative hypothesis* states that there is a significant difference in serum cholesterol levels in individual taking the two diets. (We then infer that one of the diets is more effective in reducing serum cholesterol levels.)

From the data, a statistic (or statistics) is computed. If this value falls within an accepted *level of significance*, the null hypothesis is accepted; if it falls outside the established levels of significance, the null hypothesis is rejected and the alternative hypothesis is accepted.

TABLE 21.9   Statistical Tests

Test	Characteristic
Student's *t*-test	Compares the means from two samples and involves the calculation of a *t-statistic*. The student's *t*-distribution takes into consideration the effect of sample size on variation and standard deviation; this test is often used when sample size is small.
Two sample *t*-test	Designed to test if a significant difference exists between the means of two samples where the participants were randomly assigned to the sample groups.
Paired *t*-test	Can be used for two types of experiments: so-called before and after testing where the same individual is used; matched pairs where the experimenter matches subjects.

Often, the rejection point for the null hypothesis is set at 95% and is stated as the *0.05 level of significance*. (The smaller the level of significance, the more reliable the result.) If the statistical measure falls outside the level of significance, the null hypothesis is rejected and the alternative hypothesis is accepted. Several statistics are possible, and three of these are noted in Table 21.9.

The calculated *t*-statistic was computed for the two diets, and the resultant calculated *t*-statistic is 2.65, which exceeded the critical *t*-value for a two-tailed test and the 0.05 level. The conclusion reached is that the null hypothesis is rejected and that there is a significant difference in cholesterol reduction for the two diets—diet B lowers cholesterol significantly more than diet A.

21.14.5 Statistical Inference: Validity of Assumptions and Prediction

While statistical hypothesis testing provides a mathematical mechanism for reassuring conclusions about observations, it does not materially advance the answer to two additional questions:

1. Can we confirm our statistical assumptions?

2. Can the past *predict* what we should expect from experimental outcomes of the future?

We explore two methods that can assist when addressing such questions—*Bayesian Statistics* and *Markov Analysis*.

21.14.6 Bayesian Analysis: Statistical Inference: Looking Backward

John Maynard Keynes once remarked, "When the facts change, I change my opinion. What do you do, sir?" As Sharon McGrayne (Paulos 2011) notes,

> Bayes's theorem, named after the 18th-century Presbyterian minister Thomas Bayes, addresses this selfsame essential task: How should we modify our beliefs in the light of additional information? Do we cling to old assumptions long after they've become untenable, or abandon them too readily at the first whisper of doubt? Bayesian reasoning promises to bring our views gradually into line with reality and so has become an invaluable tool for scientists of all sorts and, indeed, for anyone who wants, putting it grandiloquently, to sync up with the universe.

Bayes' theorem follows from this concept: the probability that A will occur if B occurs differs from the probability that B will occur if A occurs; this is a common error in medicine; studies in Germany and the United States show that when physicians estimate the probability that an asymptomatic woman between 40 and 50 years of age who has a positive mammogram actually has breast cancer if 7% of mammograms show cancer when there is none; also stated that the actual incidence was about 0.8% and that the false-negative rate is about 10%; in Germany, 1/3 concluded that the probability was 90% and the median estimate was 70%; in the United States 95 out of 100 physicians estimated the probability to be around 75%. From Bayesian methods, we can determine that *a positive mammogram indicates cancer in only about 9% of the cases* (Gigerenzer 2002, Mlodinow 2009).

Bayesian belief networks (BBNs) can reason with networks of propositions associated with probabilistic events. It seeks to address three issues when faced with uncertainty:

- What is the confidence of underlying beliefs?

- If the underlying belief is true, is new evidence that follows accurate?

- Independent of the belief in underlying assumptions, what is the accuracy of new evidence?

Bayes' theorem is summarized as follows (www.theriac.org 2011):

$$P(H,E) = P(H|E)\, P(E) = P(E|H)\, P(H)$$

and in its most recognized forms of these relationships:

$$P(H|E) = \frac{P(E|H)P(H)}{P(E)}$$

or

$$P(E|H) = \frac{P(H|E)P(E)}{P(H)}.$$

The symmetrical nature of this law addresses the questions noted earlier: we can compute the probability of a hypothesis (H) given the evidence (E) and vice versa (the probability of the evidence (E) given the hypothesis (H)). Conclusions can be determined from BBNs, and a simple one is shown in Figure 21.21.

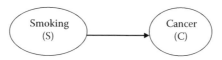

FIGURE 21.21   BBN for determining if smoking can lead to cancer.

TABLE 21.10   Causal Relations between Smoking and Cancer

Hypothesis (Smoking)		Probability	
No		0.80	
Light		0.15	
Heavy		0.05	
	P(C)/Smoking = No	P(C)/Smoking = Light	P(C)/Smoking = Heavy
P(C) = none	0.96	0.88	0.60
P(C) = benign	0.03	0.08	0.25
P(C) = malignant	0.01	0.04	0.15

Associated with the variables (smoking and cancer) as represented by the *nodes* of the net are possible outcomes. The *links* joining the nodes are considered to be *causal* conditions. Each of these outcomes has an associated probability as demonstrated in Table 21.10.

$S \in \{no, light, heavy\}$

$C \in \{none, benign, malignant\}$

where smoking can take values of none (no), light (smoker), or heavy (smoker), where definitions of *light* or *heavy* are not considered in this discussion. Cancers fall into categories of none, benign, or malignant.

Bayes evaluation is useful for investigating the following problem. We can see from Table 21.10 that if a person is a heavy smoker, there is a 0.25 chance that the individual could develop a benign tumor. However, suppose that a benign tumor has been detected: what can we say about that tumor having come from a heavy smoker. The problem to be solved is inherent in the following (Bayes) formula:

$$P(heavy\_s|benign\_c) = \frac{P(benign\_c|heavy\_s)P(heavy\_s)}{P(benign\_c)}.$$

This requires us to calculate the total probability of a benign tumor (benign_c)- the denominator in the above equation. This is determined from all the ways that the tumor could be present or:

$$p(benign\_c) = \sum_{all\_smoker\_type} p(benign|smoker\_type)p(smoker\_type).$$

Substituting from the table, $p(benign\_c) = (.03)(0.8) + (.08)(.15) + (.25)(.05) = .04$

Making this substitution into the hypothesis we are testing we get

$$P(smoke\_heavy|benign\_c) = \frac{(.25)(.05)}{.04} = .31.$$

This example originated from a relatively simple set of variables. Nets can be created with a number of added variables. For example, the tumor study mentioned earlier might be

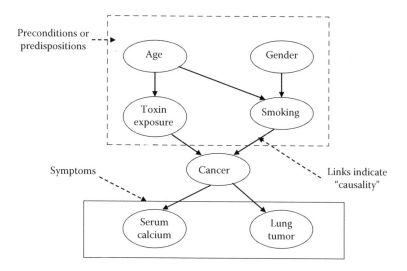

FIGURE 21.22   A more complex Bayesian network.

expanded to include a number of added instances that might lead to tumors in addition to the smoking behavior (Figure 21.22).

Features or elements of a BBN in addition to the condition being studied (cancer) have been identified in Figure 21.22: symptoms and preconditions.

21.14.7 BBN Considerations

The knowledge acquisition process for a BBN involves three steps:

1. Choosing appropriate variables

2. Deciding on the network structure

3. Obtaining data for the conditional probability tables

In general, variables should be collectively exhaustive, and mutually exclusive, thus having the following properties:

$$\overline{(\mathrm{var}\,i \wedge \mathrm{var}\,j)} \quad i \neq j.$$

Netica is a freely available software package for implementing BBN analysis but best employed for small networks. It is useful when working with BBNs and influence diagrams. It includes a graphical editor, compiler, inference engine, and other features (http://www.norsys.com/2011). Figure 21.23 is a representative diagnostic net from Netica. The top nodes indicate predispositions (i.e., a visit to Asia). These can have influence on the likelihood of the diseases, which are shown in the second row (i.e., tuberculosis, lung cancer, and bronchitis). The last row depicts the symptoms that are produced as indicators (i.e., x-ray results and Dyspnea [shortness of breath]). The organization shown in Figure 21.23 has a popular format for diagnosis: predisposition nodes (variables) at the

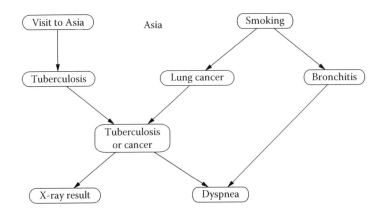

FIGURE 21.23   Sample Netica BBN net configuration for a diagnostic study from the tutorial.

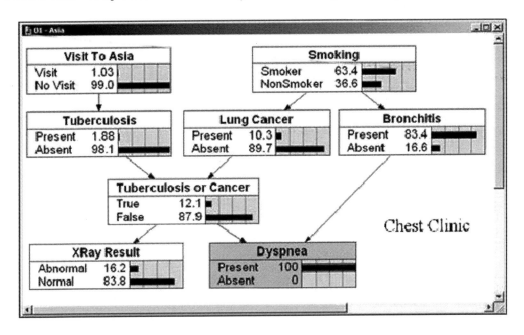

FIGURE 21.24   Representative Netica interface for diagnostic study.

top, links (causative) to internal conditions in the middle layer (as well as failure states), and finally, links to observables.

Once the net has been compiled, an interactive interface appears that permits the user to enter statistical information. The system will then reflect confidence in outcomes as well as reevaluation of assumptions regarding the causes (predispositions). A typical interface is shown in Figure 21.24.

21.14.8 Markov Chains and Analysis: Looking Forward

Markov analysis originated in the latter part of the nineteenth century but was ignored for a considerable period of time: the promulgation of IEC Standard 61508 (Functional Safety of Electrical/Electronic Safety Related Systems) reinvigorated interest. Current software tools supplement easy use of the techniques (Delone 2005, www.gap-system.org/~history/Biographies/Markov/2012).

A Markov analysis examines a sequence of events and analyzes the *tendency* of one event to be followed by another. Using this analysis, one can generate a new sequence of random but related events, which will look similar to the original. It thus provides information about the probabilities of such outcomes and can aid in drawing conclusions about observed outcomes. Such processes and analyses are useful for analyzing dependent random events—that is, events whose likelihood depends on what happened last. It would *not* be a good way to model a sequence of coin flips, for example, since every time you toss the coin, it has no memory of what happened earlier. The sequence of heads and tails is not interrelated. They are *independent events*. But many random events are affected by what happened earlier. For example, yesterday's weather does have an influence on what today's weather is. They are not independent events.

Markov analysis methods (MAMs) use stochastic modeling techniques. Stochastic modeling (Taylor and Karlin 1998): a quantitative (mathematical) description of (natural) events that predicts a set of possible outcomes determined by their likelihoods or probabilities in which time can be an important element of such prediction. A related concept is that of a *Markov chain* in which sequences of random variables (RVs) where the future variable is determined by the present variable but does not depend on how the present state came to be: *this gave rise to the theory of stochastic processes* (Taylor and Karlin 1998): *MAM* investigates a sequence of events and analyzes the tendency of one event to be followed by another. Using this analysis, we can generate a new sequence of random but related events that have probabilistic with characteristics similar to the original sequence. A key feature assumes that the future is independent of the past given the present. And the RV may be continuous or discrete.

For example, a Markov-based weather forecaster would proceed along the following steps:

1. Observe a long sequence of weather.

2. Start with today's weather.

3. Given the day's weather, use an RV to pick tomorrow's weather.

4. Make tomorrow's weather today's weather and repeat (return to step 3).

There are, of course, much more accurate ways of predicting weather, but when additional information may not be known, long run probabilities of any day being in a specific state can be useful. Such analysis has found many uses in physiology as in one such study where a Markov model was created to replicate health events over the remaining lifetime of someone newly diagnosed with glaucoma (Kymes *et al* 2010). The elements of Markov analysis are shown in Figure 21.25.

Markov models are defined by a random *probability matrix*. It is assumed to be a *memoryless* system where future states depend only on present states. The system is also characterized by stationarity wherein probabilities that govern transitions from state to state remain constant. A block diagram of a representative Markov problem is shown in Figure 21.26.

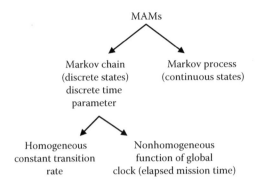

FIGURE 21.25   Elements of Markov analyses.

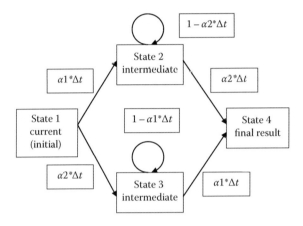

FIGURE 21.26   Block diagram for representative (two-element) Markov.

The system is a simple example of how an instrument (e.g., ECG machine, embedded processor, or instrument in physiological data collection) might function, ultimately leading to system failure (State 4). In Figure 21.26, State 1 is the current condition of the system; States 2 and 3 are some intermediate states (such as failure of part of the system). Both of these latter states might then lead to a system failure. In addition, the system represents a discrete approximation to a continuous Markov environment in which Δt is considered to be negligible. In addition, the following should be noted: $P1$ = probability of being in State 1 at time t and *not* transitioning out of that state in the time interval Δt. This is given by

$$P1(t + \Delta t) = P1(t) * [1 - (\alpha 1 + \alpha 2)*\Delta t].$$

The probability of being in State 2 at time $t + \Delta t$ is the probability of being in State 1 at time t and transitioning to State 2 (in Δt) and *not* transitioning out during Δt if it was in State 2 at the start of the computational interval. This is

$$P2(t + \Delta t) = P1(t)* \alpha 1* \Delta t + P2(t)*(1 - \alpha 2*\Delta t).$$

Continuing with this reasoning leads to a series of equations as follows after letting

$\Delta t \to 0$ as the limit from a discrete analysis to a continuous model:

$$
\begin{vmatrix} dP1(t)/dt \\ dP2(t)/dt \\ dP3(t)/dt \\ dP4(t)/dt \end{vmatrix} = \begin{vmatrix} -(\alpha 1 + \alpha 2) & 0 & 0 & 0 \\ \alpha 1 & -\alpha 2 & 0 & 0 \\ \alpha 2 & -\alpha 1 & 0 & 0 \\ 0 & \alpha 2 & \alpha 1 & 0 \end{vmatrix} * \begin{vmatrix} P1(t) \\ P2(t) \\ P3(t) \\ P4(t) \end{vmatrix}.
$$

The matrix containing the α's is the *state transition matrix* **A**, and the entire matrix equation can be solved by computer as follows:

$$
\overline{P} = \exp[A]t \cdot \overline{P(0)}.
$$

(Two methods are possible: infinite series and eigenvector; these topics are beyond the scope of this chapter.)

21.14.9 Computerized Markov Analysis Tools

Computational complexity grows rapidly with the number of states in the system. For example, with 10 elements and 2 states, the number of state computations is 20 and there are 2^{10} (or 1024) elements in matrix **A**. A system with only 10 elements is small for many applications. One simplification method is to combine states; for example, in Figure 21.26, State 2 and State 3 might be combined using the logical *OR* of these.

Two software facilities are worth noting:

1. *SHARPE*: Symbolic Hierarchical Automated Reliability and Performance Evaluator (Hirel *et al* 2000)

2. *MKV Version 3.0*: Markov Analysis Software (www.isograph-software.com 2011)

A number of application examples are available (Janssen and Limnios 1999, www.iso-graph-software.com 2011). One such demonstrates how "force generation during muscle contraction can be understood in terms of cyclical length changes in segments of actin thin filaments moving through the three-dimensional lattice of myosin thick filaments... when muscle is viewed as a Markov process, the vectorial process of chemomechanical transduction can be understood in terms of lattice dependent transitions" (Schutt and Lindberg 1998).

21.15 EMERGING DEVELOPMENTS IN PHYSIOLOGICAL INSTRUMENTATION

21.15.1 Knowledge-Based Techniques in Instrumentation

Increased capabilities in the computer have led to amplified experiments whose complexity requires highly skilled researchers. This has led to the application of *artificial intelligence* to undertake many issues associated with processing tasks that have been discussed

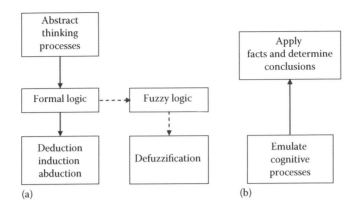

FIGURE 21.27   Approaches to developing machine-based intelligence. (a) Top-down architectures leading to Expert Systems that provide recommendations. (b) Bottom-up organization with applications in associative networks (and Neural Nets).

previously. These computer-based tasks include instrument design, calibration, and, in particular, interpretation of experimental results (Broman 1988, Kulikowski 1988, Papp *et al* 1988, Shamsolmaali 1988, Shiavi 1988, Silverman and Silver 1995, Sztipanovits and Karsai 1988).

Methodologies in AI are modeled after simplified assumptions concerning human cognitive decision making. Three elements of cognitive models include determination of facts; a system of logical rules (*IF-THEN*); and a set of processes to determine how the rules are to be applied (e.g., an *algorithm*). Two approaches employing this paradigm have found application in the scientific community. (They evolve from historical trends, which are not addressed in this chapter.) They are shown in Figure 21.27.

A well-recognized example of the Expert System approach is found on the Internet (*WebMD*) (http://www.webmd.com 2011) where a user can enter physiological symptoms and explore potential deficits. Fuzzy logic deals with circumstances where the facts and circumstances of events are uncertain (i.e., the signal was *somewhat noisy*). It has found extensive application in control systems such as robotic surgery or anesthesiology. (In this chapter, we address three aspects of machine interpretation of data significance: neural nets; BBNs; and Markov analysis.)

21.15.2 Neural Nets

Artificial networks is the general term applied to the technology known as neural networks (NNs) (Russell and Norvig 2011). Traditional computer architectures (*von Neumann design*) have solution procedures expressed in terms of logical statements, which the computer can execute. If a procedure of this kind can be specified to solve a problem, then it can be implemented on a von Neumann machines.

However, it may be difficult to specify the specific rules under which the algorithm functions. If this is the case, then it is difficult to computerize the application. For example, it is difficult to recognize soft tissue images because it is difficult to find comprehensive, explicit, logical rules to cover the myriad possibilities of distorted pictures, effects of shading, effects due to perspective, and others that impact our ability to recognize such images. For problems of this kind (e.g., voice recognition and handwriting), traditional computers have some success. Humans, however, are much better at addressing such problems because they have advantages where explicit rules are not well formulated. Humans learn to generalize from experience.

The brain does not have an explicit equivalent of the central processing unit (CPU) within the modern computer. Information processing and decision making are carried out by a network of billions of simple processing units called *neurons* (Hawkins 2004). In place of explicit rules, the brain consists of a pattern of adjustable connections between these neurons that embody the *knowledge* required for carrying out various information-processing tasks. (While the physical connections are fixed, the *strength* or intensity of the connection can be altered during *learning*.) A computer (machine) organized around a design of this kind is referred to as a *connectionist* machine (Friedenberg and Silverman 2012). Connectionist machines (e.g., NNs) do not attempt to replicate the workings of the human brain. (In fact, literal operation of an artificial neuron as a replica of a neuron is impossible.) Their objective is to achieve the strengths of the brain in regard to the brain's ability to learn from examples.

A human brain contains some 100 billion neuronal cells. (There are other types of specialized cells in the human body such as blood cells or glial cells.) A schematic representation of the neuron is shown in Figure 21.28.

- The elements of the neuron include the following: Membrane (plasma membrane): cell envelope

- Dendrites and axon (nerve fiber): projections and extensions from the soma

- Dendritic tree: network of dendrites

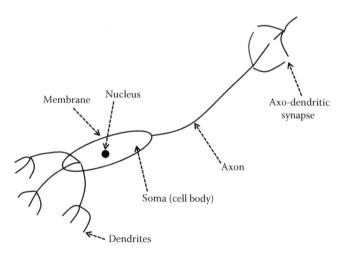

FIGURE 21.28   Schematic representation of a neuron.

Some factual information:

- Soma: 30 μm

- Dendrites: 200–300 μm

- Axon: 50 μm to several meters

- Synapse(s): points of interaction between neurons; important mechanism for information flows from one cell to another

- Synaptic cleft (space): 200–300 Å

- Presynaptic side: *input* to the junction

- Postsynaptic side: *output* side of the junction

A unique characteristic of the neuron is that it does not divide as other cells. Electrical stimulation on the presynaptic side causes transmitter chemicals to be released into the synaptic cleft. These transmitters diffuse—as in a transistor—across the cleft where receptors perform an electrochemical conversion. An *electrical* signal migrates to the soma, and if an electrical probe is inserted into the soma, the cell responses can be observed. This is shown in Figure 21.29.

The signal is referred to as an *action potential*. The soma rests at approximately −60 mV (*hyperpolarized*) and will *fire* (*depolarize*) in the presence of a large enough stimulus signal. The time between onset of the stimulus and the peak action defines the *latency* of the response. Cell firing is restricted by a *refractory interval* where cell action potentials cannot reoccur. These cell parameters result in a signal, which produces an approximately linear relationship between stimulus intensity and frequency response of the action potential. The neuron thus conveys information by *frequency modulation*.

Neuron models can be traced to the neuroscientist Warren S. McCulloch, and neuroscientist, and logician Walter Pitts. In 1943, they proposed a model that could explain how the brain could produce highly complex patterns by using many basic cells that are connected together (McCulloch and Pitts 1943). The original Mc Culloch–Pitts modes employed a transfer function with two levels; it had a step-like shape. These processing elements or artificial neurons

- Receive output signals from other neurons (as in the human model)

- Find the weighted sum of these inputs (*stimulus intensity*)

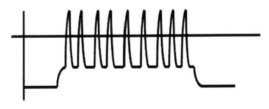

FIGURE 21.29   Neuronal cell response to electrical stimulus.

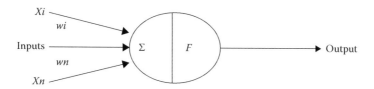

FIGURE 21.30   Schematic diagram of an artificial neuron.

- Generate an activation level—the argument of its transfer function

- Pass the activation level through a transfer function to the output

A schematic representation of an artificial neuron is shown in Figure 21.30.

The point where two neurons communicate is called a *connection*—akin to a synapse. The strength of the connection is called a *weight*, and the collection of weights for the entire net is the *weight matrix*. Of particular importance is the transfer function and while there are numerous types of functions, a widely used nonlinear function (*Sigmoid function*) is given by

$$\text{Output} = \frac{1}{\left(1 + e^{(-k^* \text{input})}\right)}.$$

NN technology is particularly well suited to applications requiring pattern recognition. (Groupings of cells into separate classifications (types) are one immediate example of NN usefulness.) There are numerous types of NNs, but all have several common features:

- The architecture of the individual neuron

- Connections between them (topology)

- Learning rules

In this chapter, we restrict ourselves to one such NN architecture characterized by several descriptors: feedforward; nonlinear; supervised; and backpropagation. (There are numerous other arrangements with complete discussion beyond the scope of this chapter.) The architecture of this model is composed of three *layers*, as shown in Figure 21.31. No neuron has an interconnection from its output back to a neuron in a previous layer—hence the *feedforward* characterization.

21.15.3 Application of NN to Physiology and Medicine

Using subsets of experimental outcomes as a group of training instances (*training set*), an NN can *learn* to identify similar outcomes (Krishna *et al* 2005, Stergious and Siganos 2011). The training can occur with or without guidance (*autoassociative*) of a *teacher*. In this chapter, we restrict the discussion to the guided formulation using the NN in Figure 21.31

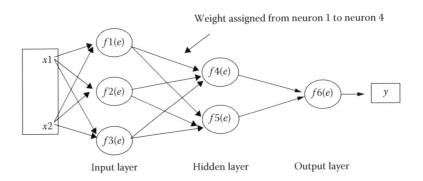

Weight assigned from neuron 1 to neuron 4

Input layer   Hidden layer   Output layer

FIGURE 21.31   Architecture of a representative three-layer NN; $x1$ and $x2$ are input signals to the NN; y is a single output, NN may have many outputs.

as representative of such nets. Such nets are particularly useful for circumstances involving pattern recognition and as such include (partial list) business, manufacturing, medicine, biology, physiology, and psychology. For example, Krishna *et al* (2005) have used such nets to create gene networks *to be used to represent regulatory relationships between genes over time* where reverse engineering of gene networks can overcome a *major obstacle in systems biology.*

21.15.4 Backpropagation (Training)

A subset of the experimental outcomes used to train the NN is provided by a sequence of signals, $x1$, and $x2$ coupled with corresponding correct outputs (z). Training is an iterative process; during each application of particular values ($x1$, $x2$, z), the weights of the NN are modified. A single iterative sequence of part of the net is shown in Figure 21.32. Figure 21.32a depicts the net output when a particular input is applied to the net. The remaining figures indicate how the error between the true output z and the net output x is reflected back (*backpropagation*) such that all weights are adjusted. These adjustments will *tend* to lead to a correct result when the signals $x1$ and $x2$ are applied in the future. The system is repeatedly tested on all training examples such that the errors reach a small but *tolerable* level. (If the net reaches a zero error level, subsequent use with the remaining experimental outcomes can lead to significant errors.) In general, training should be considered as an *art* rather than a rigid algorithm.

(1) Output of NN is computed using $x1$ and $x2$ as inputs. (2) Difference between NN output and correct output (error or δ). (3) Error is reflected back to all neurons. (4) Weights are now adjusted to new value. (Only two weight adjustments are shown.)

The adjustment equation in Figure 21.32d includes several factors in addition to the reflected error (e.g., $\delta1$) and the original weight (e.g., $w_{(x1)_1}$).

- ($df_1(e)/de$): This is the derivative of the sigmoid Function. The purpose is to *move* (adjust) the weights in the direction specified by the *slope* of the nonlinear neural function to a more *favorable* location. This mechanism is consistent with a *hill-climbing* algorithm. (The mathematics of this scheme is beyond the scope of this chapter.

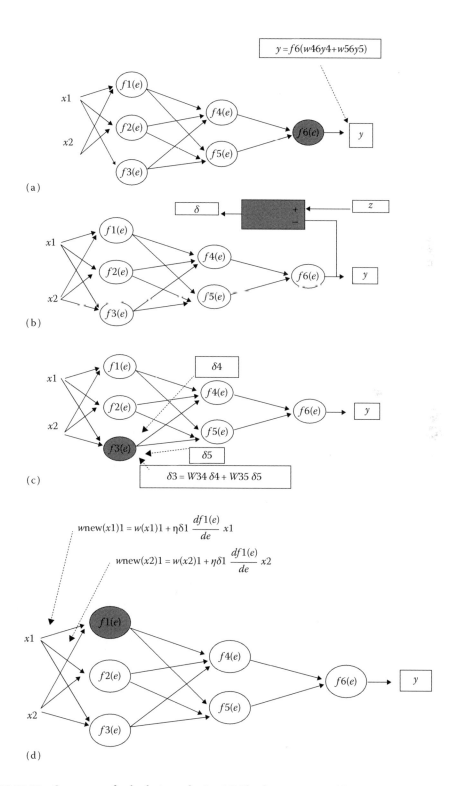

FIGURE 21.32   Sequence of calculations during NN backpropagation Training.

Mathematical evolution of this technique can be found in Stergious and Siganos [2011]). The point at which to evaluate the slope (*df/de*) is the operating point prior to weight adjustment. The sigmoid function is particularly useful because of its property that $f'(e) = f(e)(1 - f(e))k$, which makes it easy to program (evaluate).

- η is the *learning rate*. There are several methods for adjusting this: start with a large value. As weights are adjusted, decrease η so that learning *overshoots* are not encountered. Alternatively, start with a small η, then increase it as learning progresses, finally reducing it as the NN approaches its *terminal values*.

21.15.5 Summary of Knowledge-Based Technology

Knowledge-based techniques can significantly improve the services and capabilities of existing instrumentation particularly in light of the increasing complexity of contemporary experiments and higher-capacity data processing needs. Pattern recognition as implemented by NN (or Expert Systems, and Fuzzy Logic Systems) can be trained to recognize patterns (e.g., types of cancer cells or behavior). The ability of such systems to *extract the underlying theoretical relationships of experimental variables* will become an increasing measure of the power of machine-based technology.

21.15.6 Medical Record Keeping and the Paperless Medical Office

The techniques and methods described earlier will culminate in medical and research environments that will depend heavily on the computer. This has been referred to as the *paperless medical office* and is fully explored by Coiera (2003) in *Guide to Medical Informatics, the Internet and Telemedicine*. As Coiera notes "…the EMR is the computer replacement for existing medical record systems" (page 74; EMR stands for electronic medical record). The ultimate advantages of the EMR (over paper) include better use of space; durability of the records (e.g., no lost records); centralized location of data; concurrent multiuser availability; and improved efficiency of record location. Such systems can be characterized by the following:

- Captures information during clinical encounters; stores such information in a secure location and can be accessed by those *with a need to know.*

- Capabilities and resources of such systems include ordering tests and studies, image archiving, image interpretation, coding for administrative purposes, and communication between health providers.

- Portability and interoperability between platforms; standards for data representation and capture.

- For the patient: generation of clinical alerts; and avoidance of fatal drug interactions.

It is very likely that such systems will become significant in our ability to deliver health care and carry out research in the future.

BIBLIOGRAPHY

Abramson N M 1963 *Information Theory and Coding* (New York: McGraw-Hill)

Addison P S 2005 Wavelet transforms and the ECG: a review *Physiol. Meas.* **26** R155–99

Avudaiammal R 2009 *Information Coding Techniques*, 2nd edition (New York: McGraw-Hill)

Balmer L 1997 *Signals and Systems: An Introduction* (Upper Saddle River: Prentice-Hall)

Bozic S M 1994 *Digital and Kalman Filtering*, 2nd edition (London: Edward Arnold)

Broman H 1988 Knowledge-based signal processing in the decomposition of myoelectric signals *IEEE Eng. Med. Biol. Mag.* **7**(2) 24–8

Coiera E 2003 *Guide to Medical Informatics, the Internet and Telemedicine* (London: Arnold Hoder)

Cooley J W and Tukey J W 1965 An algorithm for the machine calculation of complex Fourier series *Math. Comput.* **19** 297–301

Database for the Identification of Systems (DaISy), http://www.homes.esat.kuleuven.be/~smc/daisy

David H A 2001 First occurrence of common terms in statistics and probability in H A Edwards (ed.), *Annotated Readings in the History of English* (New York: Springer)

de Weerd J P C 1981 A posteriori time-varying filtering of averaged evoked potentials I. Introduction and conceptual basis *Biol. Cybern.* **41** 211–2

Delone B N 2005 *The St Petersburg School of Number Theory* (Providence: American Mathematical Society)

Ferrara E R and Widrow B 1982 Fetal enhancement by time-sequenced adaptive filtering *IEEE Trans. Biomed. Eng.* **BME-29** 458–9

Friedenberg J and Silverman G 2012 *Cognitive Science: An Introduction to the Study of Mind*, 2nd edition (Thousand Oaks: Sage Publications)

Gibson N M, Peasgood W, Herbert J M, Woolfson M S and Crowe J A 1995 Iterative method to derive an approximate matched filter template for fetal electrocardiogram signals *Med. Eng. Phys.* **17** 188–96

Gigerenzer G 2002 *Calculated Risks: How to Know When Numbers Deceive You* (New York: Simon & Schuster) pp 40–4

Hall E C 1996 *Journey to the Moon: The History of the Apollo Guidance Computer* (Piscataway: IEEE)

Hamming R W 1980 *Coding and Information Theory* (Englewood Cliffs: Prentice-Hall)

Hamming W 1973 *Numerical Methods for Scientists and Engineers*, 2nd edition (Mineola: Dover Publications)

Hawkins J 2004 *On Intelligence* (New York: Times Books)

Haykin S 1996 *Adaptive Filter Theory*, 3rd edition (Upper Saddle River: Prentice-Hall International)

Heath S 2003 *Embedded Systems Design*, 2nd edition. EDN series for design engineers (Amsterdam: Elsevier)

Hirel C, Sahner R A, Zang X and Trivedi K S 2000 Reliability and performability modeling using SHARPE 2000 *TOOLS'00 Proceedings of 11th International Conference on Computer Performance Evaluation: Modelling Techniques Tools* (London: Springer)

Horowitz P and Hill 1989 *The Art of Electronics*, 2nd edition (Cambridge: Cambridge University Press)

http://cis.legacy.ics.tkk.fi/aapo/papers/IJCNN99_tutorialweb/IJCNN99_tutorial3.html (accessed June 27, 2011)

http://cs.pitt.edu/~kirk/CS1501/animation (accessed August 5, 2011)

http://www.britannica.com/EBchecked/topic/2011

http://www.itl.nist.gov/div898/handbook/dtoc.htm 2011

http://www.norsys.com/2011

http://www.webmd.com 2011

http://www-03.ibm.com/press/us/en/pressrelease/35251.wss 2011

Huffman D A 1952 A method for the construction of minimum-redundancy codes *Proc. IRE* **40** 1098–102

A Hyvärinen. Independent Component Analysis: Recent Advances. Philosophical Transactions of the Royal Society A, 371:20110534, 2013.

Janssen J and Limnios N (eds.) 1999 *Semi-Markov Models and Applications* (London: Springer)

Kennedy P 2011 The Cyborg in us all *New York Times*, Sunday, September 18, 2011

Krishna A, Narayanan A and Keedwell E C 2005 Neural networks and temporal gene expression data in *Lecture Notes in Computer Science* **3449** 64–73 (New York: Springer)

Kudeki E and Muson Jr D 2009 *Analog Signals and Systems* (Upper Saddle River: Pearson Education)

Kulikowski C 1988 Artificial intelligence in medical consultation systems: a review *IEEE Eng. Med. Biol. Mag.* **7** 34–9

Kymes S M, Plotzke M R, Li J Z, Nichol M B, Wu J and Fain J 2010 The increased cost of medical services for people diagnosed with primary open-angle glaucoma: a decision analytic approach *Am. J. Ophthalmol.* **150** 74–81

Lindeboom G A 2007 *Herman Boerhaave: The man and his work* (2 ed), with an updated Bibliography and an improved Edition of *Lindeboom's Bibliographia Boerhaaviana* by M J Van Liegurg Rotterdam: Erasmus Publishing

McClellan J, Schafer R and Yoder M 2003 *Signal Processing First* (Upper Saddle River: Pearson Education International)

McCulloch W S and Pitts W 1943 A logical calculus of the ideas immanent in nervous activity *Bull. Math. Biophys.* **5** 115–37

Mitra S 2011 *Digital Signal Processing* (New York: McGraw Hill)

Mlodinow L 2009 *The Drunkard's Walk* (New York: Vintage Books)

Pan J and Tompkins W J 1985 A real-time QRS detection algorithm *IEEE Trans. Biomed. Eng.* **BME-32** 230–5

Papp Z, Peceli G, Bago B and Pataki B 1988 Intelligent medical instruments *IEEE Eng. Med. Biol. Mag.* **7** 19–23

Paulos J A 2011 Book Review: *The Theory That Would Not Die: How Bayes' Rule Cracked the Enigma Code, Hunted Down Russian Submarines and Emerged Triumphant from Two Centuries of Controversy* (Sharon McGrayne: Yale University Press), *New York Times*, August 5, 2011

Russell S J and Norvig P 2011 *Artificial Intelligence: A Modern Approach*, 3rd edition (Upper Saddle River: Prentice Hall)

Schefler W C 1979 *Statistics for the Biological Sciences* (Boston: Addison-Wesley)

Schutt C E and Lindberg U 1998 Muscle contraction as a Markov process: I Energetics of the process *Acta Phys. Scand.* **163** 307–23

Shamsolmaali A, Carson E R, Collinson P O and Cramp D G June 1988 A knowledge-based system coupled to a mathematical model for interpretation of laboratory data *IEEE Eng. Med. Biol. Mag.* **7** 40–6.

Shiavi R June 1988 Factors in automated gait evaluation *IEEE Eng. Med. Biol. Mag.* **7** 29–33

Silverman G 2003 The role of neural networks in telerehabilitation *Proceedings of the First International IEEE EMBS Conference of Neural Engineering* (Capri, Italy)

Silverman G 2005a Human cognitive and perceptual motor response: an integration of idiothetic and allothetic processes *Proceedings of the Sixth International Conference on Modeling in Medicine and Biology* (Bologna, Italy)

Silverman G 2005b Impact of automation and telemedicine on the healthcare system *Proceedings of IASTED International Conference on Telehealth* (Banff, Canada)

Silverman G and Silver H 1995 *Modern Instrumentation: A Computer Approach* (Bristol: Institute of Physics Publishing)

Smith L A 2011 A tutorial on principal components analysis http://ranger.uta.edu/~guerra/Smith02tutorial.pdf (accessed July 8, 2011)

Spencer R and Ghausi M 2003 *Introduction to Electronic Circuit Design* (Upper Saddle River: Prentice Hall)

Stergious E and Siganos D 2011 Neural networks http://doc.ia.ac.uk/~nd/suprise_96/joun (accessed August 5, 2011)

Strang G and Nguyen T 1996 *Wavelets and Filter Banks*, 2nd edition (Wellesley: Wellesley-Cambridge)

Sztipanovits J and Karsai G June 1988 Knowledge-based techniques in instrumentation *IEEE Eng. Med. Biol. Mag.* **7** 13–7

Taylor H M and Karlin S 1998 *An Introduction to Stochastic Modeling*, 3rd edition (San Diego, CA: Elsevier Science Publishing Co. Inc.)

University of Michigan Chemical Engineering Process Dynamics and Controls Open Textbook 2011 https://controls.engin.umich.edu/wiki/index.php

Unser M 2000 Sampling-50 Years after Shannon *Proc. IEEE* **4** 569–87

Wilcox W 1938 The founder of statistics *Rev. Int. Statist. Inst.* **5** 321–8

Woolfson M S, Huang X B and Crowe J A 1995 Time varying Wiener filtering of the fetal ECG using the wavelet transform *IEE Colloquium on Signal Processing in Cardiography* Digest No. 1995/043

www.esat.kuleuven.ac.be/sista/daisy (accessed May 17th, 2011)

www.gap-system.org/~history/Biographies/Markov/2012

www.isograph-software.com 2011

www.theriac.org Reliability Information Analysis Center 2011

Yoon B-J 2009 Hidden Markov models and their applications in biological sequence analysis *Curr. Genomics* **10** 402–15

Zarzoso V and Nandi A K 1999 Blind source separation of independent sources for virtually any source probability density function *IEEE Trans. Biomed. Eng.* **47** 2419–32

Zarzoso V and Nandi A K 2001 Noninvasive fetal electrocardiogram extraction: blind separation versus adaptive noise cancellation *IEEE Trans. Biomed. Eng.* **48** 12–8

RECOMMENDED READING LIST

The following list is a selection of journals and magazines that contain papers and articles on various digital signal processing techniques applied to both preprocessing and analysis of biomedical and physiological data:

IEEE Transactions on Biomedical Engineering
IEEE Transactions on Signal Processing
IEEE Reviews in Biomedical Engineering
IEEE Society Instrumentation and Measurement Magazine
Signal Processing (Eurasip)
IET Signal Processing
Medical Engineering & Physics
Medical & Biological Engineering & Computing
Physiological Measurement

Index

A

Abdominal pressure, 488–489, 491, 503, 508, 510, 518, 520–521, 526
ABPM, see Ambulatory blood pressure measurement (ABPM)
ABRs, see Auditory brainstem responses (ABRs)
ACC, see Acoustic change complex (ACC)
Acoustic change complex (ACC), 318, 343–344, 346
ACS, see Air conducted sound (ACS)
Action potentials (APs), 45, 178–180, 183, 185, 212–214, 216, 221, 223–224, 227–232, 234–235, 269, 309–310, 312, 328–331, 334, 348–350, 580
Activation, 45, 47, 67–69, 156, 209, 232, 337, 529
Activation wavefronts, 45, 67–69
Active electrodes, 153, 194, 220, 224, 226–227, 329, 346
Activity, administered, 165–167
Acute respiratory distress syndrome (ARDS), 442, 449–450
ADC, 217, 273–275, 305, 423, 553–556
Adults, 16, 107, 109, 111, 113, 141–142, 160, 168, 170, 205, 207, 342–343, 352, 394, 398–399
Air-conducted sound (ACS), 274, 336–337, 467
Air flow, 424, 426–427
Air pressure, 282–283, 285, 428
Airway pressure, 423–424, 437
Airways, 2–4, 433, 436–437, 448–450
Algorithms, 17, 28, 40, 118–119, 129, 154, 203–204, 265, 388, 542, 559, 578, 582, 585
Ambulatory blood pressure measurement (ABPM), 13, 29, 31
Ambulatory urodynamics, 487, 500, 512, 520, 524, 527, 530–531
Amniotic fluid, 244–247, 250, 253
Amplifiers, 193, 197–198, 201, 214, 216–217, 220–221, 253, 255, 272, 274, 311–312, 426, 428, 430, 490–491
  differential, 197, 199, 274, 311–312

Amplitude, 38–39, 58–59, 62–63, 204–206, 216–217, 221–224, 227–229, 269–271, 274–275, 319–321, 338–339, 341–342, 344–347, 422, 545–546
  response, 460, 545–546, 549–550, 553
  spectrum, 548, 557–558
Analog filters, 221, 537, 545, 548–550, 553, 557, 559
Analog signal processing, 485–486, 498, 513
Analog signals, 216–217, 544, 548–550, 552–553, 556
Analog voltage, 554–556
An automated ABR (AABR), 323–324, 348
Angles, 135, 193, 242–243, 276, 294–296, 299, 377, 422, 463, 473–474
Antibodies, 358–359, 368–373, 375, 377–380, 390
Antigens, 358, 365, 372, 374, 377–378, 380, 526
Aortic pressure, 15, 33, 80, 108, 112
Aortic valve, 33, 79–80, 94, 103, 106–109, 111
Applanation tonometry, 14, 33–34, 37, 97
APs, see Action potentials (APs)
Area under curve (AUC), 162–165
Arrhythmias, 43–44, 54, 56–57, 65–68, 73, 399
Arterial blood, 4, 8, 82, 90, 258, 410–412, 431, 433, 443–444, 449
Arterial blood pressure, 3, 14, 16, 31, 41, 88
Arterial pressure, 15, 20, 23, 30–31, 33–35, 77, 88, 403
Arterial pulse, 14–15, 31, 33, 36
Arterial spin labeling (ASL), 82–83
Arterial stiffness, 31, 40, 75, 94–96, 99
Arteries, 3, 7, 23–24, 34–35, 37, 41, 78, 87, 89, 91, 94–97, 99, 106–107, 109, 406
  brachial, 16, 23–26, 36
  radial, 3, 34, 37–38, 90, 97–98
Artifacts
  rejection, 202–203, 320
  technical, 201, 203
Assays, 238, 357–358, 361, 369–370, 372, 374–375, 377, 379–380, 382–383, 386, 388, 390, 392
Assay sensitivity, 370, 372, 374, 377, 380, 383, 385, 387–388

Association for the Advancement of Medical Instrumentation (AAMI), 30–31
Asymmetry ratio (AR), 338
Atrial contraction, 105, 109–110, 113
Atrium
    left, 44–45, 71, 77, 79, 103, 105, 112, 114
    right, 7, 44–45, 71, 77, 79, 103, 105
AUC, *see* Area under curve (AUC)
Auditory, 209, 211–212, 230, 318, 320, 326, 329, 340, 348–349, 351–352
Auditory brainstem responses (ABRs), 317, 319–326, 329–331, 341, 343, 348–351, 353–354
    derived-band, 322–323
    stacked, 317, 321–323, 349
Auditory nerve, 323–325, 328–331, 334–335
Auditory steady-state responses (ASSRs), 326–328, 349, 351–352
Auscultation, 101–102, 110–111, 114, 118, 239
Automatic exposure control (AEC), 454, 467
AV valves, 103–104, 106–107, 109

B

Background signals, 374, 376, 378
Balloon, 71, 92, 135, 140, 143–144, 151–152, 260, 263, 424–425, 511–512, 515
Balloon catheter, 151, 425, 428
Bayesian belief networks (BBNs), 571, 573, 578
BBNs, *see* Bayesian belief networks (BBNs)
Benign prostatic hyperplasia (BPH), 489, 534
Biopotential electrodes, 47, 251, 398, 418
Bipolar recordings, 47, 67–69, 171, 200, 227
Bladder, 246, 488–490, 492, 502–503, 506–508, 510–512, 516, 519–520, 524, 527–532
    capacity, functional, 489, 502
    outlet obstruction, 489, 506, 520–521, 534
    pressure, 486, 490, 503–504, 506–508, 519, 521, 527, 532
Blind source separation (BSS), 203, 558–559, 587
Blood, 6–7, 15, 80–82, 87–91, 93, 103–107, 109, 111–112, 400, 403–410, 412, 428–429, 431–433, 440–441, 444–445
    cells, red, 90, 358–359, 413
    cooled, 93–94
    deoxygenated, 45, 87, 89
    fetal, 247–248, 257–258
    flow, 4, 7–8, 80, 86, 89, 98, 101–103, 105–107, 109, 112, 162, 191, 441, 444, 447
    flow measurements, 1, 3, 7, 99
    gas analyzer, 393, 404–405, 409–410
    gases, 404, 407, 409–410, 412, 414
    oxygenated, 45, 87, 89, 91, 440
    perfusion, 80–81

plasma, 378, 429, 431–432
pressure, 4, 6, 13–15, 17, 19, 21–23, 25, 27, 29–31, 33, 35, 39–41, 397–398, 403, 477–478
    high, 25, 40, 527
    measurement, 4, 14, 16–17, 22–25, 28, 31, 39–41, 403
    measurement of arterial, 16, 41
sample, 1, 238, 404, 406–409, 413, 440
vessels, 14, 18, 85, 87, 173, 245, 468, 470
volume of, 78, 80, 87, 106, 413
Body core temperature, 396, 415–416
Body segments, 313–314, 480
Body surface area (BSA), 165–166, 168–170
Body surface mapping, 43, 51–52
Body temperature, 4, 8, 137, 174, 211, 396, 415–416, 433–434, 478, 516
Bone-conducted vibration (BCV), 336–337, 349
Brain, 2–4, 6, 81–83, 172–175, 180–181, 189–191, 204–205, 208–209, 211–212, 214, 268–269, 343–345, 413–415, 559–560, 579–580
    function, 175, 190, 192, 206, 210–211, 352
    imaging methodologies, 190–192, 210
    stem, 173, 175, 349
Brainstem responses
    automated auditory, 348–350
    evoked auditory, 348, 351, 353
BSA, *see* Body surface area (BSA)
BSS, *see* Blind source separation (BSS)

C

Calibration, 144, 277, 404, 406, 410, 427, 450, 474, 494–497, 499, 513–514, 578
Cameras, 133, 147–148, 150, 314, 375
Cancer, 1, 168, 468, 526, 571–574
Capillaries, 75, 80–81, 85–87, 177, 290, 412
Capillary blood, 258, 261, 411–412
Carbon dioxide (CO_2), 77–78, 262, 393, 404, 408–410, 428–431, 433–434, 440, 443, 449, 516
Cardiac cycle, 32, 60, 76, 79–81, 96, 102–104, 108, 111, 118–120, 240, 242, 252
Cardiac MRI (CMR), 94
Cardiac output (CO), 7–9, 75–77, 87, 89, 91–95, 136, 408, 440–442, 444, 447–448, 450
Cardiopulmonary monitors, 398, 403, 416, 418
Cardiotocogram (CTG), 248, 250, 259, 265
Catheter, 17–18, 20–21, 67, 70–72, 92–93, 135–142, 144, 253, 260, 403–407, 424, 510–515, 517, 522, 530–531
    air-filled, 510–512, 522
    dislodged, 486, 522
    tip, 21, 70, 92, 260, 403, 424, 510–514, 517
    water-filled, 510–512, 514, 522, 530

Catheter-tip sensors, 13, 17, 20, 486, 511–512, 514, 527, 529
Cathode, 297, 301, 340, 405–406, 411, 433, 456, 463
Cavities, uterine, 247, 249–250, 254, 260, 262
Cell body, 176, 182, 215, 579
Cells, 73, 81, 176–180, 183, 186, 188, 215–216, 218, 224, 263, 268, 270, 358–359, 362–369, 580–581
Central nervous system (CNS), 171–173, 175–177, 182, 188, 190, 211–212, 214–215, 223, 330
Central processing unit (CPU), 272, 314–315, 579
Charge, 46–47, 117, 304, 364, 454–456, 463, 471
Chemiluminescent signals, 371, 379, 384, 388
Chest wall, 110, 114–115, 117, 292, 437, 449
Children, 107, 109, 111, 113, 141–142, 168, 170, 316, 325, 342–343, 349–351, 353, 393–395, 411, 416–418
    young, 333, 335, 342–343, 353, 418
Chirp stimuli, 321, 323, 349
Chromogenic in situ hybridization (CISH) assays, 386–387
Chromosome enumeration probes (CEPs), 385–386
Chronic obstructive pulmonary disease (COPD), 116, 433, 441, 445, 451, 478
CI, *see* Cochlear implant (CI)
Circuit, 18, 22, 47, 53, 89, 198–200, 242, 315, 417, 467, 523, 526
Circulation, 14, 76–78, 92–93, 98
    systemic, 87, 89, 103, 440
Clicks, 293, 320–325, 329, 332, 337, 342–343, 465
Clinical labs, 361–362, 372, 379–380, 384, 389–390
Clinical pathology, 355–357
Closure, 33, 106–107, 109, 113, 135, 207
CMAP, *see* Compound motor action potential (CMAP)
CMRR, *see* Common mode rejection ratio (CMRR)
CNS, *see* Central nervous system (CNS)
CO, *see* Cardiac output (CO)
CO_2, *see* Carbon dioxide (CO_2)
Cochlea, 319–323, 328, 331–333, 335
Cochlear implant (CI), 318, 324, 333, 335, 343, 348–351, 353, 455, 560
Collision techniques, 214, 231–233
Commercial systems, 63, 274–275, 320
Common mode rejection ratio (CMRR), 117, 221, 273–274, 423
Compound motor action potential (CMAP), 232, 309, 312
Compound nerve action potential (CNAP), 330–331
Computed tomography (CT) scanners, 287, 298–299, 301, 304–305
Concentrations, 81, 85, 89, 92, 162, 166, 178, 268, 374, 383–384, 408, 420, 428–429, 444
    antigen, 370, 378, 380

Conduction time, 56, 231, 234
Conduction velocity, 214, 232–234, 309–310
Configurations, 43, 47, 49–51, 184, 197–198, 217, 220–221, 226–227, 235, 252, 326, 424
    bipolar, 217, 220, 226–227
    tripolar, 220–221
Confocal laser scanning microscopy (CLSM), 125
Constant pressure, 18, 69, 151
Contamination, 239, 339, 361–362, 381, 384
Contractions, 45, 102, 104, 106, 132, 135, 144, 153, 158, 249–250, 253–254, 260, 262, 292, 489
    muscular, 132, 134, 144, 153, 311
    pressure, 508, 517
COPD, *see* Chronic obstructive pulmonary disease (COPD)
Core temperature, 261, 404
Cornea, 7, 34, 269, 274, 276–285
Cortex, 174, 180–182, 189–190, 196, 210, 231, 340, 348, 353
Coughs, 143, 513, 517–519, 522, 524, 528
CPU, *see* Central processing unit (CPU)
Crystal, 298, 301–303, 462, 494
CT images, 52, 84, 299
Cuff
    brachial, 16, 23, 25, 36
    deflation, 17, 23–24, 26
    electrodes, 219–221, 224, 231
    pressure, 3, 7, 16, 23, 26–28, 403
    width, 30, 403
Cystometry, 486, 490, 506–507, 518, 520–521

D

DAC, *see* Data acquisition computer (DAC)
DAC system, 299–301, 303, 306, 311
Data acquisition computer (DAC), 198, 298, 300, 303, 306, 311–312, 555
Database for the Identification of Systems (DaISy), 558, 585
Data processing, 537–539, 541, 543, 545, 547, 549, 551, 553, 555, 557, 559, 561, 563, 565, 567
Data, recorded, 29, 153–154, 184, 186
Data recorder, 146–147
Decomposition, 229, 585
Dendrites, 176–177, 181–182, 215, 579–580
Detection antibodies, 370–371, 378, 380, 390
Detectors, 126, 159, 243, 263, 282–283, 293, 297–303, 363, 366–368, 377, 412, 460–461, 469, 472
Detrusor contraction, 517–519, 524
Detrusor pressure, 486, 491, 508, 510, 518, 520, 534
Deviation, standard, 61, 63, 365, 460, 501, 566–567, 570

Devices, 29–31, 35–37, 65, 117–118, 135–138, 143, 146–148, 155–158, 217, 240, 258–259, 417–418, 425–426, 465–467, 560
water-perfused, 135–137
Diagnostic, molecular, 357–358, 391
Diaphragm, 17–18, 22–23, 115–116, 118, 423, 425–426, 508–510
Diastole, 23, 33, 76–78, 80, 94, 104–105, 107–108, 111–112, 400
Diastolic pressure (DP), 3, 7, 14–18, 23, 25–28, 32–33, 36, 39, 80, 95, 98, 333, 425–427, 577
Difference, significant, 251, 416, 569–570
Differentiator, 498–499, 523, 525
Digital filters, 498, 537, 556, 558–559
Digital signal processing (DSP), 54, 101–102, 119, 121, 203, 222, 274, 485–486, 498, 513, 544, 586–587
Dipstick, 372, 492–495, 497–498
Dipstick flowmeter, 485, 492, 494, 509
Discrete Fourier transform (DFT), 557
Diseases
infectious, 2–3, 357–358, 377, 380
processes, 292, 300–301, 488
Dislocation, 288–289, 292–293
Displacement, 16, 18, 78, 117, 239, 245–246, 254, 277, 292, 296, 313
Distance, 22, 71, 126–127, 157, 217, 221, 224, 228, 231–232, 235, 238, 271, 308–309, 455, 472
Distortion, 20, 59, 197, 201, 232, 432, 537, 541, 545, 549–550, 553, 557, 559
DNA probes, 387
Doppler ultrasound, 7, 237, 240–241, 252–253
DP, see Diastolic pressure (DP)
DSP, see Digital signal processing (DSP)
Dynamic response, 17, 421–422, 429–430, 438, 448

E

Ear, 102
ECAP, see Evoked compound action potential (ECAP)
ECG, see Electrocardiogram (ECG)
Effective refractory periods (ERPs), 44, 69–70, 206, 208–209, 318, 341–344, 352
EI, see Erythema index (EI)
Electrical activity, 3, 6, 44–46, 67, 73, 177, 182–183, 189–193, 205, 210, 224, 311–312, 320, 328, 336
Electrically, 324
Electrically evoked ABR (EABR), 324–325, 335
Electrical signals, 117, 176, 179, 183, 209, 214–215, 227, 240, 254, 259, 298, 300, 420, 423, 426
Electrical stimulus, 335, 580

Electrocardiogram (ECG), 2–3, 5–6, 44, 46, 53–55, 58–60, 62, 64–66, 73, 91, 102–103, 105–108, 200–204, 252–253, 558–559
maternal, 240–241
measurements, 54, 73
recordings, 6, 47, 53, 65–66
signals, 64–65, 102–103, 106, 108–109, 557
12-lead, 43, 47, 49, 51–52
Electrocardiography, 43, 45, 47, 49, 51, 53, 55, 57, 59, 61, 63, 65, 67, 69, 129
Electrocorticogram (ECoG), 189, 318, 328–331, 560
Electrodes, 5–6, 9, 46–47, 51–53, 67–68, 153–154, 183–186, 193–196, 198–200, 216–226, 240–241, 269, 273–274, 328–331, 398–400
array, 334, 364
ball, 329, 334
caps, 193–194, 342
design, 183–184, 218, 230
extraneural, 213, 218–219, 221, 223
ground, 51, 220, 320, 326, 329, 338
inner, 9, 464
intrafascicular, 218–219, 222, 231
locations, 46, 50, 52
microwire, 183–184
negative, 51, 68
pair, 200, 311–312, 364
pediatric, 398–399
placement, 50, 171, 193, 195–196, 328, 333
positive, 47, 49, 51, 68
regenerative, 213, 218, 220–221, 223–224
sieve, 218, 220, 223
site, 6, 198–199, 226
stimulating, 232–233, 311–312, 339–340
types, 183–184, 226
wires, 202, 311
Electroencephalogram (EEG), 2–3, 6, 185, 189–190, 192–193, 198, 201, 203–204, 206, 208, 210–212, 271, 321, 323, 342
data, 203, 211
recordings, 196, 198, 201–204, 212
signal, 198–201, 203, 209
Electrograms, 44, 46, 67–70, 189
Electromyogram (EMG), 2–3, 6, 189, 201–204, 211, 214, 216, 224, 226–227, 230, 235, 309–310, 313, 529, 533
Electroneurography (ENoG), 339–340, 348
Electrons, 6, 46, 297, 302, 405, 408, 429, 454, 456–459, 466
Electrooculogram (EOG), 189, 201–204, 269, 272, 275–276
Electrophysiology, 191, 267–268, 275, 538
Electroretinogram (ERG), 267, 269–270, 272, 275
Embedded systems, 537, 559–560, 562–563

Emotion, 174–175, 210
Endocochlear potential (EP), 328
Energy, binding, 457–458, 470
EPs, *see* Evoked potentials (EPs)
ERG signals, 270
ERPs, *see* Effective refractory periods (ERPs)
Errors, 16, 25, 28, 71, 166–168, 420, 496–497,
    511–514, 516, 527, 541–542, 554, 582
ERS, *see* Event-related synchronization (ERS)
Erythema index (EI), 124, 320, 333–335
Esophagogastric junction (EGJ), 155, 157
Esophagus, 132–133, 135, 137–140, 142, 144–149,
    155–156, 158, 423–424, 427–428
Estimation of pressure, 13, 28
Event-related desynchronization (ERD), 206, 209
Event-related synchronization (ERS), 171, 206,
    209, 437
Evoked compound action potential (ECAP), 318,
    325, 333–335, 348, 350
Evoked potentials (EPs), 209–210, 214, 230–231,
    235, 285–286, 327, 336, 340–341, 344, 346,
    348–349, 351, 353, 585
Expiration, 113, 440, 443–445
Exposure, 132, 305, 459, 461, 466–470, 474
Exposure times, 297, 466–467, 474
External anal sphincter (EAS), 142–143, 158
External sensor, 20, 34, 486, 510–512
Extracellular fluid volume (ECV), 161–163, 165–169
Extracellular space, 163–164, 183, 186, 217, 219
Extremities, 3, 5, 8, 77, 81, 288, 290, 292, 299, 307–308
Eye movements, 202, 269

F

Fast Fourier transform (FFT), 153, 438, 461, 537,
    548, 557–559
FDA, *see* Federal Drug Administration (FDA)
Federal Drug Administration (FDA), 368, 375,
    380–381, 391
Fetal
    blood sampling, 237, 257
    ECG, 240–241, 249–253, 558, 587
    heart, 239, 241–242, 251–253, 265
    membranes, 253–254, 258
    monitor, 237, 243, 252–256
    scalp electrode, 241, 251–252, 257
FFT, *see* Fast Fourier transform (FFT)
Fibers, 137, 215–216, 219, 221, 224, 231–233, 474
Filling, 77, 80, 94, 105–106, 108, 113, 145, 485–486,
    489, 500, 506, 516–517, 521–522, 529
    artifacts, 486, 521–522
    catheters, 486, 502, 517
    phase, 107, 109–110, 113, 518–519, 529

Filters, 193, 199, 202, 204, 272–273, 311–312, 363,
    467, 490–491, 513, 544–546, 549–550, 553,
    557, 559
    band-pass, 270, 545–546, 549
    optical, 366, 376–377, 379, 386
Filtration markers, 161–163, 166–167, 169
Filtration rate, glomerular, 161, 170
Flow
    current, 177–179, 198, 269
    cytometer, 362–363, 367, 375, 381, 392
        conventional, 363, 366–367
    cytometry, 359, 362–364, 367–369, 391–392
    measurements, 7, 502, 533
    meters, orifice, 426–427
    rate, 87, 93, 98, 135, 436, 438, 490–491, 494–496,
        498–501, 504, 506, 516, 518, 520, 523
        low, 424, 497, 506, 520–521
        trace, 504, 506, 519, 521
Flowmeter, 485, 491, 495–499, 503, 505, 513,
    516, 520
    load-cell, 485, 490, 495–496, 499, 503
    spinning disk, 485, 496–497
Fluid-filled catheters, 18, 20
Fluidics systems, 355, 363
Fluorescence, 263, 298, 303, 362, 367–370, 374–375,
    377, 381, 388, 406
Fluorescence resonance energy transfer (FRET),
    382–384
Fluorescence signals, 366–367, 374, 382
Fluorescent probes, 365–366, 382, 385
Fluorescent signals, 366, 372, 374, 377–378, 382–383,
    385–387
Fluorophores, 368, 374, 376–377, 382–383, 385
Forced oscillation technique (FOT), 419, 438–440,
    448, 450
Forced vital capacity (FVC), 9, 433, 445
FOT, *see* Forced oscillation technique (FOT)
Four-chambered heart, 75–76, 78–79, 98
Freezing of gait (FOG), 480–481
Frequency
    components, 20, 37, 222, 306, 326, 517, 549–550,
        552–553
    content, 111, 120, 221–222, 228
    corner, 193, 229, 523
    domain, 20, 32, 37–39, 61, 63, 154, 203
    function of, 37, 76
    modulation, 269, 327, 580
    precessional, 304, 471
    response, 13, 18, 20–21, 35, 116–119, 421–422,
        427, 438, 448, 499, 511, 515, 522, 546, 556
    spectrum, 111, 333
    test, 546–547
Frequency–volume chart, 487, 526, 532

Functional luminal imaging probe, 132, 157
Function, utricular, 336–337, 349
FVC, *see* Forced vital capacity (FVC)

G

Galvanic skin response (GSR), 211, 478
Ganzfeld stimulator, 267, 273, 275
Gases
    flowing, 426, 432, 434
    foreign, 419, 430, 443–445, 451
    soluble, 99, 428, 430, 444
GFR, *see* Glomerular filtration rate (GFR)
GI tract, 132–133, 135–136, 138, 141, 148–150, 152,
    155, 157–158
Global positioning system (GPS), 70, 478
Glomerular filtration rate (GFR), 161–170
    ECV, 166, 168–169
    uncorrected, 164–165

H

Heart rate (HR), 24, 28, 38, 45, 58, 61, 63–64, 66–67,
    69, 91, 249, 252, 397–399, 557, 559
    fetal, 240, 242–243, 248–250, 252–253, 259, 265
    instantaneous fetal, 248–251
Heart rate variability (HRV), 43, 60–61, 64, 211
Heart sounds, 80, 101–103, 106–111, 113, 115,
    117–121
    abnormal, 102, 112, 119
    fetal, 237, 239
    first, 106–108, 111
    pathological, 101–102, 114
    second, 107–109, 111
Heart valves, 103, 109–110, 118
Hemodynamics, 75–77, 79, 81, 83, 85, 87, 89, 91, 93,
    95, 97, 99
Hemoglobin, 90, 258, 376, 406, 412–413, 429,
    431–432, 440, 444, 446
Hemoglobin oxygen saturation, 258, 404, 406–407,
    409, 412–413
High-resolution manometry (HRM), 135, 139, 141,
    155, 157, 159
Holter monitoring, 44, 64–65
Holter monitors, 53, 64–65
HR, *see* Heart rate (HR)
HRM, *see* High-resolution manometry (HRM)
HRV, *see* Heart rate variability (HRV)
Human body, 2, 4, 6, 10, 114, 172, 189, 211, 215,
    268, 579
Human brain, 171–172, 176, 230, 579
Human skin, 123–125, 128, 198
Hypothesis, 268, 562, 567, 569–572

I

Images, 81–83, 125, 128, 133, 145–150, 157–158,
    244–245, 298–300, 303–305, 307, 314, 367,
    457–458, 460–462, 468–473
    digital, 460, 462
    intensifier, 453, 462, 468–469
Imaging, 11, 81–82, 99–100, 134, 147–148, 170,
    244–246, 299, 306, 461, 468, 470, 475
    ultrasonic, 237, 243–247, 454, 473
Impedance, 9, 72, 76, 131, 144–145, 156–157, 198,
    202, 205, 271, 274, 400, 402, 438–439, 447
Impedance measurements, 1, 9, 72, 145
Implanted electrode, 185–186, 223
Independent component analysis (ICA), 204, 586
Indwelling sensors, 487, 532
Information
    amount of, 187, 366, 539–540
    theory, 187–188, 542, 548, 585
Input impedance, 37, 53, 117, 171, 193, 198, 272, 274,
    438, 449
Input, inverting, 197–198
Input signal, 151, 197–198, 421–422, 544–546,
    549–550, 553, 556, 582
Instrumentation
    amplifier, 171, 199–200, 495, 510, 523, 525
    physiological measurement, 397, 418
Integrator, 467, 470, 494, 496–497, 554
Integrity, functional, 101–102, 271
International Electrotechnical Commission (IEC), 30
International Society for Clinical Electrophysiology
    of Vision (ISCEV), 272, 275–276, 286
Interstimulus interval (ISI), 232–233, 326
Intracochlear electrodes, 324, 334
Intraneural, 218–219, 222–224, 235
Intraneural electrodes, 213, 219, 221
Ion channels, 177–179

J

Joints, 288–289, 291–293, 303, 307, 311, 313, 315–316

K

Kidneys, 4, 40, 77–78, 90, 161–162, 246, 414, 489
Korotkoff sounds, 13, 16, 23, 25–26

L

LA, *see* Left atrium (LA)
Labor, 164, 167, 203, 237–238, 249–254, 256–260,
    265–266, 358
    active, 241, 250–251, 253–254, 257–260

Larmor frequency, 304, 471–472
Last image hold (LIH), 468
Lateral flow immunoassay, 355, 372–373, 390–391
LBM, *see* Lean body mass (LBM)
Lean body mass (LBM), 168–170
Left atrium (LA), 44–45, 71, 77–79, 87, 103, 105, 112, 114
Left ventricle (LV), 15, 44–45, 77–81, 87, 91, 94, 103, 110–112
LFPs, *see* Local field potentials (LFPs)
Lifetime, 76, 103, 148, 376–377, 575
Ligaments, 288, 291–292, 299, 303, 307–308, 481
Light
    absorption of, 247, 269, 406, 412, 431, 457
    intensity, 301, 303
Likelihood ratio (LR), 324
Linear amplifier, 302–303
Load cell, 482, 490–491, 495–498, 503, 509–510, 523, 525
Local field potentials (LFPs), 183, 185–186
Longitudinal intrafascicular electrodes (LIFEs), 219, 223, 225
Loop recorders, implantable, 44, 65
Lower esophageal sphincter (LES), 145, 149
Low-pass filtering, 497–498, 501, 513
Lung function, 420, 440, 445
LV, *see* Left ventricle (LV)

M

Magnetic resonance imaging (MRI), 70, 75, 78, 95, 190, 192, 209–210, 287, 289, 299, 303–304, 306, 316, 472, 475
Manometer systems, 13, 17–18
Manometry catheters, 134–135, 139–140, 144, 149
MAP, *see* Mean arterial pressure (MAP)
Mapping systems, electroanatomic, 70–71
Markov analysis methods (MAMs), 575–576
Maternal abdomen, 240, 242–243, 246–247, 249, 252–254
Mean arterial pressure (MAP), 15, 17, 23, 27–28, 32, 68–69, 76–77, 87, 403
Measure blood pressure, 22, 29, 40
Measurement method, 25, 95, 98, 264–265
Measurements
    accurate, 15, 23, 35, 40, 159, 280, 283, 427
    auscultatory, 23, 25, 28
    continuous, 4, 31, 36, 97, 404
    direct, 14, 17, 125, 260, 497
    noninvasive, 15–16, 31, 36, 76, 532
    relative, 81, 98
    site, 117, 424, 438, 509–511, 531
    systems, 41, 221, 510

techniques, indirect, 14, 125–126
Measuring blood pressure, 16, 22, 31
Measuring glomerular filtration rate, 163, 165, 167, 169
Membrane, 9, 117, 177–180, 183, 216, 234–235, 353, 372–374, 406–408, 433, 579
    composition, 372, 374
    oxygen-permeable, 404, 406, 411
MEPs, *see* Motor-evoked potentials (MEPs)
Methods, indirect, 6–7, 17, 252–253, 403
Modulation frequencies (MFs), 326–327
Modulation transfer function (MTF), 453, 460
Monitoring, home blood pressure, 15, 25, 29
Morphology, 46, 56, 62, 65, 68, 205–206, 292, 294, 558
Motility tracking system (MTS), 131, 154–155
Motor-evoked potentials (MEPs), 230–231
Motor nerves, 214, 310, 312
Motor neurons, 182, 215–216, 229, 231
Motor units, 215–216, 224, 227, 229, 232
Movement, fetal, 246, 248–249, 253
Moving average filter, 37, 39
MRI, *see* Magnetic resonance imaging (MRI)
MRI machines, 82, 287, 304–306
MR images, 303, 306
MTF, *see* Modulation transfer function (MTF)
MTS, *see* Motility tracking system (MTS)
Multiple inert gas elimination technique (MIGET), 443, 446
Murmurs, 101–102, 106, 109, 111–112, 114–120
Muscles, 2–3, 6, 80, 87, 213–217, 221, 224–230, 234–235, 288–290, 292, 298–299, 303, 307–313, 538, 545
    activity, 202, 213–214, 226–227, 229–230, 320, 335, 477
    fibers, 215–216, 224, 226–227, 232, 234–235
    recordings, 213–214, 227, 229
    skeletal, 3–4, 80, 215, 289, 310–311, 315
Myocardial blood flow, 85, 100
Myocardial contrast echocardiography (MCE), 85, 99–100

N

Natural frequency, 19, 500
    underdamped, 19, 21
Needle electrodes, 5, 193, 219, 224, 310, 320, 529
Neonatal intensive care unit (NICU), 394, 397–398, 413–414, 418
Nernst equation, 178–179, 408
Nerve
    bundle, 218, 312
    conduction, 230–231, 310

conduction velocities, 214, 233, 235, 309
facial, 339–340, 348
fibers, 215, 217, 219–221, 223, 231–233,
    321–322, 579
recordings, 213, 223–225, 227
signals, 220, 222, 227
trunk, 217–218, 231, 235, 334
Nerve conduction study (NCS), 289, 309–310
Neural activity, 171–172, 183, 210, 319–321,
    323, 342
Neural networks (NNs), 37, 480, 538, 578–579,
    581–582, 584, 586–587
Neural networks (NNs) intervals, 61
Neural responses, 185, 335, 346
Neural response telemetry (NRT), 334, 348
Neurons, 171, 173, 176–184, 186–189, 206, 215–216,
    328, 335, 560, 579–582
    artificial, 579–581
    electrical activity of, 180, 183
Neurophysiology, 171–173, 190, 538
Neurotology, 317, 319, 321, 323, 325, 327, 329, 331,
    333, 335, 337, 339, 341, 343, 347
NICU, see Neonatal intensive care unit (NICU)
Neck-shaft angle (NSA), 295–296
NNs, see Neural networks (NNs)
Noninverting amplifier, 171, 197–198
Nonspecific signal, 371, 376
Normal heart sounds, 101–102, 109, 111
Nuclear magnetic resonance (NMR), 305
Nucleic acids, 357–358, 362, 380, 383, 387–390
Nucleic acid sequences, 380–383, 387–388
Nucleus, 176, 215, 348–349, 456, 471

O

OAEs, see Oto-acoustic emissions (OAEs)
OCT, see Optical coherence tomography (OCT)
Oligonucleotide probes, 383, 387–388
Operating systems, 560, 563
Ophthalmology, 34, 267, 269, 271, 273, 275, 277–281,
    283–285
Optical coherence tomography (OCT), 10, 127–128,
    130, 132, 158–160, 286
Orthopedics, 287–291, 293, 295, 297, 299, 301, 303,
    305, 307, 309, 311, 313, 315
Oscillations, 19, 26, 106, 109, 188, 206, 424, 438
Oscillogram, 13, 16–17, 26–28, 30
Oto-acoustic emissions (OAEs), 317, 327, 331, 333
Outer hair cells (OHCs), 331, 333
Oxygen
    partial pressure of, 258, 405–406, 409–410, 431
    sensor, 258, 409–410

P

Partial pressure of carbon dioxide, 408–409
Partial pressures, 258, 398, 405–406, 411, 419,
    427–430, 432–434, 442–446
Patients
    cancer, 169, 392
    specimen, 360, 362, 369–370, 372, 376–378
    ventilated, 427–428
Pattern ERG (PERG), 269–270, 272, 275
Pattern VEP (PVEP), 271, 275
PCG signals, 118–119
PCO_2, 408–409
Pediatric intensive care units (PICUs), 393–394, 397,
    413–414
Pediatrics, 141–142, 342, 394, 397, 403–404, 410,
    413–414, 418
Perfusion, 7, 81–82, 84, 141, 261, 441–443, 468
Periods, refractory, 45, 55, 69, 180, 232, 335
Peripheral nerves, 172, 214–215, 220, 231, 308–309
PET, see Positron emission tomography (PET)
Phase, 25–26, 32, 37–39, 45, 69, 107–108, 198, 205,
    209, 300, 305, 422, 428, 438, 496
Phase response, 537, 545, 549–550
pH electrode, 258, 408–409
Phonocardiograms, 80, 108, 112–113, 118–121
Phonocardiography (PCG), 102, 105, 109, 111, 114
Photodiode, 90, 282, 284, 298, 377, 412
Photomultiplier tube (PMT), 301–303, 363, 366, 377,
    462, 467
pH sensor, 146, 258
Physiological artifacts, 201–203
PICUs, see Pediatric intensive care units (PICUs)
Piezoelectric sensors, 22, 97, 117
Piezoresistive sensors, 13, 21–22, 34, 423
Plasma, 161–162, 164, 167, 359, 444
Plasma concentration, 161, 163, 166–167
Pneumotachograph, 425–428, 434, 436, 450
PO_2, 431–432, 440, 442
Polarizable electrodes, 46–47
Position, fetal, 237, 243–244
Positive deflection, 59, 68, 270
Positron emission tomography (PET), 81, 83,
    190–192, 209–210, 560
Postsynaptic potentials (PSPs), 180, 182
Potentials, sensory-evoked, 230, 344
Preamplifier, 53, 117–118, 185, 199, 217, 255, 487,
    523, 526
Precordial, 47–49
Pregnancy, 238–240, 244, 246–248, 254,
    260–262, 266
Premature ventricular contraction (PVC), 57, 135
Pressure

atmospheric, 18, 21, 423, 434, 437, 442, 499, 508, 512–513, 533
body, 277, 280
catheters, 22, 237, 250
changes, 95, 522
diastolic, 3, 7, 14, 32, 95
esophageal, 424–425, 428, 449
gastric, 423, 425
gradients, 77, 85, 98, 101
intra-arterial, 20, 35
measured, 21, 278, 511
measurements, 1, 6, 25, 41, 131, 134, 267, 276–277, 419, 423, 506, 511–513, 517, 521–522, 532
the, 134, 520
ambulatory blood, 13, 29, 31
opening, 488–489
oscillations, 16, 27–28
peak, 32, 90, 136, 157
pleural, 423–424, 437
pulse, 15, 32–33, 36, 39, 75, 94–95, 97
carotid, 14, 36
resolution, 24, 28–29
sensors, 11, 18, 20, 151, 253, 255, 260, 263, 424–425, 486, 491, 509, 513–515, 525, 530
differential, 423, 426
external, 250, 253, 403, 490
waveform, 28, 33, 95, 97, 530
waves, 96, 519
Pressure-flow studies (PFSs), 490, 534–535
PR interval, 43, 56, 58–59
Probability, 188, 342, 345–346, 456, 465, 540, 542–543, 564, 567, 571–572, 575–576, 585
Probes, 7–8, 126, 128, 155, 184–185, 239, 242, 244–245, 254, 260, 262–263, 367–368, 381–383, 385–388, 396
whole-chromosome painting, 385–386
Pulmonary arteries (PAs), 7, 45, 78–79, 87–88, 91, 106–107, 109–110, 112, 440
Pulse wave analysis (PWA), 13–14, 31, 33, 35, 41, 97, 99
Pulse wave velocity (PWV), 36, 75, 94–96
PWA, *see* Pulse wave analysis (PWA)
P-wave, 46, 56, 59–60
P-wave onset, 56, 59
PWV, *see* Pulse wave velocity (PWV)
Pyramidal neurons, 181–182

Q

Q_{max}, maximum flow rate, 486, 501–502, 505
QRS, 46, 54–57, 59–60, 62, 104, 106, 108, 241
complexes, 54–57, 65
detection, 54, 57, 60–61

R

RA, *see* Right atrium (RA)
Radio-frequency (RF), 72, 305, 471–472
Radiographs, plain, 287, 293, 295–296, 299–300
Radius, 17–18, 30, 86–87, 96
Ratio, damping, 19, 21, 500
Reactivity, 206–208, 390
Reagents, 362, 374, 377–378, 384, 526
Real-time PCR, 382–384
Recorders, 65, 267, 272, 302
Recording
configuration, 213, 220–221, 226
electrodes, 68, 184, 190, 193, 205, 220, 231, 311–312, 328, 335, 338–340
Record pressures, 136–137
Rectum, 8, 132, 142–143, 148, 151, 396, 404, 486, 509, 512, 521, 523
Reference electrode, 153, 190, 200, 202, 220, 227, 251–252, 329, 407, 411
Reference pressure, 513–514
Rehabilitation, 211, 477, 479, 481–483, 564
Relative biological effectiveness (RBE), 455
Reporter probes, 385, 388–389
Resistance, airway, 435–436
Resonant frequency, 436, 439
Response, 18–20, 69, 142–144, 151, 208–209, 230, 270–271, 320, 322–326, 328–329, 331–335, 337–339, 343–344, 351, 503
time, 268, 396, 429
waveform, 332, 346
Retina, 10, 268–271, 275–276
Reynolds number, 86, 364–365, 437
RF coils, 305–306
Right atrium (RA), 7, 44–45, 71, 77, 87, 92, 103, 105
Rolling circle amplification (RCA) assays, 384–385
RR intervals, 55, 61

S

Sample values, 551–552
Sampling, 60, 89, 162–163, 429, 537, 541, 544, 550–553, 559, 562
Sampling frequency, 193, 548, 550, 552–553, 557, 559
Saturation, 90, 123, 201, 217, 258, 398, 429, 431–432, 447
SBC, *see* Slide-based cytometry (SBC)
SC, *see* Stratum corneum (SC)
Scalp surface, 174, 182, 189, 193
Schiotz tonometer, 267, 277–280, 285
SD, *see* Standard deviation (SD)
Selectivity, 218–219, 221, 224, 226, 374

Sensors, 17–18, 20–23, 34–35, 72–73, 117, 135–137, 141–144, 256–261, 396–397, 401–404, 408–412, 419–423, 477–478, 480, 509–514
  capacitive, 22, 507
  capacitive-based, 13, 22
  electrochemical, 405, 407
  fiber-optic, 21–22, 137–138
  linear, 422, 438
  systems, 9, 14, 17
Sensory nerve action potential (SNAP), 309, 312
Sensory nerves, 214, 309–310, 312
Signal-averaged ECG, 43, 50, 62
Signals
  amplitude, 19, 223–224, 274
  characteristics, 102, 213, 221, 226
  conditioning, 118, 222, 491, 513
  detection, 361, 375, 377, 385
  detector, 283, 285, 472
  digitized, 423, 545
  electric, 6, 216
  intensity, 82, 85, 303–304, 388, 472
  levels, 240, 270, 272
  original, 420, 548, 553
  physiological, 102, 420, 550, 559
  processing, 29, 102, 118, 198, 213–214, 222, 241, 256, 271, 403, 448, 498–499, 513
  recorded, 200–201, 203–204, 226, 235, 274–275
  sampled, 550–552
  test, 547–548
  ultrasonic, 242, 253
  underlying, 548–550, 553, 557, 559
Signal-to-noise ratio (SNR), 59, 209, 219, 222–223, 229, 235, 270, 274, 328–329, 331, 346, 376, 460, 465, 553–554
  low, 201, 384
Single-photon emission computed tomography (SPECT), 81, 83, 99, 190–192, 560
Skin, 8–9, 46, 125–130, 193–195, 198, 224, 269, 271, 274, 308, 311–312, 399, 404, 410–413, 416
  blood flow, 123, 128, 482
  electrodes, 6, 269
  irritation, 411–412
  surface, 3, 5, 126–127, 195, 226, 312, 410, 416
Slide-based cytometry (SBC), 355, 369
SNAP, see Sensory nerve action potential (SNAP)
Solid-state sensors, 25, 136–138
Solution, 202, 361, 371, 377–378, 399, 403, 405, 408–409, 428–429, 509
Sound pressure level (SPL), 116, 326, 333
Sounds, 2–3, 7, 16, 25–26, 101–102, 108, 110–112, 114–118, 246, 248, 325, 340–341, 343–344, 416–417, 472
SP, see Systolic pressure (SP)

SP and DP, 23, 27–28, 30
SPECT, see Single-photon emission computed tomography (SPECT)
Speech, 175, 192, 325, 327–328, 351, 353
Sphincter, 142
Spikes, 185–188, 194, 201, 205, 210, 229
Spontaneous OAEs (SOAEs), 317, 332
Spot film device, 466–467, 469
Standard deviation (SD), 61, 63, 365, 460–461, 501, 517, 566–567, 570
Standard electroretinography systems, 267, 276
State
  behavioral, 186–188
  physiological, 4, 211
Statistical inference, 538, 567, 570
Statistics, 501, 537–539, 541, 543, 545, 547, 549, 551, 553, 555, 557, 559, 563–567, 569–571, 585–587
Steady-state responses, 326–327, 349, 352
Steps, wash, 370, 375, 378–379, 385, 387
Stethoscope, 2–4, 7, 16, 25–26, 101–102, 110, 114, 116, 118, 239–240
Stimulation, 2, 47, 156, 219, 224, 231–232, 268–269, 273, 309, 312, 335–336, 338–339, 344, 346, 352
Stimulation intensity, 231–232, 269
Stimulators, 69, 267, 272, 275–276
Stimuli, 69–70, 202, 204, 208–209, 230, 232–233, 273, 275, 318, 323–328, 333, 336–337, 340, 342, 344–347
  click, 322–323, 329
  deviant, 342, 346
Stomach, 132–133, 140, 145, 147, 149, 152–154, 189, 423, 425
Straining, 503–506, 520–521
Stratum corneum (SC), 125–126, 194, 198, 251
Stress incontinence, 486, 489, 520, 527–528, 531
Structures, 87, 103, 172, 174, 176–177, 181–182, 190, 245–246, 262, 301, 303, 307, 458, 469, 472
Subject, 23, 25, 37–38, 154–155, 201–203, 206–208, 211, 225–226, 338, 342, 420–421, 424–425, 435, 439–440, 480–482
Surface, 46, 78–79, 214–215, 222, 224, 226, 240, 288, 308, 311–312, 377–378, 380, 387, 407, 479–480
  electrodes, 5–6, 65, 225–227, 311, 320, 329, 338, 529
  tension, 277, 280
Surface plasmon resonance (SPR), 370, 391
SYBR green, 382
Systematic errors, 28, 41
Systems
  autonomic nervous, 60, 214
  cardiovascular, 76, 98, 101

motility tracking, 131, 154
musculoskeletal, 288–289, 303, 316
second-order, 17–18, 20
vascular, 2, 4, 7–8, 10
video, 288, 314
Systole, 24, 33, 38–39, 76–78, 80, 94, 104–107, 111–112, 114, 412
Systolic pressure (SP), 3, 7, 14–17, 23–24, 26–28, 32–33, 38, 90, 328–329, 334

T

Taqman probes, 382–383
Target antigens, 358, 360, 362, 370–371, 373, 377–380
Target cells, 364–365, 369, 389
Techniques, oscillometric, 13, 23, 26–27, 29
Temperature, 4, 72–73, 92–93, 128, 146–147, 261, 396–397, 401, 404, 410, 412, 415–416, 427, 431, 433–434
    measurements, 1, 8, 393, 416
    sensors, 72, 92, 261, 264, 396, 401, 404, 411
Tendons, 288–290, 292, 299, 303, 307–308
Tests, 63–64, 141–142, 151, 169, 232, 272–275, 327, 356–357, 361–362, 372, 389–390, 443–444, 515, 526–527, 570
    procedures, 454, 474
    response, 232–233
Thickness, 125, 457–458, 461–462, 464, 473
Thoracic gas volume (VTG), 435–436, 443
Time
    amount of, 58–59, 246
    capillary flow, 372, 374
    constant, 98, 465, 472, 494, 499–500, 523
    delay, 10, 33, 109, 232, 259, 321, 323, 503
    domain, 20, 37–38, 61, 223, 327
    function of, 81–82, 85, 92, 243, 249, 252, 403
    period, 55, 58–59, 180, 223, 238, 248, 257, 443, 574
Timer, 466, 561, 563
TIPtrode, 329
Tissues, 8, 10–11, 67–69, 72, 80–82, 87, 89, 173, 175, 298–299, 303, 305–309, 412, 431, 439
    resting, 67, 69
    types, 455, 472
TM, see Tympanic membrane (TM)
TMS, see Transcranial magnetic stimulation (TMS)
Tonometer, 33–35, 97, 277–279, 281–282, 285–286
Total body water (TBW), 168–170
Tracers, 162–164, 166, 300–301
Transcranial magnetic stimulation (TMS), 230–231, 235

Transcription-mediated amplification (TMA), 384, 390
Transepidermal water loss (TWL), 125–126
Transfer function, 37–38, 116, 580–581
    inverse, 37–39
Transient-evoked otoacoustic emissions (TEOAEs), 331–332, 348
Tricuspid, 94, 103, 105–106, 110, 113
Tumors, 210, 288, 291, 293–294, 299–301, 303–304, 469, 572–573
T-wave, 46, 55–60, 62–63
TWL, see Transepidermal water loss (TWL)
Tympanic membrane (TM), 323, 328–329, 397, 415

U

Ultrasound (US), 289, 307–308
Ultrasound, intraluminal, 132, 157–158
Urethra, 488–489, 501, 504, 512, 530–531
Urethral pressure profilometry, 487, 530–531
Urine, 1–3, 161, 238, 358, 372, 414, 487, 489–490, 492–496, 503–504, 506–508, 510–511, 527, 529
    flow, 490, 501, 503
    height of, 492–493, 532
Urodynamics, 485, 487, 490, 492, 501–502, 509–510, 518, 530–535
Urodynamic system, 490, 495, 523, 525, 533
Uroflowmetry, 485–486, 490, 492, 495, 499–500, 504–506, 513, 517, 527, 529, 532, 534–535
Urology, 485, 487, 489, 491, 493, 495, 497, 499, 501, 503, 505, 507, 509, 511, 513
Uterine activity, 248–249, 251, 254, 260
Uterine contractions, 249–250, 253–254, 256, 259–260
Uterus, 238, 241, 244–245, 249, 260, 262

V

Validation, 13, 30, 41, 378, 390, 449
Values, 14, 17–18, 36–39, 123, 128, 165–168, 233, 265–266, 403–404, 421, 461, 499, 553–554, 566, 569
Valves, 78, 80, 87, 98, 101, 103, 106–107, 109–110, 112–113, 126, 242
Venous blood, 90–91
    mixed, 7, 90, 440, 444
Ventricles, right, 44–45, 79, 103, 111, 441
Ventricular contraction, 101, 106, 108–109
Ventricular pressure, 80, 106, 112
Ventricular volume, 78, 80, 104, 106–107
VEP, see Visual evoked potential (VEP)

Vessels, 7, 79, 96–97, 144, 250, 253, 292, 492–494, 497

Vestibular-evoked myogenic potentials (VEMPs), 318, 336–337, 339, 348–350, 353

Vestibular nerve, inferior, 336–337, 339

Visual evoked potential (VEP), 209, 267, 269, 271, 273, 275, 285–286

Voided volume, 485, 490–491, 497–498, 500–501, 505, 516

Voiding, 485–490, 494, 506, 508, 518, 535

Voiding disorders, 485, 489, 492

Voltage
    difference, 177–178, 199, 269
    signal, 311, 423
    source, 179, 464

Volume, 8, 77–78, 80, 91–92, 95–96, 126, 151–152, 433, 435–437, 440, 490, 492, 494–499, 513, 525–527
    filled, 491, 516, 523
    measurement, 8, 419, 425
    signal, 426–427, 498

W

Wall, luminal, 133–135, 144, 151–152, 158

Water vapor, 125–126, 430, 434, 442, 446

Wave
    delay, 323–324, 349
    output sine, 422, 438, 445, 447
    slow, 152–153, 205

Waveforms, 32–33, 37, 46, 56, 59–60, 62, 64, 96, 243, 272, 323, 332–333

Wavefront, 67–68

Wavelengths, 72, 90, 124–125, 127, 258, 263, 268, 366–367, 372, 378, 406, 412, 431–432, 454–455, 472

Wavelet transform, 120–121, 186, 223

Weights, 15, 165, 168–169, 394, 396, 425, 492, 495, 523, 581–582, 584

Wire electrodes, 224–225, 235

X

X-rays, 2–3, 132, 148, 190, 297, 299, 399, 454, 456–461, 466, 468, 473–474
    detectors, 297–299, 463, 474
    images, 69–70, 148, 287, 293, 297, 460
    machines, 67, 287, 296–297, 299
    photons, 297–298, 456–457, 464, 466
    source, 298–299, 458, 463
    tube, 296–297, 299, 456–457, 459, 463, 466, 469, 473–474

Made in United States
Orlando, FL
07 January 2023

28339863R10337